Understanding NEC® Calculations

Second Edition

Michael Holt

Delmar Publishers

I(T)P An International Thomson Publishing Company

Albany • Bonn • Boston • Cincinnati • Detroit • London • Madrid
Melbourne • Mexico City • New York • Pacific Grove • Paris • San Francisco
Singapore • Tokyo • Toronto • Washington

NOTICE TO THE READER

Publisher does not warrant or guarantee any of the products described herein or perform any independent analysis in connection with any of the product information contained herein. Publisher does not assume, and expressly disclaims, any obligation to obtain and include information other than that provided to it by the manufacturer.

The reader is expressly warned to consider and adopt all safety precautions that might be indicated by the activities herein and to avoid all potential hazards. By following the instructions contained herein, the reader willingly assumes all risks in connections with such instructions.

The publisher makes no representation or warranties of any kind, including but not limited to, the warranties of fitness for particular purpose or merchantability, nor are any such representations implied with respect to the material set forth herein, and the publisher takes no responsibility with respect to such material. The publisher shall not be liable for any special, consequential, or exemplary damages resulting, in whole or part, from the readers' use of, or reliance upon, this material.

Cover Image by Mick Brady

Delmar Staff
Publisher: Robert D. Lynch
Acquisitions Editor: Mark Huth
Developmental Editor: Jeanne Mesick
Project Editor: Tom Smith
Production Coordinator: Toni Bolognino
Art/Design Coordinator: Michael Prinzo

PNB Graphics
Design, Layout, and Typesetting: Paul Bunchuk
Internet email: pnbgraph@pnbgraphics.com

Graphic Illustrations: Mike Culbreath

email: mikeholt@gate.net
World Wide Web Site: http://www.gate.net/~mikeholt

The ITP logo is a trademark under license
Printed in the United States of America
For more information, contact:

Delmar Publishers
3 Columbia Circle, Box 15015
Albany, New York 12212-5015

International Thomson Editores
Campos Eliseos 385, Piso 7
Col Polanco
11560 Mexico D F Mexico

International Thomson Publishing Europe
Berkshire House 168 - 173
High Holborn
London WC1V 7AA
England

International Thomson Publishing GmbH
Königswinterer Strasse 418
53227 Bonn
Germany

Thomas Nelson Australia
102 Dodds Street
South Melbourne, 3205
Victoria, Australia

International Thomson Publishing Asia
221 Henderson Road
#05 - 10 Henderson Building
Singapore 0315

Nelson Canada
1120 Birchmount Road
Scarborough, Ontario
Canada M1K 5G4

International Thomson Publishing Japan
Hirakawacho Kyowa Building, 3F
2-2-1 Hirakawacho
Chiyoda-ku, Tokyo 102
Japan

3 4 5 6 7 8 9 10 XXX 99 98 97

Library of Congress Cataloging-in-Publication Data
Holt, Charles Michael.
Understanding NEC calculations / Michael Holt 2nd ed.
p. cm.
Includes index.
ISBN 0-8273-7838-6 (pbk.)
1. Electric engineering–Mathematics
2. Electric engineering—United States–Insurance requirements. I. Title.
Understanding national electrical code calculations
TK260.H65 1996 96-4104
621.319′24′076–dc20 CIP

I dedicate this book to my family.

Contents

Preface

INTRODUCTION

The purpose of this textbook is to help you understand NEC® Calculation rules as they apply to all electrical installations under 600 volts. This includes residential, commercial and industrial.

This book contains practical examples and warnings of the dangers related to improper electrical installations. In addition, it offers many helpful tips, and other suggestions.

Understanding NEC® Calculations is designed to provide you with all the tools necessary to understand and apply most electrical calculations. To understand and apply the National Electrical Code calculations, you must have a strong foundation and working knowledge of basic electrical calculations (Units 1 through 4). If you are experienced and comfortable with these calculations, you might want to skip Chapter 1 and go to Chapter 2.

This textbook is a guide to be used with the NEC®, not a replacement of the NEC®.

HOW TO USE THIS BOOK

Each unit of this book contains objectives, explanations with graphics, examples, steps on calculations, formulas and practice questions. This book is intended to be used with the 1996 National Electrical Code. As you read this book, review the author's comments, graphics, and examples with your Code Book and discuss the subjects with others. This book contains many cross references to other related Code rules. Take the time to review the cross references.

As you progress through this book, you will find some rules or some comments that you don't understand. Don't get frustrated; highlight the section in the book that you are having a problem with. Discuss it with your boss, inspector, co-worker, etc., and maybe they'll have some additional feedback. After you have completed this book, review the highlighted sections and see if you now understand those problem areas.

If you feel that I have made an error, please let me know by contacting me at 1-800-ALL-CODE.

Note: Some words are italicized to bring them to your attention. Be sure that you understand the terms before you continue with each unit.

ABOUT THE AUTHOR

Charles "Mike" Holt, Sr. of Coral Springs, Florida, has worked his way up through the electrical trade as an apprentice, journeyman, master electrician, electrical inspector, electrical contractor, electrical designer, and developer of training programs and software for the electrical industry. Formerly contributing editor to *Electrical Construction and Maintenance* magazine (*EC&M*), and *Construction Editor to Electrical Design and Installation* magazine (*EDI*). Mr. Holt is currently a contributing writer for *Electrical Contractor* magazine (*EC*). With a keen interest in continuing education, Mike Holt attended the University of Miami Master's in Business Administration Program (MBA) for Finance.

The author has provided custom in-house seminars for IAEI, NECA, ICBO, IBM, AT&T, Motorola, and the U.S. Navy, to name a few. He has taught over 1,000 classes on over 30 different electrical-related subjects ranging from alarm installations to exam preparation and voltage drop calculations. Many of Mike Holt's seminars are available on video. He continues to develop additional courses, seminars, and workshops to meet today's changing needs for individuals, organizations, vocational, and apprenticeship training programs.

Since 1982 Mike Holt has been helping electrical contractors improve the management of their business by offering business consulting, business management seminars, and computerized estimating and billing software. These software programs are used by hundreds of electrical contractors throughout the United States.

Mike Holt's extensive knowledge of the exam preparation, his hands-on experience, and his unique style of presenting information make this book a must read for those interested in passing the exam the first time.

On the personal side, Mike Holt is a national competitive barefoot water skier and has held several barefoot water ski records. He was the National Barefoot Water-ski Champion for 1988 and is currently training to regain the title by 1998. In addition to barefoot skiing, the author enjoys the outdoors, playing the guitar, reading, working with wood, and spending time with his family (he has seven children).

ACKNOWLEDGMENTS

I would like to say thank you to all the people in my life who believed in me, even those who didn't. There are many people who played a role in the development and production of this book.

I would like to thank the Culbreath Family. Mike (Master Electrician) for helping me transform my words into graphics. I could not have produced such a fine book without your help. Toni, thanks for those late nights editing the manuscript. Dawn, you're too young to know, but thanks for being patient with your parents while they worked so hard.

Next, Paul Bunchuk (PNB Graphics)—Design, Layout, and Typesetting—thank you. Paul (Master Electrician) for the layout, and editing you did. Your knowledge of computers and the NEC® has helped me put my ideas into reality.

Mike Culbreath and Paul Bunchuk, thank you both for not sacrificing quality, and for the extra effort to make sure that this book is the best that it can be.

To my family, thank you for your patience and understanding. I would like to give special thanks to my beautiful wife, Linda and my children, Belynda, Melissa, Autumn, Steven, Michael, Meghan, and Brittney.

I thank all my students; you know how much I care about you.

And thanks to all those who helped me in the electrical industry, *Electrical Construction and Maintenance* magazine for my big break, and Joe McPartland, "my mentor." Joe, you were always there to help and encourage me. I would also like to thank the following for contributing to my success: James Stallcup, Dick Lloyd, Mark Ode, D. J. Clements, Joe Ross, John Calloggero, Tony Selvestri, and Marvin Weiss.

The final personal thank you goes to Sarina, my friend and office manager. Thank you for covering the office for me the past few years while I spent so much time writing books. Your love and concern for me has helped me through many difficult times.

The author and Delmar Publishers would also like to thank those individuals who reviewed the manuscript and offered invaluable suggestions and feedback. Their assistance is greatly appreciated.

Kenneth D. Belk
Marion, Ohio

Ray Cotten, Electrical Instructor
North Tech Education Center
Palm Beach, Florida

David Figueredo, Electrical Inspector
Metro-Dade County, Florida

Mike Freiner
Florida Electrical Apprenticeship

Ronald McMurtry
Kentucky Tech – West Campus

John Mills, Master Electrician
Dade County, Florida

John A. Penley
South Central Technical College

Kurt A. Stout, Electrical Inspector
Plantation, Florida

Richard Kurtz, Consultant
Boynton Beach, Florida

Ronald Rains
Three Rivers Community College

GETTING STARTED

THE EMOTIONAL ASPECT OF LEARNING

To learn effectively, you must develop an attitude that learning is a process that will help you grow both personally and professionally. The learning process has an emotional as well as an intellectual component that we must recognize. To understand what affects our learning, consider the following:

Positive Image. Many feel disturbed by the expectations of being treated like children and we often feel threatened with the learning experience.

Uniqueness. Each of us will understand the subject matter from different perspectives and we all have some unique learning problems and needs.

Resistance To Change. People tend to resist change and resist information that appears to threaten their comfort level of knowledge. However, we often support new ideas that support our existing beliefs.

Dependence And Independence. The dependent person is afraid of disapproval, often will not participate in class discussion, and will tend to wrestle alone. The independent person spends too much time asserting differences and too little time trying to understand others' views.

Fearful. Most of us feel insecure and afraid with learning, until we understand what is going to happen and what our role will be. We fear that our performance will not match the standard set by us or others.

Egocentric. Our ego tendency is to prove someone is wrong, with a victorious surge of pride. Learning together without a win/lose attitude can be an exhilarating learning experience.

Emotional. It is difficult to discard our cherished ideas in the face of contrary facts or when overpowered by the logic of others.

HOW TO GET THE BEST GRADE ON YOUR EXAM

Studies have concluded that for students to get their best grades, they must learn to get the most from their natural abilities. It's not how long you study or how high your IQ is, it's what you do and how you study that counts. To get your best grade, you must make a decision to do your best and follow as many of the following techniques as possible.

Reality. These instructions are a basic guide to help you get the maximum grade. It is unreasonable to think that all of the instructions can be followed to the letter all of the time. Day-to-day events and unexpected situations must be taken into consideration.

Support. You need encouragement in your studies and you need support from your loved ones and employer. To properly prepare for your exam, you need to study 10 to 15 hours per week for about 3 to 6 months.

Communication With Your Family. Good communication with your family members is very important. Studying every night and on weekends may cause tension. Try to get their support, cooperation, and encouragement during this trying time. Let them know the benefits. Be sure to plan some special time with them during this preparation period; don't go overboard and leave them alone too long.

Stress. Stress can really take the wind out of you. It takes practice, but get into the habit of relaxing before you begin your studies. Stretch; do a few sit-ups and push-ups; take a 20-minute walk or a few slow, deep breaths. Close your eyes for a couple of minutes; deliberately relax the muscle groups that are associated with tension, such as the shoulders, back, neck, and jaw.

Attitude. Maintaining a positive attitude is important. It helps keep you going and helps keep you from getting discouraged.

Training. Preparing for the exam is the same as training for any event. Get plenty of rest and avoid intoxicating drugs, including alcohol. Stretch or exercise each day for at least 10 minutes. Eat light meals such as pasta, chicken, fish, vegetables, fruit, etc. Try to avoid heavy foods, such as red meats, butter, and other high-fat foods. They slow you down and make you tired and sleepy.

Eye Care. It is very important to have your eyes checked! Human beings were not designed to do constant seeing less than arm's length away. Our eyes were designed for survival, spotting food and enemies at a distance. Your eyes will be under tremendous stress because of prolonged, near-vision reading, which can result in headaches, fatigue, nausea, squinting, or eyes that burn, ache, water, or tire easily.

Be sure to tell the eye doctor that you are studying to pass an exam (bring this book and the Code Book), and you expect to do a tremendous amount of reading and writing. Prescribed nearpoint lenses can reduce eye discomfort while making learning more enjoyable and efficient.

Reducing Eye Strain. Be sure to look up occasionally, away from near tasks to distant objects. Your work area should be three times brighter than the rest of the room. Don't read under a single lamp in a dark room. Try to eliminate glare. Mixing of fluorescent and incandescent lighting can be helpful.

Sit straight, chest up, shoulders back, and weight over the seat so both eyes are an equal distance from what is being seen.

Getting Organized. Our lives are so busy that simply making time for homework and exam preparation is almost impossible. You can't waste time looking for a pencil or missing paper. Keep everything you need together. Maintain folders, one for notes, one for exams and answer keys, and one for miscellaneous items.

It is very important that you have a private study area available at all times. Keep your materials there. The dining room table is not a good spot.

Time Management. Time management and planning is very important. There simply are not enough hours in the day to get everything done. Make a schedule that allows time for work, rest, study, meals, family, and recreation. Establish a schedule that is consistent from day to day.

Have a calendar and immediately plan your homework. Follow it at all costs. Try not to procrastinate (put off something). Try to follow the same routine each week and try not to become overtired. Learn to pace yourself to accomplish as much as you can without the need for cramming.

Learn How To Read. Review the book's contents and graphics. This will help you develop a sense of the material.

Clean Up Your Act. Keep all of your papers neat, clean, and organized. Now is not the time to be sloppy. If you are not neat, now is an excellent time to begin.

Speak Up In Class. If you are in a classroom setting, the most important part of the learning process is class participation. If you don't understand the instructor's point, ask for clarification. Don't try to get attention by asking questions you already know the answer to.

Study With A Friend. Studying with a friend can make learning more enjoyable. You can push and encourage each other. You are more likely to study if someone else is depending on you.

Students who study together perform above average because they try different approaches and explain their solutions to each other. Those who study alone spend most of their time reading and rereading the text and trying the same approach time after time even though it is unsuccessful.

Study Anywhere/Anytime. To make the most of your limited time, always keep a copy of the book(s) with you. Any time you get a minute free, study. Continue to study any chance you get. You can study at the supply house when waiting for your material; you can study during your coffee break, or even while you are at the doctor's office. Become creative!

You need to find your best study time. For some it could be late at night when the house is quiet. For others, it's the first thing in the morning before things get going.

Set Priorities. Once you begin your study, stop all phone calls, TV shows, radio, snacks, and other interruptions. You can always take care of it later.

HOW TO TAKE AN EXAM

Being prepared for an exam means more than just knowing electrical concepts, the Code, and the calculations. Have you felt prepared for an exam, then choke when actually taking it? Many good and knowledgeable electricians couldn't pass their exam because they did not know how to take an exam.

Taking exams is a learned process that takes practice and involves strategies. The following suggestions are designed to help you learn these methods.

Relax. This is easier said than done, but it is one of the most important factors in passing your exam. Stress and tension cause us to choke or forget. Everyone has had experiences where they get tense and couldn't think straight. The first step is becoming aware of the tension and the second step is to make a deliberate effort to relax. Make sure you're comfortable; remove clothes if you are hot, or put on a jacket if you are cold.

There are many ways to relax and you have to find a method that works for you. Two of the easiest methods that work very well for many people follow:

• Breathing Technique: This consists of two or three slow deep breaths every few minutes. Be careful not to confuse this with hyperventilation, which is abnormally fast breathing.

• Single-Muscle Relaxation: When we are tense or stressful, many of us do things like clench our jaw, squint our eyes, or tense our shoulders without even being aware of it. If you find a muscle group that does this, deliberately relax that one group. The rest of the muscles will automatically relax also. Try to repeat this every few minutes, and it will help you stay more relaxed during the exam.

Have The Proper Supplies. First of all, make sure you have everything needed several days before the exam. The night before the exam is not the time to be out buying pencils, calculators, and batteries. The night before the exam, you should have a checklist (prepared in advance) of everything you could possibly need. The following is a sample checklist to get you started.

• Six sharpened #2H pencils or two mechanical pens with extra #2H leads. The kind with the larger leads are faster and better for filling in the answer circles.

• Two calculators. Most examining boards require quiet, paperless calculators. Solar calculators are great, but there may not be enough light to operate them.

• Spare batteries. Two sets of extra batteries should be taken. It's very unlikely you'll need them, but bring them anyway.

• Extra glasses if you use them.

• A watch for timing questions.

• All your reference materials, even the ones not on the list. Let the proctors tell you which ones are not permitted.

• A thermos of something you like to drink. Coffee is excellent.

• Some fruit, nuts, candy, aspirin, analgesic, etc.

• Know where the exam is going to take place and how long it takes to get there. Arrive at least 30 minutes early.

Note: It is also a good idea to pack a lunch rather than going out. It can give you a little time to review the material for the afternoon portion of the exam, and it reduces the chance of coming back late.

Understand The Question. To answer a question correctly, you must first understand the question. One word in a question can totally change the meaning of it. Carefully read every word of every question. Underlining key words in the question will help you focus.

Skip The Difficult Questions. Contrary to popular belief, you do not have to answer one question before going on to the next one. The irony is that the question you get stuck on is one that you will probably get wrong anyway no matter how much time you spend on it. This will result in not having enough time to answer the easy questions. You will get all stressed-out and a chain reaction is started. More people fail their exams this way than for any other reason.

The following strategy should be used to avoid getting into this situation.

• **First Pass:** Answer the questions you know. Give yourself about 30 seconds for each question. If you can't find the answer in your reference book within the 30 seconds, go on to the next question. Chances are that you'll come across the answers while looking up another question. The total time for the first pass should be 25 percent of the exam time.

• **Second Pass:** This pass is done the same as the first pass except that you allow a little more time for each question, about 60 seconds. If you still can't find the answer, go on to the next one. Don't get stuck. Total time for the second pass should be about 30 percent of the exam time.

• **Third Pass:** See how much time is left and subtract 30 minutes. Spend the remaining time equally on each question. If you still haven't answered the question, it's time to make an educated guess. Never leave a question unanswered.

• **Fourth pass:** Use the last 30 minutes of the exam to transfer your answers from the exam booklet to the answer key. Read each question and verify that you selected the correct answer on the test book. Transfer the answers carefully to the answer key. With the remaining time, see if you can find the answer to those questions you guessed at.

Guessing. When time is running out and you still have questions remaining, GUESS! Never leave a question unanswered.

You can improve your chances of getting a question correct by the process of elimination. When one of the choices is "none of these," or "none of the above," "d" is usually not the correct answer. This improves your chances from one-out-of-four (25 percent), to one-out-of-three (33 percent). Guess "All of these" or "All of the Above," and don't select the high or low number.

How do you pick one of the remaining answers? Some people toss a coin, others will count up how many of the answers were A's, B's, C's, and D's and use the one with the most as the basis for their guess.

CHECKING YOUR WORK

The first thing to check (and you should be watching out for this during the whole exam) is to make sure you mark the answer in the correct spot. People have failed the exam by ½ of a point. When they reviewed their exam, they found they correctly answered several questions on the test booklet, but marked the wrong spot on the exam answer sheet. They knew the answer was "(b) False" but marked "(d)" in error.

Another thing to be very careful of, for example, is marking the answer for question 7 in the spot reserved for question 8.

CHANGING ANSWERS

When re-reading the question and checking the answers during the fourth pass, resist the urge to change an answer. In most cases, your first choice is best and if you aren't sure, stick with the first choice. Only change answers if you are sure you made a mistake. Multiple choice exams are graded electronically, so be sure to thoroughly erase any answer that you changed. Also erase any stray pencil marks from the answer sheet.

ROUNDING OFF

You should always round your answers to the same number of places as the exam's answers. Numbers below "5" are rounded down, while numbers "5" and above are rounded up.

Example: If an exam has a multiple choice of:
(a) 2.2 (b) 2.1 (c) 2.3 (d) none of these

And your calculation comes out to 2.16, do not choose the answer (d) none of these. The correct answer is (b) 2.2, because the answers in this case are rounded off to the tenth.

Example: It could be rounded to tens, such as:
(a) 50 (b) 60 (c) 70 (d) none of these.

For this group, an answer such as 67 would be (c) 70, while an answer of 63 would be (b) 60. The general rule is to check the question's choice of answers then round off your answer to match it.

SUMMARY

• Make sure everything is ready and packed the night before the exam.

• Don't try to cram the night before the exam, if you don't know it by then.

• Have a good breakfast. Get the thermos and energy snacks ready.

• Take all your reference books. Let the proctors tell you what you can't use.

• Know where the exam is to be held and be there early.

• Bring ID and your confirmation papers from the license board if there are any.

• Review your NEC® while you wait for your exam to begin.

• Try to stay relaxed.

• Determine the time per question for each pass and don't forget to save 30 minutes for transferring your answers to the answer key.

• Remember, in the first pass answer only the easy questions. In the second pass, spend a little more time per question, but don't get stuck. In the third pass, use the remainder of the time minus 30 minutes. In the fourth pass, check your work and transfer the answers to the answer key.

THINGS TO BE CAREFUL OF

- Don't get stuck on any one question.
- Read each question carefully.
- Be sure you are marking the answer in the correct spot on the answer sheet.
- Don't get flustered or extremely tense.

To request examination copies of this book, or instructor's guide, call or write to:

Delmar Publishers
3 Columbia Circle
P.O. Box 15015
Albany, NY 12212-5015
Phone: 1-800-347-7707 • 1-518-464-3500 • Fax: 1-518-464-0301

Delmar Publishers' Online Services
To access Delmar on the World Wide Web, point your browser to:
http://www.delmar.com/delmar.html
To access through Gopher: gopher://gopher.delmar.com
(Delmar Online is part of "thomson.com", an Internet site with information on
more than 30 publishers of the International Thomson Publishing organization.)
For information on our products and services:
email: info@delmar.com

CHAPTER 1
Basic Electrical Calculations

Scope of Chapter 1

Unit 1

Electrician's Math and Basic Electrical Formulas

OBJECTIVES

After reading this unit, the student should be able to briefly explain the following concepts:

Part A - Electrician's Math
- Fractions
- Kilo
- Knowing your answer
- Multiplier
- Parentheses
- Percent increase
- Percentage
- Reciprocals
- Square root
- Squaring
- Transposing formulas

Part B - Basic Electrical Formulas
- Conductance and resistance
- Electric meters
- Electrical circuit
- Electrical circuit values
- Ohm's law
- PIE circle formula
- Power changes with the square of the voltage
- Power source
- Power wheel

After reading this unit, the student should be able to briefly explain the following terms:

Part A - Electrician's Math
- Fractions
- Kilo
- Multiplier
- Parentheses
- Percentage
- Ratio
- Reciprocal
- Rounding off
- Square root
- Squaring a number
- Transposing

Part B - Basic Electrical Formulas
- A – Ampere
- Alternating current
- Ammeter
- Ampere
- Armature
- Clamp-on ammeters
- Conductance
- Conductors
- Current
- Direct current
- Directly proportional
- E – Electromotive Force
- Electrical meters
- Electromagnetic
- Electromagnetic field
- Electron pressure
- Helically wound
- Intensity
- Inversely proportional
- Megohmeter
- Ohmmeter
- Ohms
- P – Power
- Perpendicular
- Polarity
- Polarized
- Power
- Power source
- Resistance
- Shunt bar
- Shunt meter

PART A – *ELECTRICIAN'S MATH*

1–1 *FRACTIONS*

Fractions represent a part of a number. To change a fraction to a decimal form, divide the numerator (top number of the fraction) by the denominator (bottom number of the fraction).

❑ **Examples**

$\frac{1}{6}$ = one divided by six = 0.166

$\frac{5}{4}$ = five divided by four = 1.25

$\frac{7}{2}$ = seven divided by two = 3.5

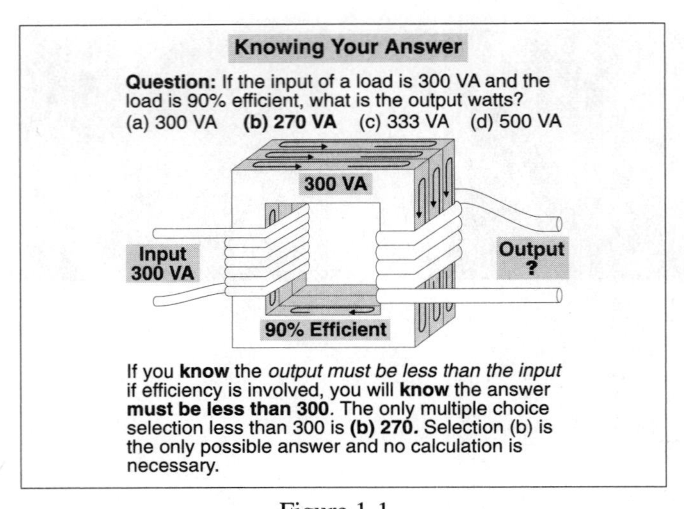

Figure 1-1

Know Your Answer Example

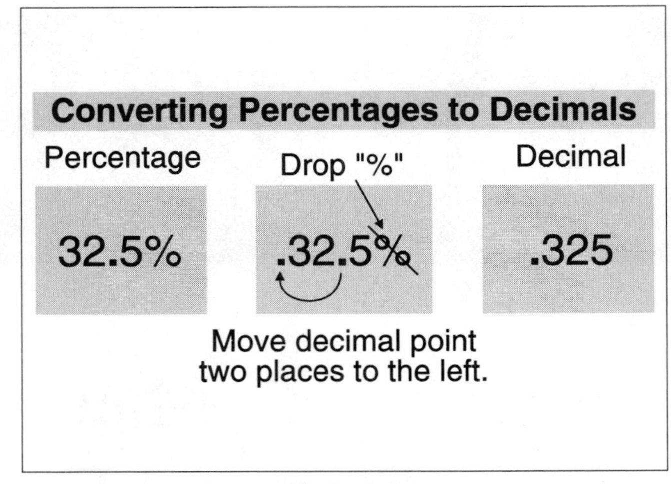

Figure 1-2

Converting Percentages to Decimals

1–2 *KILO*

The letter *k* is the abbreviation of *kilo*, which represents 1,000.

❑ **Kilo Example No. 1**

What is the wattage for an 8-kW rated range?

(a) 8 watts (b) 8,000 watts (c) 4,000 watts (d) none of these

- Answer: (b) 8,000 watts

Wattage = kW × 1,000. In this case 8 kW × 1,000 = 8,000 watts.

❑ **Kilo Example No. 2**

What is the kVA of a 300-VA load?

(a) 300 kVA (b) 3,000 kVA (c) 30 kVA (d) .3 kVA

- Answer: (d) 0.3 kVA, kW is converted to watts by dividing the watts by 1,000.

In this case; $\frac{300 \text{ VA}}{1{,}000} = 0.3 \text{ kVA}$

Note. The use of *k* is not limited to *kW*. It is used for kcmils, such as 250 kcmils.

1–3 *KNOWING YOUR ANSWER*

When working with mathematical calculations, you should know if the answer is greater than, or less than the values given.

❑ **Knowing-Your-Answer Example**

If the input of a load is 300 watts and the load is 90 percent efficient, what is the output watts? Note: Because of efficiency, the output is always less than the input, Fig. 1–1.

(a) 300 VA (b) 270 VA (c) 333 VA (d) 500 VA

- Answer: (b) 270 VA

Since the question stated that the output had to be less than the 300 watt input, the answer must be less than 300 watts. The only choice that is less than 300 watts, is 270 watts.

1–4 *PERCENTAGES*

A *percentage* is a ratio of two numbers. When changing a percent to a decimal or whole number, simply move the decimal point two places to the left, Fig. 1–2.

❑ **Percentage Example**

32.5% = .325 100% = 1.00

125% = 1.25 300% = 3.00

1–5 *MULTIPLIER*

Often a number is required to be increased or decreased by a percentage. When a percentage or fraction is used as a *multiplier*, follow these steps:

Step 1: → Convert the multiplier to a decimal form, then
Step 2: → Multiply the number by the decimal value from Step 1.

❑ **Increase By 125 Percent Example**

An overcurrent protection device (breaker or fuse) must be sized no less than 125 percent of the continuous load. If the load is 80 amperes, the overcurrent protection device would have to be sized no less than ______ amperes.

(a) 80 amperes (b) 100 amperes (c) 125 amperes (d) none of these

- Answer: (b) 100 amperes

Step 1: → Convert 125% to 1.25
Step 2: → Multiply 80 amperes by 1.25 = 100 amperes

❑ **Limit To 80 Percent Example**

The maximum continuous load on an overcurrent protection device is limited to 80 percent of the device rating. If the device is rated 50 amperes, what is the maximum continuous load?

(a) 80 amperes (b) 125 amperes (c) 50 amperes (d) 40 amperes

- Answer: (d) 40 amperes

Step 1: → Convert 80% to a decimal; 0.8
Step 2: → Multiply the number by the decimal; 50 amperes × 0.8 = 40 amperes

1–6 *PARENTHESES*

Whenever numbers are in *parentheses*, we must complete the mathematical function within the parentheses before proceeding with the problem.

❑ **Parentheses Example**

What is the voltage drop of two No. 14 conductors carrying 16 amperes for a distance of 100 feet? Use the following example for the answer:

$$VD = \frac{2 \times K \times I \times D}{CM} = \frac{(2 \text{ wires} \times 12.9 \text{ ohms} \times 16 \text{ amperes x } 100 \text{ feet})}{4{,}110 \text{ circular mils}}$$

(a) 3 volts (b) 3.6 volts (c) 10.04 volts (d) none of these

- Answer: (c) 10.04 volts

Work out the parentheses,

2 wires × 12. 9-ohms resistance × 16 amperes × 100 feet = 41,280
41,280/4,110 circular mils = 10.4 volts dropped

1–7 *PERCENT INCREASE*

Increasing a number by a specific *percentage* is accomplished by:

Step 1: → Converting the percentage to decimal form.
Step 2: → Adding one to the decimal value from Step 1.
Step 3: → Multiplying the number to be increased by the multiplier from Step 2.

❑ **Percent Increase Example**

Increase the whole number 45 by 35%.

(a) 61 (b) 74 (c) 83 (d) 104

- Answer: (a) 61

Step 1: → Convert 35% into .35.
Step 2: → Add one to the decimal value from Step 1; 1 + .35 = 1.35
Step 3: → Multiply the number by the multiplier; 45 × 1.35 = 60.75

1–8 *PERCENTAGE RECIPROCALS*

A *reciprocal* is a whole number converted into a fraction, with the number one as the top number. This fraction is then converted to a decimal form.

Step 1: → Convert the number to a decimal.
Step 2: → Divide the number into one.

❑ **Reciprocal Example No. 1**

What is the reciprocal of 80 percent?

(a) .80 percent (b) 100 percent (c) 125 percent (d) none of these

- Answer: (c) 1.25 or 125 percent

Step 1: → Convert the number to a decimal; 80% = .8

Step 2: → Divide the number into one; $1/0.8 = 1.25$ or 125%

❑ **Reciprocal Example No. 2**

A continuous load requires an overcurrent protection device sized no smaller than 125 percent of the load. What is the maximum continuous load permitted on a 100-ampere overcurrent protection device?

(a) 100 amperes (b) 125 amperes (c) 80 amperes (d) none of these

- Answer: (c) 80 amperes

Step 1: → Convert the number to a decimal; 125% = 1.25

Step 2: → Divide the number into one; $1/1.25 = 0.8$ or 80%

If the overcurrent device is sized no less than 125 percent of the load, the load is limited to 80 percent of the overcurrent protection device (reciprocal). Therefore, the maximum load is limited to 100 amperes× .8 = 80 amperes

1–9 *SQUARING*

Squaring a number is simply multiplying the number by itself such as: $23^2 = 23 \times 23 = 529$

❑ **Squaring Example No. 1**

What is the power consumed, in watts, of a No. 12 conductor that is 200 feet long and has a resistance of 0.4 ohm? The current flowing in the circuit is 16 amperes. Formula: Power = $I^2 \times R$

(a) 50 watts (b) 150 watts (c) 100 watts (d) 200 watts

- Answer: (c) 100 watts

$P = I^2 \times R$, I = 16 amperes, R = 0.4 ohm

$P = 16 \text{ amperes}^2 \times 0.4$ ohm

P = 102.4 watts; answers are rounded to 50s.

❑ **Squaring Example No. 2**

What is the area, in square inches, of a one-inch raceway whose diameter is 1.049 inches?
Use the formula: Area = $\pi \times r^2$, $\pi = 3.14$, r = radius, radius is ½ the diameter.

(a) 1 square inch (b) .86 square inch (c) .34 square inch (d) .5 square inch

- Answer: (b) 0.86 square inch

Raceway area = $\pi \times r^2 = 3.14 \times (½ \times 1.049)^2 = 3.14 \times .5245^2$

$3.14 \times (.5245 \times .5245) = 3.14 \times .2751 = 0.86$ square inch

1–10 *SQUARE ROOT*

The *square root* of a number is the opposite of squaring a number. For all practical purposes, to determine the square root of any number, you must use a calculator with a square root key. For exam preparation purposes, the only number you need to know the square root of is 3, which is 1.732. To multiply, divide, add, or subtract a number by a square root value, simply determine the square root value first, and then perform the math function. The steps to determine the square root of a number follow.

Step 1: → Enter number in calculator.

Step 2: → Press the √_ key of the calculation.

❑ **Example**

What is the $\sqrt{3}$?

(a) 1.55 (b) 1.73 (c) 1.96 (d) none of these

- Answer: (b) 1.732

Step 1: → Type 3:

Step 2: → Press the √_ key = 1.732.

❑ **Square Root Example No. 1**

$\frac{36{,}000 \text{ watts}}{(208 \text{ volts} \times \sqrt{3})}$ is equal to _____amperes?

(a) 120 (b) 208 (c) 360 (d) 100

- Answer: (d) 100

Step 1: → Determine the $\sqrt{3} = 1.732$
Step 2: → Multiply 208 volts × 1.732 = 360 volts
Step 3: → Divide 36,000 watts/360 volts = 100 amperes

❑ **Square Root Example No. 2**

The phase voltage is equal to $\left(\frac{208 \text{ volts}}{\sqrt{3}}\right)$ _____?

(a) 120 volts (b) 208 volts (c) 360 volts (d) none of these

- Answer: (a) 120 volts

Step 1: → Determine the $\sqrt{3} = 1.732$.
Step 2: → Divide 208 volts by 1.732 = 120 volts.

1–11 *ROUNDING OFF*

Numbers below 5 are rounded down, while numbers 5 and above are rounded up. *Rounding* to three significant figures should be sufficient for most calculations, such as

.12459 – the fourth number is 5 or above = .125 rounded up
1.6744 – the fourth number is below 5 = 1.67 rounded down
21.996 – the fourth number is 5 or above = 22 rounded up
367.28 – the fourth number is below 5 = 367 rounded down

Rounding For Exams

You should always round off your answer in the same manner as the answers to an exam question. Do not choose any of these in an exam until you have checked the answers for rounding off. If after rounding off your answer to the exam format there is no answer, choose "none of these."

❑ **Rounding Example**

The sum of 12, 17, 28, and 40 is equal to _______?

(a) 80 (b) 90 (c) 100 (d) none of these

- Answer: (c) 100

The answer is actually 97, but there is no 97 as a choice. Do not choose "none of these" in an exam until you have checked how the choices are rounded off. The choices in this case are all rounded off to the nearest tens.

1–12 *TRANSPOSING FORMULAS*

Transposing is an algebraic function used to rearrange formulas. The formula $I = P/E$ can be transposed to $P = I \times E$ or $E = P/I$, Fig. 1–3.

❑ **Transpose Example**

Transpose the formula $CM = \frac{(2 \times K \times I \times D)}{VD}$, to find the voltage drop of the circuit, Fig. 1–4.

(a) $VD = CM$

(b) $VD = (2 \times K \times I \times D)$

(c) $VD = (2 \times K \times I \times D) \times CM$

(d) none of these

- Answer: (d) none of these

$$VD = \frac{(2 \times K \times I \times D)}{CM}$$

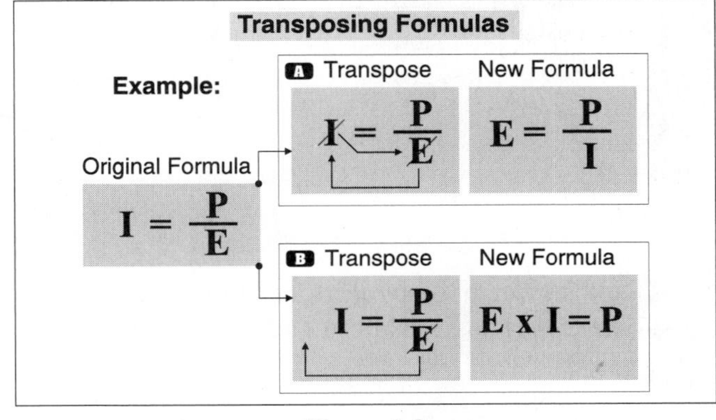

Figure 1-3

Transpose Formulas

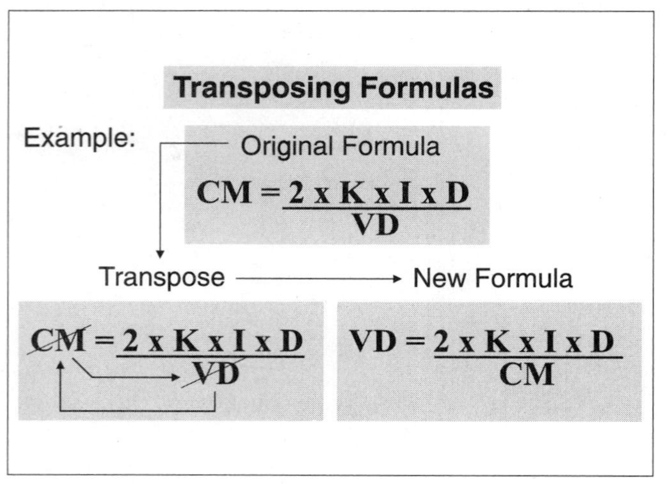

Figure 1-4

Transpose Example

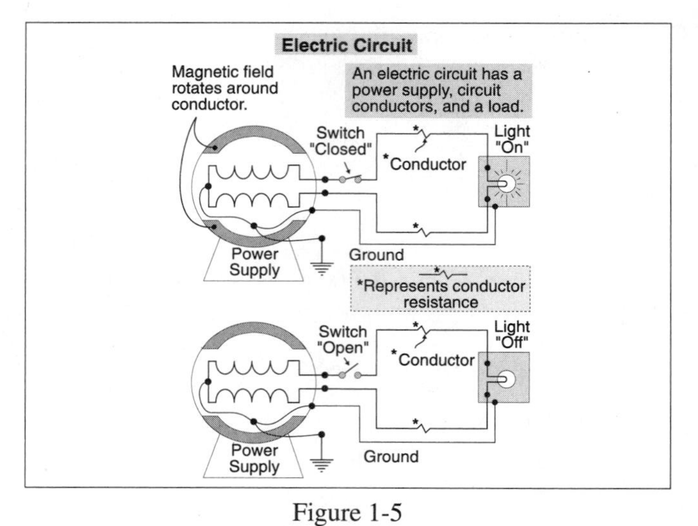

Figure 1-5

The Electrical Circuit

PART B – *BASIC ELECTRICAL FORMULAS*

1–13 *ELECTRICAL CIRCUIT*

An electric circuit consists of the power source, the conductors, and the load. For current to travel in the circuit, there must be a complete path from one terminal of the power supply, through the conductors and the load, back to the other terminal of the power supply, Fig. 1–5.

1–14 *ELECTRON FLOW*

Inside a direct current power source (such as a battery) the electrons travel from the positive terminal to the negative terminal. However, outside of the power source, electrons travel from the negative terminal to the positive terminal, Fig. 1–6.

1–15 *POWER SOURCE*

In any completed circuit, it takes a force to push the electrons through the power source, conductor, and load. The two most common types of power sources are *direct current* and *alternating current.*

Direct Current

The polarity from direct current power sources never changes; that is, the current flows out of the negative terminal of the power source always in the same direction. When the power supply is a *battery*, the polarity and the magnitude remain the same, Fig. 1–7.

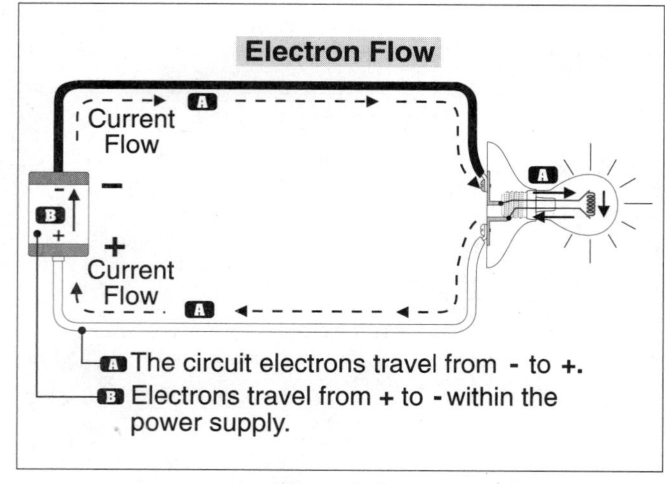

Figure 1-6

Electron Flow

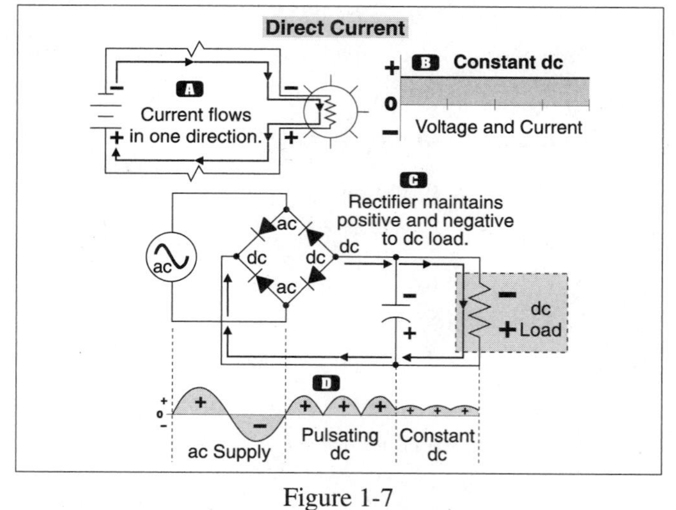

Figure 1-7

Direct Current – Constant Polarity

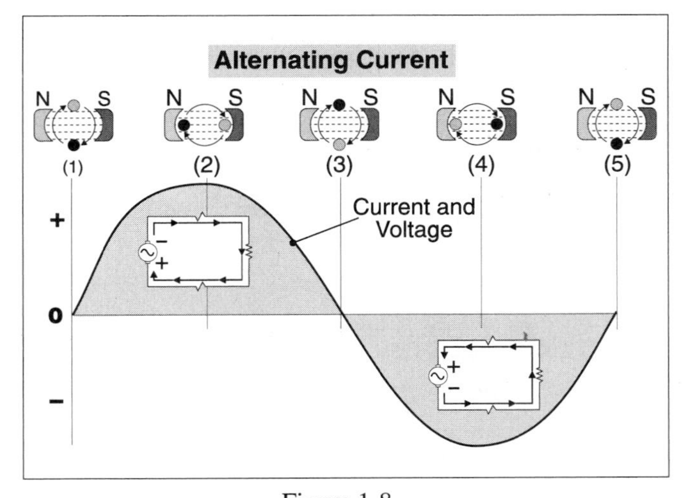

Figure 1-8

Alternating Current – Alternating Polarity

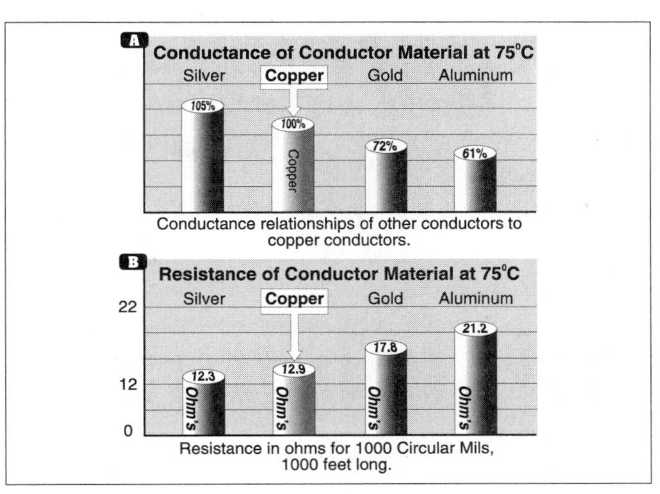

Figure 1-9

Conductance and Resistance

Alternating Current

Alternating current power sources produce a voltage and current that has a constant change in polarity and magnitude at a constant frequency. Alternating current flow is produced by a *generator* or an *alternator*. Fig. 1–8.

1–16 *CONDUCTANCE AND RESISTANCE*

Conductance

Conductance is the property of metal that permits current to flow. The best conductors, in order of their conductivity are: silver, copper, gold, and aluminum. Although silver is a better conductor of electricity than copper, copper is used more frequently because it is less expensive, Fig. 1–9 Part A.

Resistance

Resistance is the opposite of conductance. It is the property that opposes the flow of electric current. The resistance of a conductor is measured in ohms according to a standard length of 1,000 feet , Fig. 1–9 Part B. This value is listed in the National Electrical Code, Chapter 9, Table 8, for direct current circuits and Chapter 9, Table 9, for alternating current circuits.

1–17 *ELECTRICAL CIRCUIT VALUES*

In an electrical circuit there are four circuit values that we must understand. They are voltage, resistance, current and power, Fig. 1–10.

Voltage

Electron pressure is called *electromotive force* and is measured by the unit *volt*, abbreviated by the letter *E* or *V*. Voltage is also a term used to described the difference of potential between any two points.

Resistance

The friction opposition to the flow of electrons is called *resistance*, and the unit of measurement is the *ohm*, abbreviated by the letter *R*. Every component of an electric circuit contains resistance including the power supply.

Current

Free electrons moving in the same direction in a conductor produce an electrical *current* sometimes called *intensity*. The rate at which electrons move is measured by the unit called *ampere*, abbreviated by the letter *I* or *A*.

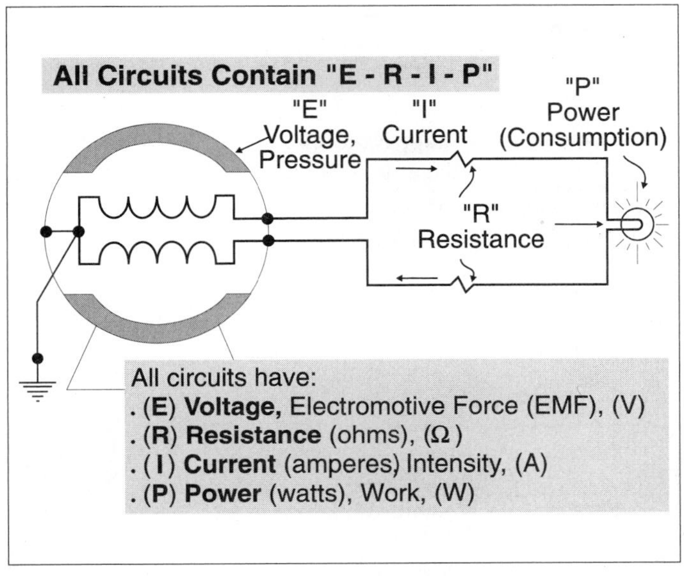

Figure 1-10

Electrical Circuit Values

Power

The rate of work that can be produced by the movement of electrons is called *power*, and the unit is the *watt.* It is very common to see the *W* symbol for watts used instead of *P* for power.

Note: A 100 watt lamp will consume 100 watts per hour.

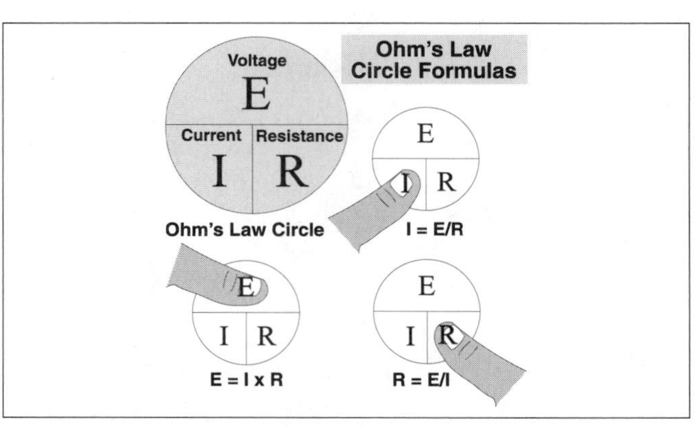

Figure 1-11

Ohm's Law Formulas

1–18 *OHM'S LAW I = E/R*

Ohm's law, $I = E/R$, demonstrates the relationship between current, voltage, and resistance in a direct current, or an alternating current circuit that supplies only resistive loads, Fig. 1–11.

$$I = \frac{E}{R} \qquad E = I \times R \qquad R = \frac{E}{I}$$

This law states that:

1. **Current is directly proportional to the voltage.** This means that if the voltage is increased by a given percentage, the current would increase by that same percentage; or if the voltage is decreased by a given percentage, the current will decrease by the same percentage, Fig. 1–12 Part A.
2. **Current is inversely proportional to the resistance.** This means that an increase in resistance will result in a decrease in current, and a decrease in resistance will result in an increase in current, Fig. 1–12 Part B.

Opposition To Current Flow

In a direct current circuit, the physical resistance of the conductor opposes the flow of electrons. In an alternating current circuit, there are three factors that oppose current flow. Those factors are *conductor resistance*, *inductive reactance*, and *capacitive reactance*. The opposition to current flow, due to a combination of resistance and *reactance*, is called *impedance*, measured in ohms , and abbreviated with the letter Z. Impedance will be covered later, so for now assume that all circuits have very little or no reactance.

❑ **Ohm's Law Ampere Example**

A 120-volt power source supplies a lamp that has a resistance of 192 ohms. What is the current flow of the circuit, Fig. 1–13.

(a) 0.6 ampere (b) 0.5 ampere (c) 2.5 amperes (d) 1.3 amperes

- Answer: (a) .6 amperes

Step 1: → What is the current, I?
Step 2: → What do you know? E = 120 volts, R = 192 ohms
Step 3: → The formula is $I = E/R$.
Step 4: → The answer is

$$I = \frac{120 \text{ volts}}{192 \text{ ohms}} = 0.625 \text{ ampere}$$

❑ **Ohm's Law Voltage Example**

What is the voltage drop of two No. 12 conductors that supply a 16-ampere load located 50 feet from the power supply? The total resistance of both conductors is 0.2 ohm, Fig. 1–14.

(a) 16 volts (b) 32 volts
(c) 1.6 volts (d) 3.2 volts

- Answer: (d) 3.2 volts

Step 1: → What is the question? It is, what is voltage drop, E?
Step 2: → What do you know about the conductors? I = 16 amperes, R = 0.2 ohm.
Step 3: → The formula is $E = I \times R$.
Step 4: → The answer is $E = 16 \text{ amperes} \times .2 \text{ ohm}$, $E = 3.2$ volts.

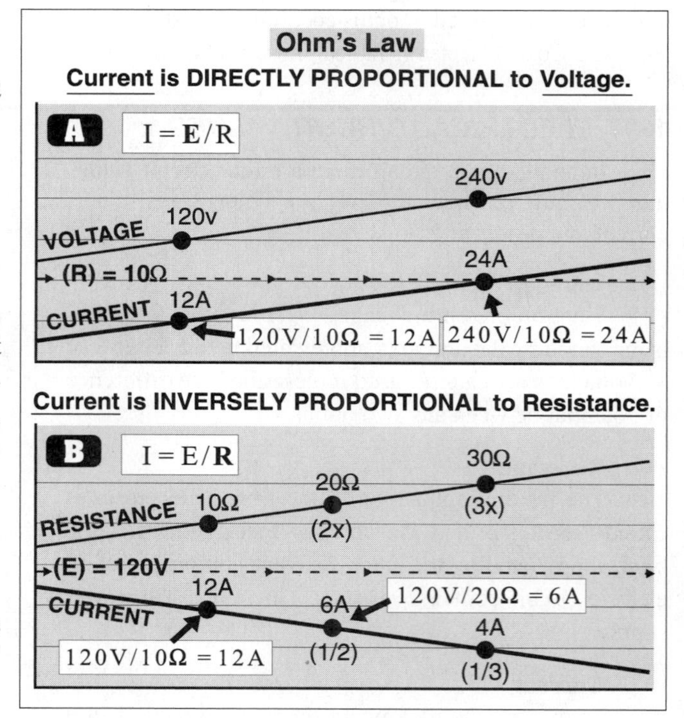

Figure 1-12

Part A – Current Proportional to Voltage
Part B – Current Inversely to Resistance

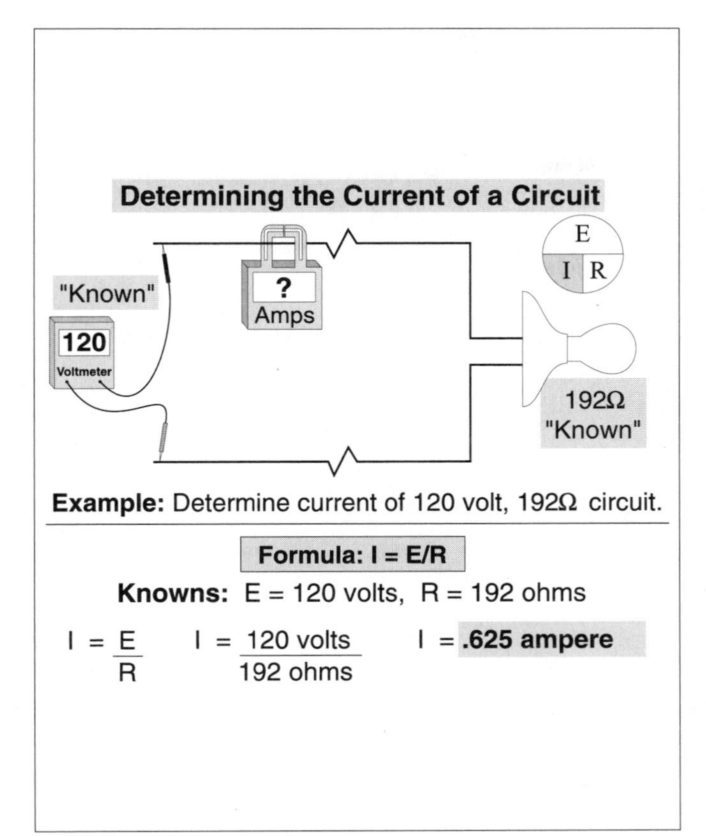

Figure 1-13

Determining the Current of a Circuit

Determining Voltage Drop With Ohm's Law

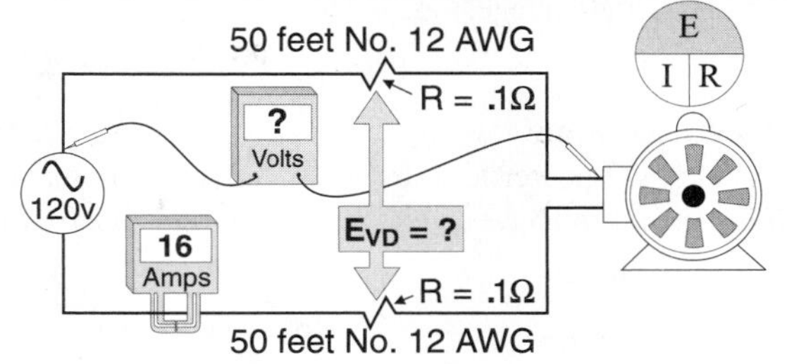

Example: Determine conductor VD on a 120 volt circuit.

Formula: $E_{VD} = I \times R$

To determine the voltage drop of conductors, use resistance of conductors.

Known: I = 16 Amperes (given)
Known: R of each Conductor = .1 ohm

$E_{VD} = I \times R$ E_{VD} = 16 amps x .1 ohm

E_{VD} = **1.6 volts per conductor**

Note: Voltage drop of both conductors = 16 amperes x .2 ohm = 3.2 volts
Note: Load operates at 120 volts - 3.2 vd = 116.8 volts

Figure 1-14

Voltage Example

❑ Ohms Law Resistance Example

What is the resistance of the circuit conductors when the conductor voltage drop is 3 volts and the current flowing through the conductors is 100 amperes, Fig. 1–15?

(a) .03 ohm
(b) 0.2 ohm
(c) 3 ohms
(d) 30 ohms

• Answer: (a) .03 ohm

Step 1: → What is the question? It is, what is resistance, R?

Step 2: → What do you know about the conductors?
E = 3 volts drop,
I = 100 amperes

Step 3: → The formula is $R = E/I$

Step 4: → The answer is

$$R = \frac{3 \text{ volts}}{100 \text{ amperes}} = R = .03 \text{ ohm}$$

Determining Resistance of Conductors

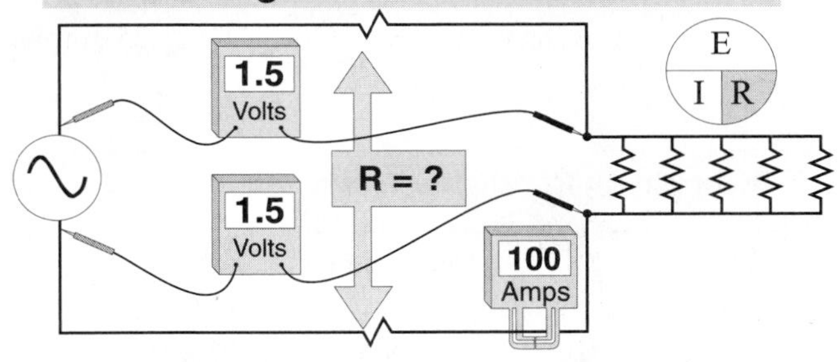

Example: Determine resistance of conductors.

Formula: R = E/I

Known: E_{VD} = 1.5 volts per conductor
Known: I = 100 amperes

$R = \frac{E}{I}$ $R = \frac{1.5 \text{ volts}}{100 \text{ amp}}$

R = **.015 ohm per conductor**

R = .015Ω x 2 conductors = **.03 ohm for both conductors**

OR... $R = \frac{3 \text{ volts}}{100 \text{ amps}}$ = **.03 ohm for both conductors**

Figure 1-15

Resistance Example

1–19 *PIE CIRCLE FORMULA*

The PIE circle formula, shows the relationships between power, current, and voltage, Fig. 1–16.

P = E × I **I = P/E** **E = I × R**

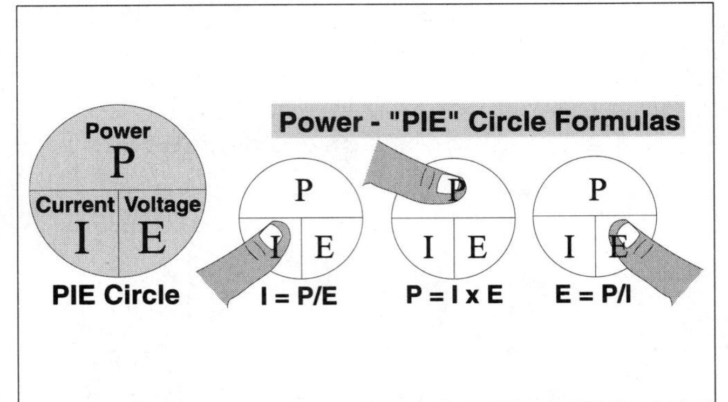

Figure 1-16

PIE Formula Circle

❑ **Power Example**

What is the power loss, in watts, for two conductors that carry 12 amperes and have a voltage drop of 3.6 volts, Fig. 1–17?

(a) 4.3 watts (b) 43 watts

(c) 432 watts (d) none of these

- Answer: (b) 43 watts

Step 1: → What is the question? It is, what is the power, P?

Step 2: → What do you know?
I = 12 amperes, E = 3.6 volts drop

Step 3: → The formula is P = I × E.

Step 4: → The answer is
P = 12 amperes × 3.6 volts = 43.2 watts per hour.

❑ **Current Example**

What is the current flow, in amperes, in the circuit conductors that supply a 7.5-kW heat strip rated 240 volts when connected to a 240-volt power supply, Fig. 1–18?

(a) 25 amperes (b) 31 amperes (c) 39 amperes (d) none of these

- Answer: (b) 31 amperes

Step 1: → What is the question? It is, what is the current, I?

Step 2: → What do you know?
P = 7,500 watts, E = 240 volts

Step 3: → The formula is I = P/E.

Step 4: → The answer is $I = \frac{7{,}500 \text{ watts}}{240 \text{ volts}} = 31.25$ amperes.

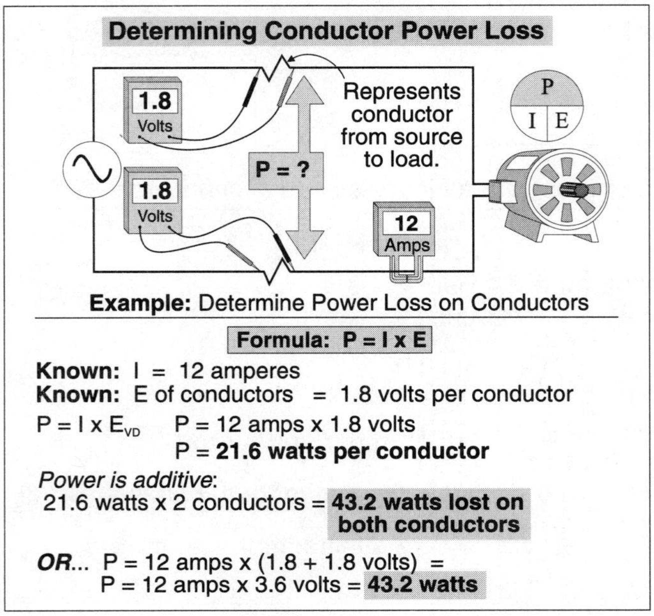

Determining Conductor Power Loss

1.8 Volts

1.8 Volts

P = ?

Represents conductor from source to load.

P I E

12 Amps

Example: Determine Power Loss on Conductors

Formula: P = I x E

Known: I = 12 amperes
Known: E of conductors = 1.8 volts per conductor

$P = I \times E_{VD}$ P = 12 amps x 1.8 volts
P = **21.6 watts per conductor**

Power is additive:
21.6 watts x 2 conductors = **43.2 watts lost on both conductors**

OR... P = 12 amps x (1.8 + 1.8 volts) =
P = 12 amps x 3.6 volts = **43.2 watts**

Figure 1-17

Power Example

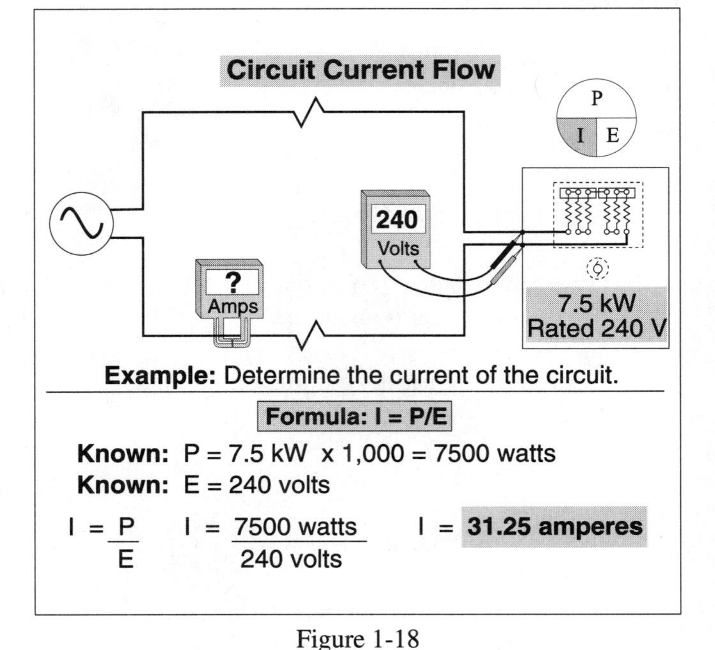

Figure 1-18

Current Example

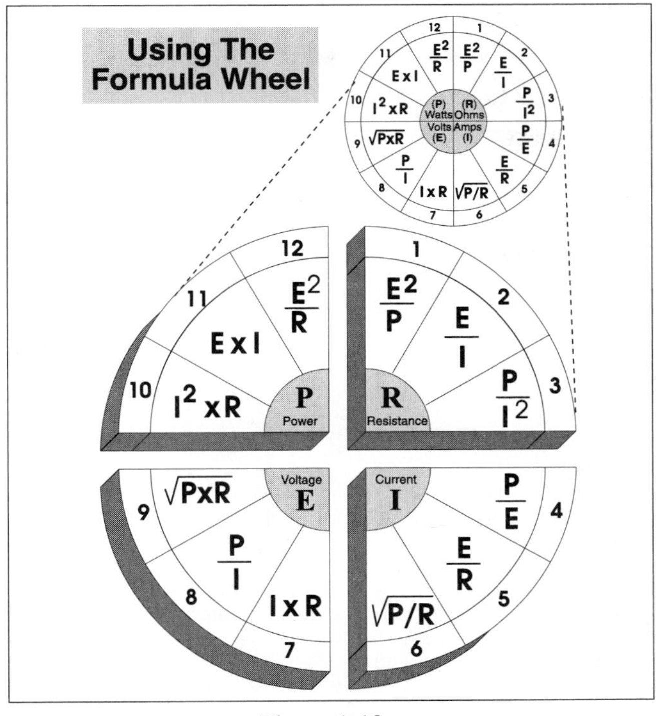

Figure 1-19

Power Wheel

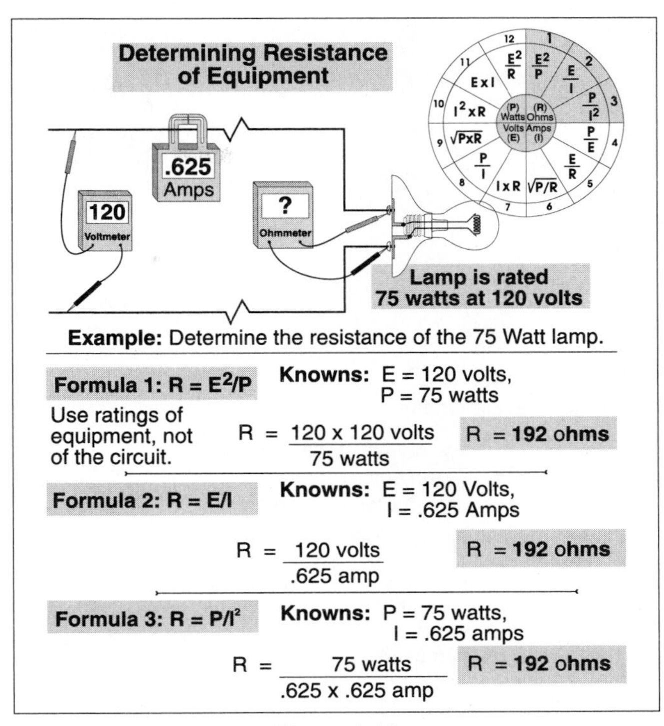

Figure 1-20

Resistance Example

1–20 *FORMULA WHEEL*

The formula wheel combines the Ohm's law and the PIE formulas. The formula wheel is divided into four sections with 3 formulas in each section, Fig. 1–19.

❑ **Resistance Example**

What is the resistance of a 75-watt light bulb that is rated 120 volts, Fig. 1–20?

(a) 100 ohms (b) 192 ohms

(c) 225 ohms (d) 417 ohms

- Answer: (b) 192 ohms

  $R = E^2/P$

  E = 120-volt rating

  P = 75 watt rating

  R = 120 volts^2/75 watts

  R = 192 ohms

❑ **Current Example**

What is the current flow of a 10-kW heat strip connected to a 230-volt power supply, Fig. 1–21?

(a) 13 amperes (b) 26 amperes

(c) 43 amperes (d) 52 amperes

- Answer: (c) 43 amperes

  I = P/E

  P = 10,000 watts

  E = 230 volts

  I = 10,000 watts/230 volts

  I = 43 amperes

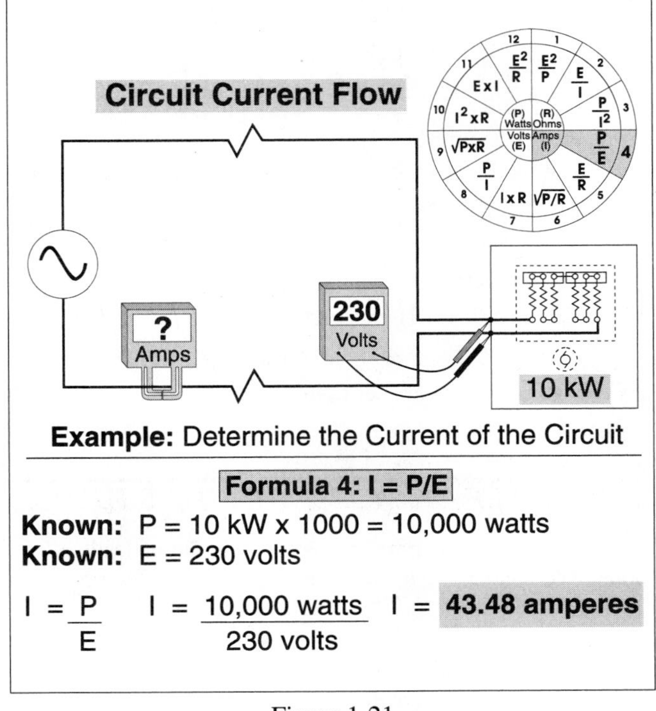

Figure 1-21

Current Example

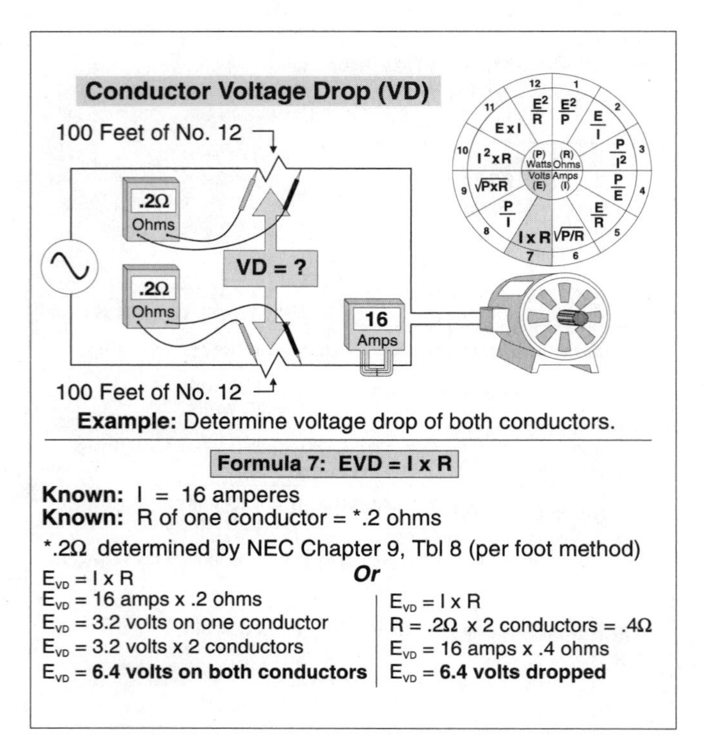

Figure 1-22

Voltage Example

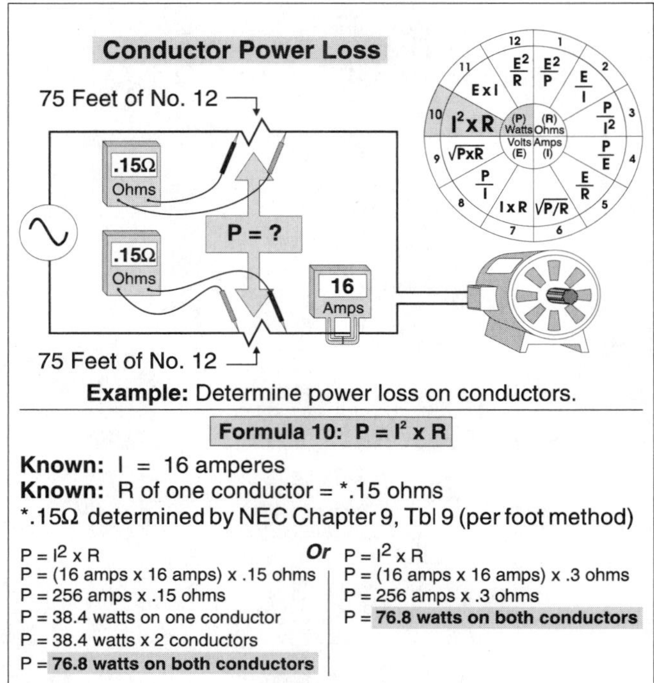

Figure 1-23

Power Example

❑ **Voltage Example**

What is the voltage drop of 200 feet of No. 12 conductor that carries 16 amperes, Fig. 1–22?

Note: The resistance of No. 12 conductor is 2-ohms per 1,000 feet.

(a) 1.6 volts drop (b) 2.9 volts drop (c) 3.2 volts drop (d) 6.4 volts drop

- Answer: (d) 6.4 volts drop

  $E = I \times R$, $I = 16$ amperes, $R = 0.2$ ohm, 2 ohms/1,000 = .002-ohm per foot × 200 = .4 ohm

  $E = 16$ amperes $\times$ 0.4 ohm

  $E = 6.4$ volts drop

❑ **Power Example**

The total resistance of two No. 12 conductors, 75 feet long, is 0.3 ohm (.15 ohm for each conductor). The current of the circuit is 16 amperes. What is the power loss of the conductors in watts per hour, Fig. 1–23?

(a) 19 watts (b) 77 watts (c) 172.8 watts (d) none of these

- Answer: (b) 77 watts

  $P = I^2R$, $I = 16$ amperes, $R = 0.3$ ohm

  $P = 16$ amperes$^2 \times 0.3$ ohm

  $P = 76.8$ watts per hour

1–21 *POWER CHANGES WITH THE SQUARE OF THE VOLTAGE*

The *power* consumed by a resistor is dramatically affected by the voltage applied. The power is affected by the square of the voltage and directly to the resistance, Fig. 1–24.

$$P = \frac{E^2}{R}$$

❑ **Power Changes With Square Of Voltage Example**

What is the power consumed of a 9.6-kW heat strip rated 230 volts connected to 115-, 230- and 460-volt power supplies? The resistance of the heat strip is 5.51 ohms, Fig. 1–25.

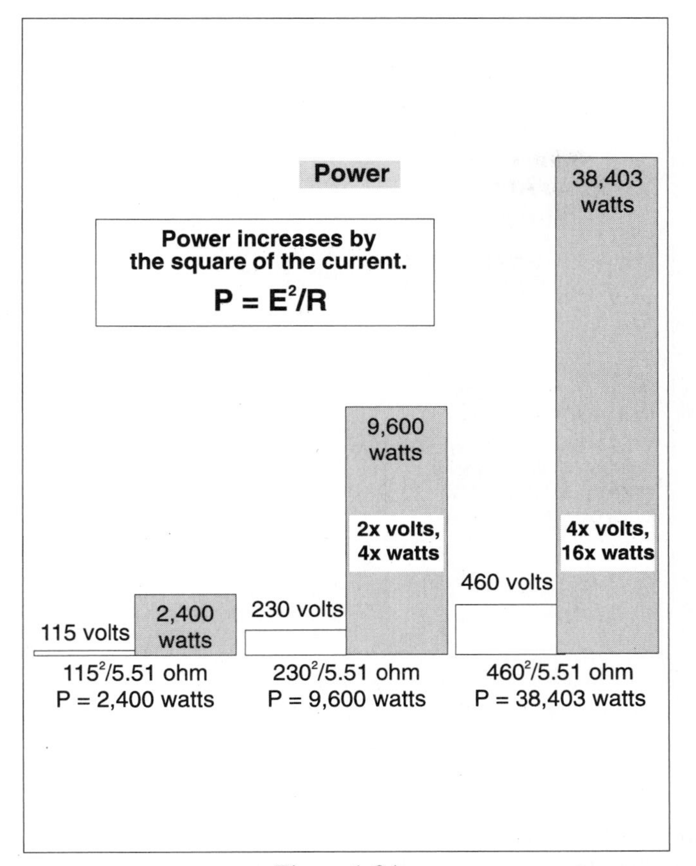

Figure 1-24

Power Calculations

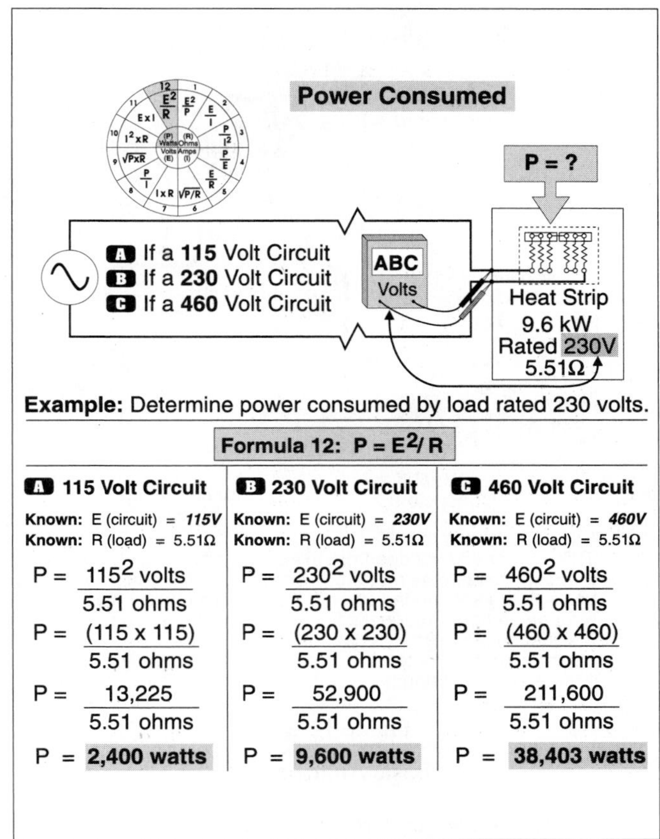

Figure 1-25

Power Changes with the Square of the Voltage

Step 1: → What is the question? It is, what is the power consumed, P?

Step 2: → What do you know about the heat strip?
E = 115 volts, 230 volts and 460 volts, R = 5.51 ohms

Step 3: → The formula to determine power is, $P = E^2/R$.
E = 115 volts, 230 volts, and 460 volts, R = 5.51 ohms
P = 115 volts^2/5.51 ohms, P = 2,400 watts
P = 230 volts^2/5.51 ohms, P = 9,600 watts (2 times volts = 4 times power)
P = 460 volts^2/5.51 ohms, P = 38,403 watts (4 times volts = 16 times power)

1–22 *ELECTRIC METERS*

Basic electrical meters use a helically wound coil of conductor, called a *solenoid*, to produce a strong electromagnetic field to attract an iron bar inside the coil. The iron bar that moves inside the coil is called an *armature*. When a meter has positive (+) and negative (–) shown for the meter leads, the meter is said to be *polarized*. The negative (–) lead must be connected to the negative terminal, and the positive (+) lead must be connected to the positive terminal of the power source.

Ammeter

An ammeter is a meter that has a *helically* wound coil, and it directly utilizes the circuit energy to measure direct current. As current flows through the meter's helically wound coil, the combined electromagnetic field of the coil draws in the iron bar. The greater the current flow through the meter's coil, the greater the electromagnetic field, and the further the armature is drawn into the coil. Ammeters are connected in series with the circuit and are used to measure only direct current, Fig. 1–26.

Ammeters are connected in *series* with the power supply and the load. If the ammeter is accidentally connected in *parallel* to the power supply, the current flow through the meter will be extremely high. The excessive high current through the meter will destroy the meter due to excessive heat.

If the ammeter is not connected to the proper *polarity* when measuring direct current, the meter's needle will quickly move in the reverse direction and possibly damage the meter's calibration.

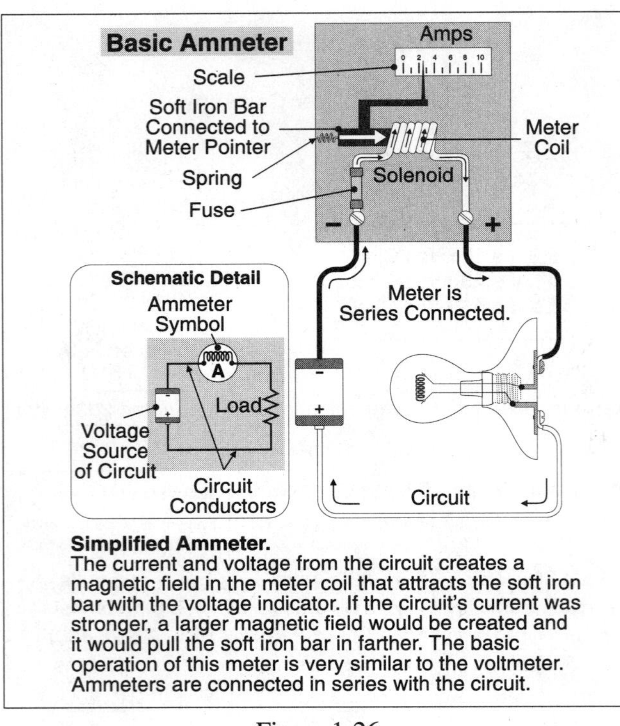

Figure 1-26

Basic Ammeter

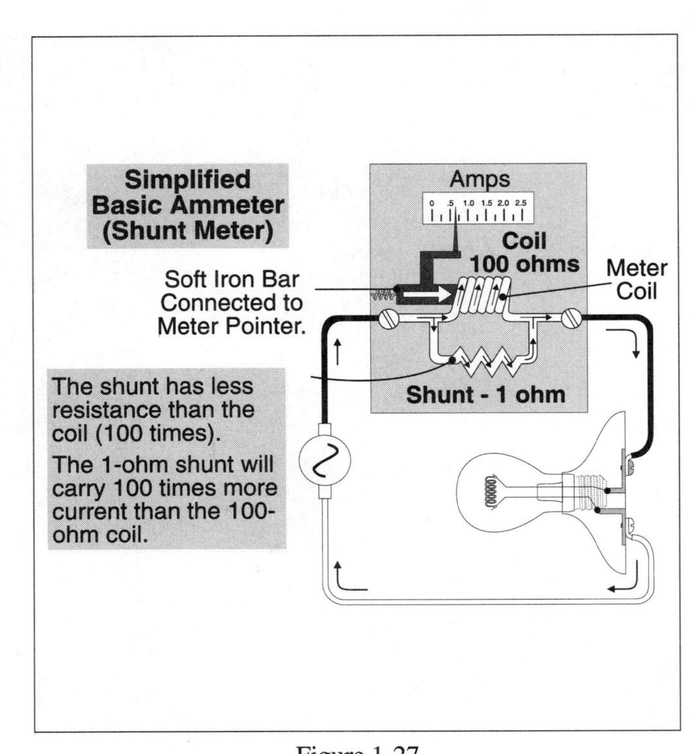

Figure 1-27

Shunt Ammeter Current Example

Ammeters that measure currents larger than 10/1,000 amperes (10 milliamperes, 10 mA) often contain a device called a *shunt*. The shunt bar is placed in parallel with the meter coil. This permits the current flow to divide between the meter's coil and the shunt bar. The current through the meter coil depends on the resistance of the shunt bar.

❑ **Shunt Ammeter Current Example**

What is the current flow through the meter if the shunt bar is 1 ohm and the coil is 100 ohms, Fig. 1–27?

(a) the same as the shunt
(b) 10 times less than the shunt
(c) 100 times less than the shunt
(d) 1,000 times less than the shunt

• Answer: (c) 100 times less

Since the shunt is 100 times less resistant than the meter's coil, the shunt bar will carry 100 times more current than the meter's coil, or the meter's coil will carry 100 times less current than the shunt bar.

Clamp-on Ammeter

Clamp-on ammeters are used to measure alternating current. They are connected *perpendicular* around the conductor (90 degrees) without breaking the circuit. A clamp-on ammeter indirectly utilizes the circuit energy by *induction* of the electromagnetic field. The clamp on ammeter is actually a transformer (sometimes called a *current transformer*). The primary winding is the phase conductor, and the secondary winding is the meter's coil, Fig. 1–28.

The electromagnetic field around the phase conductor expands and collapses, which causes electrons to flow in the meter's circuit. As current flows through the meter's coil, the electromagnetic field of the meter draws in the armature. Since the phase conductor serves as the primary (one turn), the current to be measured must be high enough to produce an electromagnetic field that is strong enough to cause the meter to operate.

Ohmmeter and Megohmeter

Ohmmeters are used to measure the resistance of a circuit or component and can be used to locate open circuits or shorts. An ohmmeter has an armature, a coil and it's own power supply, generally a battery. Ohmmeters are always connected to deenergized circuits and polarity is not required to be observed. When an ohmmeter is used, current flows through the meter's coil causing an electromagnetic field around the coil and drawing in the armature. The greater the current flow, as a result of lower resistance ($I = E/R$), the greater the magnetic field and the further the armature is drawn into the coil. A short circuit will be indicated by a reading of zero and a circuit that is opened will be indicated by infinity (∞), Fig. 1–29.

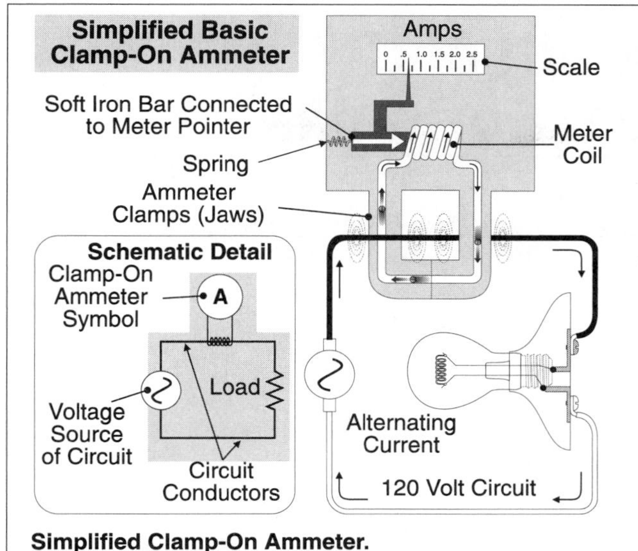

Simplified Clamp-On Ammeter.
The current from the circuit induces a current in the clamps of the ammeter, which creates current in the coil of the meter. This creates a magnetic field in the meter coil that attracts the soft iron bar with the ammeter indicator. If the circuit's current was stronger, a larger magnetic field would be created and it would pull the soft iron bar in farther. The clamps of the ammeter must be perpendicular to the circuit conductor to get an accurate reading.

Figure 1-28

Clamp–on Ammeters

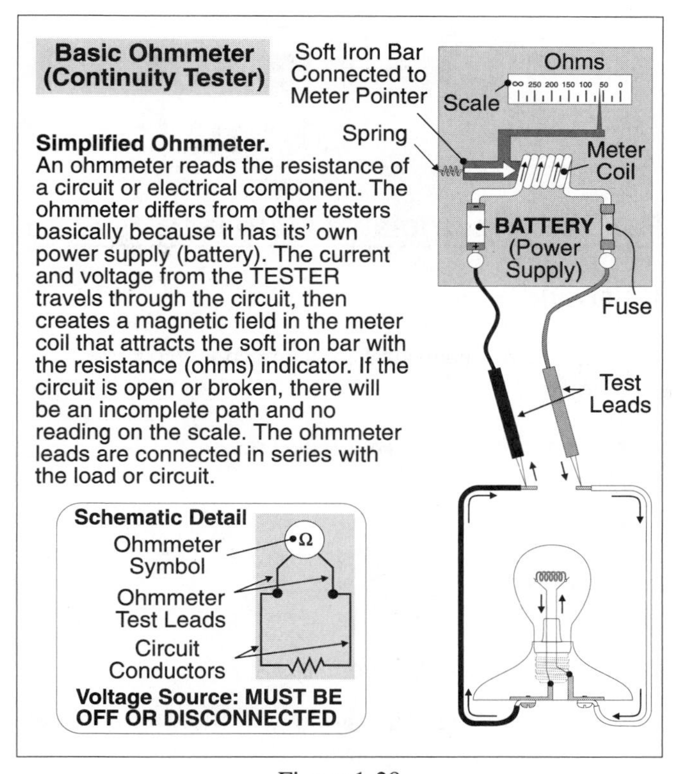

Figure 1-29

Ohm Meter

The *megger*, also called *megohmmeter* or *megohmer* is an instrument designed to measure very high resistances, such as those found in cable insulation between motor or transformer windings.

Voltmeter

Voltmeters are used to measure both direct current and alternating current voltage. A voltmeter contains a *resistor* in *series* with the coil and utilizes the circuit energy for its operation. The purpose of the resistor in the meter is to reduce the current flow through the meter. As current flows through the meter's helically wound coil, the combined electromagnetic field of the coil draws in the iron bar. The greater the circuit voltage, the greater the current flow through the meter's coil ($I=E/R$). The greater the current flow through the meter's coil, the greater the electromagnetic field and the further the armature is drawn into the coil, Fig. 1–30.

Polarity must be observed when connecting voltmeters to direct current circuits. If the meter is not connected to the proper polarity, the meter's needle will quickly move in the reverse direction and damage the meter's calibration. Polarity is not required when connecting voltmeters to alternating current circuits.

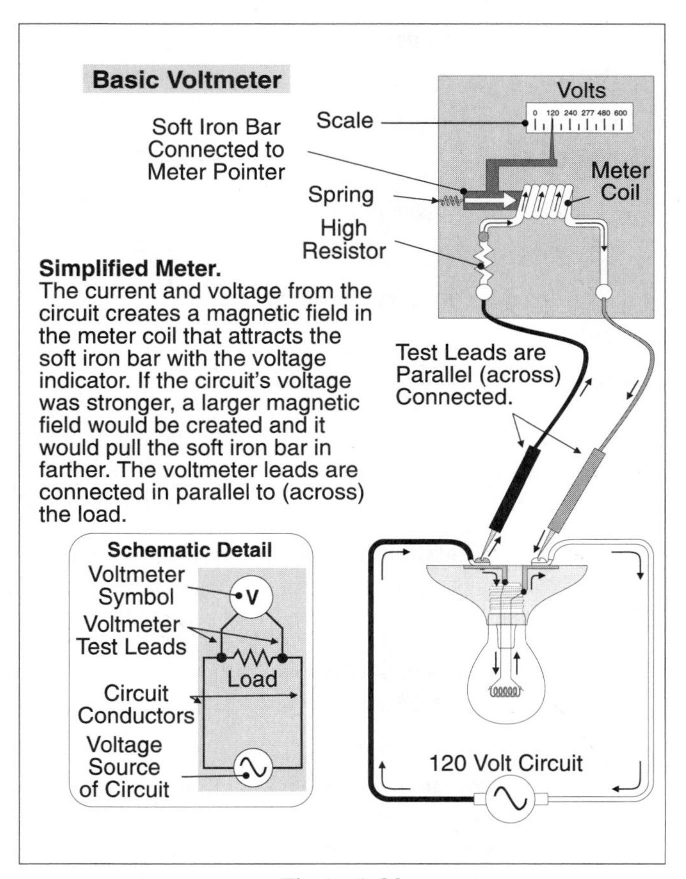

Figure 1-30

Voltmeter

Unit 1 – Electrician's Math and Basic Electrical Formulas

Part A – Electrician's Math (• Indicates that 75% or less get the question correct)

1–1 Fractions

1. The decimal equivalent for the fraction ½ is ________.
(a) .5 (b) 5 (c) 2 (d) .2

2. The decimal equivalent for the fraction $4/18$ is ________.
(a) 4.5 (b) 1.5 (c) 2.5 (d) .2

1–2 Kilo

3. What is the kW of a 75-watt load?
(a) 75 kW (b) 7.5 kW (c) .75 kW (d) .075 kW

1–3 Knowing Your Answer

4. • The output VA of a transformer is 100 VA. The transformer efficiency is 90 percent. What is the transformer input? Note: Because of efficiency, input is always greater than output.
(a) 90 watts (b) 110 watts (c) 100 watts (d) 125 watts

1–4 Percentages

5. When changing a percent value to a decimal or whole number, simply move the decimal point two places to the ________.
(a) right (b) left (c) depends (d) none of these

6. The decimal equivalent for 75 percent is ________.
(a) .075 (b) .75 (c) 7.5 (d) 75

7. The decimal equivalent for 225 percent is________.
(a) 225 (b) 22.5 (c) 2.25 (d) .225

8. The decimal equivalent for 300 percent is ________.
(a) .03 (b) .3 (c) 3 (d) 30.0

1–5 Multiplier

9. The method of increasing a number by another number is called the ________.
(a) percentage (b) decimal
(c) fraction (d) multiplier

10. An overcurrent protection device (breaker or fuse) is required to be sized no less than 115 percent of the load. If the load is 20 amperes, the overcurrent protection device would have to be sized at no less than ________.
(a) 20 amperes (b) 23 amperes (c) 17 amperes (d) 30 amperes

11. The maximum continuous load on an overcurrent protection device is limited to 80 percent of the device rating. If the device is rated 90 amperes, the maximum continuous load is ________ amperes.
(a) 72 (b) 90 (c) 110 (d) 125

12. A 50-ampere rated wire is required to be adjusted for temperature. If the correct multiplier is .80, which of the following statements is/are correct?
(a) the answer will be less than 50 amperes
(b) 80 percent of the ampacity (50) can be used
(c) the formula is 50 amperes × .8
(d) all the above

1–6 Percent Increase

13. The feeder demand load for an 8-kW load, increased by 20 percent is ________ kVA.
(a) 8 (b) 9.6 (c) 6.4 (d) 10

1–7 Percentage Reciprocals

14. What is the reciprocal of 125 percent?
(a) .8 (b) 100 percent
(c) 125 percent (d) none of these

15. A continuous load requires an overcurrent protection device sized no smaller than 125 percent of the load. What is the maximum continuous load permitted on a 100-ampere overcurrent protection device?
(a) 100 amperes (b) 125 amperes
(c) 80 amperes (d) 110 amperes

1–8 Squaring

16. What is the power consumed in watts of a No. 12 conductor that is 100 feet long and has a resistance of (R) 0.2 ohm, when the current (I) in the circuit is 16 amperes? Formula: Power $= I^2 \times R$
(a) 75 watts (b) 50 watts
(c) 100 watts (d) 200 watts

17. • What is the area in square inches of a 2-inch raceway?
Formula: Area $= \pi r^2$, $\pi = 3.14$, r = radius, which equals ½ the diameter.
(a) 1 square inch (b) 2 square inches
(c) 3 square inches (d) 4 square inches

18. The numeric equivalent of 4^2 is ________.
(a) 2 (b) 8 (c) 16 (d) 32

19. The numeric equivalent of 12^2 is ________.
(a) 3.46 (b) 24 (c) 144 (d) 1,728

1–9 Square Root

20. What is the square root of 1,000 ($\sqrt{1000}$)?
(a) 3 (b) 32 (c) 100 (d) 500

21. The square root of the number 3 is ________.
(a) 1.732 (b) 9 (c) 729 (d) 1.5

1–10 Rounding Off

22. • The sum of 5, 7, 8 and 9 is ________ approximately.
(a) 20 (b) 25 (c) 30 (d) 35

1–11 Parentheses

23. What is the distance of two No. 14 conductors carrying 16 amperes with a voltage drop of 10 volts? Formula:
$$D = \frac{(4{,}100 \text{ circularmils} \times 10 \text{ volts drop})}{(2 \text{ wires} \times 12.9 \text{ amperes} \times 16 \text{ ohms})}$$
(a) 50 feet (b) 75 feet (c) 100 feet (d) 150 feet

24. What is the current in amperes of a three-phase, 18-kW, 208-volt load? Formula: $I = \frac{W}{E \times \sqrt{3}}$
(a) 25 amperes (b) 50 amperes
(c) 100 amperes (d) 150 amperes

1–12 Transposing Formulas

25. • Transpose the formula $CM = \frac{(2 \times K \times I \times D)}{VD}$ to find the distance of the circuit.

(a) $D = VD \times CM$
(b) $D = \frac{(2 \times K \times I \times C \times M)}{VD}$
(c) $D = (2 \times K \times I) \times CM$
(d) $D = \frac{CM \times VD}{2 \times K \times I}$

26. If $I = P/E$, which of the following statements contain the correct transposed formula?

(a) $P = E/I$
(b) $P = I/E$
(c) $P = I \times E$
(d) $P = I^2E$

Part B – Basic Electrical Formulas

1–13 Electrical Circuits

27. An electric circuit consists of the ________.

(a) power source
(b) conductors
(c) load
(d) all of these

1–14 Electron Flow

28. Inside the power source, electrons travel from the positive terminal to the negative terminal.

(a) True
(b) False

1–15 Power Source

29. The polarity of ________ current power sources never change. One terminal is always negative and the other is always positive. ________ current flows out of the negative terminal of the power source at the same polarity.

(a) static, Static
(b) direct, Direct
(c) alternating, Alternating
(d) all of the above

30. ________ current power sources produce a voltage that has a constant change in polarity and magnitude in one direction exactly the same as it does in the other.

(a) Static
(b) Direct
(c) Alternating
(d) All of the above

1–16 Conductance And Resistance

31. Conductance is the property of metal that permits current to flow. The best conductor, in order of conductivity, are: ________.

(a) gold, silver, copper, aluminum
(b) copper, gold, copper, aluminum
(c) gold, copper, silver, aluminum
(d) silver, copper, gold, aluminum

1–17 Electrical Circuit Values

32. The ________ is the pressure required to force one ampere of electrons through a one ohm resistor.

(a) ohm
(b) watt
(c) volt
(d) ampere

33. All conductors have resistance that opposes the flow of electrons. Some materials have more resistance than others. ________ has the lowest resistance and ________ is more resistant than copper.

(a) Silver, gold
(b) Gold, aluminum
(c) Gold, silver
(d) None of these

34. Resistance is represented by the letter *R*, and it is expressed in ________.

(a) volts
(b) impedance
(c) capacitance
(d) ohms

35. The opposition to the flow of current can be thought of as restricting the flow of electrons in the circuit. Every component of an electric circuit contains resistance, except the generator or transformer.
(a) True (b) False

36. In electrical systems, the volume of electrons that moves through a conductor is called the ________ of the circuit.
(a) intensity (b) voltage (c) power (d) resistance

37. The rate of work that can be produced by the movement of electrons is called ________.
(a) voltage (b) current (c) power (d) none of these

1–18 Ohm's Law ($I = E/R$)

38. The Ohms's law formula, $I = E/R$ demonstrates that current is ________ proportional to the voltage, and ________ proportional to the resistance.
(a) indirectly, inversely (b) inversely, directly
(c) inversely, indirectly (d) directly, inversely

39. In an alternating current circuit, which factors oppose current flow?
(a) resistance (b) capacitance reactance
(c) induction reactance (d) all of these

40. The opposition to current flow due in an alternating current circuit is called ________ and is often represented by the letter Z.
(a) resistance (b) capacitance (c) induction (d) impedance

41. • What is the voltage drop of two No. 12 conductors supplying a 16-ampere load, located 100 feet from the power supply?
Formula: $E_{VD} = I \times R$, I = 16 amperes, R = 200 feet of No. 12 wire = 0.4 ohm
(a) 6.4 volts (b) 12.8 volts (c) 1.6 volts (d) 3.2 volts

42. What is the resistance of the circuit conductors when the conductor voltage drop is 7.2 volts and the current flow is 50 amperes?
(a) 0.14 ohm (b) 0.3 ohm (c) 3 ohms (d) 14 ohms

1–19 Pie Circle Formula

43. What is the power loss in watts for a conductor that carries 24 amperes and has a voltage drop of 7.2 volts?
(a) 175 watts (b) 350 watts (c) 700 watts (d) 2,400 watts

44. • What is the power of a 10-kW heat strip rated 240 volts, when connected to a 208-volt circuit?
(a) 8 kW (b) 9 kW (c) 12 kW (d) 15 kW

1–20 Formula Wheel

45. • The formulas listed in the formula wheel apply to ________.
(a) direct current circuits only
(b) alternating current circuits with unity power factor
(c) a and b
(d) none of these

46. When working any formula, the key to getting the correct answer, is following these four simple steps:
Step 1: → Know what the question is asking.
Step 2: → Determine the knowns of the circuit or resistor.
Step 3: → Select the formula.
Step 4: → Work out the formula calculation.
(a) True (b) False

47. The total resistance of two No. 12 conductors, 150 feet long is 0.6 ohm, and the current of the circuit is 16 amperes. What is the power loss of the conductors in watts per hour?
(a) 75 watts (b) 150 watts (c) 300 watts (d) 600 watts

48. • What is the conductor power loss in watts for a 120-volt circuit that has a 3 percent voltage drop and carries a current flow of 12 amperes? The load operates 24 hours per day, 365 days each year.
(a) 43 watts (b) 86 watts (c) 172 watts (d) 722 watts

49. • What does it cost per year (8.6 cents per kW) for the power loss of a conductor? The No. 12 circuit conductor resistance is 0.3 ohm and the current flow is 12 amperes.
(a) $33.00 (b) $13.00 (c) $130.00 (d) $1,300.00

1–21 Power Changes With The Square Of The Voltage

50. • What is the power consumed of a 10-kW heat strip rated 230 volts, connected to a 115-volt circuit?
(a) 10 kW (b) 2.5 kW (c) 5 kW (d) 20 kW

51. A(n) ________ is connected in series with the load.
(a) watt-hour meter (b) voltmeter (c) power meter (d) ammeter

1–22 Electrical Meters

52. • Ammeters are used to measure ________ current, are connected in series with the circuit, and are said to shunt the circuit.
(a) direct current (b) alternating current
(c) a and b (d) none of these

53. Clamp on ammeters have one coil connected ________ around the circuit conductor.
(a) series (b) parallel
(c) series-parallel (d) at right angles (perpendicular)

54. An ohmmeter has a ________ connected in series with the resistor. As current flows through the meter coil, the magnetic field around the coil draws in the soft iron bar. The greater the current flow through the circuit, the greater the magnetic field and the further the iron bar is drawn into the coil.
(a) coil and resistor (b) coil and power supply
(c) two coils (d) none of these

Challenge Questions

1–2 Kilo

55. • kVA is equal to ______.
(a) 100 VA (b) 1,000 volts (c) 1,000 watts (d) 1,000 VA

1–16 Conductance And Resistance

56. • ______ is not an insulator.
(a) Bakelite (b) Oil (c) Air (d) Salt water

1–17 Electrical Circuit Values

57. • ______ is not the force that moves electrons.
(a) EMF (b) Voltage (c) Potential (d) Current

58. Conductor resistance varies with ______.
(a) material (b) voltage (c) current (d) power

1–18 Ohm's Law ($I = E/R$)

59. • If the contact resistance of a connection increases and the current remains the same, the voltage drop across the connection will ______.
(a) increase (b) decrease (c) remain the same (d) cannot be determined

60. • To double the current of a circuit when the voltage remains constant, the *R* (resistance) must be ______.
(a) doubled (b) reduced by half (c) increased (d) none of these

61. • An ohmmeter is being used to test a relay coil. The equipment instructions indicate that the resistance of the coil should be between 30 and 33 ohms. The ohmmeter indicates that the actual resistance is less than 22 ohms. This reading would most likely indicate ______.
(a) the coil is okay (b) an open coil
(c) a shorted coil (d) a meter problem

1–20 Formula Wheel

62. • To calculate the power consumed by a resistive appliance, one needs to know ______.
(a) voltage and current (b) current and resistance
(c) voltage and resistance (d) any of these

63. • The number of watts of heat given off by a resistor is expressed by the formula $I^2 \times R$. If 10 volts is applied to a 5 ohm resistor, then _____ watts of heat will be given off.
(a) 500 (b) 250 (c) 50 (d) 20

64. • Power loss in a circuit because of heat can be determined by the formula ______.
(a) $P = R \times I$ (b) $P = I \times R$ (c) $P = I^2 \times R$ (d) none of these

65. • If current remains the same and resistance increases, the circuit will consume ______ power.
(a) higher (b) lower

66. When a lamp that is rated 500-watts at 115 volts is connected to a 120-volt power supply, the current of the circuit will be ______ amperes. *Tip:* Does power remain the same when voltage is changed?
(a) 3.8 (b) 4.5 (c) 2.7 (d) 5.5

1–21 Power Changes With The Square Of The Voltage

67. A toaster will produce less heat at low voltage. As a result of the low voltage, ______.
(a) its total watt output will decrease (b) the current flow will decrease
(c) the resistance is not changed (d) all of these

68. • When a resistive load is operated at a voltage 10 percent higher than the nameplate rating of the appliance, the appliance will ______.
(a) have a longer life (b) draw a lower current
(c) use more power (d) none of these

69. • A 1,500-watt heater rated 230 volts is connected to a 208-volt supply. The power consumed for this load is ______ watts. *Tip:* When the voltage is reduced, will the power be greater or less!
(a) 1,625 (b) 1,750 (c) 1,850 (d) 1,225

70. • The total resistance of a circuit is 12 ohms; the load is 10 ohms, and the wire 2 ohms. If the current of the circuit is 3 amperes, then the power consumed by the circuit conductors is ______ watts.
(a) 28 (b) 18 (c) 90 (d) 75

1–22 Electric Meters

71. • The best instrument for detecting an electric current is a(n) ______.
(a) ohmmeter (b) voltmeter (c) ammeter (d) wattmeter

72. The polarity of a circuit being tested must be observed when connecting an ohmmeter to ______.
(a) an alternating current circuit
(b) a direct current circuit
(c) any circuit
(d) polarity doesn't matter because the circuit is not energized

73. • When the test leads of an ohmmeter are shorted together, the meter will read ______ on the scale.
(a) 0 (zero ohms) (b) high (c) either a or b (d) both a and b

74. • A short circuit is indicated by a reading of ______ when tested with an ohmmeter.
(a) zero (b) ohms (c) infinity (d) R

75. Voltmeters are used to measure ______.
(a) voltages to ground (b) voltage differences
(c) AC voltages only (d) DC voltages only

76. Voltmeters must be connected in ______ with the circuit component being tested.
(a) series (b) parallel (c) series-parallel (d) multiwire

77. To measure the voltage across a load, you would connect a(n) ______.
(a) voltmeter across the load
(b) ammeter across the load
(c) voltmeter in series with the load
(d) ammeter in series with the load

78. A voltmeter is connected in ______ to the load.
(a) series (b) parallel (c) series-parallel (d) none of these

79. • In the course of normal operation, the least effective instrument in indicating that a generator may overheat because it is overloaded is a(n) ______.
(a) ammeter (b) voltmeter (c) wattmeter (d) none of these

80. • A direct-current voltmeter (not a digital meter) can be used to measure ______.
(a) power (b) frequency (c) polarity (d) power factor

81. • Polarity must be observed when connecting a voltmeter to ______ current circuit.
(a) an alternating (b) a direct
(c) any (d) polarity doesn't matter

82. The minimum number of wattmeters necessary to measure the power in the load of a balanced three-phase, four-wire system is ______.
(a) 1 (b) 2 (c) 3 (d) 4

Unit 2

Electrical Circuits

OBJECTIVES

After reading this unit, the student should be able to briefly explain the following concepts:

- Series circuit calculations
- Parallel circuit calculations
- Series-parallel circuit
- Calculations
- Multiwire circuit calculations
- Neutral current calculations
- Dangers of multiwire circuits

After reading this unit, the student should be able to briefly explain the following terms:

- Amp-hour
- Equal resistor method
- Kirchoff's law
- Multiwire circuits
- Neutral conductor
- Nonlinear loads
- Parallel circuits
- Phases
- Pigtailing
- Product of the sum method
- Reciprocal method
- Series circuit
- Series-parallel circuit
- Unbalanced current
- Ungrounded conductors (hot wires)

PART A – *SERIES CIRCUITS*

INTRODUCTION TO SERIES CIRCUITS

A *series circuit* is a circuit in which the current leaves the voltage source and flows through every electrical device with the same intensity before it returns to the voltage source. If any part of a series circuit is opened, the current will stop flowing in the entire circuit, Fig. 2–1.

For most practical purposes, series (*closed loop*) circuits are not used for building wiring, but they are important for the operation of many *control* and *signal circuits*, Fig. 2–2. Motor control circuit stop switches are generally wired in series with the starter's coil and the line conductors. Dual-rated motors, such as 460/230 volts, will have their winding connected in series when supplied by the higher voltage and connected in parallel when supplied by the lower voltage.

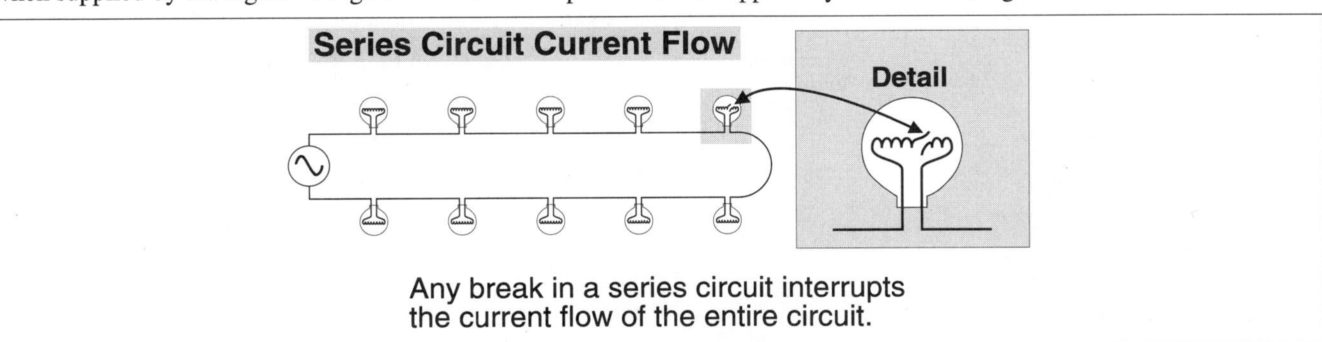

Figure 2-1

Series Circuit Current Flow

2–1 *UNDERSTANDING SERIES CALCULATIONS*

It is important that you understand the relationship between resistance, current, voltage, and power of series circuits, Fig. 2–3.

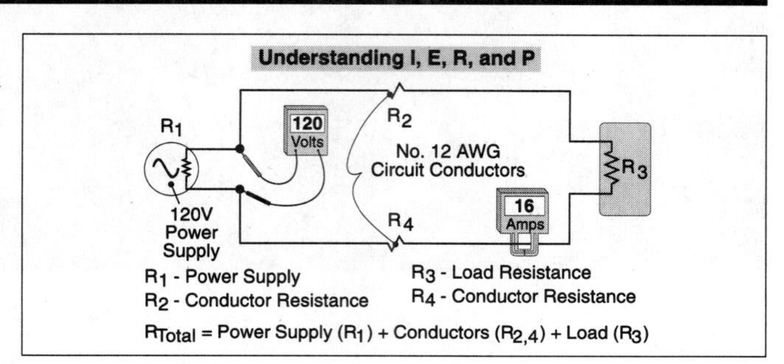

Figure 2-3

Understanding I, E, R, and P

Calculating Resistance Total

In a series circuit, the total resistance of the circuit is equal to the sum of all the series resistor's resistance, according to the formula, $\mathbf{R_T = R_1 + R_2 + R_3 + R_4}$

❑ **Total Resistance Example**

What is the total resistance of the loads in Figure 2–4?

(a) 2.5 ohms (b) 5.5 ohms

(c) 7.5 ohms (d) 10 ohms

- Answer: (c) 7.5 ohms

R_1 – Power Supply	0.05 ohm
R_2 – Conductor No. 1	0.15 ohm
R_3 – Appliance	7.15 ohms
R_4 – Conductor No. 2	0.15 ohm
Total Resistance:	7.50 ohms

Calculating Voltage Drop

The result of current flowing through a resistor is voltage lost across the resistor which is called *voltage drop*. In a closed loop circuit, the sum of the voltage drops of all the loads is equal to the voltage source, Fig. 2–5. This is known as *Kirchoff's First Law*, of the *voltage law*: The voltage drop (E_{VD}) of each resistor can be determined by the formula:

$\mathbf{E_{VD} = I \times R}$ **I = Current of the circuit** **R = Resistance of the resistor**

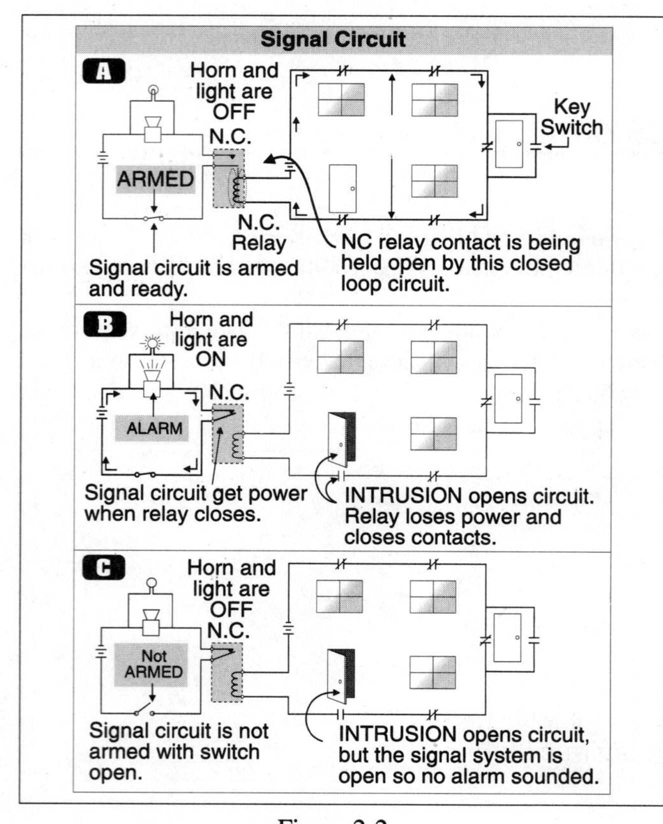

Figure 2-2

Control and Signal Circuits

Resistance is Additive - Series Circuit

Conductor 1

R1 .05Ω

R1 = Power Supply
R2 & R4 = Conductors
R3 = Load

R2 .15Ω

R3 7.15Ω

R4 .15Ω

120V Power Supply

Appliance (Load)

Conductor 2

Example: Determine total resistance (R_T) of the circuit.

Formula: $R_T = R1 + R2 + R3 + R4...$

$R1 = .05$ ohm $R2 = .15$ ohm

$R3 = 7.15$ ohms $R4 = .15$ ohm

$R_T = .05\Omega + .15\Omega + 7.15\Omega + .15\Omega$

$R_T =$ **7.5 ohms**

Figure 2-4

Total Resistance Example

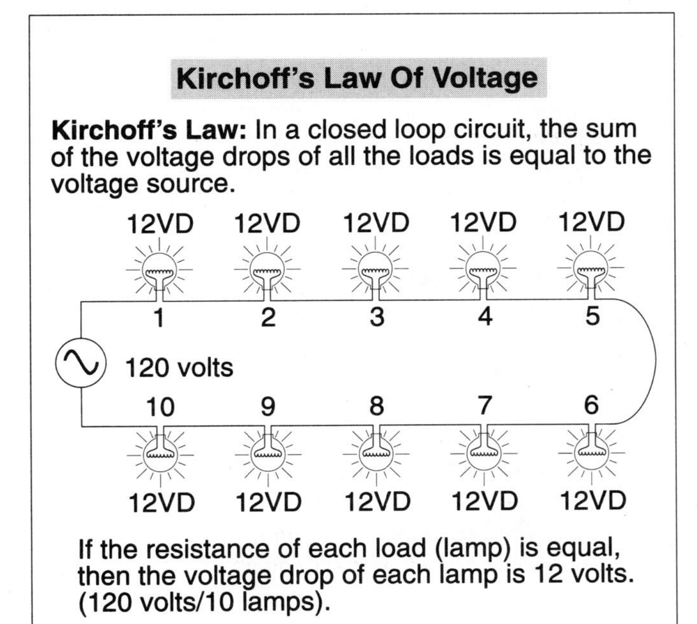

Figure 2-5

Voltage Drop Distribution

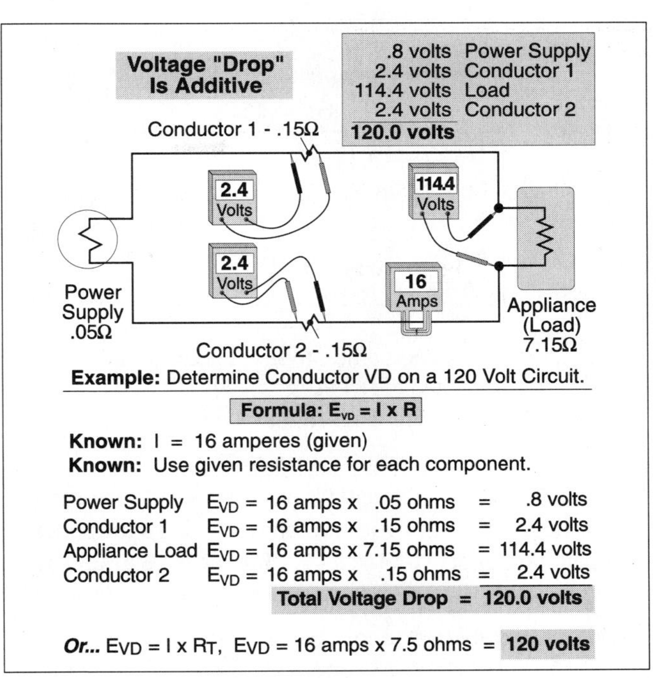

Figure 2-6

Series – Voltage Drop Example

❑ **Voltage Drop Example**

What is the voltage drop across each resistor in Figure 2–6 using the formula $E_{VD} = I \times R$?

- Answer: $E_{VD} = I \times R$

R_{1vd} – Power Supply	16 amperes × 0.05 ohm	=	0.8 volt drop
R_{2vd} – Conductor No. 1	16 amperes × 0.15 ohm	=	2.4 volts drop
R_{3vd} – Appliance	16 amperes × 7.15 ohms	=	144.4 volts drop
R_{4vd} – Conductor No. 2	16 amperes × 0.15 ohm	=	2.4 volts drop
Total Voltage Drop	16 amperes × 7.50 ohms	=	120 volts drop

Note: Due to rounding off, the sum of the voltage drops might be slightly different than the voltage source.

Voltage of Series Connected Power Supplies

When *power supplies* are connected in series, the voltage of each power supply will add together, providing that all the polarities are connected properly.

❑ **Series Connected Power Supplies Example**

What is the total voltage output of four 1.5-volt batteries connected in series, Fig. 2–7?

(a) 1.5 volts (b) 3.0 volts
(c) 4.5 volts (d) 6.0 volts

- Answer: (d) 6 volts

Current of Resistor or Circuit

In a series circuit, the *current* throughout the circuit is constant and does not change. The current through each resistor of the circuit can be determined by the formula:

I = E/R
E = Voltage drop of the load or circuit
R = Resistance of the load or circuit

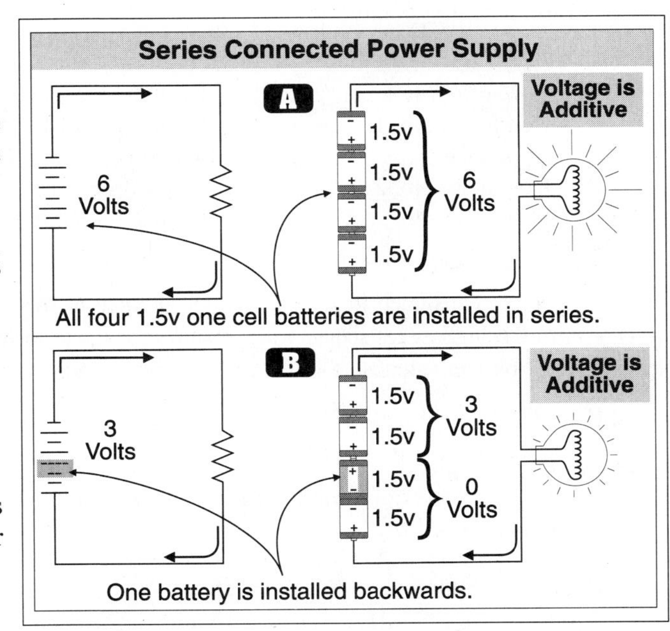

Figure 2-7

Series Power Supplies Example

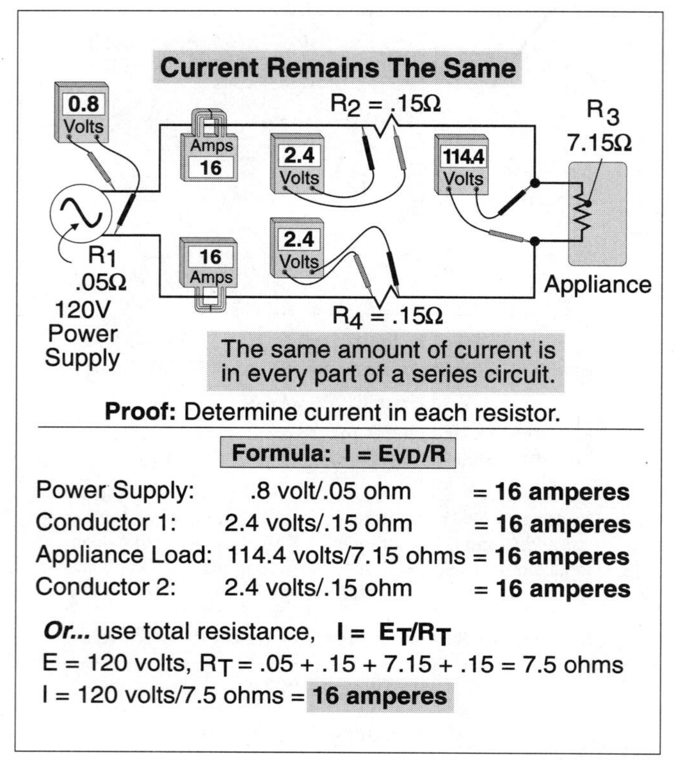

Figure 2-8

Series Current Example

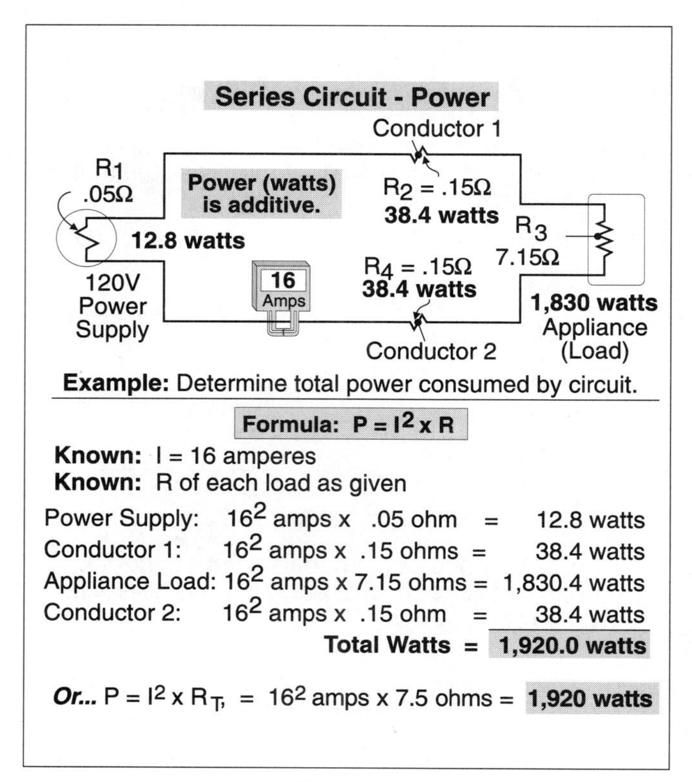

Figure 2-9

Series Power Example

Note: If the resistance is not given, you can determine the resistance of a resistor (if you know the nameplate voltage and power rating of the load) by the formula: $R = E^2/P$, E = Nameplate voltage rating (squared), P = Nameplate power rating.

❑ **Current Example**

What is the current flow through the series circuit in Figure 2–8?

(a) 4 amperes (b) 8 amperes (c) 12 amperes (d) 16 amperes

- Answer: (d) 16 amperes

$I = E/R$

R_1 – Power Supply	0.8 volt drop/0.05 ohm	=	16 amperes
R_2 – Conductor No. 1	2.4 volts drop /0.15 ohm	=	16 amperes
R_3 – Appliance	114.4 volts drop/7.15 ohms	=	16 amperes
R_4 – Conductor No. 2	2.4 volts drop/0.15 ohm	=	16 amperes
Circuit Current	120 volts/7.5 ohms	=	16 amperes

Power of Resistor or Circuit

The *power* consumed in a series circuit is equal to the sum of the power of all of the resistors in the series circuit. The resistor with the highest resistance will consume the most power, and the resistor with the smallest resistance will consume the least power. You can calculate the power consumed (watts) of each resistor or of the circuit by the formula:

$P = I^2R$ **I^2 = Current of the circuit (squared)** **R = Resistance of circuit or resistor**

❑ **Power Example**

What is the power consumed of each resistor in Figure 2–9?

- Answer: $P = I^2R$

R_1 – Power Supply	16 amperes^2 × 0.05 ohm	=	12.8 watts
R_2 – Conductor No. 1	16 amperes^2 × 0.15 ohm	=	38.4 watts
R_3 – Appliance	16 amperes^2 × 7.15 ohms	=	1,830.4 watts
R_4 – Conductor No. 2	16 amperes^2 × 0.15 ohm	=	38.4 watts

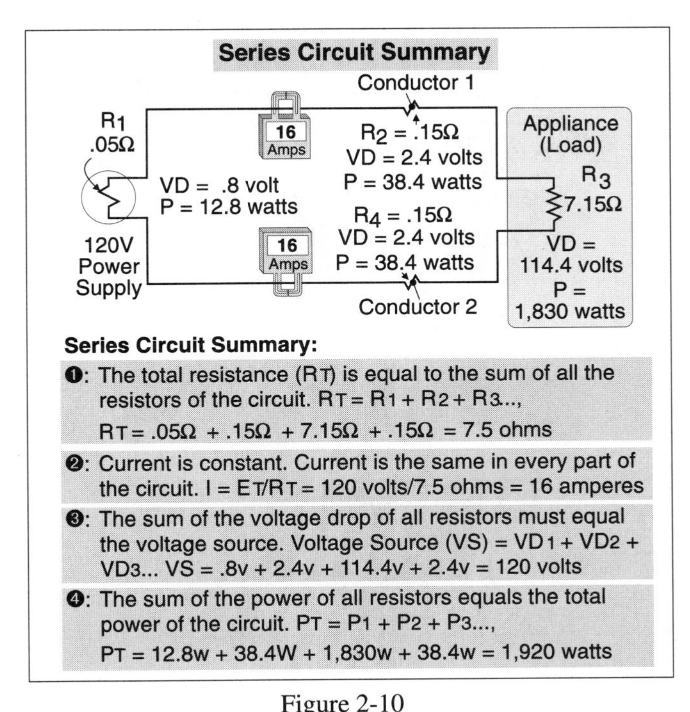

Figure 2-10

Series Circuit Summary

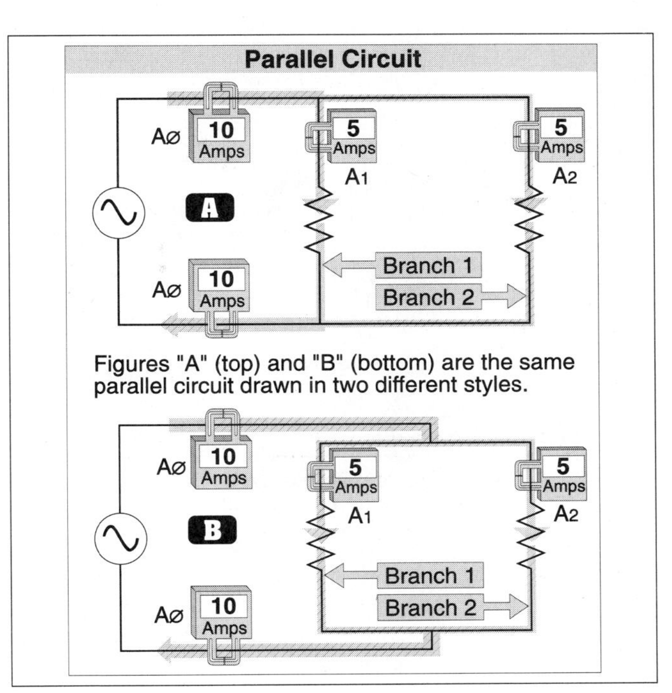

Figure 2-11

Parallel Circuit – Current Flow

2–2 *SERIES CIRCUIT SUMMARY*

Figure 2–10, Series Circuit Summary.

Note 1: → The total resistance of the series circuit is equal to the sum of all the resistors of the circuit.

Note 2: → Current is constant.

Note 3: → The sum of the voltage drop of all resistors must equal the voltage source.

Note 4: → The sum of the power of all resistors equals the total power of the circuit.

PART B – *PARALLEL CIRCUIT*

INTRODUCTION TO PARALLEL CIRCUITS

A *parallel circuit* is a circuit in which current leaves the voltage source, branches through different parts of the circuit, and then returns to the voltage source, Fig. 2–11. Parallel is a term used to describe a method of connecting electrical components so that the current can flow through two or more different branches of the circuit.

2–3 *PRACTICAL USES OF PARALLEL CIRCUITS*

Parallel circuits are commonly used for building wiring; in addition, parallel (*open loop*) circuits are often used for fire alarm systems and the internal wiring of many types of electrical equipment, such as motors and transformers. Dual-rated motors, such as 460/230 volts, have their winding connected in parallel when supplied by the lower voltage, and in series when supplied by the higher voltage, Fig. 2–12.

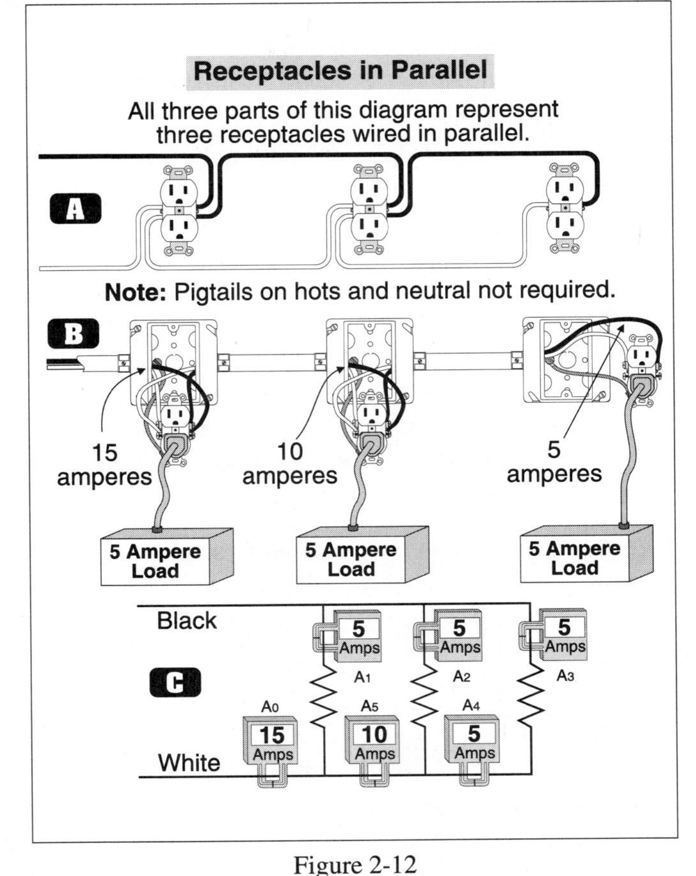

Figure 2-12

Parallel Wiring of Equipment

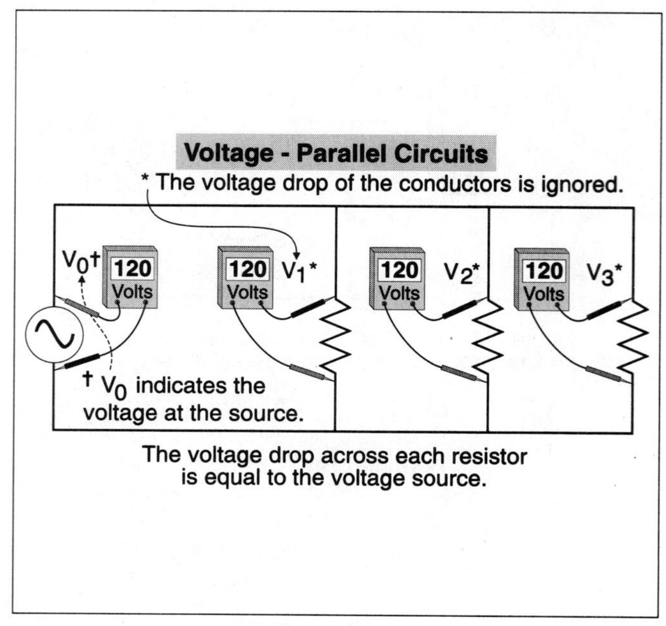

Figure 2-13

Parallel Voltage of Each Branch

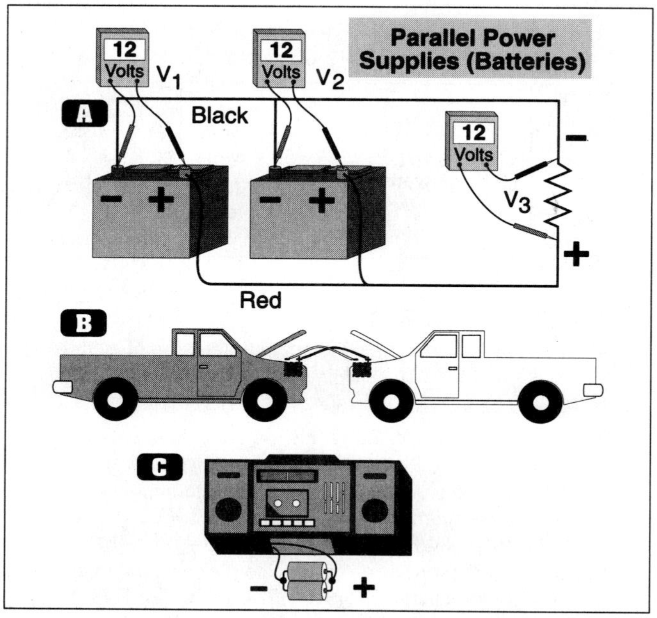

Figure 2-14

Parallel Power Supply Example

2–4 *UNDERSTANDING PARALLEL CALCULATIONS*

It is important that you understand the relationship between voltage, current, power, and resistance of a parallel circuit.

Voltage of Each Branch

The *voltage drop* across loads connected in parallel is equal to the voltage that supplies each parallel branch, Fig. 2–13.

Power Supplies Connected in Parallel

When *power supplies* are connected in parallel, voltage remains the same but the current, or amp-hour, capacity is increased. When connecting batteries in parallel, always connect batteries of the same voltage with the proper polarity.

❑ **Parallel Connected Power Supplies Example**

If two 12-volt batteries are connected in parallel, what is the output voltage, Fig. 2–14?

(a) 3 volts (b) 6 volts (c) 12 volts (d) 24 volts

- Answer: (c) 12 volts, voltage remains the same when connected in parallel

Note: Two 12-volt batteries connected in parallel will result in an output voltage of 12 volts, but the amp-hour capacity will be increased resulting in longer service.

Current Through Each Branch

The *current* from the power supply is equal to the sum of the branch circuits. The current in each branch depends on the branch voltage and branch resistance and can be calculated by the formula:

I = E/R **E = Voltage of Branch** **R = Resistance of Branch**

❑ **Current Through Each Branch Example**

What is the current of each appliance in Figure 2–15?

- Answer: $I = E/R$.

R_1 – Coffee Pot .... 120 volts/16 ohms .... =. . 7.50 amperes
R_2 – Skillet .... 120 volts/13 ohms .... =. . 9.17 amperes
R_3 – Blender .... 120 volts/36 ohms .... =. . 3.33 amperes
Total Current = 7.5 amperes + 9.17 amperes + 3.33 amperes =. 20.00 amperes

Power Consumed Of Each Branch

The total *power* consumed of any circuit is equal to the sum of the branch powers. Each branch power depends on the branch current and resistance. The power can be found by the formula:

$P = I^2R$ **I = Current of Each Branch** **R = Resistance of Each Branch**

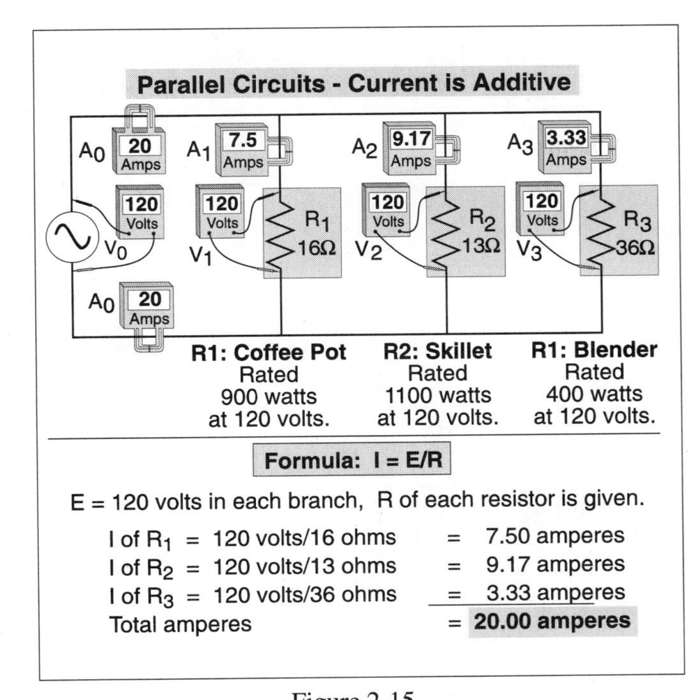

Figure 2-15

Parallel Circuits – Current Is Additive

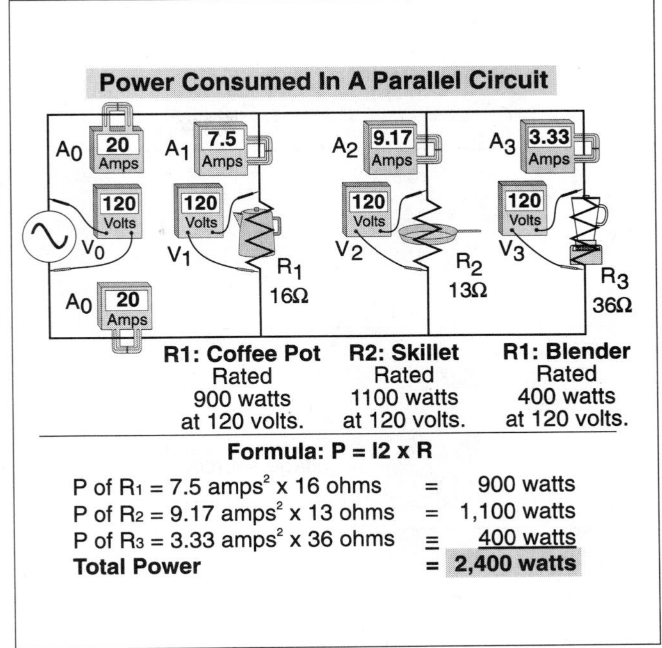

Figure 2-16

Parallel Circuits Summary – Power Of Each Branch

❑ **Power of Each Branch Example**

What is the power consumed of each appliance in Figure 2–16?

- Answer: $P = I^2R$

R_1 – Coffee Pot .... 7.5 amperes^2 × 16 ohms ... = ... 900 watts

R_2 – Skillet .... 9.17 amperes^2 × 13 ohms .. = ... 1,093 watts, round to 1,100 watts

R_3 – Blender .... 3.33 amperes^2 × 36 ohms .. = ... 399 watts, round to 400 watts

Total Power = 900 watts + 1,100 watts + 400 watts ... = ... 2,400 watts

2–5 *PARALLEL CIRCUIT RESISTANCE CALCULATIONS*

In a parallel circuit, the total circuit *resistance* is always less than the smallest resistor and can be determined by one of three methods:

Equal Resistor Method

The *equal resistor method* can be used when all the resistors of the parallel circuit have the same resistance. Simply divide the resistance of one resistor by the number of resistors in parallel.

R_T = R/N **R = Resistance of One Resistor** **N = Number of Resistors**

❑ **Equal Resistors Method Example**

The total resistance of three, 10-ohm resistors is ______, Fig. 2–17.

(a) 10 ohms (b) 20 ohms (c) 30 ohms (d) none of these

- Answer: (d) none of these

$$R_T = \frac{\text{Resistance of One Resistor}}{\text{Number of Resistors}} = \frac{10 \text{ ohms}}{3 \text{ resistors}} = 3.33 \text{ ohms}$$

Product Of The Sum Method

The product of the sum method can be used to calculate the resistance of two resistors.

$$R_T = \frac{R_1 \times R_2 \text{ (Product)}}{R_1 + R_2 \text{ (Sum)}}$$

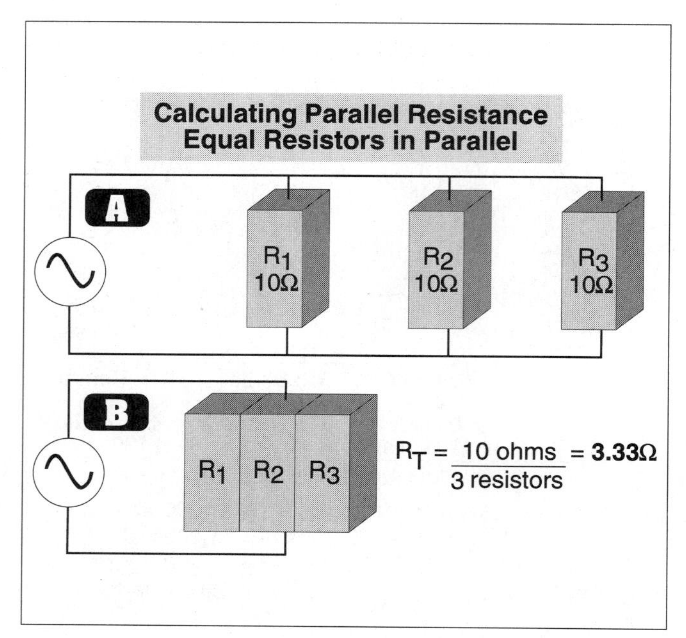

Figure 2-17

Parallel Resistance – Equal Resistors Example

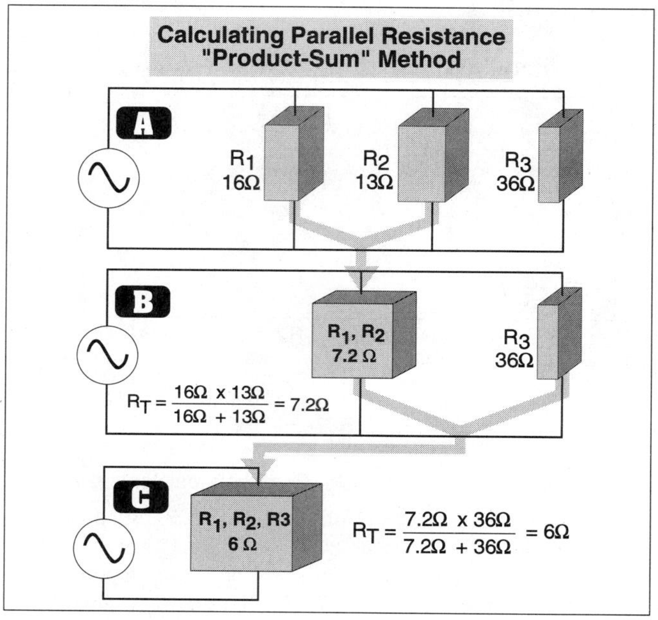

Figure 2-18

Parallel Resistance – Product/Sum Method Example

The term *product* means the answer of numbers that are multiplied together. The term *sum* is the answer to numbers that are added together. The product of the sum method can be used for more than two resistors, but only two can be calculated at a time.

❑ **Product Of The Sum Method Example**

What is the total resistance of a 16 ohm coffee pot and a 13 ohm skillet connected in parallel, Fig. 2–18?

(a) 16 ohms (b) 13.09 ohms (c) 29.09 ohms (d) 7.2 ohms

- Answer: 7.2 ohms

The total resistance of parallel circuit is always less than the smallest resistor (13 ohms).

$$R_T = \frac{R_1 \times R_2}{R_1 + R_2} = \frac{16 \text{ ohms} \times 13 \text{ ohms}}{16 \text{ ohms} + 13 \text{ ohms}} = 7.20 \text{ ohms}$$

Reciprocal Method

The advantage of the *reciprocal method* is that this formula can be used for an unlimited number of parallel resistors.

$$\mathbf{R_T} = \frac{1}{1/R_1 + 1/R_2 + 1/R_3 \ldots}$$

❑ **Reciprocal Method Example**

What is the resistance total of a 16-ohm, 13-ohm, and 36-ohm resistor connected in parallel, Fig. 2–19?

(a) 13 ohms (b) 16 ohms (c) 36 ohms (d) 6 ohms

- Answer: (d) 6 ohms

$$R_T = \frac{1}{1/16 \text{ ohm} + 1/13 \text{ ohm} + 1/36 \text{ ohm}} = R_T = \frac{1}{0.0625 \text{ ohm} + 0.0769 \text{ ohm} + 0.0278 \text{ ohm}}$$

$$R_T = \frac{1}{0.1672 \text{ ohm}} = R_T = 6 \text{ ohms}$$

2–6 *PARALLEL CIRCUIT SUMMARY*

Note 1: → The total resistance of a parallel circuit is always less than the smallest resistor, Fig. 2–20.
Note 2: → Current total of a parallel circuit is equal to the sum of the currents of the individual branches.
Note 3: → Power total is equal to the sum of the power in all the individual branches.

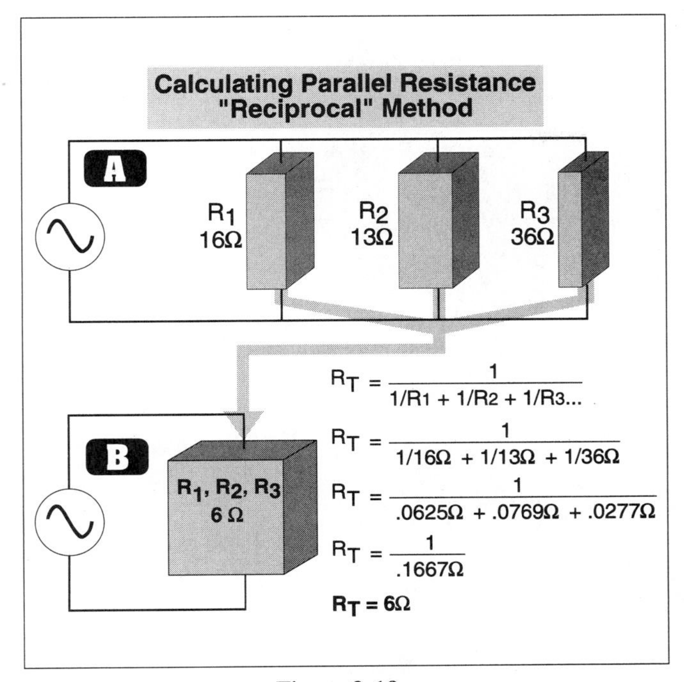

Figure 2-19

Parallel Resistance Reciprocal Method Example

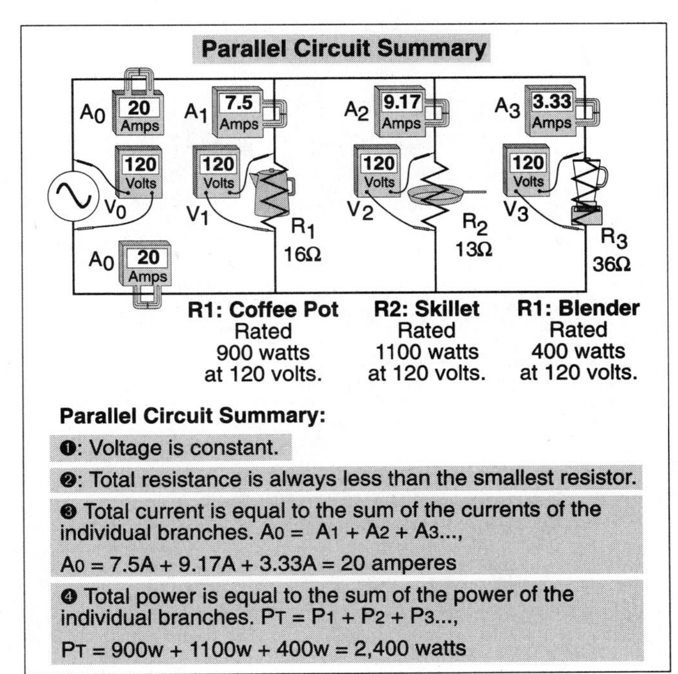

Figure 2-20

Parallel Summary

PART C – *SERIES-PARALLEL AND MULTIWIRE CIRCUITS*

INTRODUCTION TO SERIES-PARALLEL CIRCUITS

A *series-parallel* circuit is a circuit that contains some resistors in series and some resistors in parallel to each other, Fig. 2–21.

2–7 *REVIEW OF SERIES AND PARALLEL CIRCUITS*

To have a better understanding of series-parallel circuits, let's review the rules for series and parallel circuits. That portion of the circuit that contains resistors in series must comply with the rules of series circuits, and that portion of the circuit that is connected in parallel must comply with the rules of parallel circuits, Fig. 2–22.

Series Circuit Rules:

Note 1: → Resistance is additive.
Note 2: → Current is constant.
Note 3: → Voltage is additive.
Note 4: → Power is additive.

Parallel Circuit Rules

Note 1: → Resistance is less than the smallest resistor.
Note 2: → Current is additive.
Note 3: → Voltage is constant.
Note 4: → Power is additive.

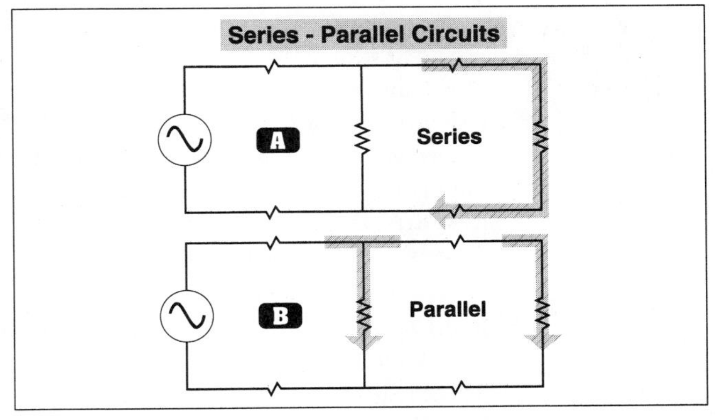

Figure 2-21

Series-Parallel Circuit

2–8 *SERIES-PARALLEL CIRCUIT RESISTANCE CALCULATIONS*

When determining the *resistance total* of a series-parallel circuit, it is best to redraw the circuit so you can see the series components and the parallel branches. Determine the resistance of the series components first or the parallel components depending on the circuit, then determine the resistance total of all the branches. Keep breaking the circuit down until you have determined the total effective resistance of the circuit as one resistor.

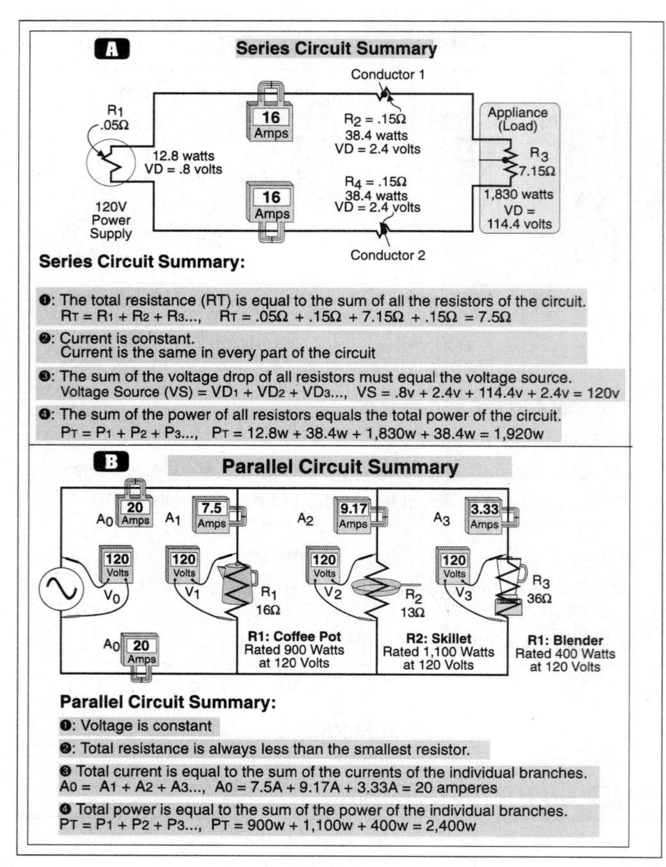

Figure 2-22

Series and Parallel Review

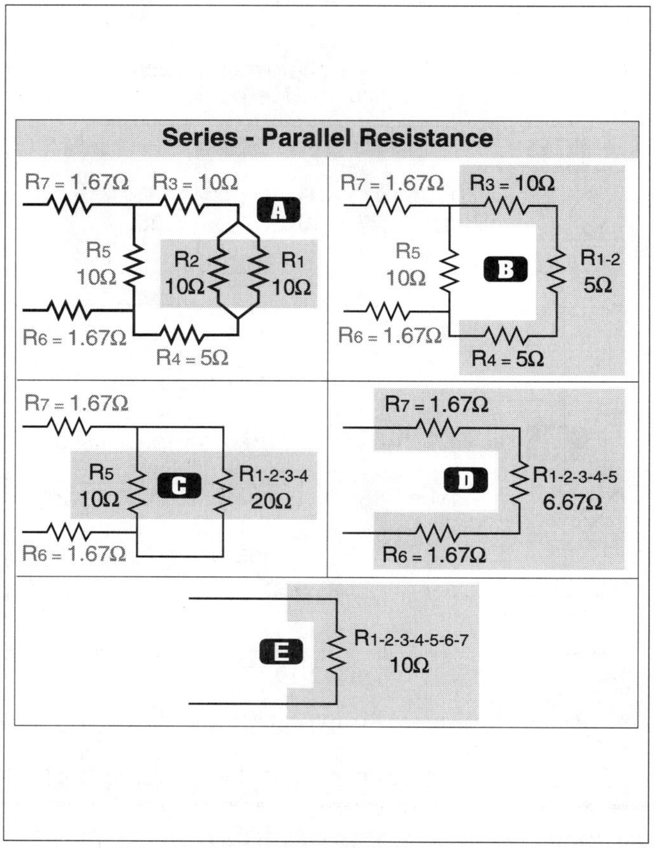

Series - Parallel Resistance

A: R7 = 1.67Ω, R3 = 10Ω, R5 10Ω, R2 10Ω, R1 10Ω, R6 = 1.67Ω, R4 = 5Ω

B: R7 = 1.67Ω, R3 = 10Ω, R5 10Ω, R1-2 5Ω, R6 = 1.67Ω, R4 = 5Ω

C: R7 = 1.67Ω, R5 10Ω, R1-2-3-4 20Ω, R6 = 1.67Ω

D: R7 = 1.67Ω, R1-2-3-4-5 6.67Ω, R6 = 1.67Ω

E: R1-2-3-4-5-6-7 10Ω

Figure 2-23

Series – Parallel Resistance

❑ Series-Parallel Circuit Resistance Example

What is the resistance total of the circuit shown in Figure 2–23, Part A?

- Answer:

Step 1: → Determine the resistance of the equal parallel resistors - R_1 and R_2 – Figure 2–23 Part A:
Using the *Equal Resistor Method*
The *equal resistor method* can be used when all the resistors of the parallel circuit have the same resistance. Simply divide the resistance of one resistor by the number of resistors in parallel.
$R_T = R/N$, R = Resistance of one resistor, N = Number of resistors
R_T = 10 ohms/2 resistors, R_T = 5 ohms

Step 2: → Redraw the circuit – Figure 2–23 Part B, now determine the series resistance of $R_{1/2}$ (5 ohms), plus R_3 (10 ohms), plus R_4 (5 ohms).
Series resistance total is equal to $R_{1/2} + R_3 + R_4$
R_T = 5 ohms + 10 ohms + 5 ohms, R_T = 20 ohms

Step 3: → Redraw the circuit – Figure 2–23 Part C, now determine the parallel resistance of R_5, plus the resistance of $R_{1,2,3,4,}$. Remember the resistance total of a parallel circuit is always less than the smallest parallel branch (10 ohms). Since we have only two parallel branches, the resistance can be determined by the product of the sum method.

$$R_T = \frac{R_1 \times R_2 \text{ (Product)}}{R_1 + R_2 \text{ (Sum)}} \qquad R_T = \frac{(10 \text{ ohms x } 20 \text{ ohms})}{(10 \text{ ohms} + 20 \text{ ohms})} \qquad R_T = 6.67 \text{ ohms}$$

Step 4: → Redraw the circuit – Figure 2–23 Part D, now determine the series resistance total of R_7, plus R_6, plus $R_{1,2,3,4,5}$
Series resistance total is equal to $R_7 + R_6 + R_{1,2,3,4,5}$
R_T = 1.67 ohms + 1.67 ohms + 6.67 ohms, R_T = 10 ohms

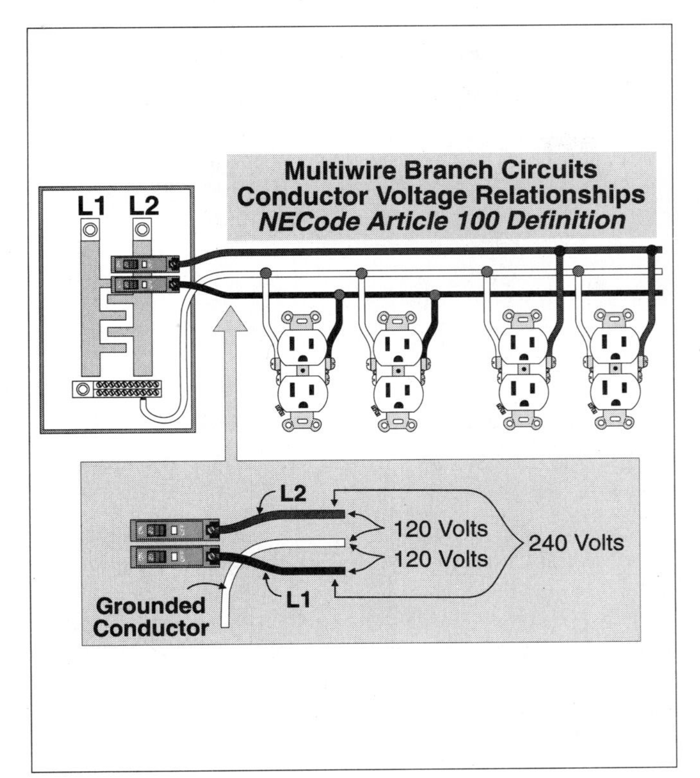

Figure 2-24

Multiwire Branch Circuit

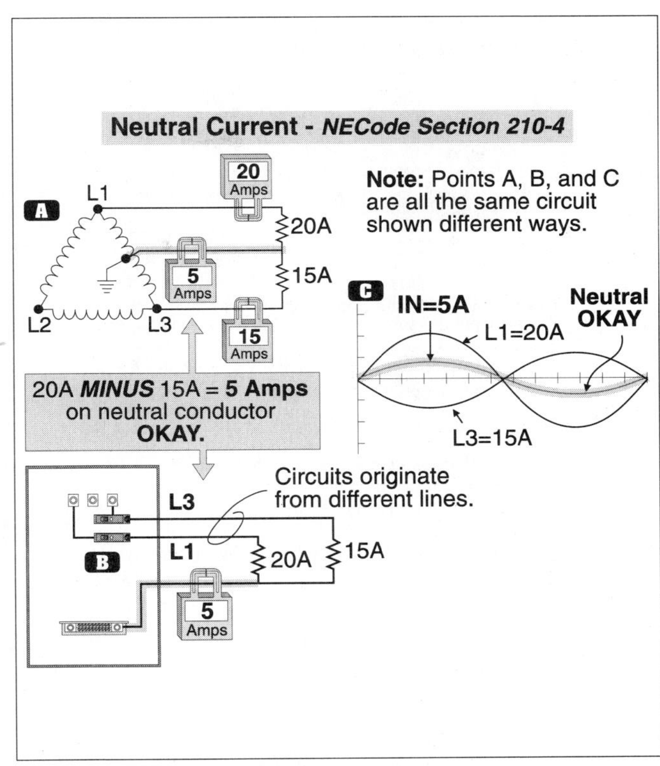

Figure 2-25

Neutral Current – 120/240 Volt, 3-Wire Example

PART D – *MULTIWIRE CIRCUITS*

INTRODUCTION TO MULTIWIRE CIRCUITS

A multiwire circuit is a circuit consisting of two or more *ungrounded* conductors (hot conductors) having a *potential difference* between them, and an equal difference of potential between each hot wire and the *neutral* or *grounded conductor,* Figure 2–24.

Note: The National Electrical Code contains specific requirements on multiwire circuits, see Article 100 definition of multiwire circuit, Section 210–4 branch circuit requirements, and Section 310–13(b) requirements on pigtailing of the neutral conductor.

2–9 *NEUTRAL CURRENT CALCULATIONS*

When current flows on the *neutral* conductor of a multiwire circuit, the current is called *unbalanced current.* This current can be determined according to the following:

120/240-Volt, 3-Wire Circuits

A 120/240-volt, 3-wire circuit, consisting of two hot wires and a neutral, will carry no current when the circuit is balanced. However, the neutral (or *grounded*) conductor will carry unbalanced current when the circuit is not balanced. The neutral current can be calculated as:

Line 1 Current less Line 2 Current

❑ **120/240-Volt, 3-Wire Neutral Current Example**

What is the neutral current if $line_1$ current is 20 amperes and $line_2$ current is 15 amperes, Fig. 2–25?

(a) 0 amperes (b) 5 amperes (c) 10 amperes (d) 35 amperes

- Answer: (b) 5 amperes

  $Line_1$ current = 20 amperes

  $Line_2$ current = 15 amperes

  Unbalance = 5 amperes

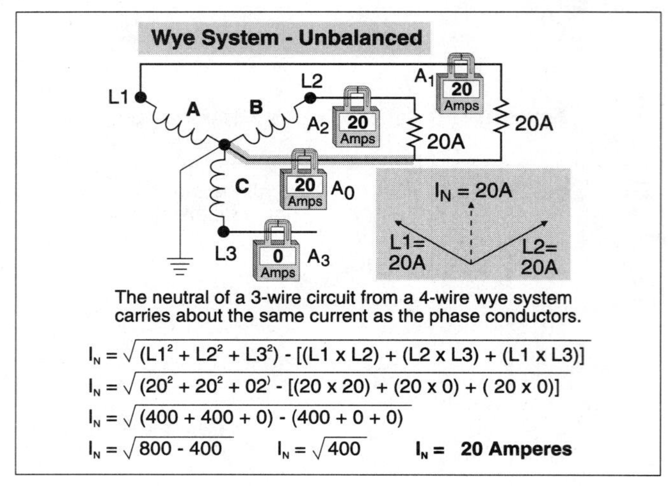

Figure 2-26

Neutral Current – 120/208 Volt, 3-Wire Example

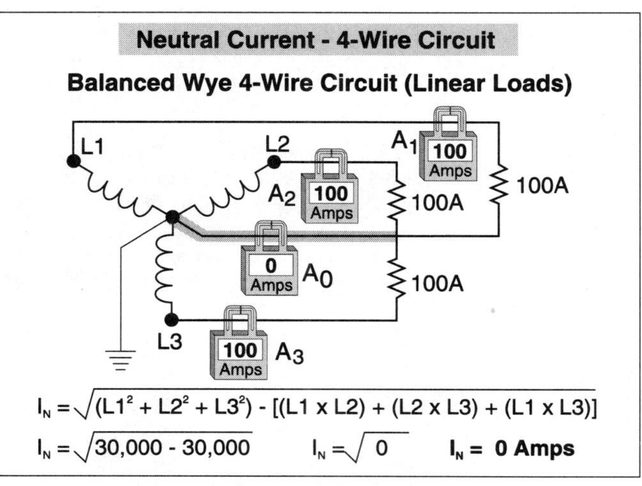

Figure 2-27

Neutral Current – Wye 4-Wire Balanced Example

Wye 3-Wire Circuit Neutral Current

A 3-wire wye, 208Y/120- or 480Y/277-volt circuit, consisting of two phases and a neutral always carries unbalanced current. The current on the neutral or grounded conductor is determined by the formula:

$$I_N = \sqrt{(\text{Line } 1^2 + \text{Line } 2^2) - (\text{Line } 1 \times \text{Line } 2)}$$

Line 1 = Current of one phase

Line 2 = Current of other phase

❑ **Three-wire Wye Circuit Neutral Current Example**

What is the neutral current for a 20 ampere, 3-wire circuit (two hots and a neutral)? Power is supplied from a 208Y/120-volt feeder, Fig. 2–26?

(a) 40 amperes (b) 20 amperes (c) 60 amperes (d) 0 amperes

- Answer: (b) 20 amperes

$I_N = \sqrt{(20 \text{ amperes}^2 + 20 \text{ amperes}^{2)} - (20 \text{ amperes} \times 20 \text{ amperes})} = \sqrt{400}$

$I_N = 20$ amperes

Wye 4-wire Circuit Neutral Current

A 4-wire wye, 120/208Y- or 277/480Y-volt circuit will carry no current when the circuit is balanced but will carry unbalanced current when the circuit is not balanced.

$I_N = \sqrt{(\text{Line } 1^2 + \text{Line } 2^2 + \text{Line } 3^2) - [(\text{Line } 1 \times \text{Line } 2) + (\text{Line } 2 \times \text{Line } 3) + (\text{Line } 1 \times \text{Line } 3)]}$

❑ **Example Balanced Circuits**

What is the neutral current for a 4-wire, 208Y/120-volt feeder where L_1 = 100 amperes, L_2 = 100 amperes, and L_3 = 100 amperes, Fig. 2–27?

(a) 50 amperes (b) 100 amperes (c) 125 amperes (d) 0 amperes

- Answer: (d) 0 amperes

$I_N = \sqrt{(100 \text{ Amps}^2 + 100 \text{ Amps}^2 + 100 \text{ Amps}^2) - [(100A \times 100A) + (100A \times 100A) + (100A \times 100A)]} = \sqrt{0}$

$I_N = 0$ amperes

❑ **Unbalanced Circuits – Example**

What is the neutral current for a 4-wire, 208Y/120-volt feeder where L_1 = 100 amperes, L_2 = 100 amperes, and L_3 = 50 amperes, Fig. 2–28?

(a) 50 amperes (b) 100 amperes (c) 125 amperes (d) 0 amperes

- Answer: (a) 50 amperes

$I_N = \sqrt{(100A^2 + 100A^2 + 50A^2) - [(100A \times 100A) + (100A \times 50A) + (100A \times 50A)]} = \sqrt{2{,}500}$

$I_N = 50$ amperes

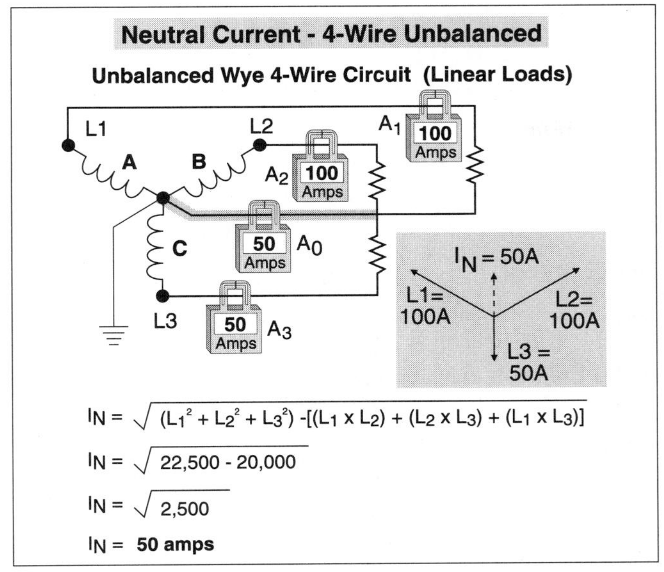

Figure 2-28

Neutral Current – Wye 4-Wire Unbalanced

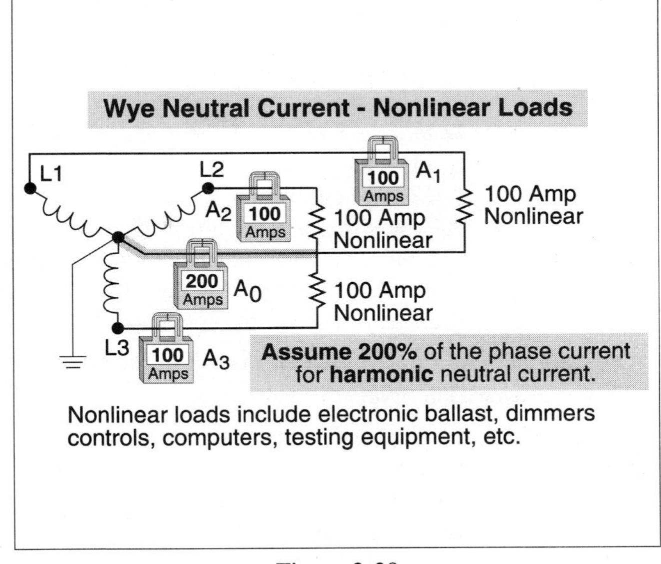

Figure 2-29

Neutral Current – Nonlinear Loads

Nonlinear Load Neutral Current

The neutral conductor of a balanced wye 4-wire circuit will carry current when supplying power to *nonlinear* loads, such as computers, copy machines, laser printers, fluorescent lighting, etc. The current can be as much as two times the phase current, depending on the harmonic content of the loads.

❑ Nonlinear Loads – Example

What is the neutral current for a 4-wire, 208Y/120-volt feeder supplying power to a nonlinear load, where L_1 = 100 amperes, L_2 = 100 amperes, and L_3 = 100 amperes? Assume that the harmonic content results in neutral current equal to 200 percent of the phase current, Fig. 2–29.

(a) 80 amperes (b) 100 amperes (c) 125 amperes (d) 200 amperes

- Answer: 200 amperes

2–10 *DANGERS OF MULTIWIRE CIRCUITS*

Improper wiring, or mishandling of multiwire circuits, can cause excessive neutral current (*overload*) or destruction of electrical equipment because of *overvoltage* if the neutral conductor is opened.

Overloading of the Neutral Conductor

If the ungrounded conductors (hot wires) of a multiwire circuit are connected to different phases, the current on the grounded conductor (neutral) will cancel. If the ungrounded conductors are not connected to different phases, the current from each phase will add on the neutral conductor. This can result in an *overload* of the neutral conductor, Fig. 2–30.

Note: Overloading of the neutral conductor will cause the insulation to look discolored. Now you know why the neutral wires sometimes look like they were burned.

Overvoltage To Electrical Equipment

If the neutral conductor of a multiwire circuit is opened, the multiwire circuit changes from a parallel circuit into a series circuit. Instead of two 120-volt circuits, we now have one 240-volt circuit, which can result in fires and the destruction of the electric equipment because of *overvoltage,* Fig. 2–31.

To determine the operating voltage of each load in an open multiwire circuit, use the following steps:

Step 1: → Determine the resistance of each appliance,

$R = E^2/P$

E = Appliance voltage nameplate rating

P = Appliance power nameplate rating

Step 2: → Determine the circuit resistance, $R_T = R_1 + R_2$

Step 3: → Determine the current of the circuit,
$I_T = E_S/R_T$,
E_S = Voltage, phase to phase
R_T = Resistance total, from Step 2

Step 4: → Determine the voltage for each appliance,
$E = I_T \times R_X$.
I_T = Current of the circuit
R_X = Resistance of each resistor

Step 5: → Determine the power consumed of each appliance; $P = E^2/R_X$
E = Voltage the appliance operates at (squared)
Rx = Resistance of the appliance

❑ **Example – At what voltage does each of the loads operate at if the neutral is opened,** Fig. 2–31.?

- Answer: Hair dryer = 77 volts, T.V. = 163 volts

Step 1: → Determine the resistance of each appliance, $R = E^2/P$
E = Appliance voltage nameplate rating
P = Appliance power nameplate rating
Hair dryer rated 1,275 watts at 120 volts.
$R = E^2/P$, $R = 120\ \text{volts}^2/1{,}275$ watts = 11.3 ohms
Television rated 600 watts at 120 volts.
$R = E^2/P$, $R = 120\ \text{volts}^2/600$ watts = 24 ohms

Step 2: → Determine the circuit resistance, $R_t = R_1 + R_2$
R_T = 11.3 ohms + 24 ohms = R_T = 35.3 ohms

Step 3: → Determine the current of the circuit, $I_T = E_S/R_T$.
I_T = Current of the Circuit, R_T = ResistanceTotal

$$I_T = \frac{240 \text{ volts}}{35.3 \text{ ohms}} = 6.8 \text{ amperes}$$

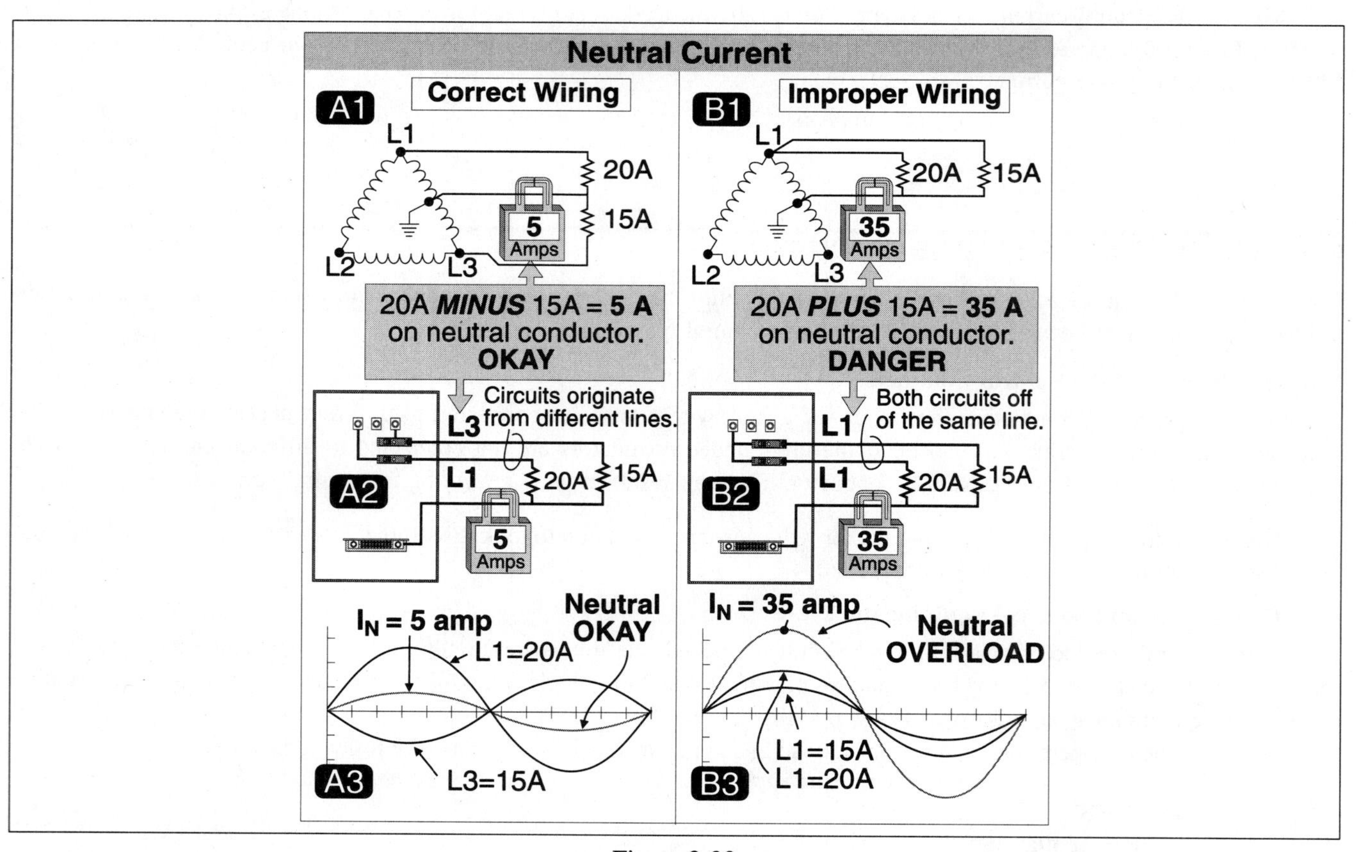

Figure 2-30

Neutral Overload

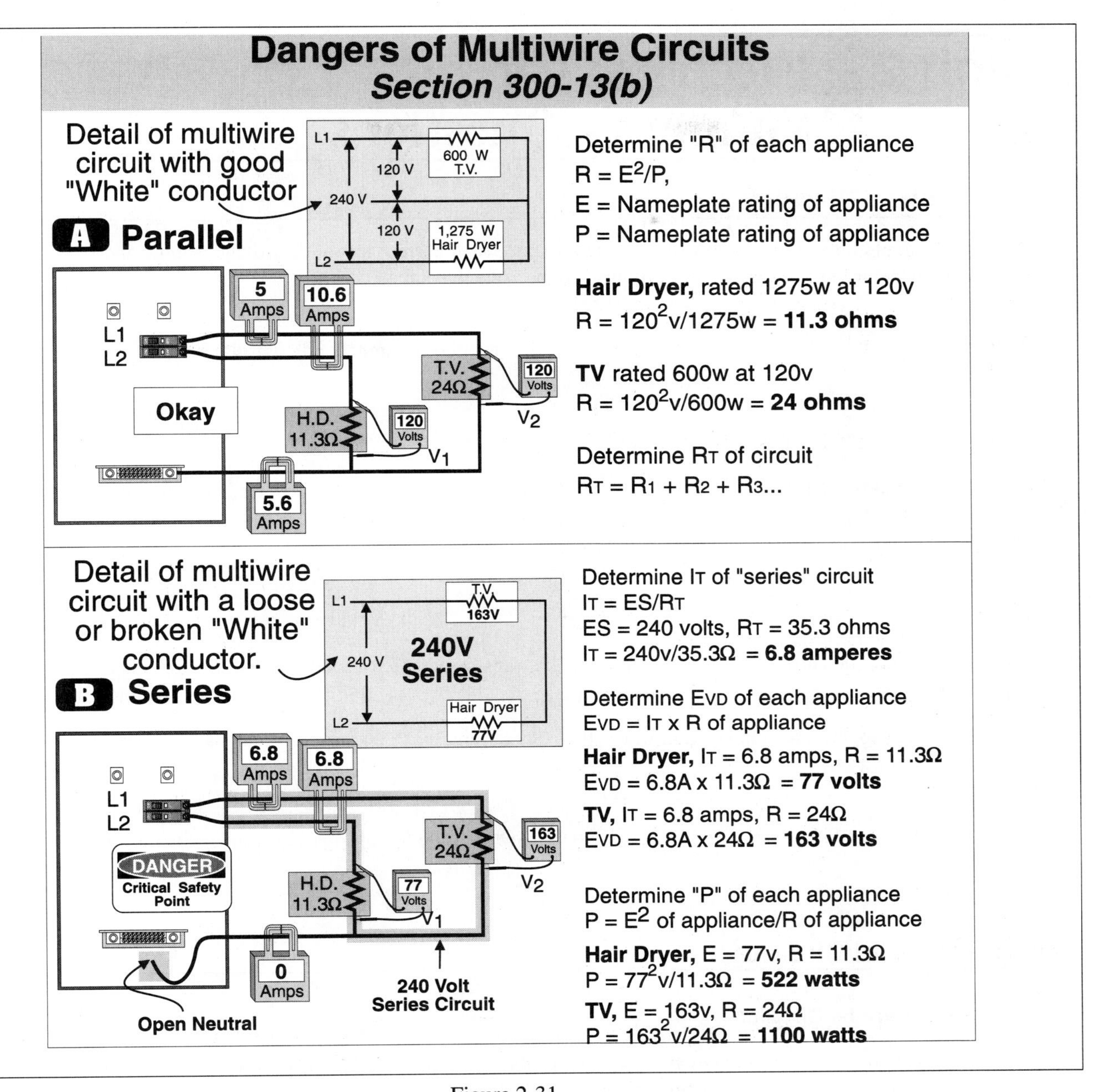

Figure 2-31

Overvoltage – Open Neutral

Step 4: → Determine the voltage for each appliance, $E = I_T \times R_x$

I_T = Current of the circuit, R = Resistance of each resistor

Hair dryer: 6.8 amperes × 11.3 ohms = 76.84 volts

Television: 6.8 amperes × 24 ohms = 163.2 volts

The 120-volt rated TV in the split second before it burns up or explodes is operating at 163.2 volts.

Step 5: → Determine the power consumed of each appliance, $P = E^2/R$

E^2 = Voltage the appliance operates at (squared)

R = Resistance of the appliance

Hair Dryer: $P = 76.8\ \text{volts}^2/11.3$ ohms = 522 watts

Television: $P = 163.2\ \text{volts}^2/24$ ohms = 1,110 watts

Note: The 600-watt, 120-volt rated TV will operate at 163 volts and consume 1,110 watts. Because of the dangers associated with multiwire branch-circuits, don't use them for sensitive or expensive equipment, such as computers, stereos, research equipment, etc.

Unit 2 – Electrical Circuits Summary Questions

Part A – Series Circuits

1. A closed-loop circuit is a circuit in which a specific amount of current leaves the voltage source and flows through every electrical device in a single path before it returns to the voltage source.
(a) True (b) False

2. For most practical purposes, closed-loop system circuits are used for signal and control circuits.
(a) True (b) False

2–1 Understanding Series Calculations

3. Resistance opposes the flow of electrons. In a series circuit, the total circuit resistance is equal to the sum of all the series resistor.
(a) True (b) False

4. The opposition to current flow results in ________.
(a) current (b) voltage (c) voltage drop (d) none of these

5. In a series circuit, the current is ________ through the transformer, the conductors, and the appliance.
(a) proportional (b) distributed (c) additive (d) constant

6. • When power supplies are connected in series, the voltage remains the same, provided that all the polarities are connected properly.
(a) True (b) False

7. The power consumed in a series circuit is equal to the power of the largest resistor in the series circuit.
(a) True (b) False

Part B – Parallel Circuits

Introduction To Parallel Circuits

8. A ________ circuit is a circuit in which current leaves the voltage source, branches through different parts of the circuit in different magnitudes, and then branches back to the voltage source.
(a) series (b) parallel (c) series-parallel (d) multiwire

2–4 Understanding Parallel Calculations

9. • The power supply provides the pressure needed to move the electrons; however, the ________ oppose(s) the current flow.
(a) power supply (b) conductors
(c) appliances (d) all of these

10. When power supplies are connected in parallel, the amp-hour capacity remains the same.
(a) True (b) False

11. The total current of a parallel circuit is equal to the sum of the branch currents. The current in each branch can be calculated by the formula, $I = E/R$.
(a) True (b) False

12. When current flows through a resistor, power is consumed. The power consumed of each branch can be determined by the formula, $P = I^2 \times R$. The total power consumed in a parallel circuit is equal to the largest branch power.
(a) True (b) False

2–5 Parallel Circuit Resistance Calculations

13. The basic method(s) of calculating total resistance, R_T of a parallel circuit is/are ________.
(a) equal resistor method (b) product of the sum method
(c) reciprocal method (d) all of these

14. The total resistance of three 6-ohm resistors in parallel is ________.
(a) 6 ohms (b) 12 ohms (c) 18 ohms (d) none of these

15. The circuit resistance of a 600-watt coffee pot and a 1,000-watt skillet is ________ ohms when connected to a 120 volt parallel circuit.
(a) 24 (b) 14.4 (c) 38.4 (d) 9

16. The resistance total of a 20-ohm, 20-ohm, and 10-ohm resistor in parallel is ________ ohms.
(a) 5 (b) 20 (c) 30 (d) 50

2–6 Parallel Circuit Summary

17. • Which of the following statements is/are true about parallel circuits?
I. The total resistance of a parallel circuit is less than the smallest resistor of the circuit.
II. Current total is equal to the sum of the branch currents.
III. The power of all resistors is equal to the sum of the branch powers.
(a) I and II (b) I, II, and III (c) II and III (d) I and III

Part C – Series–Parallel And Multiwire Circuits

Introduction To Series-Parallel Circuits

18. A ________ is a circuit that contains some resistors in series and some resistors in parallel to each other.
(a) parallel circuit (b) series circuit (c) series-parallel circuit (d) none of these

Part D – Multiwire Branch Circuits

Introduction To Multiwire Circuits

19. A multiwire circuit has two or more ungrounded conductors having a potential difference between them and an equal difference of potential between each ungrounded conductor and the grounded conductor.
(a) True (b) False

2–9 Neutral Current Calculations

20. • The current on the grounded conductor of a 2-wire circuit will be ________ percent of the current on the ungrounded conductor.
(a) 50 (b) 70 (c) 80 (d) 100

Three-Wire Circuits

21. • A balanced 3-wire, 120/240-volt circuit is connected so the ungrounded conductors are from different transformer phases (Line$_1$ and Line$_2$). The current on the grounded conductor will be ________ percent of the ungrounded conductor current.
(a) 0 (b) 70 (c) 80 (d) 100

22. A 3-wire, 120/240-volt circuit will carry 10 amperes unbalanced neutral current if Line$_1$ = 20 amperes and Line$_2$ = 10 amperes.
(a) True (b) False

23. • What is the neutral current for a 20 ampere, 3-wire, 208Y/120-volt circuit?
(a) 0 amperes (b) 10 amperes (c) 20 amperes (d) 40 amperes

Four-Wire Circuits

24. Wye 4-wire Circuits: The neutral of a 4-wire, 208Y/120- or 480Y/277-volt circuit will carry the unbalanced current when the circuit is balanced.
(a) True (b) False

25. What is the neutral current for a 4-wire, 208Y/120-volt circuit, L_1 = 20 amperes, L_2 = 20 amperes, L_3 = 20 amperes?
(a) 0 amperes (b) 10 amperes (c) 20 amperes (d) 40 amperes

26. • The grounded conductor (neutral) of a balanced wye 4-wire circuit will carry no current when supplying power to balanced *nonlinear loads.*
(a) True (b) False

27. • A three-phase, 4-wire, 208Y/120-volt circuit supplying nonlinear load, L_1 = 20 amperes, L_2 = 20 amperes, L_3 = 20 amperes. The neutral conductor will carry as much current as ________ amperes.
(a) 0 (b) 10 (c) 15 (d) 40

2–10 Dangers Of Multiwire Circuits

28. Improper wiring or mishandling of multiwire branch circuits can cause ________ connected to the circuit.
(a) overloading of the ungrounded conductors
(b) overloading of the grounded conductors
(c) destruction of equipment because of overvoltage
(d) b and c

29. • Because of the dangers associated with the open neutral (grounded conductor), the continuity of the ________ conductor cannot be dependent on the receptacle. (Code Rule in Article 300)
(a) ungrounded (b) grounded (c) a and b (d) none of these

Challenge Questions

Part A – Series Circuits

30. • Two resistors, one 4 ohms, and one 8 ohms, are connected in series. If the voltage dropped across both resistors is 12 volts, then the current that would pass through the 4-ohm resistor is ______ amperes.
(a) 1 (b) 2 (c) 4 (d) 8

31. • A series circuit has four 40-ohm resistors and the power supply is 120 volts. The voltage drop of each resistor would be ______.
(a) one-quarter of the source voltage (b) 30 volts
(c) the same across each resistor (d) all of these

32. • The power consumed in a series circuit is ______.
(a) the sum of the power consumed of each load
(b) determined by the formula $P_T = I^2 \times R_T$
(c) determined by the formula $P_T = E \times I$
(d) all of these

33. • The reading on voltmeter 2 (V_2) is ______, Fig. 2–32.
(a) 5 volts
(b) 7 volts
(c) 10 volts
(d) 6 volts

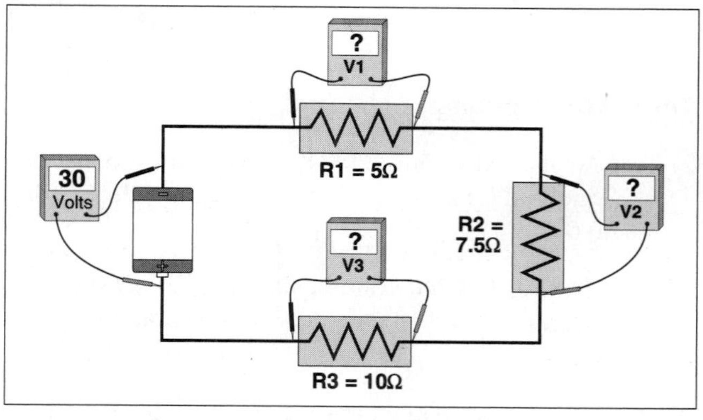

Figure 2-32

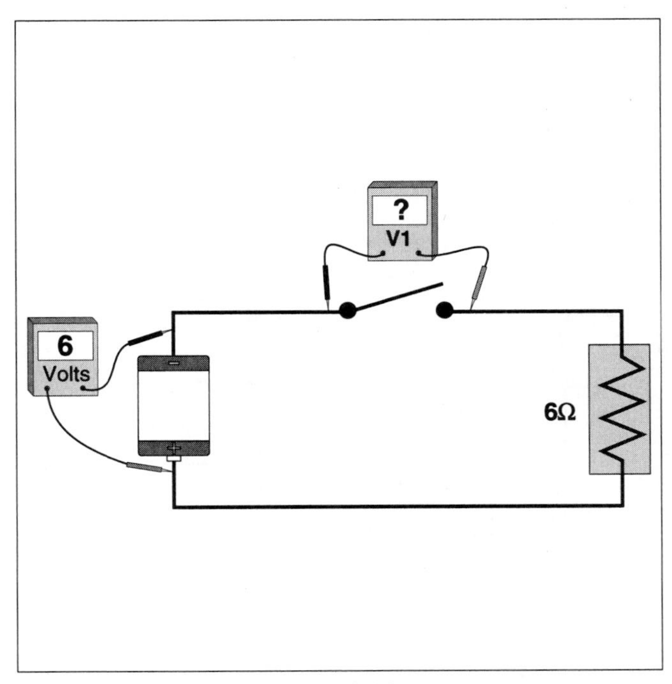

Figure 2-33

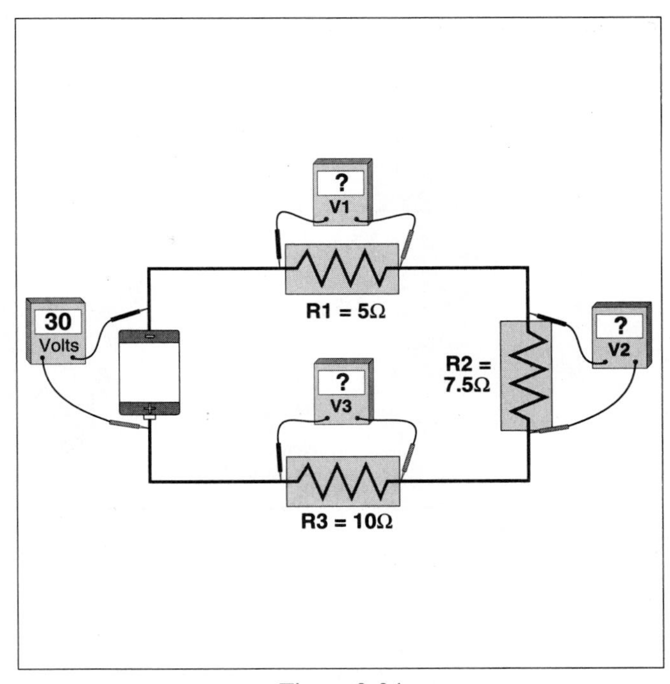

Figure 2-34

34. The voltmeter connected to the circuit on the right would read ______, Fig. 2–33.
(a) 3 volts (b) 12 volts (c) 6 volts (d) 18 volts

Part B – Parallel Circuits

35. • In general, when multiple light bulbs are wired in a single fixture, they are connected in ______ to each other.
(a) series (b) series-parallel
(c) parallel (d) order of wattage

36. • A single-phase, dual-rated, 120/240-volt motor will have its winding connected in ______ when supplied by 120 volts.
(a) series
(b) parallel
(c) series-parallel
(d) parallel-series

37. • The voltmeters shown in Figure 2–34 are connected ______ each of the loads.
(a) in series to (b) across from
(c) in parallel to (d) b and c

38. • If the supply voltage remains constant, four resistors will consume the most power when they are connected ______, Fig. 2–34.
(a) all in series (b) all in parallel
(c) with two parallel pairs in series (d) with one pair in parallel and the other two in series

Circuit Resistance

39. A parallel circuit has three resistors. One resistor is 2 ohms, one is 3 ohms, and the other is 7 ohms. The total resistance of the parallel circuit would be ______ ohm(s).
Remember the total resistance of any parallel circuit is always less than the smallest resistor.
(a) 12
(b) 1
(c) 42
(d) 1.35

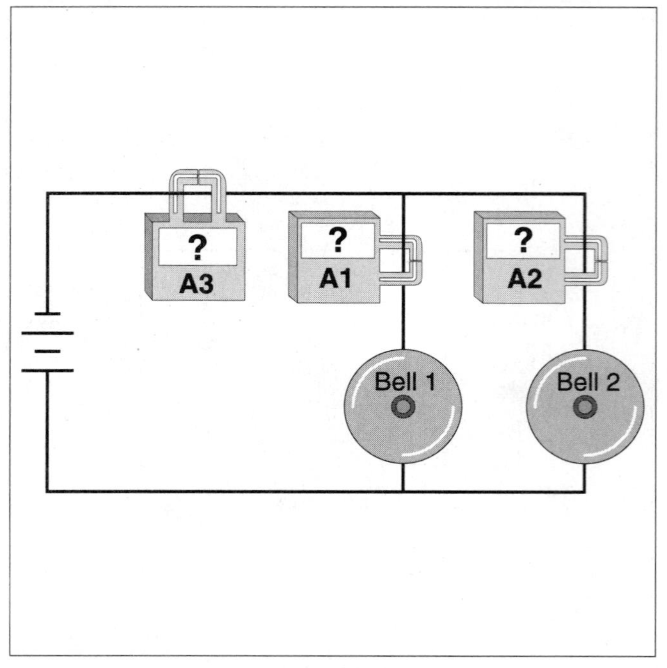

Figure 2-35

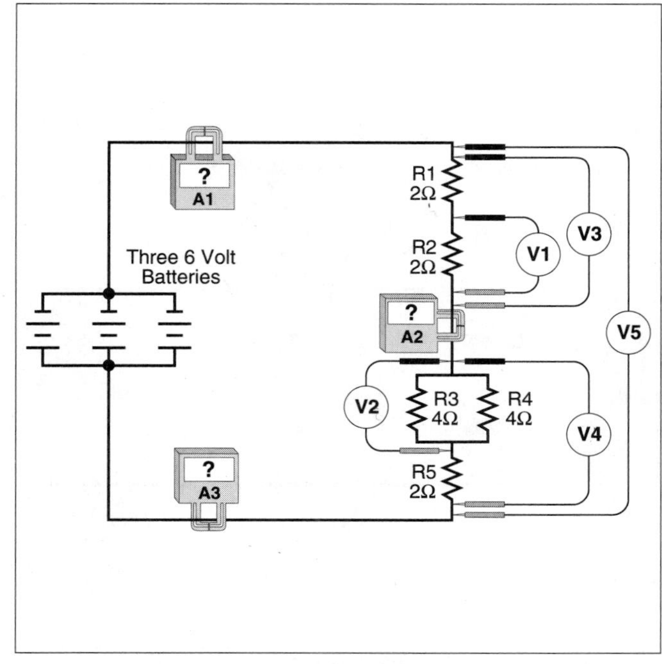

Figure 2-36

Figure 2–35 applies to the next three questions.

40. • The total current of the circuit can be measured by ammeter ______ only, Fig. 2–35.
(a) 1 (b) 2 (c) 3 (d) none of these

41. • If bell 2 consumed 12 watts of power when supplied by two 12-volt batteries (connected in series), the resistance of this bell would be ____ ohms, Fig. 2–35.
(a) 9.6 (b) 44 (c) 576 (d) 48

42. Determine the total circuit resistance of the parallel circuit based on the following facts, Figure 2–35.
1. The current on ammeter 1 reads .75 ampere.
2. The voltage of the circuit is 30 volts.
3. Bell 2 has a resistance of 48 ohms.
Tip: Resistance total of a parallel circuit is always.
(a) 22 ohms (b) 48 ohms (c) 1920 ohms (d) 60 ohms

Part C – Series-Parallel Circuits

Figure 2–36 applies to the next three questions.

43. • The total current of this circuit can be read on ______, Fig. 2–36.
I. Ammeter 1 II. Ammeter 2
III. Ammeter 3
(a) I only (b) II only (c) III only (d) I, II and III

44. • The voltage reading of V_2 is ______ volts, Fig. 2–36.
(a) 1.5 (b) 4 (c) 4 (d) 8

45. • The voltage reading of V_4 is ______ volts, Fig. 2–36.
(a) 1.5 (b) 3 (c) 5 (d) 8

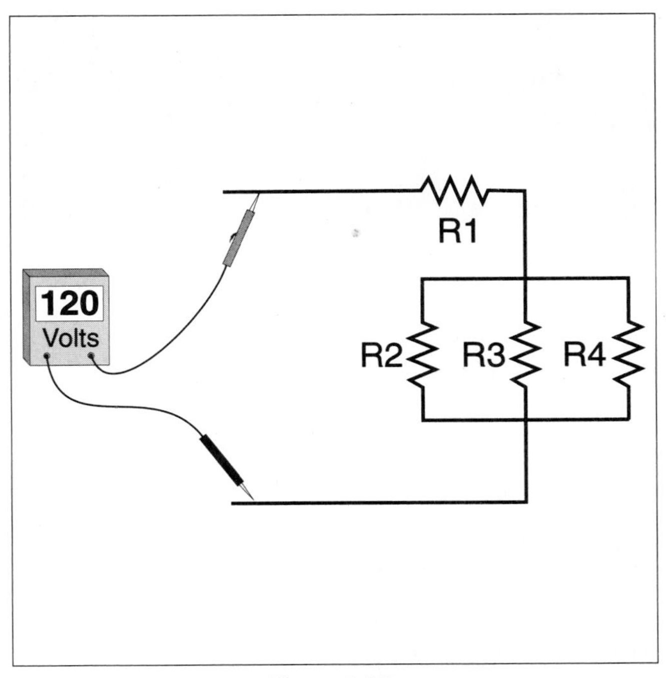

Figure 2-37

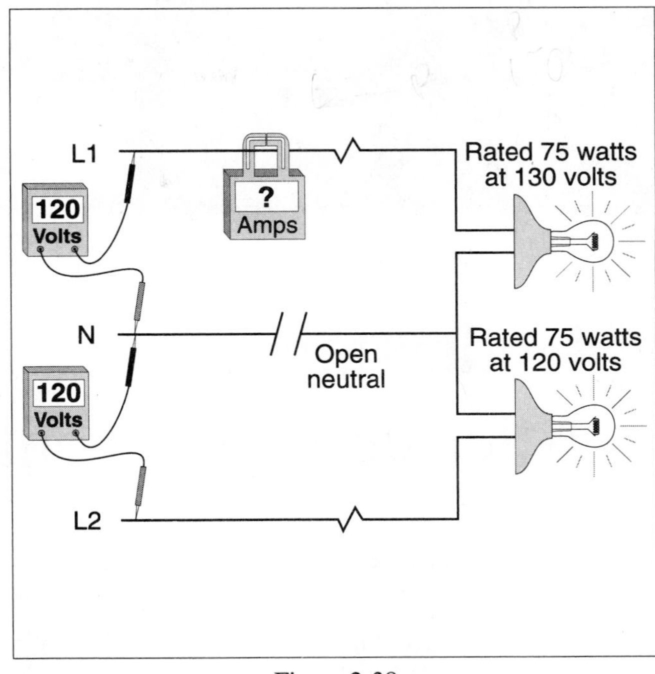

Figure 2-38

Figure 2–37 applies to the next two questions.

46. Resistor R_1 has a resistance of 5 ohms, resistors R_2, R_3, and R_4 have a resistance of 15 ohms each. The total resistance of this series-parallel circuit is ______ ohms, Fig. 2–37.
(a) 50
(b) 35
(c) 25
(d) 10

47. • What is the voltage drop across R_1? R_1 is 5 ohms and total resistance of R_2, R_3, and R_4 is 5 ohms, Fig. 2–37.
(a) 60 volts
(b) 33 volts
(c) 40 volts
(d) 120 volts

Part D – Multiwire Circuits

48. • If the neutral of the circuit in the diagram is opened, the circuit becomes one series circuit of 240 volts. Under this condition, the current of the circuit is ______ ampere(s). *Tip:* Determine the resistance total, Fig. 2–38.
(a) .67
(b) .58
(c) 2.25
(d) .25

Unit 3

Understanding Alternating Current

OBJECTIVES

After reading this unit, the student should be able to briefly explain the following concepts:

Part A – Alternating Current Fundamentals

- Alternating current
- Armature turning frequency
- Current flow
- Generator
- Magnetic cores
- Phase differences in degrees
- Phase– In and Out
- Values of alternating current
- Waveform

Part B - Induction And Capacitance

- Charge, discharging and testing of capacitors
- Conductor impedance
- Conductor shape
- Induced voltage and applied current
- Magnetic cores
- Use of capacitors

Part C - Power Factor And Efficiency

- Apparent power (volt-amperes)
- Efficiency
- Power factor
- True power (watts)

After reading this unit, the student should be able to briefly explain the following terms:

Part A – Alternating Current Fundamentals

- Ampere-turns
- Armature speed
- Coil
- Conductor cross-sectional area
- Effective (RMS)
- Effective to peak
- Electromagnetic field
- Frequency
- Impedance
- Induced voltage
- Magnetic field
- Magnetic flux lines
- Peak
- Peak to effective
- Phase relationship
- RMS
- Root-mean-square
- Self-inductance
- Skin effect
- Waveform

Part B And C - Induction And Capacitance

- Back-EMF
- Capacitance
- Capacitive reactance
- Capacitor
- Coil
- Counterelectromotive force
- Eddy currents
- Farads
- Frequency
- Henrys
- Impedance
- Induced voltage
- Induction
- Inductive reactance
- Phase relationship
- Self-inductance
- Skin effect

Part D - Power Factor And Efficiency

- Apparent power
- Efficiency
- Input watts
- Output watts
- Power factor
- True power
- Volts-amperes
- Wattmeter

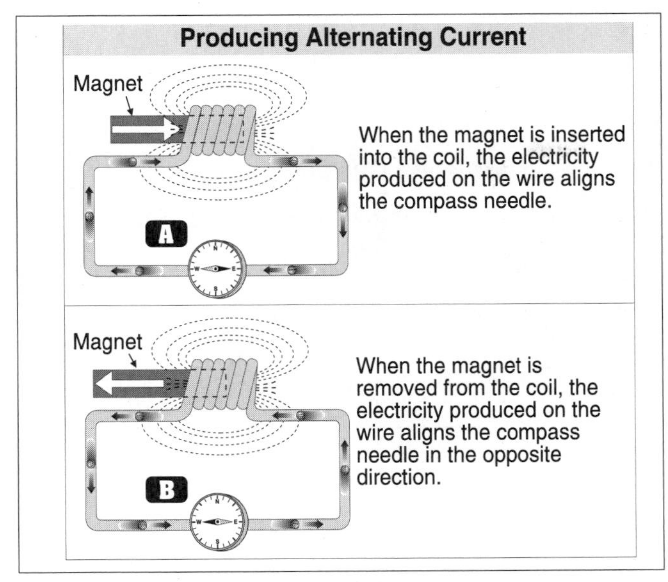

Figure 3-1

Alternating Current

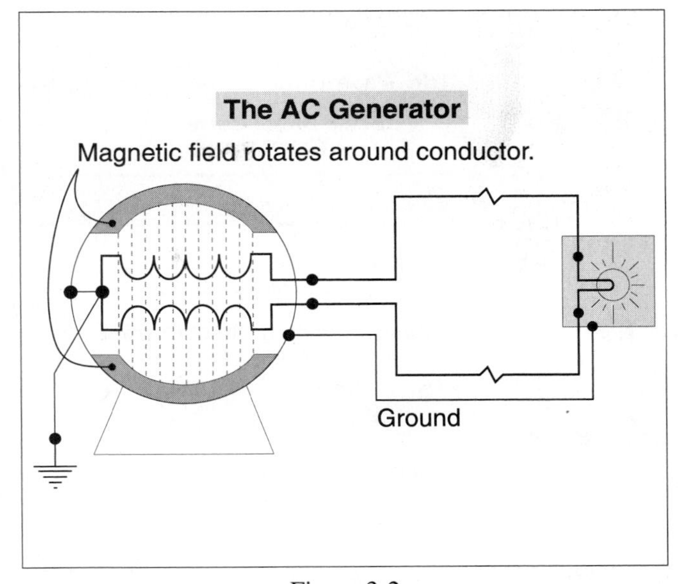

Figure 3-2

Alternating Current Generator

PART A – *ALTERNATING CURRENT FUNDAMENTALS*

3–1 *CURRENT FLOW*

For current to flow in a circuit, the circuit must have a *closed loop* and the power supply must push the electrons through the completed circuit. The transfer of electrical energy can be accomplished by electrons flowing in one constant direction (*direct current*), or by electrons flowing in one direction and then reversing in the other direction (*alternating current*).

3–2 *ALTERNATING CURRENT*

Alternating current is generally used instead of direct current for electrical systems and building wiring. This is because alternating current can easily have voltage variations (*transformers*). In addition alternating current can be transmitted inexpensively at *high voltage* over long distances resulting in reduced *voltage drop* of the power distribution system as well as smaller power distribution wire and equipment. Alternating current can be used for certain applications for which direct current is not suitable.

Alternating current is produced when electrons in a conductor are forced to move because there is a moving magnetic field. The lines of force from the magnetic field cause the electrons in the wire to flow in a specific direction. When the lines of force of the magnetic field move in the opposite direction, the electrons in the wire will be forced to flow in the opposite direction, Fig. 3–1. We must remember that electrons will flow only when there is relative motion between the conductors and the magnetic field.

3–3 *ALTERNATING CURRENT GENERATOR*

An alternating current *generator* consists of many loops of wire that rotates between the *flux lines* of a magnetic field. Each conductor loop travels through the magnetic lines of force in opposite directions, causing the electrons within the conductor to move in a specific direction, Fig. 3–2. The force on the electrons caused by the *magnetic flux lines* is called *voltage*, or *electromotive force*, and is abbreviated as *EMF*, E or V.

The magnitude of the electromotive force is dependent on the number of turns (wraps) of wire, the strength of the magnetic field, and the speed at which the coil rotates. The rotating conductor loop mounted on a shaft is called a *rotor*, or *armature*. Slip, or collector rings, and carbon brushes are used to connect the output voltage from the generator to an external circuit.

3–4 *WAVEFORM*

A *waveform* is a pictorial view of the shape and magnitude of the level and direction of current or voltage over a period of time. The waveform represents the magnitude and direction of the current or voltage overtime in relationship with the generators armature.

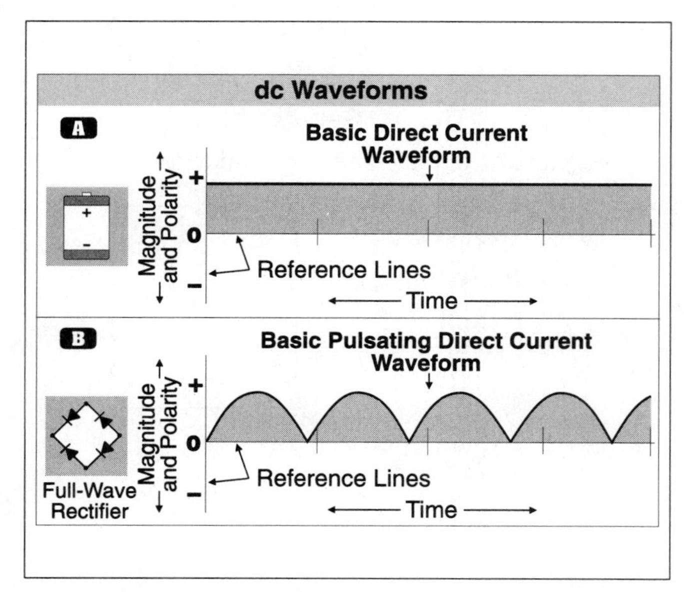

Figure 3-3

Direct Current Waveform

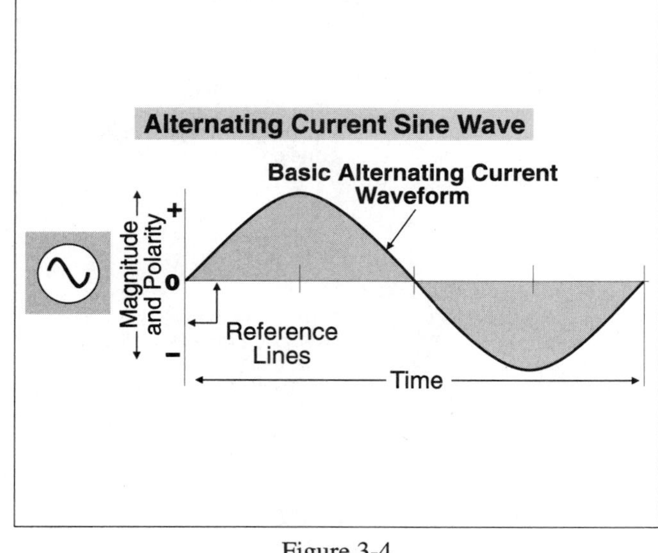

Figure 3-4

Alternating Current Sine Wave

Direct Current Waveform

The *polarity* of the voltage of direct current is positive, and the current flows through the circuit in the same direction at all times. In general, direct current voltage and current will remain the same magnitude, particularly when supplied from batteries or capacitors, Fig. 3–3.

Alternating Current Waveform

The waveform for alternating current displays the level and direction of the current and voltage for every instant of time for one full revolution of the *armature*. When the waveform of an alternating current circuit is symmetrical with positive above, and negative below the zero reference level, the waveform is called a *sine wave*, Fig. 3–4.

3–5 *ARMATURE TURNING FREQUENCY*

Frequency is a term used to indicate the number of times the generator's armature turns one full revolution (360 degrees) in one second, Fig. 3–5. This is expressed as *hertz*, or *cycles per second*, and is abbreviated as *Hz* or cycles per second (*cps*). In the United States, frequency is not a problem for most electrical work because it remains constant at 60 hertz.

Armature Speed

A current or voltage that is produced when the generator's armature makes one cycle (360 degrees) in 1/60th of a second is called 60-Hz. This is because the armature travels at a speed of one cycle every 1/60th of a second. A 60-Hz circuit will take 1/60th of a second for the armature to complete one full cycle (360 degrees), and it will take 1/120th of a second to complete 1/2 cycle (180 degrees), Fig. 3–6.

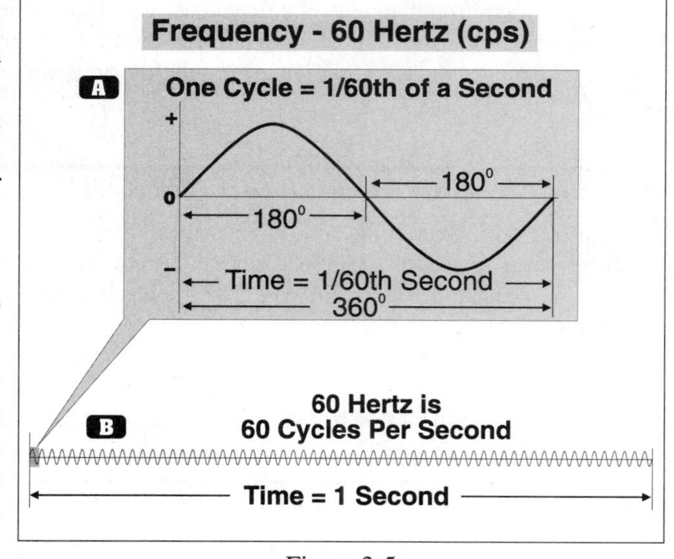

Figure 3-5

Armature Turn Frequency

3–6 *PHASE – IN AND OUT*

Phase is a term used to indicate the time relationship between two waveforms, such as voltage to current, or voltage to voltage. *In phase* means that two waveforms are in step with each other, that is, they cross the horizontal zero axis at the same time, Fig. 3–7. The terms *lead* and *lags* are used to describe the relative position, in degrees, between two waveforms. The waveform that is ahead, *leads*, the waveform behind, *lags*, Fig. 3–8.

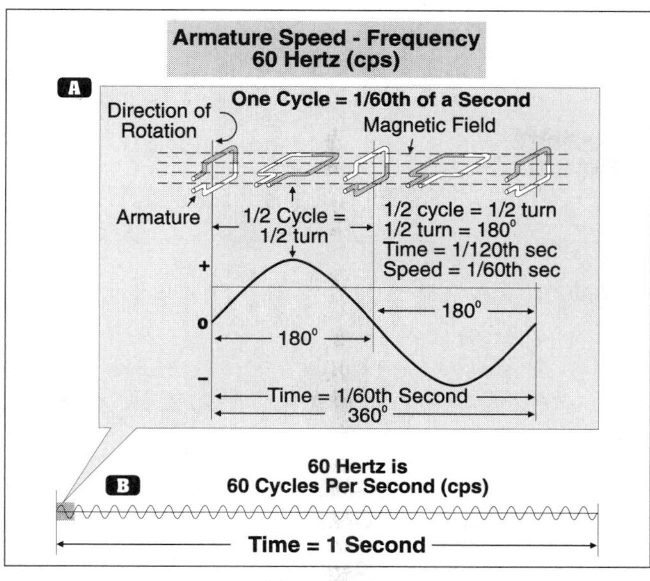

Figure 3-6

Armature Speed

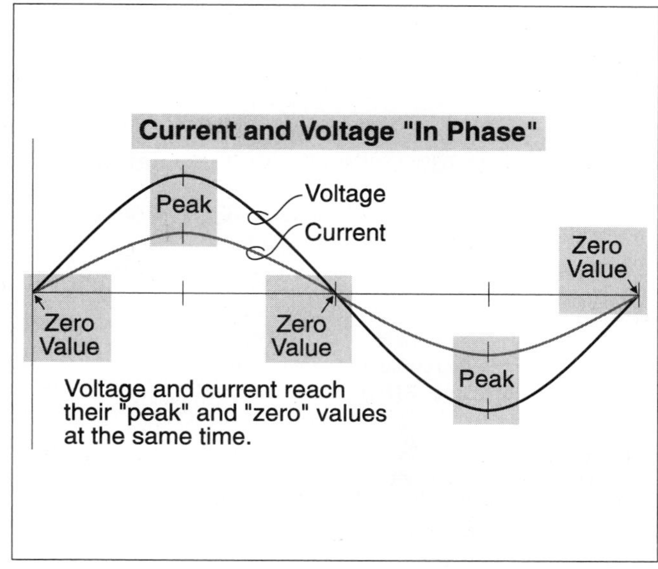

Figure 3-7

In Phase

3–7 *PHASE DIFFERENCES IN DEGREES*

Phase differences are expressed in *degrees*, one full revolution of the armature is equal to 360 degrees. Single-phase, 120/240-volt power is out of phase by 180 degrees, and three-phase power is out of phase by 120 degrees, Figure 3–9.

3–8 *VALUES OF ALTERNATING CURRENT*

There are many different types of values in alternating current, the most important are *peak* (maximum) and *effective*. In alternating current systems, effective is the value that will cause the same amount of heat to be produced as in a direct current circuit containing only resistance. Another term for effective is *RMS*, which is *root-mean-square*. In the field, whenever you measure voltage, you are always measuring effective voltage. Newer types of clamp-on ammeters have the ability to measure peak as well as effective current, Fig. 3–10.

Note: The values of direct current generally remain constant and are equal to the effective values of alternating current.

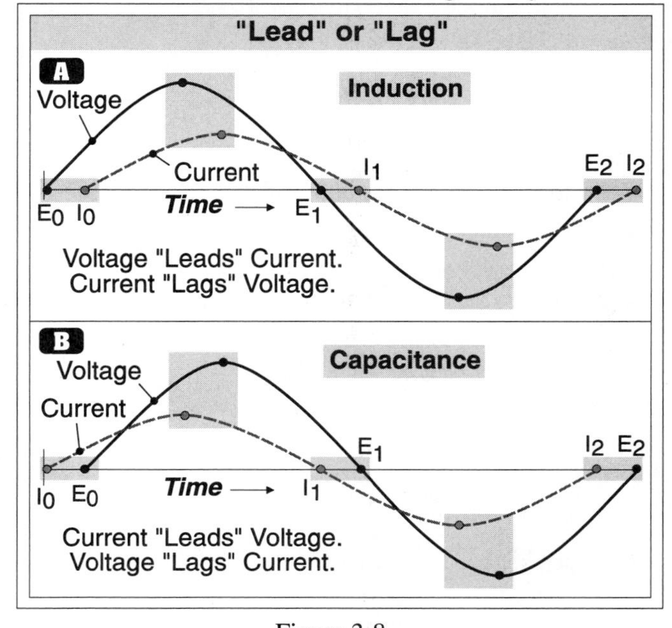

Figure 3-8

Out of Phase – Lead or Lag

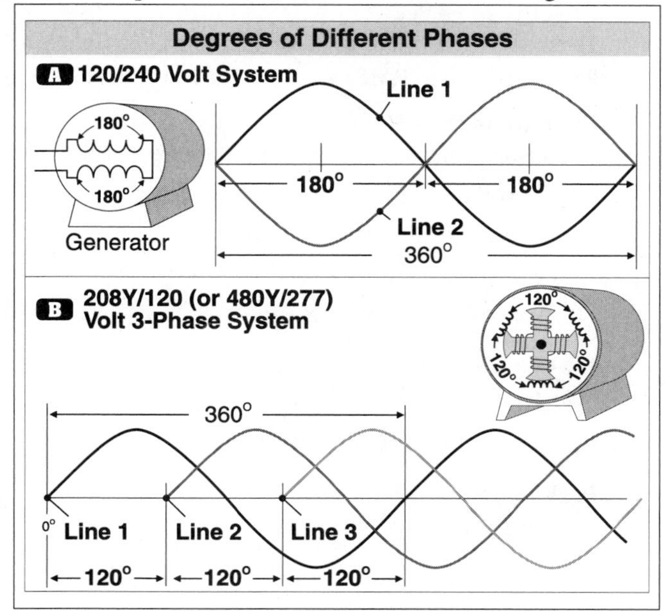

Figure 3-9

Phase Differences in Degrees

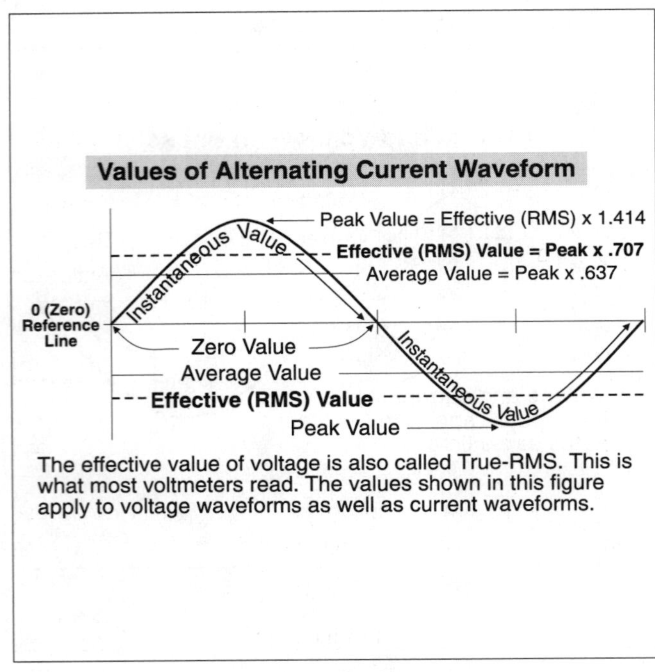

Figure 3-10

Values of Alternating Current

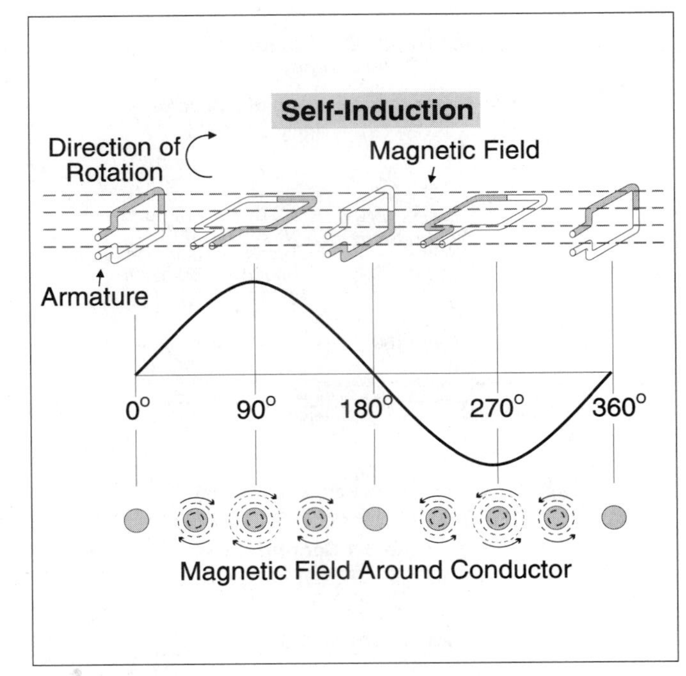

Figure 3-11

Electromagnetic Field Within Conductor

❑ **Alternating Current Versus Direct Current Heating Example**

Twelve amperes effective alternating current will have the same heating effect as ______-amperes direct current.

(a) 7 (b) 9 (c) 11 (d) 12

- Answer: (d) 12

❑ **Voltage Example**

One Hundred and twenty volt effective alternating current will produce the same heating effect as _____ volts direct current.

(a) 70 (b) 90 (c) 120 (d) 170

- Answer: = 120

Effective to Peak

To convert effective to peak, or vise versa, the following formulas should be helpful:

Peak = Effective (RMS) × 1.414 times or Peak = Effective (RMS)/0.707

❑ **Effective to Peak Example**

What is the peak voltage of a 120-volt effective circuit?

(a) 120 volts (b) 170 volts (c) 190 volts (d) 208 volts

- Answer: (b) 170 volts

  Peak Volts = Effective (RMS) volts × 1.411 Effective (RMS) = 120 volts

  Peak Volts = 120 volts × 1.414 Peak Volts = 170 volts

Peak to Effective

Peak to effective can be determined by the formula:

Effective (RMS) = Peak × .707

❑ **Peak to Effective Example**

What is the effective current of 60 amperes peak?

(a) 42 amperes (b) 60 amperes (c) 72 amperes (d) none of these

- Answer: (a) 42 amperes

  Effective Current = Peak Current × 0.707 Peak Current = 60 amperes

  Effective Current = 60 amperes × 0.707 Effective Current = 42 amperes

PART B – *INDUCTION*

INDUCTION INTRODUCTION

When a magnetic field is moved through a conductor, the electrons in the conductor will move. The movement of electrons caused by electromagnetism is called *induction.* In an alternating current circuit, the current movement of the electrons increases and decreases with the rotation of the *armature* through the magnetic field. As the current flow through the conductor increases or decreases, it causes an expanding and collapsing electromagnetic field within the conductor, Fig. 3–11. This varying electromagnetic field within the conductor causes the electrons in the conductor to move at 90 degrees to the flowing electrons within the conductor. This movement of electrons from the conductor's electromagnetic field is called *self-induction*. Self-induction (induced voltage) is commonly called *counterelectromotive-force*, *CEMF*, or *back-EMF*.

3–9 *INDUCED VOLTAGE AND APPLIED CURRENT*

The *induced voltage* in the conductor from the conductor's electromagnetism (CEMF) is always 90 degrees *out of phase* with the applied current, Fig. 3–12. When the current in the conductor increases, the induced voltage tries to prevent the current from increasing by storing some of the electrical energy in the *magnetic field* around the conductor. When the current in the conductor decreases, the induced voltage attempts to keep the current from decreasing by releasing the the energy from the magnetic field back into the conductor. The opposition to any change in current because of induced voltage is called *inductive reactance*, which is measured in ohms, abbreviated as X_L, and is known as *Lentz's Law.*

Inductive reactance changes proportionally with frequency and can be found by the formula:

$X_L = 2\pi fL$ $\pi = 3.14$, f = frequency L= Inductance in henrys

❑ **Inductive Reactance Example**

What is the inductive reactance of a conductor in a transformer that has an inductance of 0.001 henrys? Calculate at 60-Hz, 180-Hz, and 300-Hz, Fig. 3–13.

- Answers:

  $X_L = 2\pi fL$

  $\pi = 3.14$, f = 60-Hz, 180-Hz and 300-Hz, L= Inductance measured in henrys, 0.001

  X_L at 60-Hz = 2 × 3.14 × 60-Hz × 0.001 henrys, X_L = 0.377 ohms

  X_L at 180-Hz = 2 × 3.14 × 180-Hz × 0.001 henrys, X_L = 1.13 ohms

  X_L at 300-Hz = 2 × 3.14 × 300-Hz × 0.001 henrys, X_L = 1.884 ohms

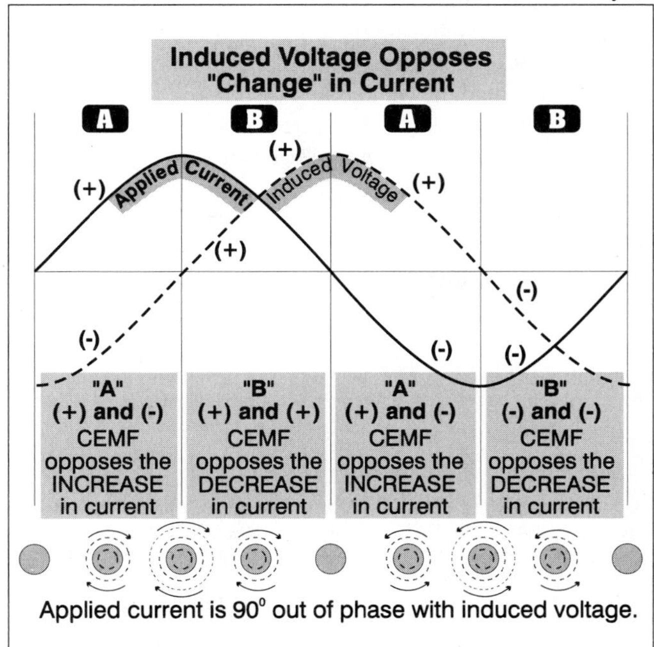

Figure 3-12

Induced Voltage versus Applied Voltage

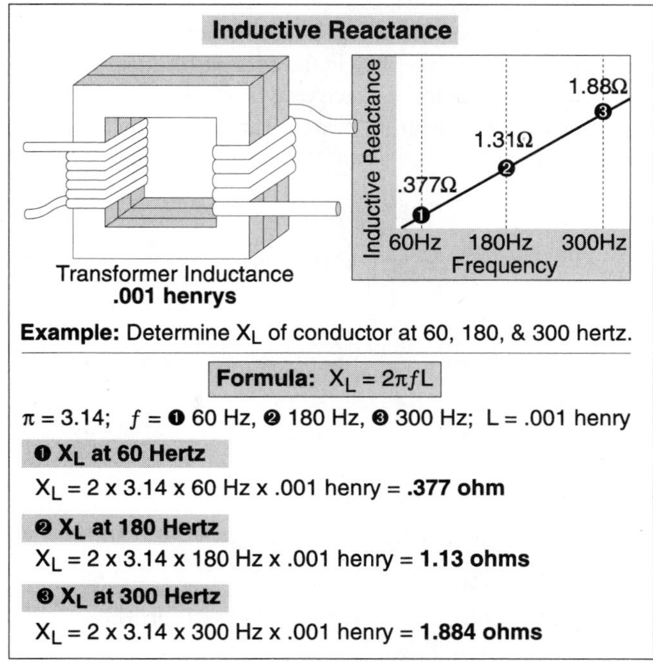

Figure 3-13

Inductive Reactance Example

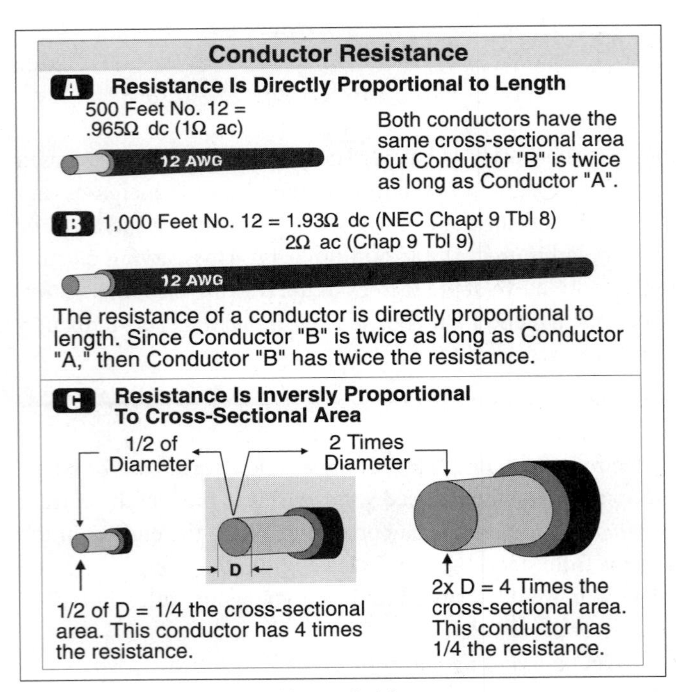

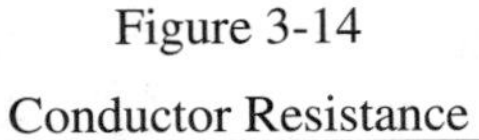

Figure 3-14

Conductor Resistance

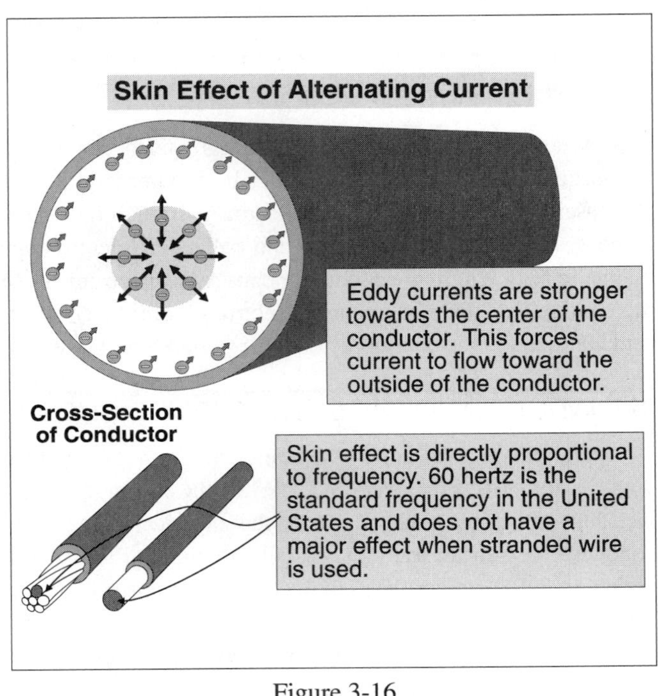

Figure 3-16

Skin Effect

3–10 *CONDUCTOR IMPEDANCE*

The *resistance* of a conductor is a physical property of the conductors that oppose the flow of electrons. This resistance is directly proportional to the conductor length, and is inversely proportional to the conductor cross-sectional area. This means that the longer the wire, the greater the conductor resistance; the smaller the wire, the greater the conductor resistance, Fig. 3–14.

The opposition to alternating current flow due to *inductive reactance* and conductor resistance is called *impedance (Z).* Impedance is only present in alternating current circuits and is measured in ohms. The opposition to alternating current (impedance - Z) is greater than the resistance (R) of a direct current circuit, because of counterelectromotive-force, *eddy currents*, and *skin effect.*

Eddy Currents and Skin Effect

Eddy currents are small independent currents that are produced as a result of the expanding and collapsing magnetic field of alternating current flow, Fig. 3–15. The expanding and collapsing magnetic field of an alternating current circuit also induces a counterelectromotive-force in the conductors that repels the flowing electrons towards the surface of the conductor, Fig. 3–16. The counterelectromotive-force causes the circuit current to flow near the surface of the conductor rather than at the conductors center. The flow of electrons near the surface is known as *skin effect*. Eddy currents and skin effect decrease the effective conductor cross-sectional area for current flow, that result in an increase in the the conductor's *impedance* which is measured in ohms.

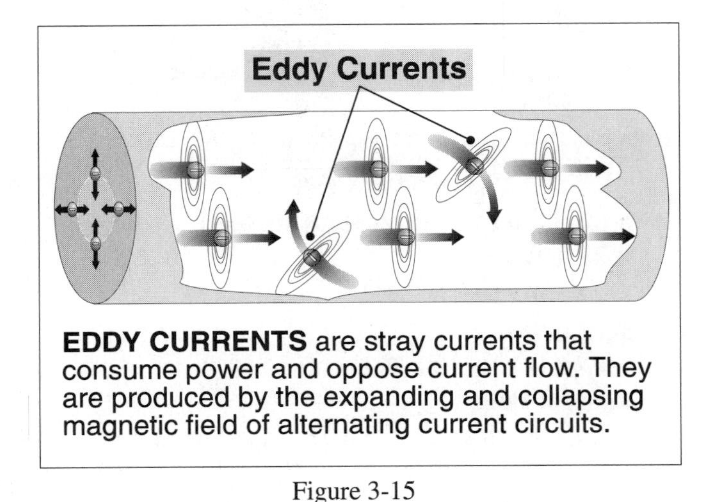

Figure 3-15

Eddy Currents

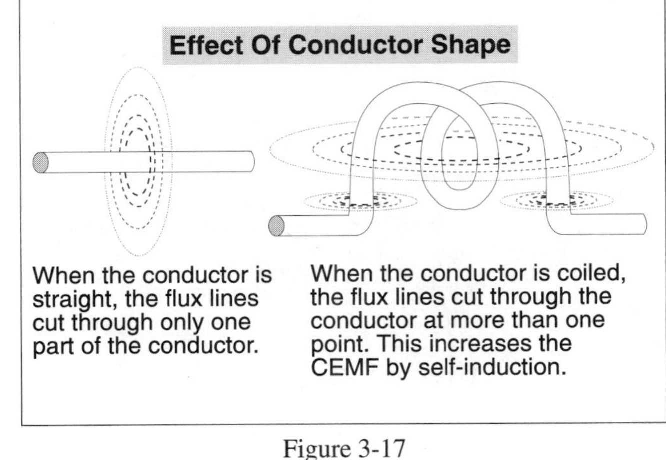

Figure 3-17

Induction and Conductor Loops

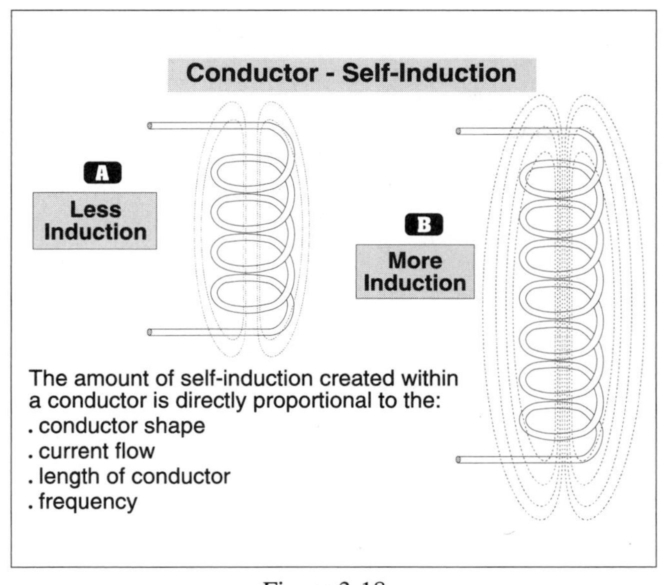

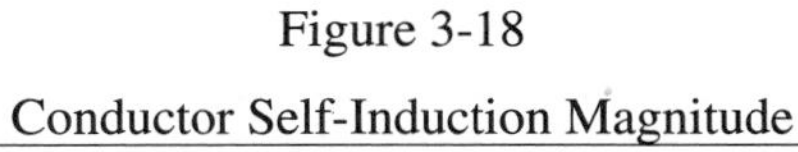
Figure 3-18

Conductor Self-Induction Magnitude

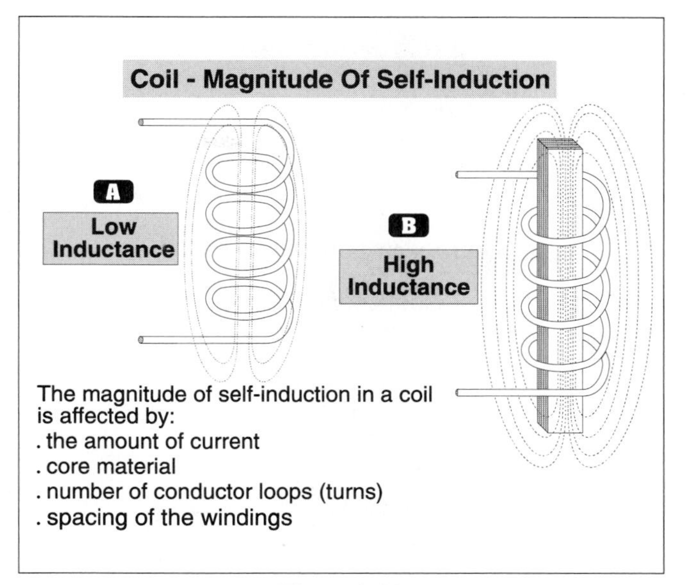

Figure 3-19

Coil Self-Induction Magnitude

3–11 *INDUCTION AND CONDUCTOR SHAPE*

The physical shape of a conductor affects the amount of *self-induced voltage* in the conductor. If a conductor is coiled into adjacent loops *(winding)*, the electromagnetic field (*flux lines*) of each conductor loop adds together to create a stronger overall magnetic field, Fig. 3–17. The amount of self-induced voltage created within a conductor is directly proportional to the current flow, the length of the conductor, and the frequency at which the magnetic fields cut through the conductors, Fig. 3–18.

3–12 *INDUCTION AND MAGNETIC CORES*

Self-inductance in a *coil* of conductors is affected by the winding current magnitude, the core material, the number of conductor loops (*turns*), and the spacing of the winding, Fig. 3–19. *Iron cores* within the conductor winding permits an easy path for the electromagnetic flux lines which produce a greater counterelectromotive-force than air core windings. In addition, conductor coils are measured in *ampere-turns*, that is, the current in amperes times the number of conductor loops or turns.

Ampere-Turns = Coil Amperes × Number of Coil Turns

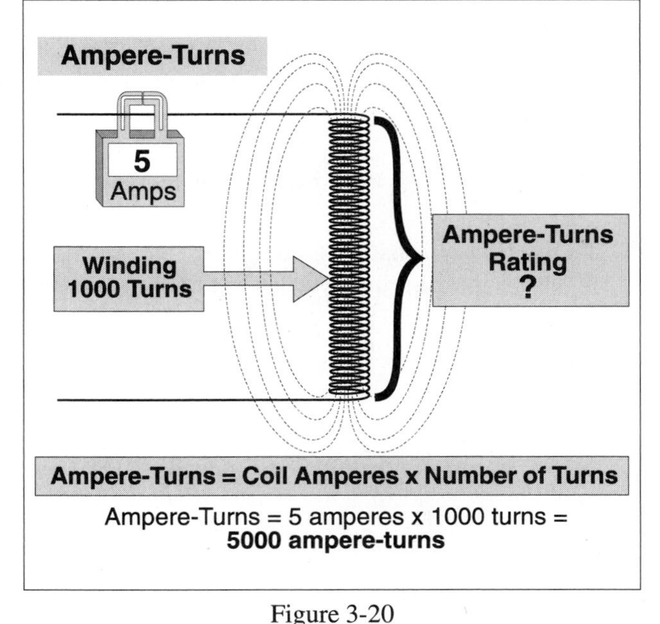

Figure 3-20

Ampere Turn Example

❑ **Ampere-Turns Example**

A magnetic coil in a transformer has 1,000 turns and carries 5 amperes. What is the ampere-turns rating of this transformer, Fig. 3–20?

(a) 5,000 ampere-turns
(b) 7,500 ampere-turns
(c) 10,000 ampere-turns
(d) none of these

- Answer: (a) 5,000 ampere-turns

  Ampere-Turns = Amperes × Number of Turns

  Ampere-turns = 5 amperes × 1,000 turns

  Ampere-turns = 5,000

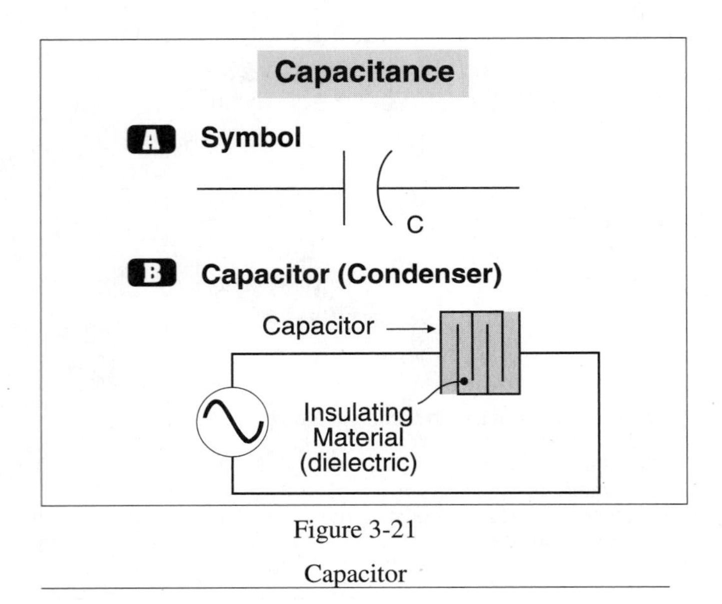

Figure 3-21

Capacitor

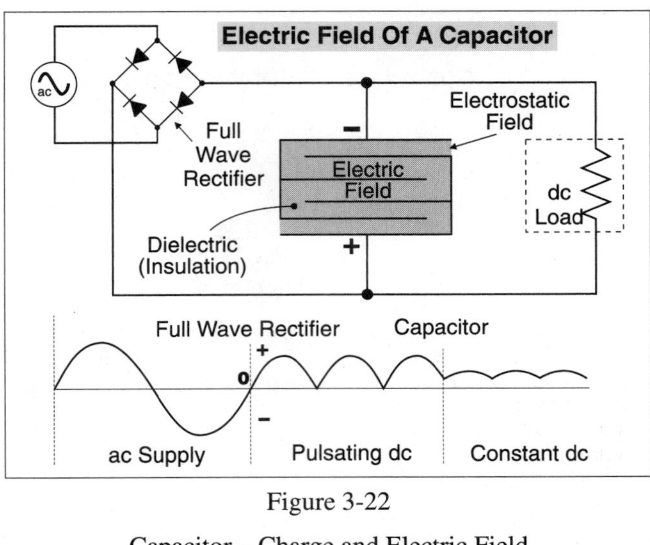

Figure 3-22

Capacitor – Charge and Electric Field

PART C – *CAPACITANCE*

CAPACITANCE INTRODUCTION

Capacitance is the property of an electric circuit that enables the storage of electric energy by means of an *electrostatic field*, much like a spring stores mechanical energy, Fig. 3–21. Capacitance exists whenever an insulating material separates two objects that have a difference of potential. A capacitor is simply two metal foils separated by wax paper rolled together. Devices that intentionally introduce capacitance into circuits are called *capacitors* or *condensers.*

3–13 *CHARGE, TESTING AND DISCHARGING*

When a capacitor has a potential difference between the plates, it is said to be *charged.* One plate has an excess of free electrons (–), and the other plate has a lack of electrons (+). Because of the insulation (*dielectric*) between the capacitor foils, electrons cannot flow from one plate to the other. However, there are electric lines of force between the plates, and this force is called the *electric field*, Fig. 3–22.

Factors that determine capacitance are the surface area of the plates, the distance between the plates, and the *dielectric* insulating material between the plates. Capacitance is measured in *farads*, but the opposition offered to the flow of current by a capacitor is called *capacitive reactance*, which is measured in ohms. Capacitive reactance is abbreviated X_C*, is directly proportional to frequency and can be calculated by the equation:*

$$X_C = \frac{1}{2\pi fC}$$ **π = 3.14, f = frequency, C= Capacitance measure in farads**

❑ **Capacitive Reactance Example**

What is the capacitive reactance of a 0.000001 farad capacitor supplied by 60-Hz, 180-Hz, and 300-Hz, Fig. 3–23?

- Answer: $X_C = \frac{1}{2\pi fC}$.

  $\pi = 3.14$

  f=frequency; 60-Hz, 180-Hz, and 300-Hz

  C=Capacitancemeasuredinfarads,0.0000001

  X_C at 60–Hz $= \frac{1}{2 \times 3.14 \times 60\text{–hertz} \times 0.000001 \text{ farads}}$, X_C = 2,654 ohms

  X_C at 180–Hz $= \frac{1}{2 \times 3.14 \times 180\text{–hertz} \times 0.000001 \text{ farads}}$, X_C = 885 ohms

  X_C at 300–Hz $= \frac{1}{2 \times 3.14 \times 300\text{–hertz} \times 0.000001 \text{ farads}}$, X_C = 530.46 ohms

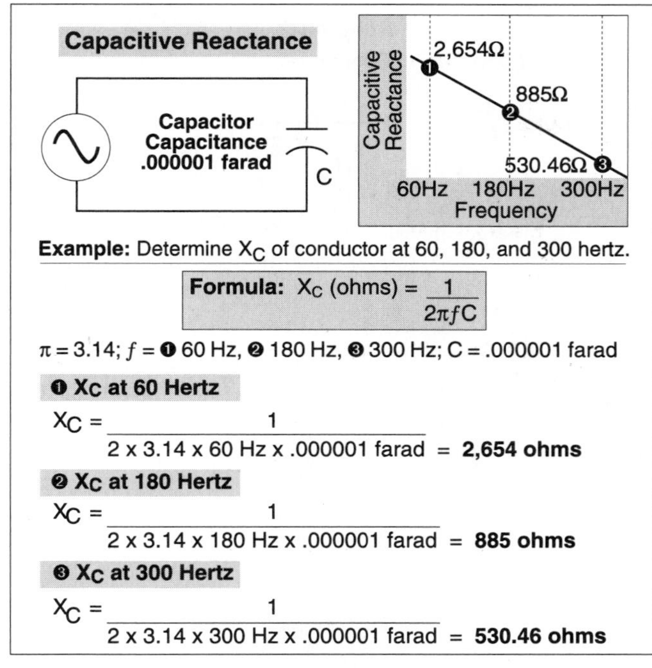

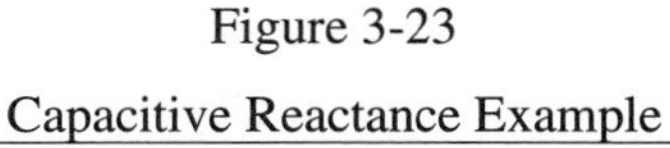

Figure 3-23

Capacitive Reactance Example

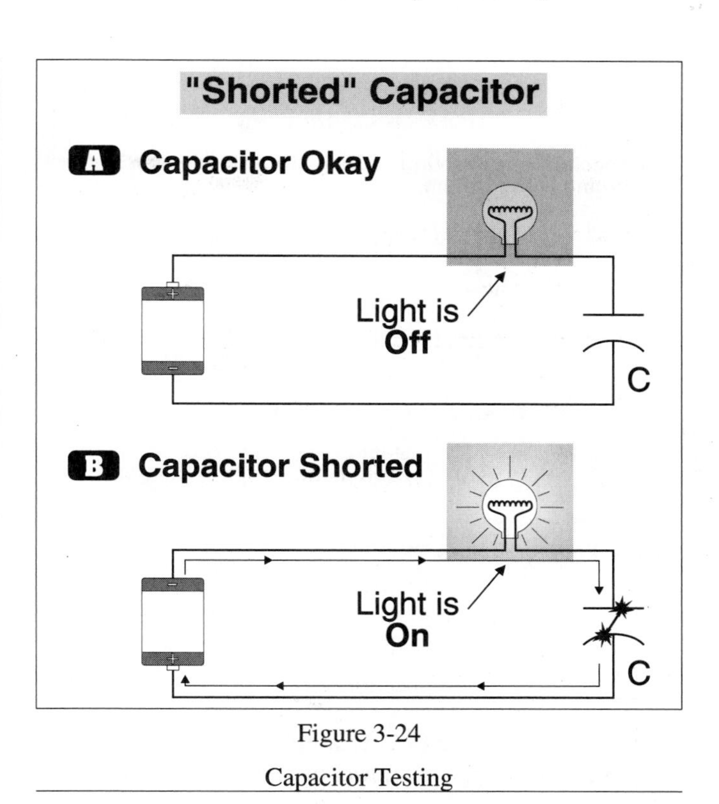

Figure 3-24

Capacitor Testing

Capacitor Short and Discharge

The more the capacitor is charged, the stronger the electric field. If the capacitor is overcharged, the electrons from the negative plate could be pulled through the dielectric insulation to the positive plate, resulting in a short of the capacitor. To discharge a capacitor, all that is required is a conducting path between the capacitor plates.

Testing a Capacitor

If a capacitor is connected in series with a direct current power supply and a test lamp, the lamp will be continuously illuminated if the capacitor is shorted, but will remain dark if the capacitor is good, Fig. 3–24.

3–14 *USE OF CAPACITORS*

Capacitors (condensers) are used to prevent arcing across the *contacts* of low ampere switches and direct current power supplies for electronic loads such as computer, Fig. 3–25. In addition, capacitors cause the current to lead the voltage by as much as 90 degrees and can be used to correct poor power factor due to inductive reactive loads such as motors. Poor power factor, because of induction, causes the current to lag the voltage by as much as 90 degrees. Correction of the power factor to *unity* (100%) is known as *resonance*, and occurs when inductive reactance is equal to capacitive reactance in a series circuit; $X_L = X_C$.

PART D – *POWER FACTOR AND EFFICIENCY*

POWER FACTOR INTRODUCTION

Alternating current circuits develop inductive and capacitive reactance which causes some power to be stored temporarily in the electromagnetic field of the inductor, or in the electrostatic field of the capacitor, Fig. 3–26. Because of the temporary storage of energy, the voltage and current of inductive and capacitive circuits are not in phase. This out-of-phase relationship between voltage and current is called *power factor*, Figure 3–27. Power factor affects the calculation of power in alternating current circuits, but not direct current circuits.

3–15 *APPARENT POWER (VOLT-AMPERES)*

In alternating current circuits, the value from multiplying volts times amperes equals *apparent power* and is commonly referred to as *volt–amperes*, *VA*, or *kVA*. Apparent power (VA) is equal to or greater than *true power* (*watts*) depending on the *power factor*. When sizing circuits or equipment, always size the circuit equipment according to the apparent power (volt-ampere), not the true power (watt).

Apparent Power = Volts × Amperes, or Apparent Power = True Power/Power Factor

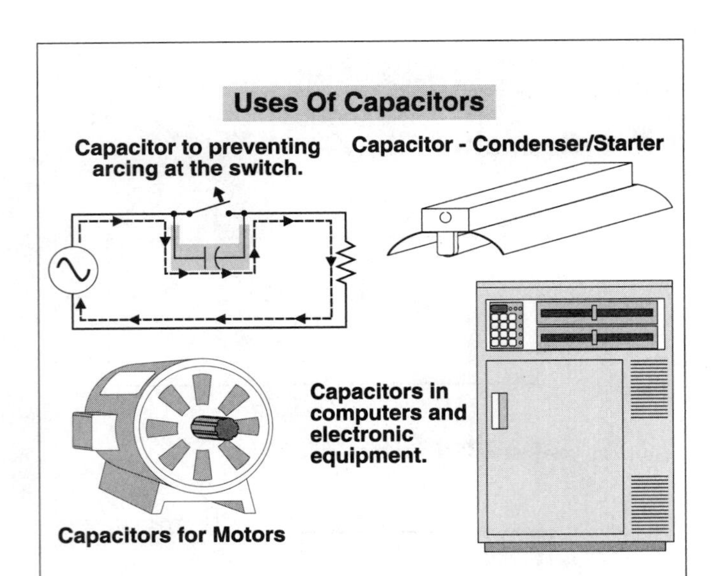

Figure 3-25

Capacitor Uses

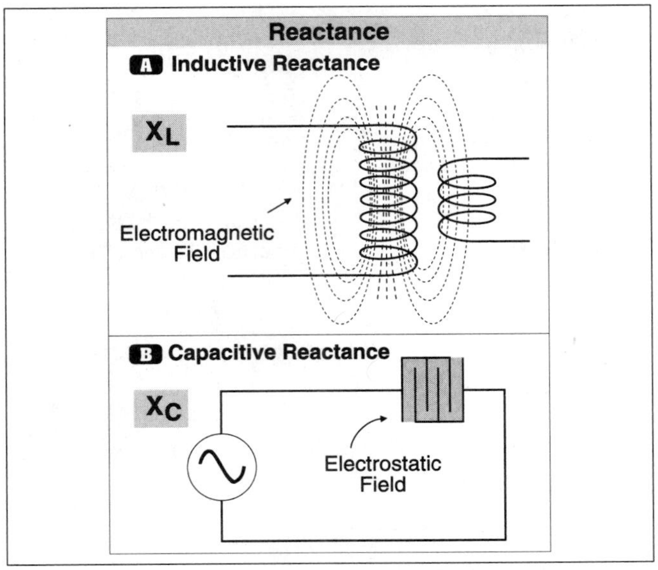

Reactance

A Inductive Reactance

X_L

Electromagnetic Field

B Capacitive Reactance

X_C

Electrostatic Field

Figure 3-26

Reactance

❑ **Apparent Power (VA) Example No. 1**

What is the apparent power of a 16-ampere load operating at 120 volts with a power factor of 80 percent, Fig. 3–28?

(a) 2,400 VA (b) 1,920 VA (c) 1,632 VA (d) none of these

- Answer: (b) 1,920 VA

  Apparent Power = Volt × Amperes, VA = 120 volts × 16 amperes = 1,920 VA

❑ **Apparent Power Example No. 2**

What is the apparent power of a 250-watt fixture that has a power factor of 80 percent, Fig. 3–29?

(a) 200 VA (b) 250 VA (c) 313 VA (d) 375 VA

- Answer: (c) 313 VA

  Apparent Power = True Power/Power Factor

  True Power = 250 watts, Power Factor = 80% or 0.8

  Apparent Power = 250 watts/.8

  Apparent Power = 313 volt-amperes

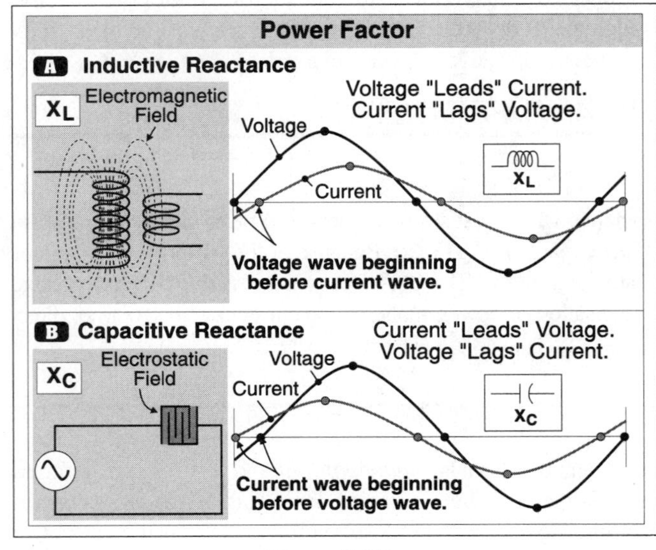

Figure 3-27

Power Factor

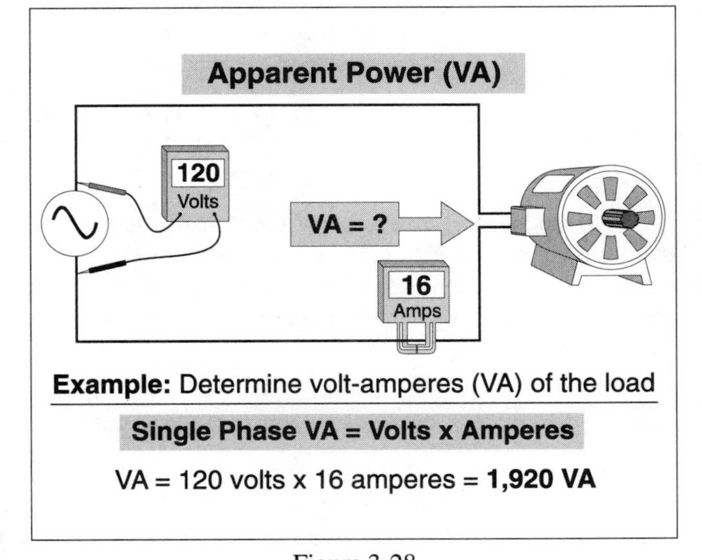

Figure 3-28

Apparent Power Example

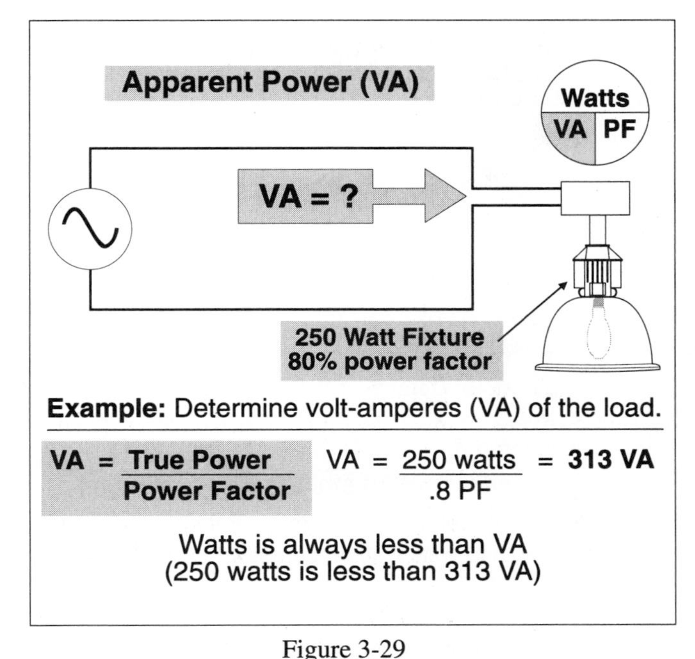

Figure 3-29

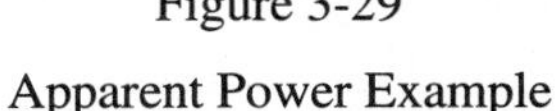
Apparent Power Example

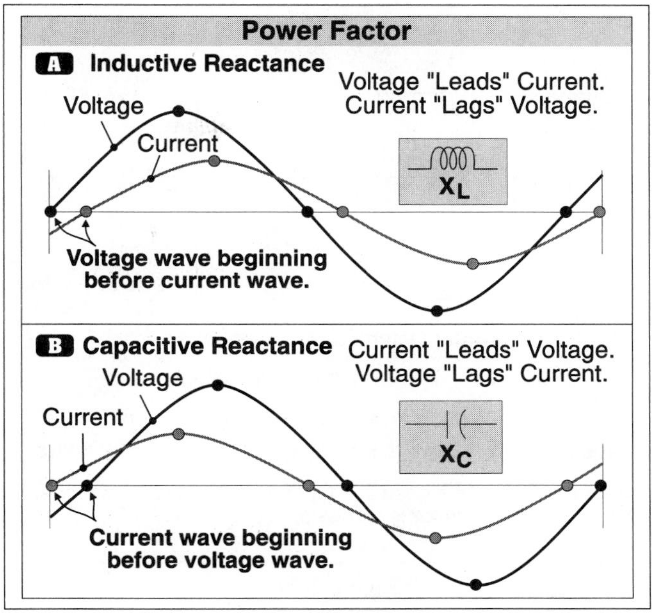

Power Factor
A Inductive Reactance
Voltage "Leads" Current.
Current "Lags" Voltage.
Voltage
Current
X_L
Voltage wave beginning before current wave.
B Capacitive Reactance
Current "Leads" Voltage.
Voltage "Lags" Current.
Voltage
Current
X_C
Current wave beginning before voltage wave.

Figure 3-30

Power Factor Example

3–16 *POWER FACTOR*

Power factor is the ratio of active power (R) to apparent power (Z) expressed as a percent that does not exceed 100 percent. Power factor is a form of measurement of how far the voltage and current are out of phase with each other. In an alternating current circuit that supplies power to resistive loads, such as incandescent lighting, heating elements, etc., the circuit voltage and current will be *in phase*. The term in phase means that the voltage and current reach their zero and peak values at the same time, resulting in a power factor of 100 percent or *unity*. Unity can occur if the circuit only supplies resistive loads or capacitive reactance (X_C) is equal to inductive reactance(X_L).

The formulas for determining power factor are:

Power Factor = True Power/Apparent Power, or
Power Factor = Watts/Volt-Amperes , or
Power Factor = Resistance/Impedance
Power Factor = R/Z, Resistance (Watts)/Impedance (VA)

The formulas for current flow are:

I = Watts/(Volts Line to Line times Power Factor) Single-Phase
I = Watts/(Volts Line to Line times 1.732 times Power Factor) Three-Phase

❑ **Power Factor Example**

What is the power factor of a fluorescent lighting fixture that produces 160 watts of light and has a ballast rated for 200 volt-amperes, Fig. 3–30?

(a) 80 percent (b) 90 percent (c) 100 percent (d) none of these

- Answer: (a) 80 percent

$$\text{Power Factor} = \frac{\text{True Power}}{\text{VA}}, = \frac{\text{160 watts}}{\text{200 volt–amperes}}, = 0.80 \text{ or } 80\%$$

❑ **Current Example**

What is the current flow for three 5-kW, 115-volt, single-phase loads that have a power factor of 90 percent?

(a) 72 amperes (b) 90 amperes (c) 100 amperes (d) none of these

- Answer: (a) 72 amperes

I = Watts/(Volts Line to Line times Power Factor) Single-Phase

I = 15,000 Watts/(230 volts × 0.9)

I = 72 amperes

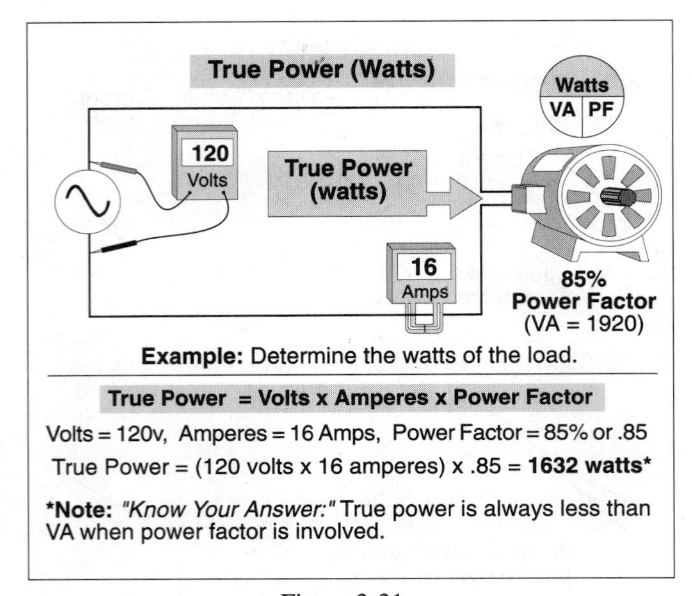

Figure 3-31

Alternating Current – True Power Example

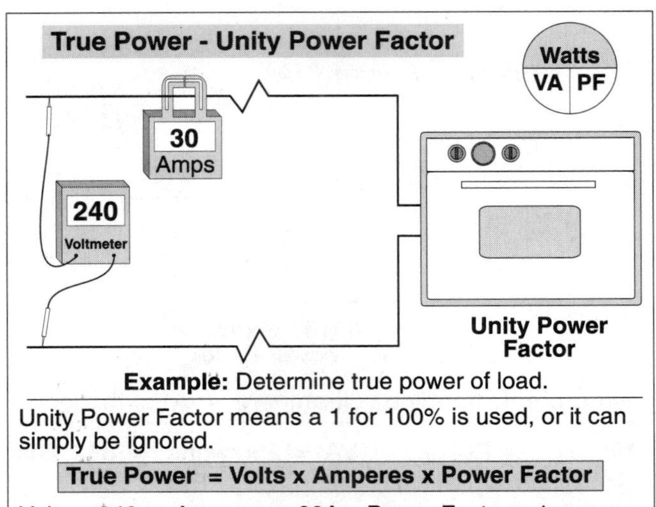

Figure 3-32

Direct Current – True Power Example

3–17 *TRUE POWER (WATTS)*

Alternating Current

True power is the energy consumed for work, expressed by the unit called the *watt*. Power is measured by a wattmeter which is connected series-parallel [series (measures amperes) and parallel (measures volts)] to the circuit conductors. The true power of a circuit that contains inductive or capacitance reactance can be calculated by the use of one of the following formulas:

True Power (alternating current) = Volts × Amperes × Power Factor, or

True power (alternating current) = Volts × Amperes × θ

θ (Cosine) is the symbol for power factor

Note: True power for a direct current circuit is calculated as Volts × Amperes.

❑ **Alternating Current – True Power Example**

What is the true power of a 16-ampere load operating at 120 volts with a power factor of 85 percent, Fig. 3–31?

(a) 2,400 watts (b) 1,920 watts (c) 1,632 watts (d) none of these

- Answer: (c) 1,632 watts

  True Power = Volts-Amperes × Power Factor

  Watts = (120 volts × 16 amperes) × 0.85, Watts = 1,632 watts per hour

Note: True power is always equal to or less than apparent power.

Direct Current

In direct current or alternating current circuits at unity power factor, the true power is simply:

True Power Direct Current = Volts × Amperes

Note: True power for alternating current with reactance = Volts × Amperes × θ Cosine

❑ **Direct Current – True Power Example**

What is true power of a 30-ampere, 240-volt resistive load that has unity power factor, Fig. 3–32?

(a) 4,300 watts (b) 7,200 watts (c) 7,200 VA (d) none of these

- Answer: (b) 7,200 watts, not VA

  True Power = Volts × Amperes

  Volts = 240 volts, Amperes = 30 amperes

  True Power = 240 volts × 30 amperes

  True Power = 7,200 watts

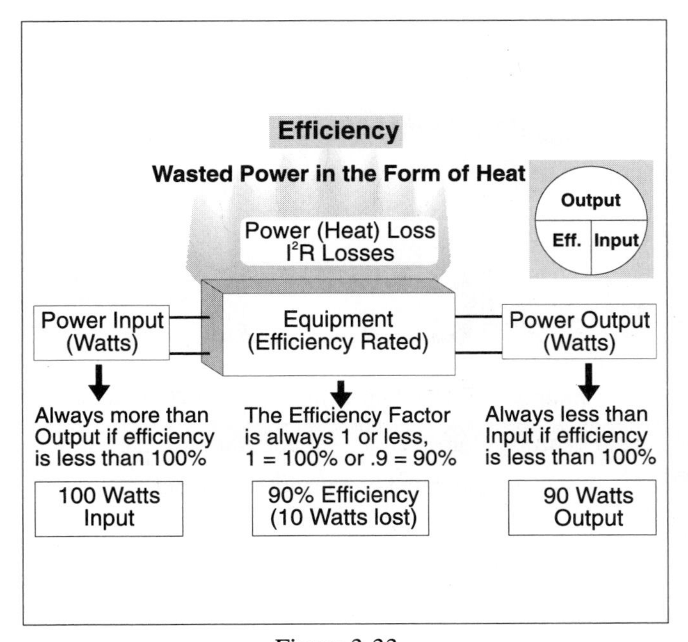

Figure 3-33

Efficiency

Calculating Efficiency

Equipment

Power Input 800 Watts

Power Output 640 Watts

Output / Eff. / Input

Efficiency ?

Example: Determine efficiency of equipment.

Formula: **Efficiency = Output Watts / Input Watts**

Answer must always be 1 or less.

Efficiency = 640 Watts Output / 800 Watts Input

Efficiency = **.8 or 80%**

Figure 3-34

Efficiency Example

3–18 *EFFICIENCY*

Efficiency has nothing to do with power factor. Efficiency is the ratio of input power to output power, where power factor is the ratio of true power (watts) to apparent power (volt-amperes). Energy that is not used for its intended purpose is called power loss. Power losses can be caused by conductor resistance, friction, mechanical loading, etc. Power losses of equipment are expressed by the term efficiency, which is the *ratio* of the input power to the output power, Fig. 3–33. The formulas for efficiency are:

Efficiency = Output watts/Input watts

Input watts = Output watts/Efficiency

Output watts = Input watts × Efficiency

❑ **Efficiency Example**

If the input of a load is 800 watts and the output is 640 watts, what is the efficiency of the equipment, Fig. 3–34?

(a) 60 percent (b) 70 percent (c) 80 percent (d) 100 percent

- Answer: (c) 80 percent

  Efficiency is always less than 100 percent,

  $$\text{Efficiency} = \frac{\text{Output}}{\text{Input}}$$

  $$\text{Efficiency} = \frac{640 \text{ watts}}{800 \text{ watts}}$$

  Efficiency = 0.8 or 80%

❑ **Input Example**

If the output of a load is 250 watts and the equipment is 88 percent efficient, what are the input watts, Fig. 3–35?

(a) 200 watts (b) 250 watts (c) 285 watts (d) 325 watts

- Answer: (c) 285 watts

  Input is always greater than the output,

  $$\text{Input} = \frac{\text{Output}}{\text{Efficiency}}$$

  $$\text{Input} = \frac{250 \text{ watts}}{0.88 \text{ efficiency}}$$

  Input = 284 watts

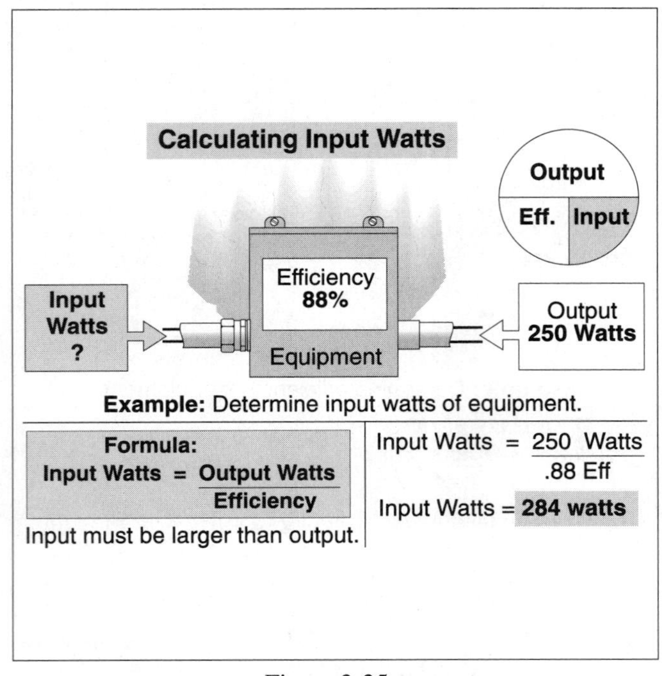

Figure 3-35

Input Example

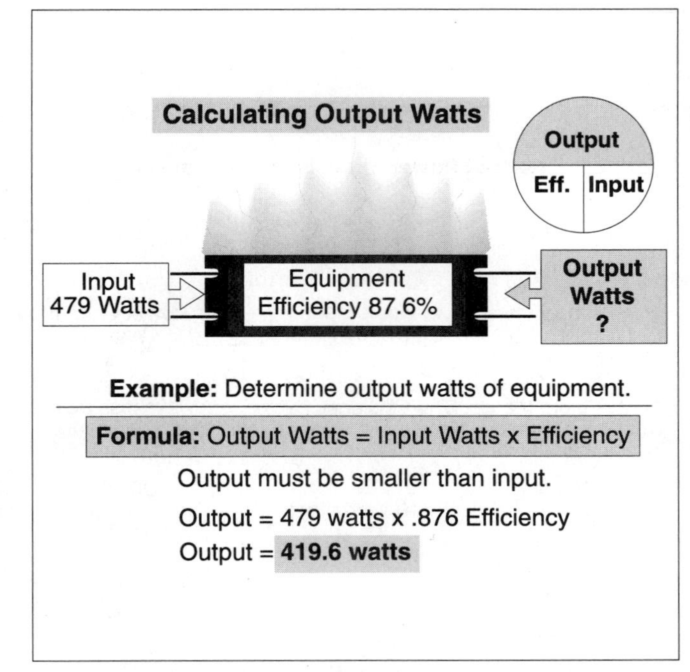

Figure 3-36

Output Example

❑ Output Example

If we have a load that is 87.6 percent efficient, for every 479 watts of input we will have________ watts of output, Fig. 3–36.

(a) 440 (b) 480 (c) 390 (d) 420

- Answer: (d) 420 watts

  Output is always less than input.

  $\text{Output} = \text{Input} \times \text{Efficiency}$

  $\text{Output} = 479 \text{ watts} \times 0.876$

  $\text{Output} = 419.6 \text{ watts}$

Unit 3 – Understanding Alternating Current Summary Questions

Part A – Alternating Current Fundamentals

3–1 Current Flow

1. The effects of electron movement is the same regardless of the direction of the current flow.
(a) True (b) False

3–2 Alternating Current

2. Alternating current is primarily used because it can be transmitted inexpensively and because it can be used for certain applications for which direct current is not suitable.
(a) True (b) False

3. Faradays experiments revealed that when a magnetic field moves through a coil of wire, the lines of force of the magnetic field makes the electrons in the wire flow in a specific direction. When the magnetic field moves in the opposite direction, electrons in the wire flow continue to flow in the same direction.
(a) True (b) False

3–3 Alternating Current Generator

4. A simple alternating current generator consists of a rotating loop of wire between the lines of force of opposite poles of a magnet. The magnitude of the voltage is dependent on the ________.
(a) number of turns of wire
(b) strength of the magnetic field
(c) speed at which the coil rotates
(d) all of these

3–4 Waveform

5. • A waveform is used to display the level and direction of the current and voltage. The waveform for ________ circuits displays the level and direction of the current and voltage for every instant of time for one full revolution of the armature.
(a) direct current (b) alternating current
(c) a and b (d) none of these

6. For alternating current circuits, the waveform is called the ________.
(a) frequency (b) cycle
(c) degree (d) none of these

3–5 Armature Turns Frequency

7. The number of times the armature turns in one second is called the frequency. Frequency is expressed as ________ or cycles per second.
(a) degrees (b) sign wave
(c) phase (d) hertz

3–6 Phase

8. When two waveforms are in step with each other, they are said to be in phase. In a purely resistive alternating current circuit, the current and voltage are in phase. This means that they both reach their zero and ________ values at the same time.
(a) peak (b) effective
(c) average (d) none of these

3–7 Degrees

9. Phase differences are expressed in ________.
(a) sign (b) phase (c) hertz (d) degrees

10. The terms ________ and ________ are used to describe the relative positions in time of two waveforms (voltage or current).
(a) hertz, phase (b) frequency, phase
(c) sine, degrees (d) lead and lag

3–8 Values Of Alternating Current

11. • ________ is the value of the voltage or current at any one particular moment of time.
(a) Peak (b) Root mean square
(c) Effective (d) Instantaneous

12. ________ is the maximum value that alternating current or voltage reaches, both for positive and negative.
(a) Peak (b) Root mean square
(c) Instantaneous (d) None of these

13. • Effective is the alternating current voltage or current value that produces the same amount of heat in a resistor that would be produced by the same amount of direct current voltage or current. ________ is the same as effective.
(a) Peak (b) Root mean square
(c) Instantaneous (d) None of these

Part B – Induction

Induction Introduction

14. The movement of electrons because of electromagnetism is called ________.
(a) flux lines (b) voltage (c) induction (d) magnetic field

15. The induction of voltage in a conductor because of expanding and collapsing magnetic fields is know as ________.
(a) flux lines (b) power
(c) self-induced voltage (d) magnetic field

3–9 Induced Voltage And Applied Current

16. A change in current produces a magnetic field through the conductor which produces an induced voltage that always opposes the change in current. The induced voltage that opposes the change in current is called ________.
(a) cemf (b) counter electromotive force
(c) back-emf (d) all of these

17. When the conductor current increases, the direction of the induced voltage (cemf) in the conductor is opposite to the direction of the conductor current and tries to prevent the current from ________.
(a) decreasing (b) increasing

18. When the current in the conductor decreases, the direction of the induced voltage in the conductor attempts to keep the current from decreasing by releasing the energy from the magnetic field back into the conductor.
(a) True (b) False

3–10 Conductor Impedance

19. In direct current circuits, the only property that effects current and voltage flow is ________ which is a physical property of the conductors that oppose current flow.
(a) voltage (b) cemf (c) back-emf (d) none of these

20. Conductor resistance is directly proportional to the conductor length and inversely proportional to conductor cross sectional area.
(a) True (b) False

21. Alternating currents produce a cemf that is set up inside the conductor which increases the effective resistance of the conductor because of eddy currents and skin effect.
(a) True (b) False

22. The opposition to current flow in a conductor because of resistance and induction is called ________.
(a) resistance (b) capacitance
(c) induction (d) impedance

23. • Eddy currents are small independent currents that are produced as a result of direct current. Eddy currents flow erratic through a conductor, consume power and increase the effective conductor resistance by opposing the current flow.
(a) True (b) False

24. The expanding and collapsing magnetic field of the conductors current induces a voltage in the conductors which repels the flowing electrons towards the surface of the conductor. This has the effect of decreasing the effective conductor cross-sectional area which causes an increased in the conductor impedance. The flow of electrons near the surface is known as ________.
(a) eddy currents (b) induced voltage
(c) impedance (d) skin effect

3–11 Conductor Shape

25. The amount of the self-induced voltage created within the conductor is directly proportional to the current flow, the length of the conductor, and the frequency at which the magnetic fields cut through the conductors.
(a) True (b) False

3–12 Magnetic Cores

26. • Self-inductance (cemf) in a coil is affected by the ________.
(a) winding length and shape (b) core material
(c) frequency (d) all of these

Part C – Capacitance

Capacitance Introduction

27. ________ is a property of an electric circuit that enables it to store electric energy by means of an electrostatic field and to release this energy at a later time.
(a) Capacitance (b) Induction
(c) Self-induction (d) None of these

3–13 Charged, Testing, And Discharging

28. When a capacitor has a potential difference between the plates, it is said to be ________. One plate has an excess of free electrons, and the other plate has a lack of them.
(a) induced (b) charged (c) discharged (d) shorted

29. If the capacitor is overcharged, the electrons from the negative plate could be pulled through the insulation to the positive plate. The capacitor is said to have ________.
(a) charged (b) discharged
(c) induced (d) shorted

30. To discharge a capacitor, all that is required is a ________ path between the capacitor plates.
(a) conducting (b) insulating
(c) isolating (d) none of these

3–14 Uses Of Capacitors

31. What helps prevent arcing across the contacts of electric switches?
(a) springs (b) condenser
(c) inductor (d) resistor

32. The current in a purely capacitive circuit ________.
(a) leads the applied voltage by 90 degrees
(b) lags the applied voltage by 90 degrees
(c) leads the applied voltage by 180 degrees
(d) lags the applied voltage by 180 degrees

33. Circuits containing inductive or capacitive reactance temporarily store power in the electromagnetic field of induction and the electrostatic field of capacitors.
(a) True (b) False

Part D – Power Factor And Efficiency

3–15 Apparent Power

34. If you measure voltage and current in an inductive or capacitive circuit and then multiply them together, you would obtain the circuits ________.
(a) true power (b) power factor (c) apparent power (d) power loss

35. Apparent power is equal to or greater than true power depending on the power factor.
(a) True (b) False

36. When sizing circuits or equipment, always size the circuit components and transformers according to the apparent power, not the True power.
(a) True (b) False

3–16 Power Factor

37. Power factor is a measurement of how far the voltage and current are out of phase with each other. Power factor is the ratio between True Power (resistive load) to Apparent Power (reactive load). Power factor can be express by the formula ________ .
(a) P/E (b) R/Z (c) I^2R (d) Z/R

38. • When current and voltage are in phase, the power factor is ________ .
(a) 100 percent (b) unity (c) 90 degrees (d) a and b

39. In an alternating current circuit that supplies power to resistive loads such as incandescent lighting, heating elements, etc., the circuit voltage and current are said to be ________, resulting in a power factor of unity.
(a) out of phase (b) leading by 90 degrees
(c) 90 degrees out of phase (d) none of these

3–17 True Power (Watts)

40. True power is the energy consumed for work expressed by the term watts. To determine the true power of a circuit that contains inductive or capacitance reactance, we must multiply the volts times the current times the ________.
(a) efficiency (b) sine wave (c) power factor (d) none of these

41. In direct current circuits, the voltage and current are constant and the true power is simply, volts times amperes.
(a) True (b) False

42. What does it cost per year (9 cents per kWh) for 10 - 150 watt recessed fixtures to operate if they are on 6 hours per day?
(a) $150 (b) $300 (c) $500 (d) $800

Power Factor Examples

43. A 2 × 4 recessed fixture contains four 34 watt lamps, and the ballast is rated 1.5 amperes at 120 volts. What is the power factor of the ballast assuming 100 percent efficiency?
(a) 55 percent (b) 65 percent (c) 70 percent (d) 75 percent

44. • What is the apparent power of a 20 ampere load operating at 120 volts, power factor 85 percent?
(a) 2,400 watts (b) 1,920 watts (c) 1,632 watts (d) 2,400 VA

45. • What is the true power of a 20 ampere load operating at 120 volts, power factor 85 percent?
(a) 2,400 watts (b) 1,920 watts (c) 1,632 watts (d) none of these

46. • Since power factor cannot be greater than 100 percent, true power is equal to or less than the apparent power. Because of power factor, the VA of the load is greater than the watts, which results in less loads per circuit, a greater number of circuits, and larger transformers.
(a) True (b) False

47. • What is the true power of a 10-ampere circuit operating at 120 volts with unity power factor?
(a) 1,200 VA (b) 2,400 VA (c) 1,200 watts (d) 2,400 watts

48. What size transformer is required for a 125-ampere, 240 volt single-phase load?
(a) 3 kVA (b) 30 kVA (c) 12.5 kVA (d) 15 kVA

49. What size transformer is required for a 30 kW-load that has a power factor of 85 percent?
(a) 12.5 kVA (b) 35 kVA (c) 7.5 kVA (d) 15 kVA

50. • How many 20-ampere, 120-volt circuits are required for 42 - 300 watt recessed fixtures (noncontinuous load)? Tip: Only so many fixtures are permitted on a circuit.
(a) 3 circuits (b) 4 circuits (c) 5 circuits (d) 6 circuits

51. • How many 20-ampere, 120-volt circuits are required for 42 - 300 watt recessed fixtures (noncontinuous load) with a power factor of 85 percent?
(a) 5 circuits (b) 6 circuits (c) 7 circuits (d) 8 circuits

3–18 Efficiency

52. Efficiency is the ratio of the input power to output power.
(a) True (b) False

53. If the output is 1,320 watts and the input is 1,800 watts, what is the efficiency of the equipment?
(a) 62 percent (b) 73 percent (c) 0 percent (d) 100 percent

54. If the output is 160 watts and the equipment is 88 percent efficient, what is the input amperes at 120 volts?
(a) .75 amperes (b) 1.500 amperes (c) 2.275 amperes (d) 3.250 amperes

55. If we have a transformer that is 97 percent efficient, for every 1 kW input, we have ________ watts output.
(a) 970 watts (b) 1,000 watts (c) 1,030 watts (d) 1,300 watts

Challenge Questions

Part A – Alternating Current Fundamentals

3–2 Alternating Current

56. The primary reason(s) for high voltage transmission lines is/are ______.
I. reduced voltage drop II. smaller wire III. smaller equipment
(a) I only (b) II only
(c) III only (d) all of these

57. One of the advantages of a higher voltage system as compared to a lower voltage system (for the same wattage loads) is ______.
(a) reduced voltage drop (b) reduced power use
(c) large currents (d) lower electrical pressure

58. The advantage of alternating current over direct current is that alternating current provides for ______.
(a) better speed control (b) ease of voltage variation
(c) lower resistance at high currents (d) none of these

3–4 Waveforms

59. A waveform represents ______.
(a) the magnitude and direction of current or voltage
(b) how current or voltage can vary with time
(c) how output voltage can vary with the generator armature
(d) all of these

3–5 Armature Turning Frequency

60. Frequency, of an alternating current waveform, is the number of times the current or voltage goes through 360º in ______.
(a) 1/10th of a second (b) 5 seconds (c) 1 second (d) 60 seconds

61. How much time does it take for 60 Hz alternating current to travel through 180 degrees?
(a) 1/120 of a second (b) 1/40 of a second
(c) 1/180 of a second (d) none of these

3–8 Values Of Alternating Current

62. • The heating effects of 10 amperes alternating current as compared with 10 amperes direct current is ______.
(a) the same (b) less (c) greater (d) none of these

63. If the maximum value of an alternating current system is 50 amperes, the RMS value would be approximately ______.
(a) 25 amperes (b) 30 amperes (c) 35 amperes (d) 40 amperes

64. • The maximum value of 120 volt direct current is equal to the maximum value of an equivalent alternating current.
(a) True (b) False

65. • You are getting a 120 volt reading on your voltmeter, this is an indication of the ______ value of the voltage source.
(a) average (b) peak (c) effective (d) instantaneous

Part B – Induction

3–9 Voltage And Applied Current

66. A change in current that produces a counter electromotive force whose direction is such that it opposes the change in current is known as ______ Law.
(a) Kirchoff's 2nd (b) Kirchoff's 1st (c) Lenz's (d) Hertz

67. Inductive reactance is abbreviated as ______.
(a) IR (b) L_X (c) X_L (d) Z

68. Inductive reactance changes proportionally with frequency.
(a) True (b) False

69. Inductive reactance is measured in ______.
(a) farads (b) watts (c) ohms (d) coulombs

70. • If the frequency is constant, the inductive reactance of a circuit will ______.
(a) remain constant regardless of the current and voltage changes
(b) vary directly with the voltage
(c) vary directly with the current
(d) not affect the impedance

3–10 Conductor Impedance

71. The total opposition to current flow in an alternating current circuit is expressed in ohms and is called ______.
(a) impedance (b) conductance (c) reluctance (d) none of these

72. Impedance is present in ______ type circuit(s).
(a) resistance (b) direct current (c) alternating current (d) none of these

73. Conductor resistance to alternating current flow is ______ the resistance to direct current.
(a) higher than (b) lower than

Part C – Capacitance

3–13 Charge, Testing And Discharging

74. If a test lamp is placed in series with a capacitor with a direct current voltage source and the lamp is continuously illuminated, it is an indication that the capacitor is ______.
(a) fully charged (b) shorted (c) fully discharged (d) open-circuited

75. The insulating material between the surface plates of a capacitor is called the ______.
(a) inhibitor (b) electrolyte (c) dielectric (d) regulator

76. Capacitors are measured in ______.
(a) watts (b) volts (c) farads (d) henrys

77. Capacitive reactance is measured in ______.
(a) ohms (b) volts (c) watts (d) henrys

3–14 Uses Of Capacitors

78. • In a circuit that has only capacitive reactance (X_C), the voltage and current are said to be out of phase to each other because the voltage ______.
(a) leads the current by 90° (b) lags the current by 90°
(c) leads the current (d) none of these

79. Resonance occurs when ______.
(a) only resistance occurs in the system
(b) the power factor is equal to 0
(c) $X_L = X_C$ are in a series circuit
(d) when R = Z

Part D – Power Factor And Efficiency

3–15 Apparent Power

80. If you multiply the voltage times the current in an inductive or capacitive circuit, the answer you obtain will be the ______ of the circuit.
(a) watts (b) true power (c) apparent power (d) all of these

81. • The apparent power of a 19.2 ampere, 120 volt load is ______.
(a) 2,304 kVA (b) 2.3 kVA (c) 2.3 VA (d) 230 kVA

3–16 Power Factor

82. Power factor in an alternating current circuit will be unity (100%), if the circuit contains only ______.
(a) induction motors (b) transformers
(c) reactance coils (d) resistive load such as electric space heater coils

83. • Three 8 kW electric discharge lighting bank circuits, which have a 92 percent power factor, are connected to a 230-volt, three-phase source. The current flow of these lights is ______.
(a) 37 amperes (b) 50 amperes (c) 65 amperes (d) 75 amperes

3–17 True Power

84. • A wattmeter is connected in ______ in the circuit.
(a) series (b) parallel (c) series-parallel (d) none of these

85. • True power is always voltage times current for _______ .
(a) all alternating current circuits
(b) direct current circuits
(c) alternating current circuits at unity power factor
(d) b and c

86. Power consumed in either a single-phase alternating current or direct current system is always equal to ______.
(a) E × I cos (b) E × R (c) $I^2 \times R$ (d) E/(I × R)

87. • The power consumed on a 76 ampere, 208 volt three-phase circuit that has a power factor of 89 percent is ______.
(a) 27,379 watts (b) 35,808 watts (c) 24,367 watts (d) 12,456 watts

88. • The true power of a single phase 2.1 kVA load with a power factor of 91 percent is ______.
(a) 2.1 kW (b) 1.91 kW (c) 1.75 kW (d) 1,911 kW

3–18 Efficiency

89. • Motor efficiency can be determined by which of the following formulas?
(a) hp × 746 (b) hp × 746/VA Input
(c) hp × 746/Watts Input (d) hp × 746/kVA Input

90. The efficiency ratio of a 4,000 VA transformer with a secondary VA of 3,600 VA is ______.
(a) 80 (b) 9 (c) .9 (d) 1.1

Unit 4

Motors and Transformers

OBJECTIVES

After reading this unit, the student should be able to briefly explain the following concepts:

Part A - Motors
Alternating current motors
Dual voltage motors
Horsepower/watts
Motor speed control
Nameplate amperes
Reversing DC motors
Reversing AC motors
Volt-ampere calculations

Part B - Transformers
Transformer primary vs. secondary
Transformer current
Transformer kVA rating
Transformer power losses
Transformer turns ratio

After reading this unit, the student should be able to briefly explain the following terms:

Part A - Motors
Armature winding
Commutator
Dual voltage
Field winding
Horsepower ratings
Magnetic field
Motor full load current
Nameplate amperes
Torque
Watts rating

Part B - Transformers
Auto transformers
Circuit Impedance
Conductor losses
Core losses
Counter-electromotive-force
Current transformers (ct)
Eddy current losses
Excitation current
Flux leakage loss
Hysteresis losses
Kilo volt-amperes
Line current
Ratio
Self-excited
Step-down transformer
Step-up transformers

PART A – *MOTORS*

MOTOR INTRODUCTION

The *electric motor* operates on the principle of the attracting and repelling forces of magnetic fields. One magnetic field (permanent or electromagnetic) is *stationary* and the other magnetic field (called the *armature* or *rotor*) rotates between the stationary magnetic field, Fig. 4–1. The turning or repelling forces between the magnetic fields is called *torque*. Torque is dependent on the strength of the stationary magnetic field, the strength of the armatures magnetic field, and the physical construction of the motor.

The repelling force of like magnetic polarities, and the attraction force of unlike polarities, causes the armature to rotate in the electric motor. For a direct current motor, a device called a *commutator* is placed on the end of the conductor loop. The purpose of the commutator is to maintain the proper polarity of the loop, so as to keep the armature or rotor turning.

The *armature winding* of a motor will carry short-circuit current when voltage is first applied to the motor windings. As the armature turns, it cuts the lines of force of the *field winding* resulting in an increase in counter-electromotive-force, Fig. 4–2. The increased counter-electromotive-force results in an increase in inductive reactance, which will increase the circuit impedance. The increased circuit impedance causes a decrease in the motor armature running current, Fig. 4–3.

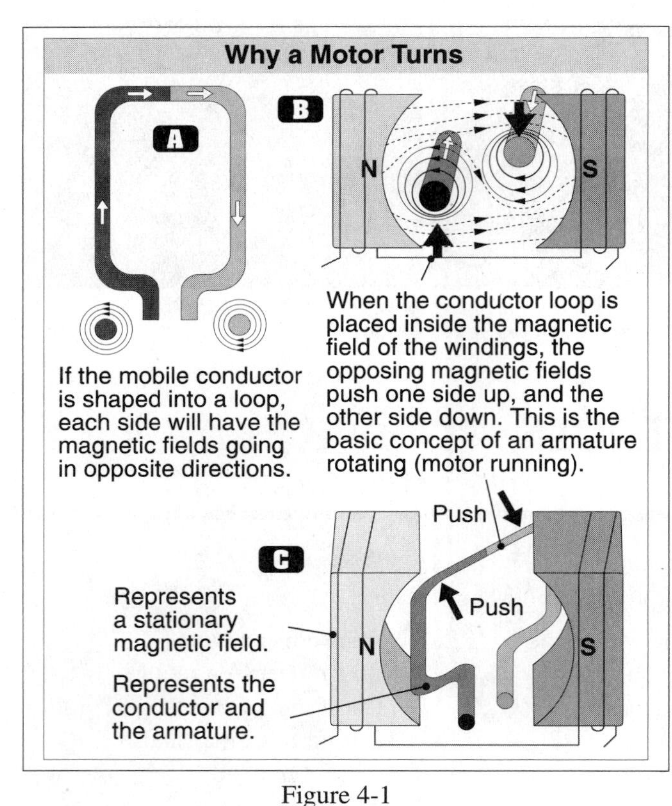

Figure 4-1

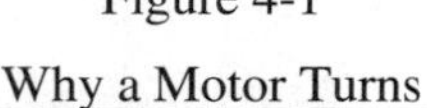
Why a Motor Turns

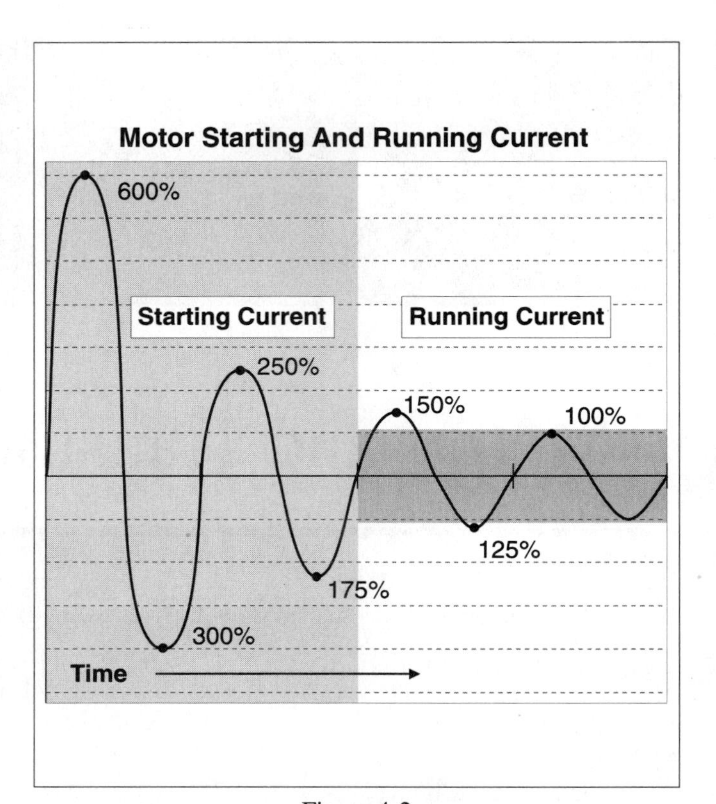

Figure 4-2

Motor Starting Current

4–1 *MOTOR SPEED CONTROL*

One of the advantages of the *direct current motor* over the alternating current motor is the motor's ability to maintain a constant speed. But, series direct current motors are susceptible to run away when not connected to a load.

If the speed of a direct current motor is increased, the armature winding will be cut by the magnetic field at an increasing rate, resulting in an increase of the armature's counter-electromotive-force. The increased counter-electromotive-force in the armature acts to cut down on the armature current, resulting in the motor slowing down. Placing a load on a direct current motor causes the motor to slow down, which reduces the rate at which the armature winding is cut by the magnetic field flux lines. A reduction of the armature flux lines results in a decrease in the armature's counter-electromotive-force and an increase in motor speed.

4–2 *REVERSING A DIRECT CURRENT MOTOR*

To reverse a direct current motor, you must reverse either the *magnetic field* of the field winding or the magnetic field of the armature. This is accomplished by reversing either the field or armature current flow, Fig. 4–4. Because most direct current motors have the field and armature winding connected to the same direct current power supply, reversing the polarity of the power supply changes both the field and armature simultaneously. To reverse the rotation of a direct current motor, you must reverse either the field or armature leads, but not both.

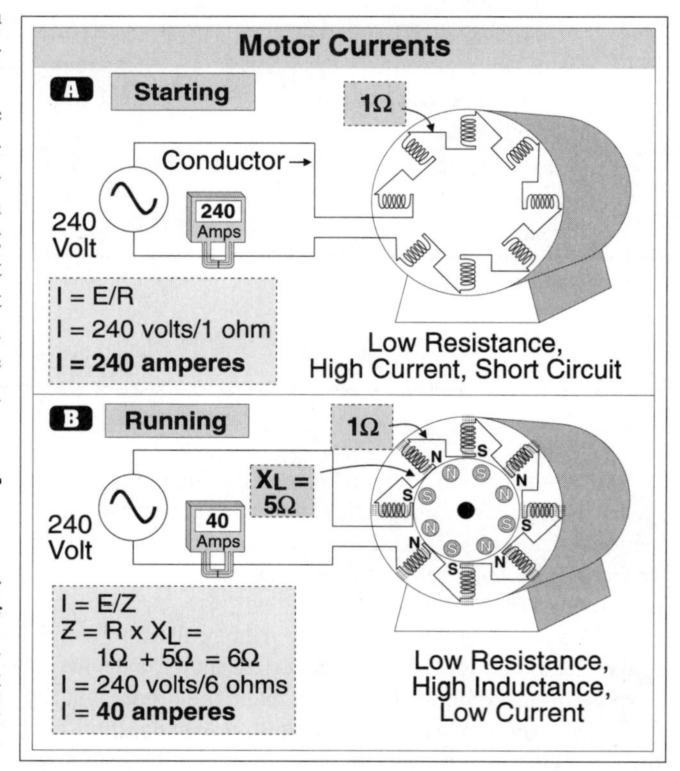

Figure 4-3

Motor Current and Impedance

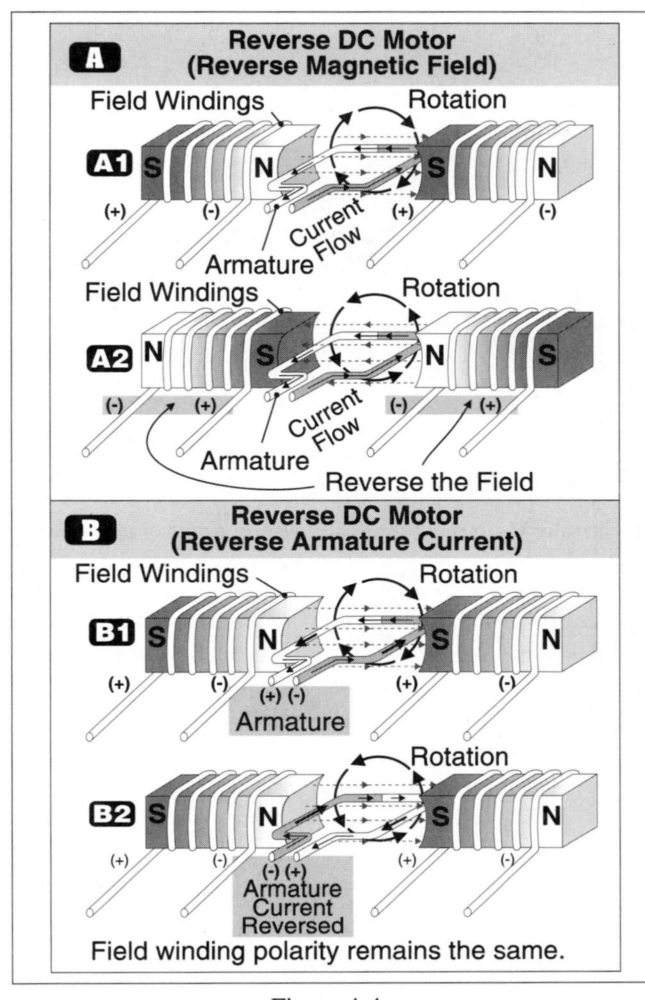

Figure 4-4

Reversing Direct Current Motors

Reversing Three-Phase AC Motors

A

L1
L3
L2
T1
T2
T3
C A
B
Rotation A-B-C

B

L1
L3
L2
T1
T2
T3
C A
B
Rotation C-B-A

Figure 4-5

Reversing Three-Phase Alternating Current Motors

4–3 *ALTERNATING CURRENT MOTORS*

Fractional horsepower motors that can operate on either alternating or direct current are called *universal motors*. A motor that will not operate on direct current is called an induction motor.

The purest form of an alternating current motor is the induction motor, which has no physical connection between its rotating member (*rotor*), and stationary member (*stator*). Two common types of alternating current motors are synchronous and wound rotor motors.

Synchronous Motors

The synchronous motor's rotor is locked in step with the rotating stator field. Synchronous motors maintain their speed with a high degree of accuracy and are used for electric clocks and other timing devices.

Wound Rotor Motors

Because of their high starting torque requirements, wound rotor motors are used only in special applications. In addition, wound rotor motors only operate on three-phase alternating current power.

4–4 *REVERSING ALTERNATING CURRENT MOTORS*

Three-phase alternating current motors can be reversed by reversing any two of the three line conductors that supply the motor, Fig. 4–5.

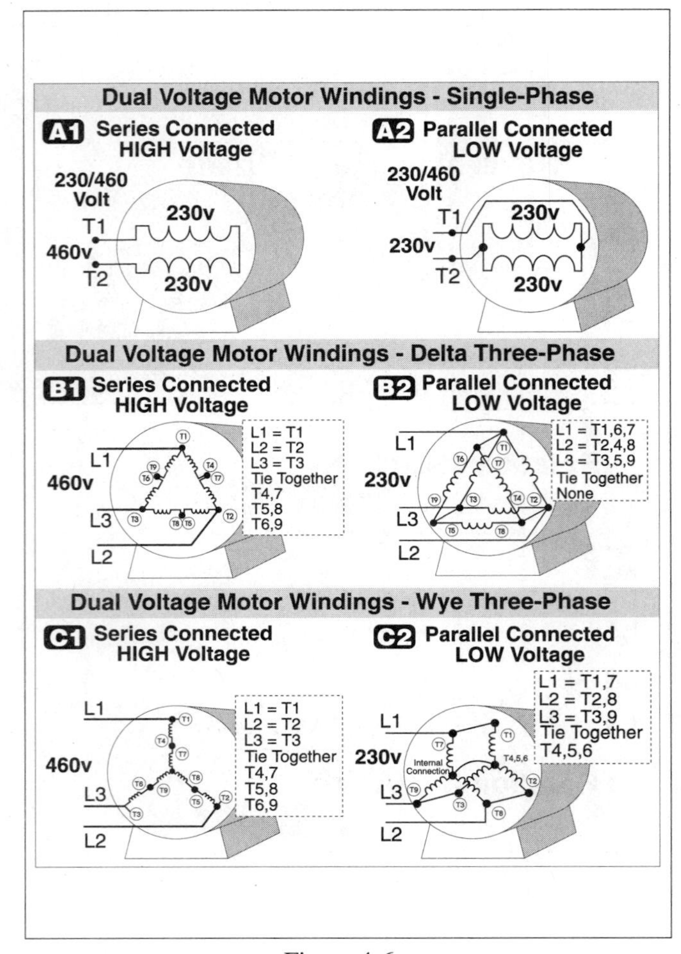

Figure 4-6

Dual Voltage Motors

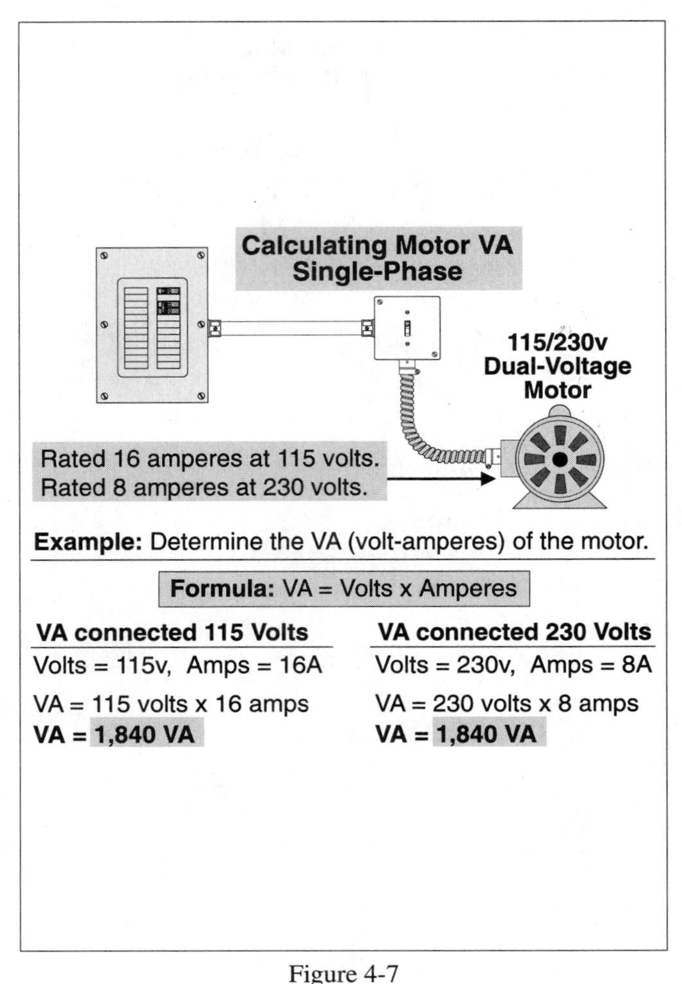

Figure 4-7

Motor VA – Single-Phase Example

4–5 *MOTOR VOLT-AMPERE CALCULATIONS*

Dual voltage motors are made with two field windings, each rated for the lower voltage marked on the nameplate of the motor. When the motor is wired for the lower voltage, the field windings are connected in parallel; when wired for the higher voltage, the motor windings are connected in series, Fig. 4–6.

Motor Input VA

Regardless of the voltage connection, the power consumed for a motor is the same at either voltage. To determine the *motor input* apparent power (VA), use the following formulas:

Motor VA (single-phase) = Volts × Amperes **Motor VA (three-phase) = Volts × Amperes × $\sqrt{3}$**

❑ Motor VA Single-Phase Example

What is the motor VA of a single-phase 115/230-volt, 1-horsepower motor that has a current rating of 16 amperes at 115-volts and 8 amperes at 230-volts, Fig. 4–7?

(a) 1,450 VA (b) 1,600 VA (c) 1,840 VA (d) 1,920 VA

- Answer: (c) 1,840 VA

  Motor VA = Volts × Amperes

  Volts = 115 or 230, Amperes = 16 or 8 amperes

  Motor VA = 115 volts × 16 amperes, or

  Motor VA = 230 volts × 8 amperes

  Motor VA = 1,840 VA

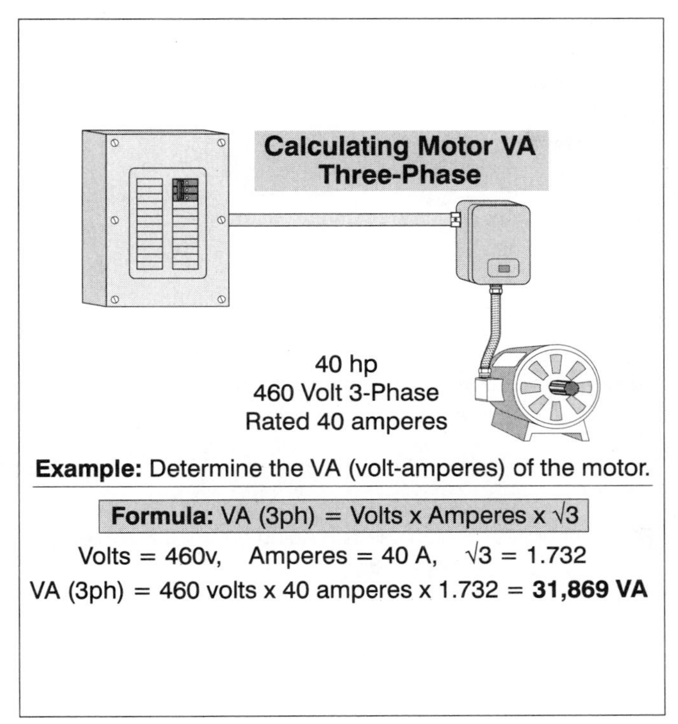

Figure. 4-8

Motor VA – Three-Phase Example

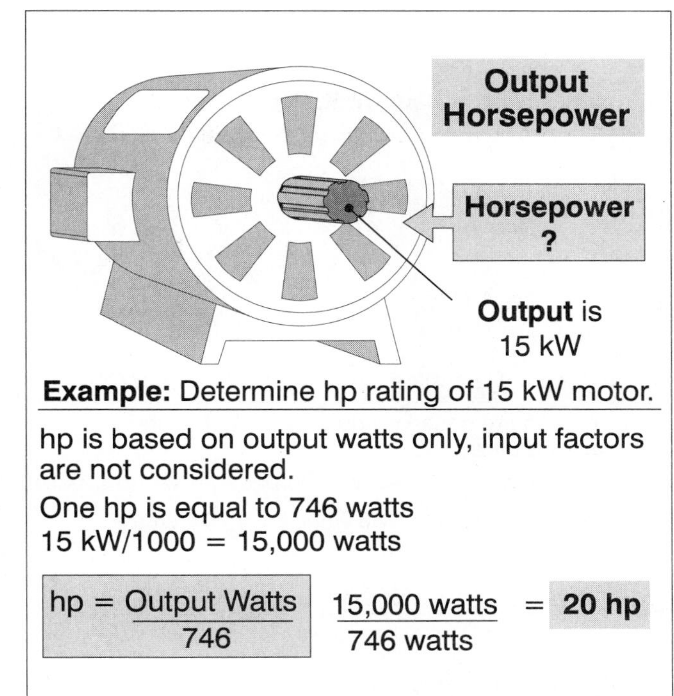

Figure. 4-9

Motor Horsepower Example

❑ **Motor VA Three-Phase Example**

What is the motor VA of a three-phase, 230/460-volt, 30-horsepower motor that has a current rating of 40 amperes at 460-volts, Fig. 4–8?

(a) 41,450 VA (b) 31,600 VA (c) 21,840 VA (d) 31,869 VA

- Answer: (d) 31,869 VA

  Motor VA = Volts × Amperes × $\sqrt{3}$

  Volts = 460 volts, Amperes = 40 amperes, $\sqrt{3}$ = 1.732

  Motor VA = 460 volts × 40 amperes × 1.732

  Motor VA = 31,869 VA

4–6 *MOTOR HORSEPOWER/WATTS*

The mechanical work (output) of a motor is rated in *horsepower* and can be converted to electrical energy as 746-watts per horsepower.

Horsepower Size = Output Watts/746 Watts **Motor Output Watts = Horsepower × 746**

❑ **Motor Horsepower Example**

What size horsepower motor is required to produce 15-kW output watts, Fig. 4–9?

(a) 5 hp (b) 10 hp (c) 20 hp (d) 30 hp

- Answer: (c) 20 horsepower

  $$\text{Horsepower} = \frac{\text{Output Watts}}{746}$$

  $$\text{Horsepower} = \frac{15{,}000 \text{ watts}}{746 \text{ watts}}$$

  Horsepower = 20 horsepower

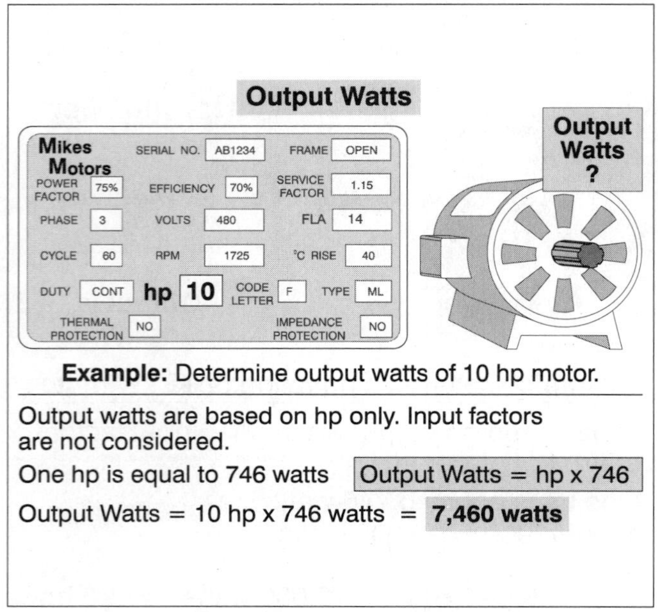

Figure 4-10

Motor Output Watts Example

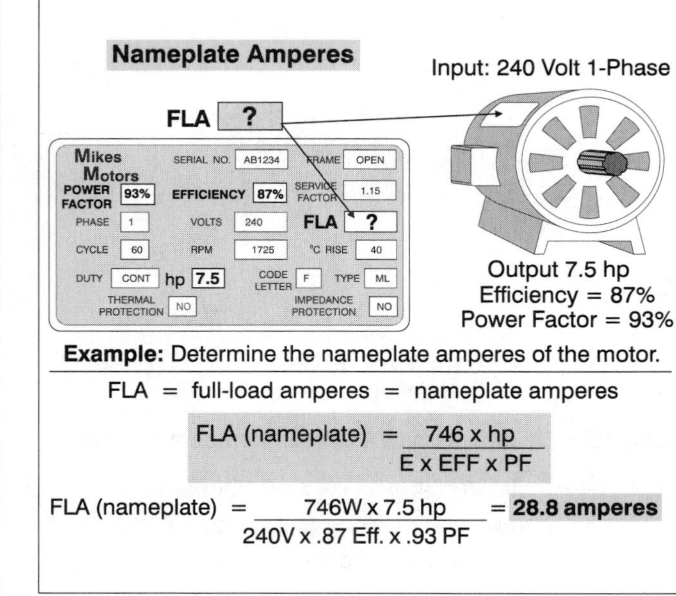

Figure 4-11

Nameplate Amperes – Single-Phase Example

❑ **Motor Output Watt Example**

What is the output watt rating of a 10-hp motor, Fig. 4–10?

(a) 3 kW (b) 2.2 kVA (c) 3.2 kW (d) 7.5 kW

Answer: (d) 7.5 kW

Output watts = horsepower × 746 watts

Output watts = 10 horsepower × 746 watts = 7,460 watts

$$\text{Output watts} = \frac{7{,}460 \text{ watts}}{1{,}000} = 7.46 \text{ kW}$$

4–7 *MOTOR NAMEPLATE AMPERES*

The motor nameplate indicates the motor operating voltage and current. The actual current drawn by the motor depends on how much the motor is loaded. It is important not to overload a motor above its rated horsepower, because the current of the motor will increase to a point that the motor winding will be destroyed from excess heat. The nameplate current can be calculated by the following formulas:

$$\textbf{Single-Phase Motor Nameplate Amperes} = \frac{\textbf{Motor Horsepower} \times \textbf{746 Watts}}{(\textbf{Volts} \times \textbf{Efficiency} \times \textbf{Power Factor})}$$

$$\textbf{Three-Phase Motor Nameplate Amperes} = \frac{\textbf{Motor Horsepower} \times \textbf{746 Watts}}{(\textbf{Volts} \times \sqrt{3} \times \textbf{Efficiency} \times \textbf{Power Factor})}$$

❑ **Nameplate Amperes – Single-Phase Example**

What is the nameplate amperes for a 7.5-horsepower motor rated 240-volts, single-phase? The efficiency is 87 percent and the power factor is 93 percent, Fig. 4–11.

(a) 19 amperes (b) 24 amperes (c) 19 amperes (d) 29 amperes

- Answer: (d) 29 amperes

$$\text{Nameplate amperes} = \frac{\text{Horsepower} \times 746 \text{ Watts}}{(\text{Volts} \times \text{Efficiency} \times \text{Power Factor})}$$

$$\text{Nameplate amperes} = \frac{7.5 \text{ Horsepower} \times 746 \text{ Watts}}{(240 \text{ volts} \times 0.87 \text{ Efficiency} \times 0.93 \text{ Power Factor})}$$

Nameplate amperes= 28.81 amperes

❑ **Nameplate Amperes Three-Phase Example**

What it the nameplate amperes of a 40-horsepower motor rated 208-volts, three-phase? The efficiency is 80 percent and the power factor is 90 percent, Fig. 4–12.

(a) 85 amperes (b) 95 amperes (c) 105 amperes (d) 115 amperes

- Answer: (d) 115 amperes

$$\text{Nameplate amperes} = \frac{\text{Horsepower} \times 746 \text{ Watts}}{(\text{Volts} \times \sqrt{3} \times \text{Efficiency} \times \text{Power Factor})}$$

$$\text{Nameplate amperes} = \frac{40 \text{ Horsepower} \times 746 \text{ Watts}}{(208 \text{ volts} \times 1.732 \times 0.8 \text{ Efficiency} \times 0.9 \text{ Power Factor})}$$

Nameplate amperes = 29,840/259.4

Nameplate amperes = 115 amperes

PART B – *TRANSFORMER BASICS*

TRANSFORMER INTRODUCTION

A *transformer* is a stationary device used to raise or lower voltage. Transformers have the ability to transfer electrical energy from one circuit to another by mutual induction between two conductor coils. *Mutual induction* occurs between two conductor coils *(windings)* when electromagnetic lines of force within one winding induces a voltage into a second winding, Fig. 4–13.

Current transformers use the circuit conductors as the primary winding and steps the current down for metering. Often, the *ratio* of the current transformer is 1,000 to 1. This means that if there are 400 amperes on the phase conductors, the current transformer would step the current down to 0.4 ampere for the meter.

The *magnetic coupling* (*magnetomotive force, mmf*) between the primary and secondary winding can be increased by increasing the winding *ampere-turns*. Ampere-turns can be increased by increasing the number of coils and/or the current through each coil. When the current in the core of a transformer is raised to a point where there is high *flux density*, additional increases in current will produce few additional flux lines. The transformer metal iron core is said to be *saturated*.

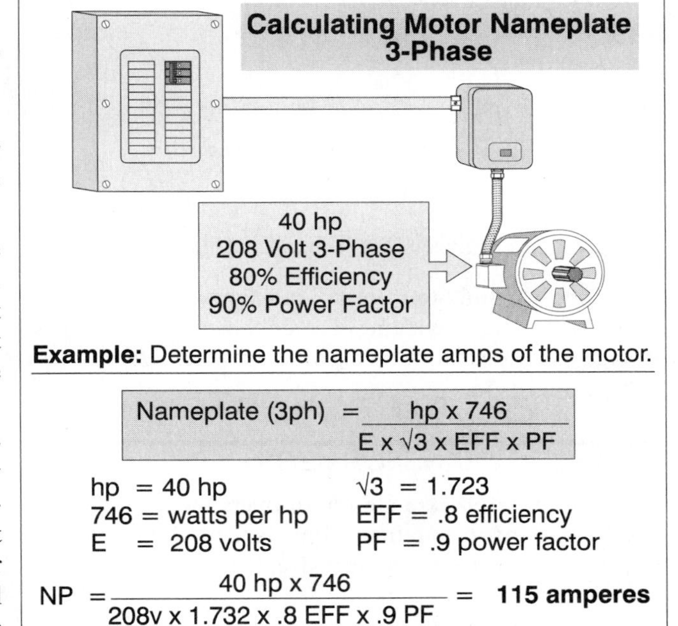

Figure 4-12

Nameplate Amperes – Three-Phase Example

4–8 *TRANSFORMER PRIMARY versus SECONDARY*

The transformer *winding* connected to the source is called the *primary winding*. The transformer winding connected to the load is called the *secondary winding*. Transformers are reversible, that is, either winding can be used as the primary or secondary.

4–9 *TRANSFORMER SECONDARY AND PRIMARY VOLTAGE*

Voltage induced in the secondary winding of a transformer is equal to the sum of the voltages induced in each *loop* of the secondary winding. The voltage induced in the secondary of a transformer depends on the number of secondary conductor turns cut by the primary magnetic flux lines. The greater the number of secondary conductor loops, the greater the secondary voltage, Fig. 4–14.

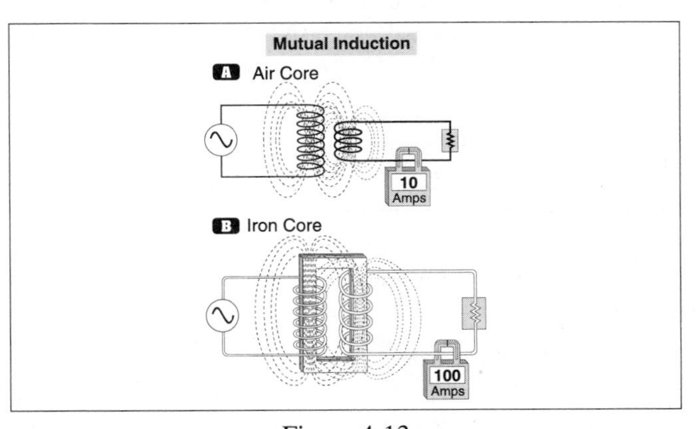

Figure 4-13

Transformer – Mutual Induction

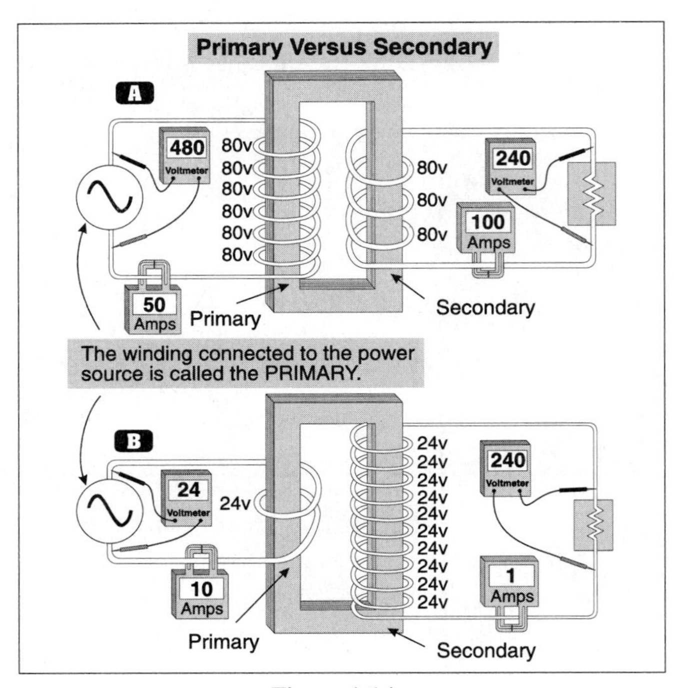

Figure 4-14

Secondary and Primary Voltage

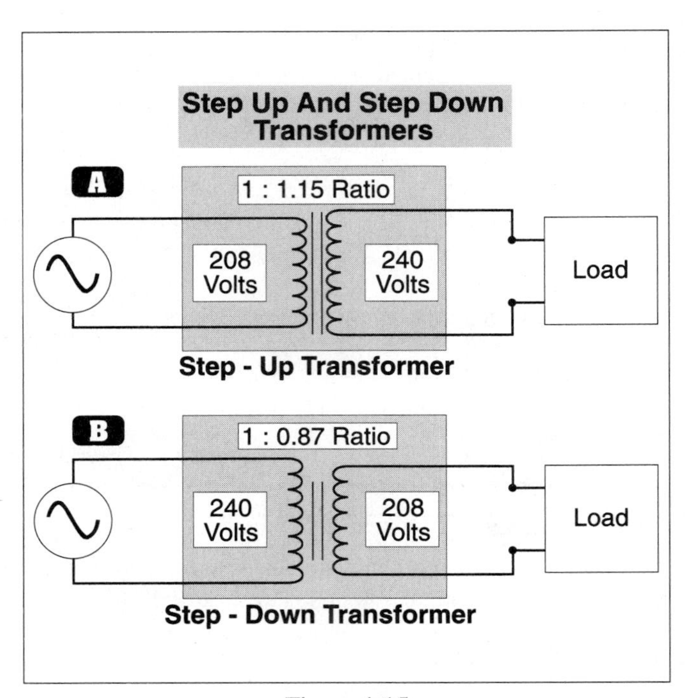

Figure 4-15

Step-Up and Step-Down Transformers

Step-Up and Step-Down Transformers

The secondary winding of a *step-down transformer* has fewer turns than the primary winding, resulting in a lower secondary voltage. The secondary winding of a *step-up transformer* has more turns than the primary winding, resulting in a higher secondary voltage, Fig. 4–15.

4–10 *AUTO TRANSFORMERS*

Auto transformers are transformers that use the same common winding for both the primary and the secondary. The disadvantage of an auto transformer is the lack of isolation between the primary and secondary conductors, but they are often used because they are less expensive, Fig. 4–16.

4–11 *TRANSFORMER POWER LOSSES*

When current flows through the winding of a transformer, power is dissipated in the form of heat. This loss is referred to as conductor I^2R loss. In addition, losses include flux leakage, core loss from eddy currents, and hysteresis heating losses, Fig. 4–17.

Conductor Resistance Loss

Transformer windings are made of many turns of wire. The resistance of the conductors is directly proportional to the length of the conductors, and inversely proportional to the cross-sectional area of the conductor. The more turns there are, the longer the conductor is, and the greater the conductor resistance. Conductor losses can be determined by the formula: $P = I^2R$.

Flux Leakage Loss

The *flux leakage loss* represents the electromagnetic flux lines between the primary and secondary winding that are not used to convert electrical energy from the primary to the secondary. They represent wasted energy.

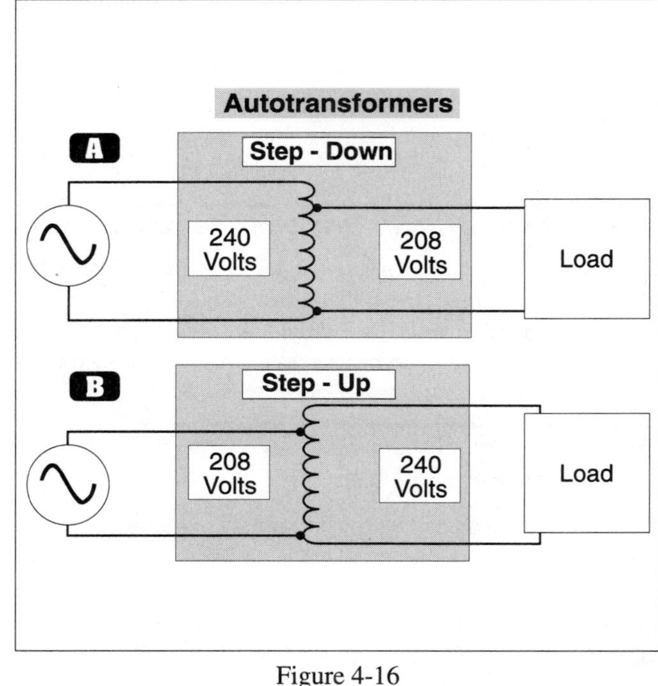

Figure 4-16

Auto Transformer

Core Losses

Iron is the only metal used for transformer cores because it offers low resistance to magnetic flux lines (*low reluctance*). Iron cores permit more flux lines between the primary and secondary winding, thereby increasing the magnetic coupling between the primary and secondary windings. However, alternating currents produce electromagnetic fields within the windings that induce a circulating current in the iron core. These circulating currents (*eddy currents*) flow within the iron core producing power losses that cannot be transferred to the secondary winding. Transformer iron cores are laminated to have a small cross-sectional area to reducing the eddy currents and their associated losses.

Hysteresis Losses

Each time the primary magnetic field expands and collapses, the transformer's iron core molecules realign themselves to the changing polarity of the electromagnetic field. The energy required to realign the iron core molecules to the changing electromagnetic field is called *hysteresis losses.*

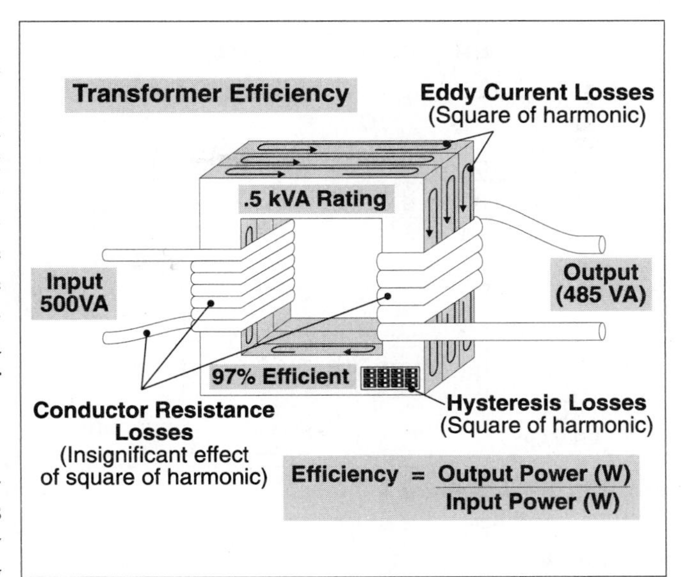

Figure 4-17

Transformer Losses

Heating By The Square Of The Frequency

Hysteresis and eddy current losses are affected by the square of the alternating current frequency. For this reason, care must be taken when iron core transformers are used in applications involving high frequencies and nonlinear loads. If the transformer operates at the third harmonic (180-Hz), the losses will by the square of the frequency. At 180-Hz, the frequency is three times the fundamental frequency (60-Hz). Therefore, the heating effect will be $3^2 = 3 \times 3 = 9$.

❑ **Transformer Losses By Square Of Frequency Example**

What is the effect on 60-Hz rated transfer for third harmonic loads (180 Hz)?

(a) heating increases three times (b) heating increases six times

(c) heating increases nine times (d) no significant heating

- Answer: (c) heating increases nine times

4–12 *TRANSFORMER TURNS RATIO*

The relationship of the primary winding voltage to the secondary winding voltage is the same as the relation between the number of primary turns as compared to the number of secondary turns. This relationship is called turns ratio or *voltage ratio.*

❑ **Delta Winding Turns Ratio Example**

What is the turns ratio of a delta/delta transformer? The primary winding is 480-volts, and the secondary winding voltage is 240-volts, Fig. 4–18?

(a) 4:1 (b) 1:4 (c) 2:1 (d) 1:2

- Answer: (c) 2:1

The primary winding voltage is 480 and the secondary winding voltage is 240. This results in a ratio of 480:240 or 2:1.

❑ **Wye Winding Ratio Example**

What is the turns ratio of a delta/wye transformer? The primary winding is 480-volts, and the secondary winding is 120-volts, Fig. 4–19?

(a) 4:1 (b) 1:4 (c) 2.3:1 (d) 1:2.3

- Answer: (a) 4:1

  The primary winding voltage is 480, and the secondary winding voltage is 120. This results in a ratio of 480:120 or 4:1.

4–13 *TRANSFORMER kVA RATING*

Transformers are rated in kilo volt-amperes, abbreviated as kVA.

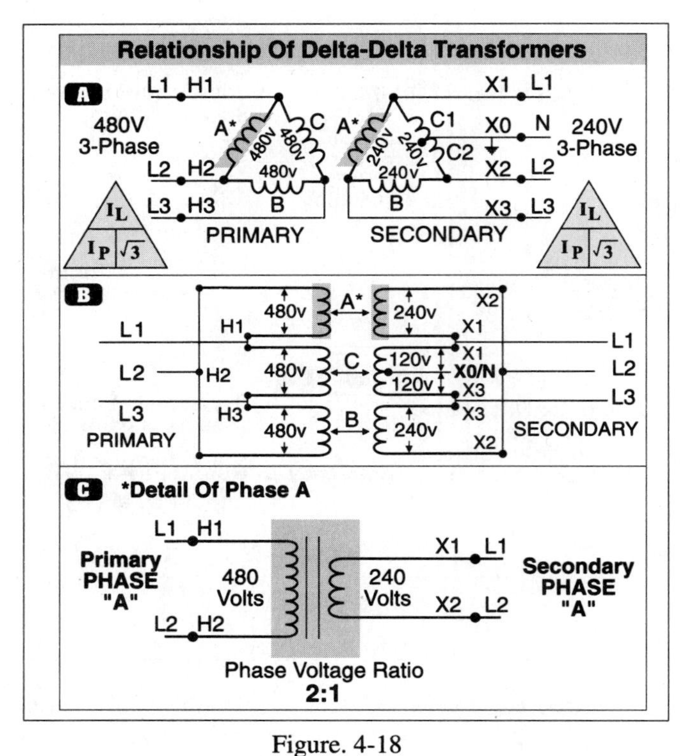

Figure. 4-18

Delta Winding Ratio Example

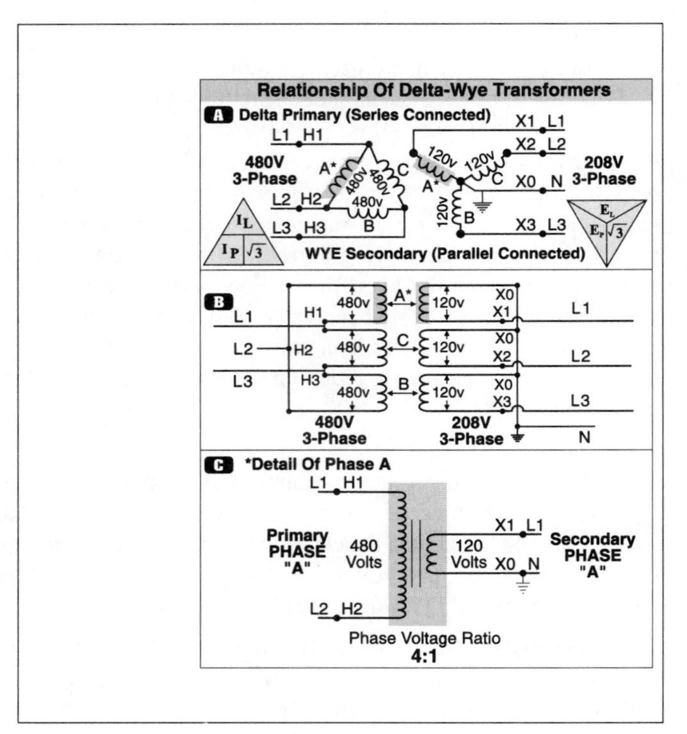

Figure. 4-19

Wye Winding Ratio Example

4–14 *TRANSFORMER CURRENT*

Whenever the number of primary turns is greater than the number of secondary turns, the secondary voltage will be less than the primary voltage. This will result in secondary current being greater than the primary current because the power remains the same, but the voltage changes. Since the secondary voltage of most transformers is less than the primary voltage, the secondary conductors will carry more current than the primary, Fig. 4–20. Primary and secondary line current can be calculated by:

Current (Single-Phase) = Volt-Amperes/Volts

Current (Three-Phase) = Volt-Amperes/(Volts × 1.732)

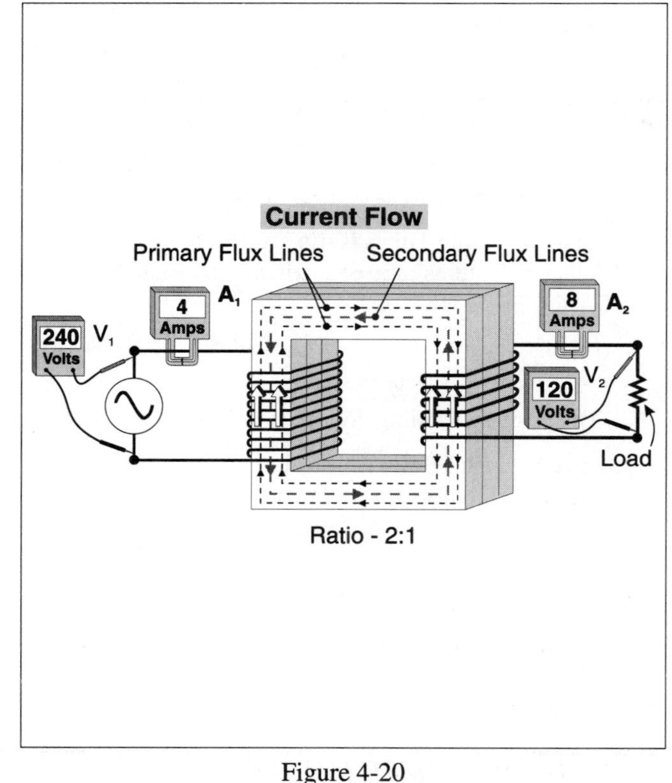

Figure 4-20

Secondary Current Greater Than Primary

❑ Single-Phase Transformer Current Example

What is the primary and secondary line current for a single-phase 25-kVA transformer, rated 480-volts primary and 240-volts secondary? Which conductor is larger, Fig. 4–21?

(a) 52/104 amperes (b) 104/52 amperes

(c) 104/208 amperes (d) 208/104 amperes

• Answer: (a) 52/104 amperes. The secondary conductor is larger.

Primary current =

$$VA/E = \frac{25{,}000\ VA}{480\ volts} = 52\ amperes$$

Secondary current =

$$VA/E = \frac{25{,}000\ VA}{240\ volts} = 104\ amperes$$

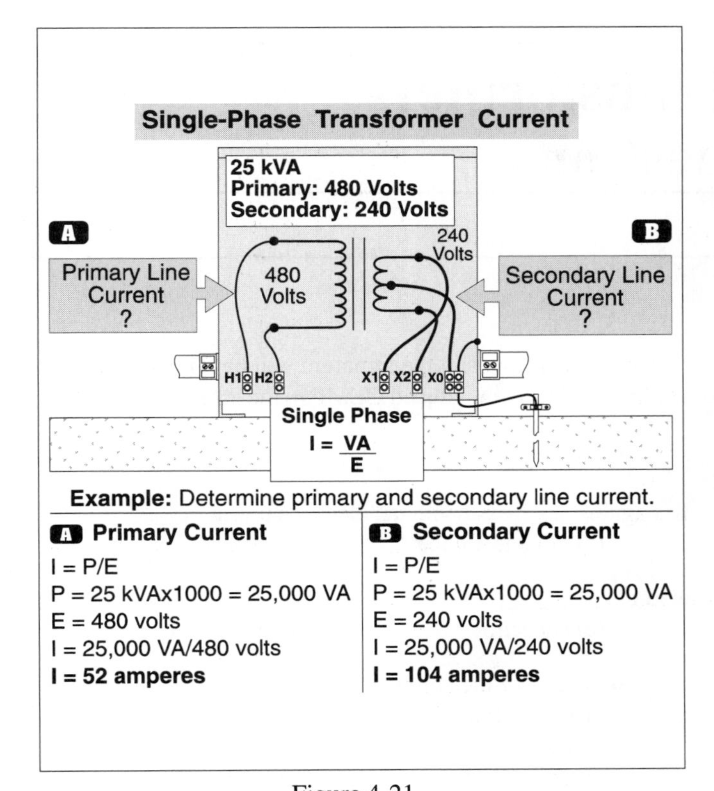

Figure 4-21

Single Phase Current Example

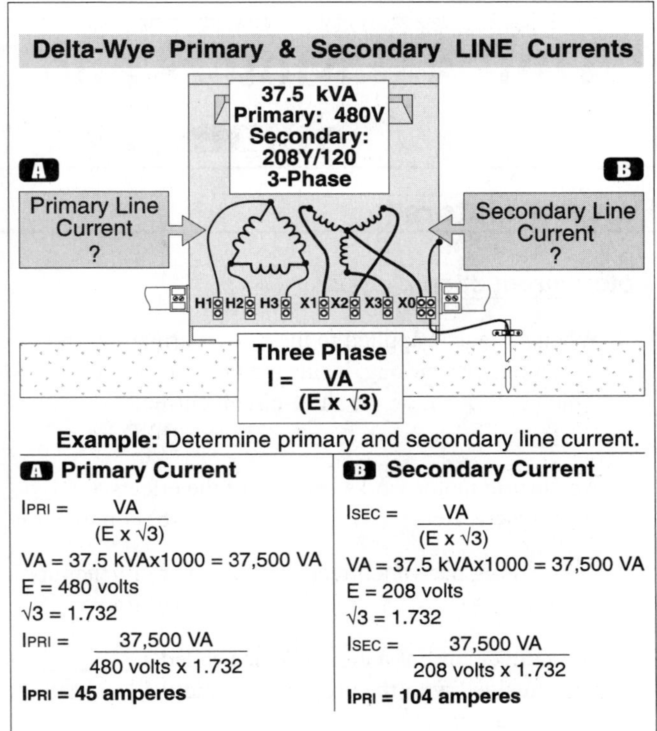

Figure 4-22

Three Phase Transformer Current Example

❑ Three-Phase Transformer Current Example

What is the primary and secondary line current for a three-phase 37.5-kVA transformer, rated 480-volts primary and 208-volts secondary, Fig. 4–22?

(a) 45/104 amps (b) 104/40 amps (c) 208/140 amps (d) 140/120 amps

• Answer: (a) 45/104 amperes

The secondary conductor is larger

$$\text{Primary current} = \frac{VA}{E \times \sqrt{3}} = \frac{37{,}500 \text{ VA}}{(480 \text{ volts} \times 1.732)} = 45 \text{ amperes}$$

$$\text{Secondary current} = \frac{VA}{E \times \sqrt{3}} = \frac{37{,}500 \text{ VA}}{(208 \text{ volts} \times 1.732)} = 104 \text{ amperes}$$

Unit 4 – Motors and Transformers Summary Questions

Part A – Motors

Motor Introduction

1. When voltage is applied to the motors armature, short-circuit current will flow and the armature will begin to turn. As the armature starts turning, it cuts the lines of force of the field winding, resulting in induced CEMF in the armature conductor, which cuts down on the short-circuit current.
(a) True (b) False

2. An electric motor works because of the effects a ________ has against a wire carrying an electric current.
(a) magnetic field (b) commutator (c) voltage source (d) none of these

3. The rotating part of a direct current motor or generator is called the ________.
(a) shaft (b) rotor (c) capacitor (d) field

4. For a direct current motor, a device called a ________ is placed on the end of the conductor loop. The polarity of the loop is maintained with the proper magnetic field to keep the opposing magnetic fields pushing each other. This in turn keeps the loop turning.
(a) coil (b) resistor (c) commutator (d) none of these

4–1 Motor Speed Control

5. • One of the great advantages of the direct current motor is the motor's ability to maintain a constant speed. If the speed of a direct current motor is increased, the armature will be cut through the field winding at an increasing rate resulting in a lower CEMF that acts to cut down on the increased armature current, which slows the motor back down.
(a) True (b) False

6. • Placing a load on a direct current motor causes the motor to slow down, which increases the rate at which the field flux lines are being cut by the armature. As a result, the armature CEMF increases, resulting in an increase in the applied armature voltage and current. The increase in current results in increase in motor speed.
(a) True (b) False

4–2 Reversing A Direct Current Motor

7. To reverse a direct current motor, you must reverse the direction of the ________.
(a) field current (b) armature current (c) a or b (d) a and b

4–3 Alternating Current Motors

8. In a(n) ________ motor, the rotor is actually locked in step with the rotating stator field and is dragged along at the synchronous speed of the rotating magnetic field. ________ motors maintain their speed with a high degree of accuracy and are used for electric clocks and other timing devices.
(a) alternating current (b) universal (c) wound rotor (d) synchronous

9. ________ rotor motors are used only as special applications because of their high starting torque requirements and only operate on three-phase alternating current power.
(a) Alternating current (b) Universal
(c) Wound rotor (d) Synchronous

10. ________ motors are fractional horsepower motors that operate equally well on alternating current and direct current and are used for vacuum cleaners, electric drills, mixers, and light household appliances.
(a) Alternating current (b) Universal
(c) Wound rotor (d) Synchronous

4–4 Reversing Alternating Current Motors

11. • Three-phase alternating current motors can be reversed by changing the wiring from the line wiring from ABC phase configuration to________.
(a) BCA (b) CAB (c) CBA (d) ABC

4–5 Motor Volt-Ampere Calculations

12. Dual voltage 277/480-volt motors are made with two field windings, each rated at 277 volts. The field windings are connected in parallel for ________-volt operation and in series for ________-volt operation.
(a) 277, 480 (b) 480, 277 (c) 277, 277 (d) 480, 480

4–6 Motor Horsepower/watts

13. What size horsepower motor is required to produce 30-kW output?
(a) 20 horsepower (b) 30 horsepower (c) 40 horsepower (d) 50 horsepower

14. What are the output watts of a 15 horsepower?
(a) 11 kW (b) 15 kVA (c) 22 kW (d) 31 kW

15. What are the output watts of a 5-horsepower, alternating current, three-phase 480 volt motor, efficiency 75 percent and power factor 70 percent?
(a) 3.75 kW (b) 7.5 kVA (c) 7.5 kW (d) 10 kW

4–7 Motor Nameplate Amperes

16. • For all practical purposes you will not need to calculate the motor nameplate current; but you should understand how it is calculated.
(a) True (b) False

17. What are the nameplate amperes for a 5-horsepower motor, 240volt, single-phase, efficiency 93 percent, power factor 87 percent?
(a) 9 amperes (b) 19 amperes (c) 28 amperes (d) 31 amperes

18. What are the nameplate amperes of a 20-horsepower, 208volt, three-phase motor, power factor 90 percent and efficiency 80 percent?
(a) 50 amperes (b) 58 amperes (c) 65 amperes (d) 80 amperes

Part B – Transformer Basics

Transformer Introduction

19. A __________ is a stationary device used to raise or lower the voltage and has the ability to transfer electrical energy from one circuit to another, with no physical connection between the two.
(a) capacitor (b) motor (c) relay (d) transformer

20. Transformers operate on the principle of ________.
(a) magnetoelectricity (b) triboelectric effect
(c) thermocouple (d) mutual induction

4–8 Primary Versus Secondary

21. The transformer winding that is connected to the source is called the ________ winding and the transformer winding that is connected to the load is called the ________. Transformers are reversible; that is, either winding can be used as the primary or secondary.
(a) secondary, primary (b) primary, secondary
(c) depends on the wiring (d) none of these

22. Voltage induced in the secondary winding of a transformer is dependent on the number of secondary turns as compared to the number of primary turns.
(a) True (b) False

23. The secondary winding of a step-down transformer has ________ turns than the primary, resulting in a ________ secondary voltage as compared to the primary.
(a) less, higher (b) more, lower (c) less, lower (d) more, higher

24. The secondary winding of a step-up transformer has ________ turns than the primary, resulting in a ________ secondary voltage as compared to the primary.
(a) less, higher (b) more, lower (c) less, lower (d) more, higher

4–10 Auto Transformers

25. Auto transformers use the same winding for both the primary and secondary. The disadvantage of an autotransformer is the lack of ________ between the primary and secondary conductors.
(a) power (b) voltage (c) isolation (d) grounding

4–11 Transformer Power Losses

26. The most common causes of power losses for transformers are ________.
(a) conductor resistance (b) eddy currents (c) hysteresis (d) all of these

27. The leakage of the electromagnetic flux lines between the primary and secondary winding represents wasted energy.
(a) True (b) False

28. • The expanding and collapsing electromagnetic field of the transformer also induces a voltage in the transformer core. The induced voltage causes ________ to flow within the core, which removes energy from the transformer winding and represents wasted power.
(a) eddy currents (b) flux (c) inductive (d) hysteresis

29. Eddy currents can be reduced by dividing the core into many flat sections or laminations. Because the laminations have a ________ cross-sectional area, the resistance offered to the eddy currents is greatly increased.
(a) round (b) porous (c) large (d) small

30. As current flows through the transformer, the iron-core is temporarily magnetized by the electromagnetic field created by the alternating current. Each time the primary magnetic field expands and collapses, the core molecules realign themselves to the changing polarity of the electromagnetic field. The energy required to realign the core molecules to the changing electromagnetic field is called the ________ loss of the core.
(a) eddy current (b) flux (c) inductive (d) hysteresis

4–12 Transformer Turns Ratio

31. The relationship of the primary winding voltage to the secondary winding voltage is the same as the relation between the number of conductor turns on the primary as compared to the secondary. This relationship is called ________.
(a) ratio (b) efficiency (c) power factor (d) none of these

32. • The primary phase voltage is 240 and the secondary phase is 480. This results in a ratio of ________.
(a) 1:2 (b) 2:1 (c) 4:1 (d) 1:4

4–13 Transformer kVA Rating

33. Transformers are rated in ________.
(a) VA (b) kW (c) watts (d) kVA

4–14 Transformer Current

34. • The current flow in the secondary transformer winding creates a electromagnetic field that opposes the primary electromagnetic field resulting in less primary CEMF. The primary current automatically increases in direct proportion to the secondary current.
(a) True (b) False

35. • The transformer winding with the ________ number of turns will have the lower current and the winding with the ________ number of turns will have the higher current.
(a) least, most (b) most, most (c) least, least (d) most, least

Challenge Questions

Part A – Motors

4–1 Motor Speed Control

36. A(n) ______ type of electric motor tends to run away if it is not always connected to its load.
(a) direct current series
(b) direct current shunt
(c) alternating current induction
(d) alternating current synchronous

37. • A(n) ______ motor has a wide speed range.
(a) alternating current (b) direct current (c) synchronous (d) induction

4–2 Reversing A Direct Current Motor

38. • If the two line (supply) leads of a direct current series motor are reversed, the motor will ______.
(a) not run (b) run backwards
(c) run the same as before (d) become a generator

39. • To reverse a direct current motor, one may simply reverse the supply wires.
(a) True (b) False

4–3 Alternating Current Motors

40. The ______ induction motor is used only in special applications and is always operated on three-phase alternating current power.
(a) compound (b) synchronous (c) split phase (d) wound rotor

41. • The rotating part of a direct current motor or generator is called the ______.
(a) shaft (b) rotor (c) armature (d) b or c

4–5 Motor Volt-Ampere Calculations

42. The input volt-amperes of a 5-horsepower (15.2-amperes), 230-volt, three-phase motor is closest to ______ VA.
(a) 7,500 (b) 6,100 (c) 5,300 (d) 4,600

43. The VA of a 1-horsepower (16-amperes), 115-volt motor is ______ VA.
(a) 2,960 (b) 1,840 (c) 3,190 (d) 1,650

Part B – Transformer Basics

4–11 Transformer Power Losses

44. • When the current in the core of a transformer has risen to a point where high flux density has been reached, and additional increases in current produce few additional flux lines, the metal core is said to be ______.
(a) maximum (b) saturated (c) full (d) none of these

45. • Magnetomotive force (mmf) can be increased by increasing the ______.
(a) number of ampere-turns (b) current in the coils
(c) number of coils (d) all of these

4–12 Transformer Turns Ratio

46. • The secondary current of a transformer that has a turns ratio of 2:1 is ______ than the primary current.
(a) higher (b) lower (c) the same (d) none of these

4–13 Transformer kVA Rating

47. The primary kVA rating for a transformer that is 100 percent efficient, with a secondary of 12 volts (E_s) and with secondary current (I_s) of 5 amperes is ______ kVA.
(a) 600 (b) 30 (c) 6 (d) 0.06

4–14 Transformer Current

48. • The primary a transformer has 100 turns (N_p = 100) and the secondary has 10 turns (N_s = 10). What is the primary current of the transformer if the secondary current (I_s) of 5 amperes is _______ amperes?
(a) 25 (b) 10 (c) 5 (d) 0.5

49. • Which winding of a current transformer will carry more current? *Note:* A clamp-on ammeter is a current transformer, the meter is the secondary.
(a) primary (b) secondary (c) interwinding (d) tertiary

50. If a transformer primary winding has 900 turns (N_p = 900) and the secondary winding has 90 turns (N_s = 90 turns). Which winding of the transformer has a larger a larger conductor?
(a) primary (b) secondary (c) interwinding (d) none of these

51. If the primary of a transformer is 480 volts, and the secondary is 240 volts. Which winding of this transformer has a larger conductor?
(a) tertiary (b) secondary (c) primary (d) windings are equal

52. If the transformer voltage turns ratio is 5:1, and the secondary has 10 ampere-turns, the primary current would be ______ if the secondary current was 10 amperes.
(a) 25 amperes (b) 10 amperes (c) 2 amperes (d) cannot be determined

53. The transformer primary is 240 volts, the secondary is 12 volts, and the load is two 100-watt lamps. The transformer is 92 percent efficient. The secondary current of this transformer is ______.
(a) 1 amperes (b) 17 amperes (c) 28 amperes (d) none of these

54. The transformer primary is 240 volts, the secondary is 120 volts, and the load is 1,500-watt lamps. The transformer is 92 percent efficient. The primary current of this transformer is ______ amperes.
(a) 6.8 (b) 8.6 (c) 9.9 (d) 7.8

Figure 4–23 applies to the next two questions.

55. • The primary kVA of this transformer is ______ kVA, Fig. 4–23?
(a) 72.5
(b) 42
(c) 30
(d) 21

56. • The primary current of this transformer is ______ amperes, Fig. 4–23?
(a) 90
(b) 50
(c) 25
(d) 12

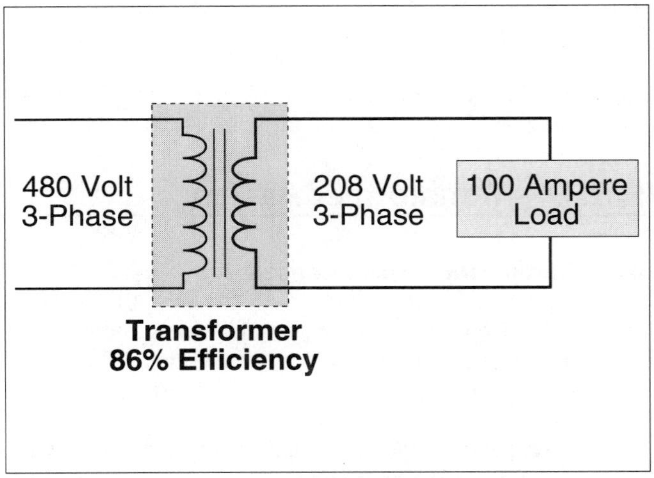

Figure 4-23

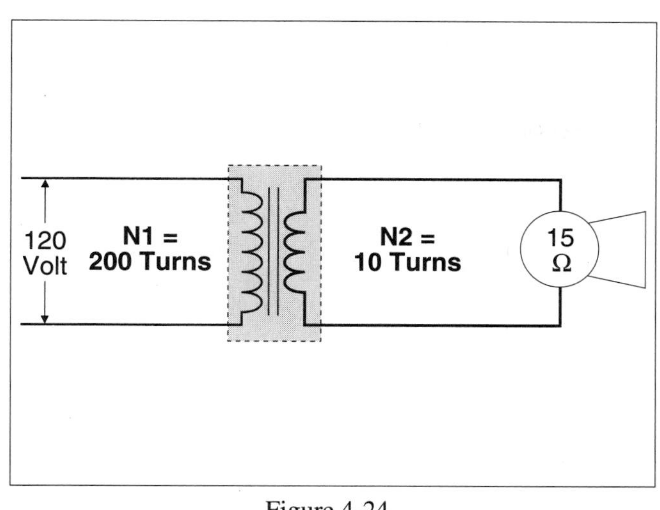

Figure 4-24

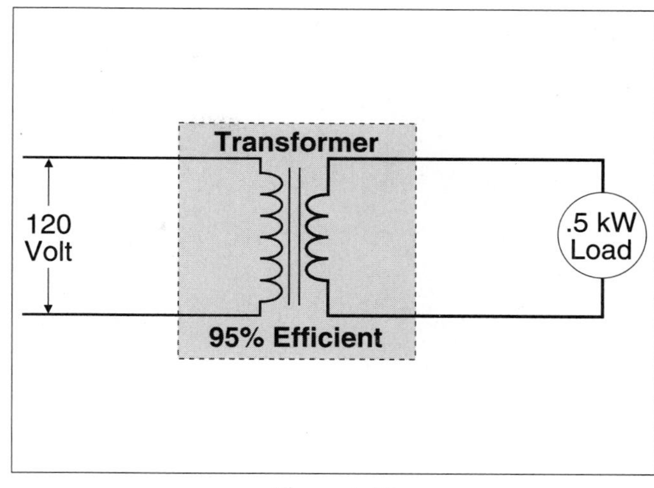

Figure 4-25

57. • The secondary voltage of this transformer is ______ volts, Fig. 4–24?
(a) 6 (b) 12
(c) 24 (d) 30

Figure 4–35 applies to the next two questions.

58. The primary current of the transformer is ______ ampere(s), Fig. 4–25?
(a) 0.416 (b) 4.38
(c) 3.56 (d) 41.6

59. The primary power for the transformer is ______ VA, Fig. 4–25?
(a) 526 (b) 400
(c) 475 (d) 550

Transformer Miscellaneous

60. The transformer primary is 240 volts, the secondary is 12 volts, and the load is two 100 watt lamps. The transformer is 92 percent efficient. The secondary VA of this transformer ______.
(a) is 200 VA
(b) is the same as the primary VA
(c) cannot be calculated
(d) none of these

61. • The secondary of a transformer is 24 volts with a load of 5 amperes. The primary is 120 volts and the transformer is 100 percent efficient. The secondary VA of this transformer ______.
(a) 120 VA
(b) is the same as the primary VA
(c) cannot be calculated
(d) a and b

62. The transformer primary is 240 volts, the secondary is 12 volts, and the load is two 100 watt lamps. The transformer is 92 percent efficient. The primary VA of this transformer ______.
(a) is 185 VA
(b) is 217 VA
(c) is 0.217 VA
(d) cannot be determined

63. The output of a generator is 20 kW. The input kVA is ______ kVA if the efficiency rating is 65 percent.
(a) 11 (b) 20 (c) 31 (d) 33

CHAPTER 2
NEC® Calculations And Code Questions

Scope of Chapter 2

Unit 5

Raceway, Outlet Box, And Junction Boxes Calculations

OBJECTIVES

After reading this unit, the student should be able to briefly explain the following concepts:

Part A – Raceway Calculations
Existing raceway calculation
Raceway sizing
Raceway properties
Understanding The National Electrical Code Chapter 9

Part B – Outlet Box Calculations
Conductor equivalents
Sizing box – conductors all the same size
Volume of box

Part C – Pull And Junction Box Calculations
Depth of box and conduit body sizing
Pull and junction box size calculations

After reading this unit, the student should be able to briefly explain the following terms:

Part A – Raceway Calculations
Alternating current conductor
Resistance
Bare conductors
Bending radius
Compact aluminum building
Wires
Conductor properties
Conductor fill
Conduit bodies
Cross-sectional area of insulated conductors
Expansion characteristics of PVC
Fixture wires
Grounding conductors
Lead-covered conductor
NEC® errors
Nipple size
Raceway size
Spare space area

Part B – Outlet Box Calculations
Cable clamps
Conductor terminating in the box
Conductor running through the box
Conduit bodies
Equipment bonding jumpers
Extension rings
Fixture hickey
Fixture stud
Outlet box
Pigtails
Plaster rings
Short radius conduit bodies
Size outlet box
Strap
Volume
Yoke

Part C – Pull And Junction Box Calculations
Angle pull calculation
Distance between raceways
Horizontal dimension
Junction boxes

PART A – *RACEWAY FILL*

5–1 *UNDERSTANDING THE NATIONAL ELECTRICAL CODE, CHAPTER 9*

Chapter 9 – Tables And Examples, Part A – Tables

Table 1 – Conductor Percent Fill

The maximum percentage of conductor fill is listed in Table 1 of Chapter 9 and is based on common conditions where the length of the conductor and number of raceway bends are within reasonable limits [Fine Print Note after Note 9], Fig. 5–1.

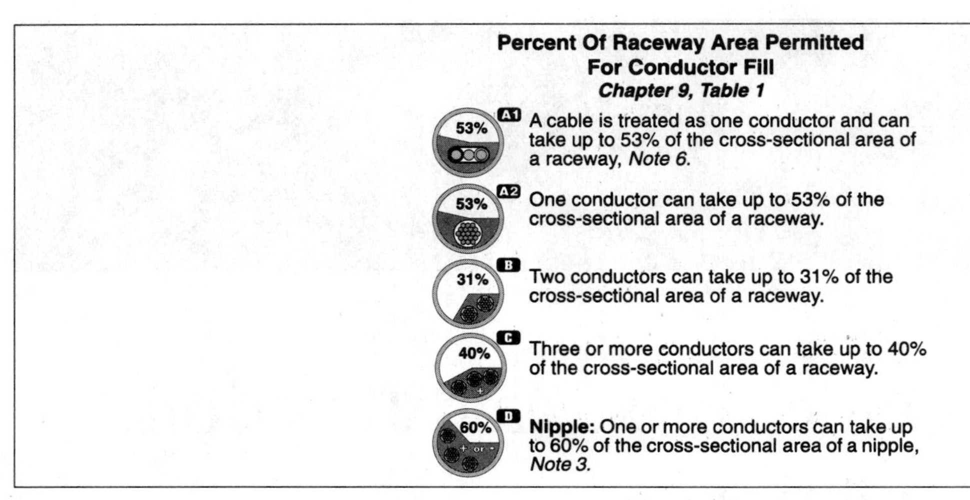

Figure 5-1

Percent Of Raceway Area Permitted For Conductor Fill

Table 1 of Chapter 9, Maximum Percent Conductor Fill	
Number of Conductors	**Percent Fill Permitted**
1 conductor	53% fill
2 conductors	31% fill
3 or more conductors	40% fill
Raceway 24 inches or less	60% fill Chapter 9, Note 3

Table 1, Note 1 – Conductors All The Same Size And Insulation

When all of the conductors are the same size and insulation, the number of conductors permitted in a raceway can be determined simply by looking at the tables located in Appendix C – Conduit and Tubing Fill Tables for Conductors and Fixture Wires of the Same Size.

Tables C1 through C13 are based on maximum percent fill as listed in Table 1 of Chapter 9.

Table C1 – Conductors and fixture wires in electrical metallic tubing
Table C1A – Compact conductors and fixture wires in electrical metallic tubing
Table C2 – Conductors and fixture wires in electrical nonmetallic tubing
Table C2A – Compact conductors and fixture wires in electrical metallic tubing
Table C3 – Conductors and fixture wires in flexible metal conduit
Table C3A – Compact conductors and fixture wires in flexible metal conduit
Table C4 – Conductors and fixture wires in intermediate metal conduit
Table C4A – Compact conductors and fixture wires in intermediate metal conduit
Table C5 – Conductors and fixture wires in liquidtight flexible nonmetallic conduit (gray type)
Table C5A – Compact conductors and fixture wires in liquidtight flexible nonmetallic conduit (gray type)
Table C6 – Conductors and fixture wires in liquidtight flexible nonmetallic conduit (orange type)
Table C6A – Compact conductors and fixture wires in liquidtight flexible nonmetallic conduit (orange type)

Note: The appendix does not have a table for liquidtight flexible nonmetallic conduit of the black type
Table C7 – Conductors and fixture wires in liquidtight flexible metallic conduit
Table C7A – Compact conductors and fixture wires in liquidtight flexible metal conduit
Table C8 – Conductors and fixture wires in rigid metal conduit
Table C8A – Compact conductors and fixture wires in rigid metal conduit
Table C9 – Conductors and fixture wires in rigid PVC Conduit Schedule 80
Table C9A – Compact conductors and fixture wires in rigid PVC Conduit Schedule 80
Table C10 – Conductors and fixture wires in rigid PVC Conduit Schedule 40

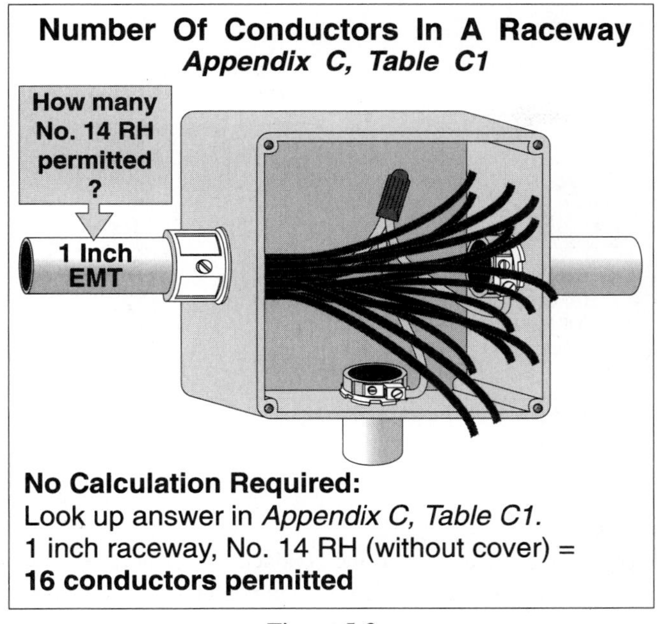

Figure 5-2

Number of Conductors In A Raceway

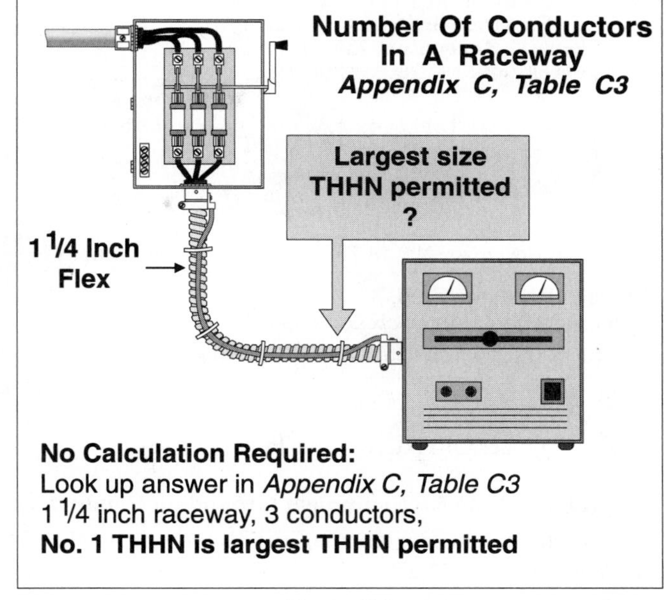

Figure 5-3

Number Of Conductors In A Raceway

Table C10A – Compact conductors and fixture wires in rigid PVC Conduit Schedule 40
Table C11 – Conductors and fixture wires in Type A, PVC Conduit
Table C11A – Compact conductors and fixture wires in Type A, PVC Conduit
Table C12 – Conductors and fixture wires in Type EB, PVC Conduit
Table C12A – Compact conductors and fixture wires in Type EB, PVC Conduit

❑ **Appendix C – Table C1 Example**

How many No. 14 RHH conductors (without cover) can be installed in a one inch electrical metallic tubing, Fig. 5–2?

(a) 25 conductors (b) 16 conductors (c) 13 conductors (d) 19 conductors

- Answer: (b) 16 conductors, Appendix C, Table C1

❑ **Appendix C – Table C2A – Compact Conductor Example**

How many compact No. 6 XHHW conductors can be installed in a 1¼ inch nonmetallic tubing?

(a) 10 conductors (b) 6 conductors (c) 16 conductors (d) 13 conductors

- Answer: (a) 10 conductors, Appendix C, Table C2A

❑ **Appendix C – Table C3 Example**

If we have a 1¼ inch flexible metal conduit and we want to install three THHN conductors (not compact), what is the largest conductor permitted to be installed, Fig. 5–3?

(a) No. 1 (b) No. 1/0 (c) No. 2/0 (d) No. 3/0

- Answer: (a) No. 1, Appendix C, Table C3

❑ **Appendix C – Table C4 Example**

How many No. 4/0 RHH conductors (with outer cover) can be installed in two inch intermediate metal conduit?

(a) 2 conductors (b) 1 conductors (c) 3 conductors (d) 4 conductors

- Answer: (c) 3 conductors, Appendix C, Table C4

❑ **Appendix C – Table C5 – Fixture Wire Example**

How many No. 18 TFFN conductors can be installed in a ¾ inch liquidtight flexible nonmetallic conduit gray type, Fig. 5–4?

(a) 40 (b) 26 (c) 30 (d) 39

- Answer: (d) 39, Appendix C, Table 5

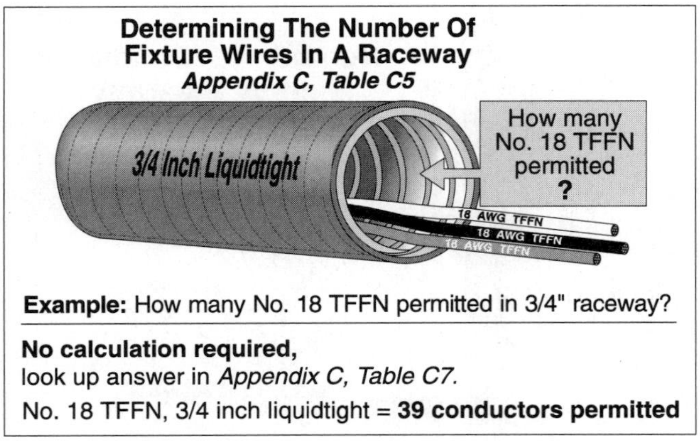

Determining The Number Of Fixture Wires In A Raceway
Appendix C, Table C5

3/4 Inch Liquidtight

How many No. 18 TFFN permitted ?

18 AWG TFFN
18 AWG TFFN
18 AWG TFFN

Example: How many No. 18 TFFN permitted in 3/4" raceway?

No calculation required,
look up answer in *Appendix C, Table C7.*
No. 18 TFFN, 3/4 inch liquidtight = **39 conductors permitted**

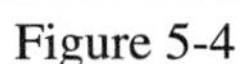

Figure 5-4

Number Of Fixture Wires In A Raceway

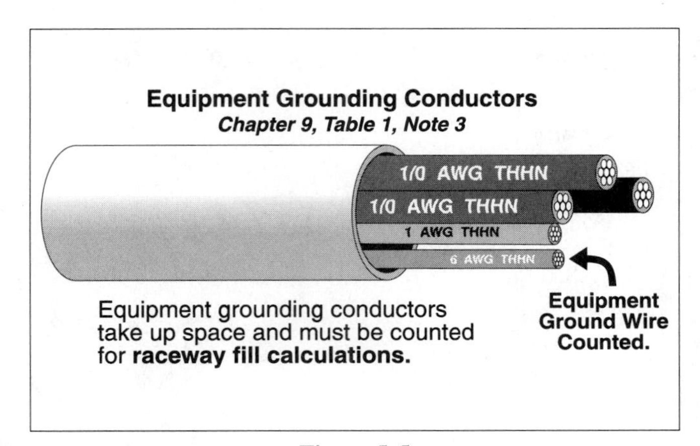

Figure 5-5

Equipment Grounding Conductors

Table 1, Note 3 – Equipment Grounding Conductors

When equipment grounding conductors are installed in a raceway, the actual area of the conductor must be used when calculating raceway fill. Chapter 9, Table 5 can be used to determine the cross-sectional area of insulated conductors, and Chapter 9, Table 8 can be used to determine the cross-sectional area of bare conductors [Note 8 or Table 1, Chapter 9], Fig. 5–5.

Table 1, Note 4 – Nipples, Raceways Not Exceeding 24 Inches

The cross-sectional areas of conduit and tubing can be found in Table 4 of Chapter 9. When a conduit or tubing raceway does not exceed 24 inches in length, it is called a nipple. Nipples are permitted to be filled to 60% of their total cross-sectional area, Fig. 5–6.

Table 1, Note 7

When the calculated number of conductors (all of the same size and insulation) results in 0.8 or larger, the next whole number can be used. But, be careful, this only applies when the conductors are all the same size and insulation. In effect, this note only applies for the development of Appendix C – Tables 1 through 13, and should not be used for exam purposes.

Table 1, Note 8

The dimensions for bare conductor are listed in Table 8 of Chapter 9.

Chapter 9, Table 4 – Conduit And Tubing Cross-Sectional Area

Table 4 of Chapter 9 lists the dimensions and cross-sectional area for conduit and tubing. The cross-sectional area of a conduit or tubing is dependent on the raceway type (cross-sectional area of the area) and the maximum percentage fill as listed in Table 1 of Chapter 9.

❑ **Conduit Cross-sectional Area Example**

What is the total cross-sectional area in square inches of a 1¼ inch rigid metal conduit that contains three conductors, Fig. 5–7?

(a) 1.063 square inches

(b) 1.526 square inches

(c) 1.098 square inches

(d) any of these

- Answer: (b) 1.526 square inches
  Chapter 9, Table 4

Chapter 9, Table 5 – Dimensions Of Insulated Conductors And Fixture Wires

Table 5 of Chapter 9 lists the cross-sectional area of insulated conductors and fixture wires.

Raceway Fill - "Nipples" At 60%
Chapter 9, Table 1, Note 4

A **nipple** is a raceway that is 24 inches or less. Nipples are permitted a 60% conductor fill.

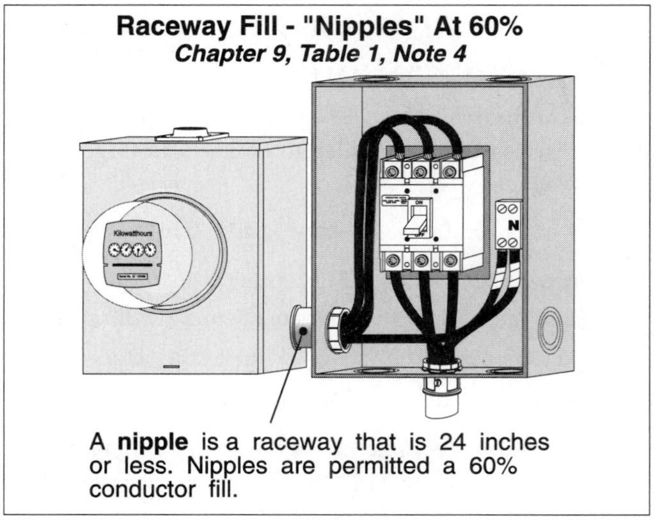

Figure 5-6

Raceway Fill – Nipples At 60%

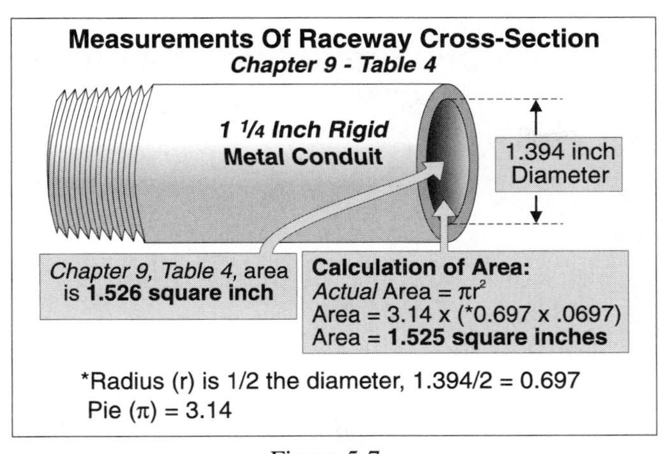

Figure 5-7

Measurements Of Raceway Cross-section

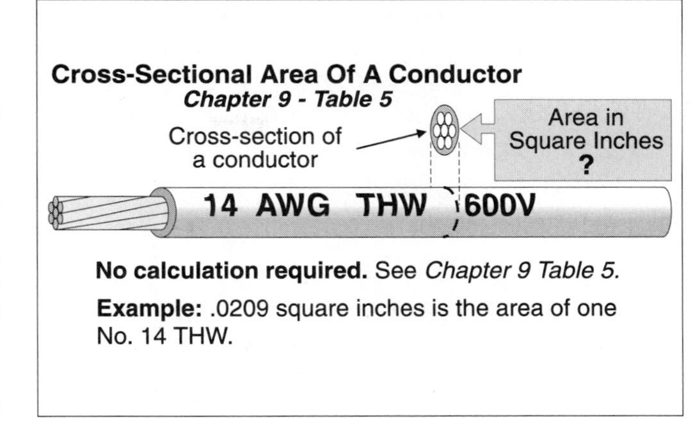

Figure 5-8

Conductor Cross-Sectional Area

Table 5 of Chapter 9 – Conductor Cross-Sectional Area							
	RH	RHH/RHW With *Cover*	RHH/RHW With *out Cover* or THW	TW	THHN THWN TFN	XHHW	BARE *Stranded Conductors*
Size AWG/kcmil	Approximate Cross-Sectional Area – Square Inches						
Column 1	Column 2	Column 3	Column 4	Column 5	Column 6	Column 7	Chapter 9, Table 8
14	.0209	.0293	.0209	.0139	.0097	.0139	**.004**
12	.0260	.0353	.0260	.0181	.0133	.0181	**.006**
10	.0437	.0437	.0333	.0243	.0211	.0243	**.011**
8	.0835	.0835	.0556	.0437	.0366	.0437	**.017**
6	.1041	.1041	.0726	.0726	.0507	.0590	**.027**
4	.1333	.1333	.0973	.0973	.0824	.0814	**.042**
3	.1521	.1521	.1134	.1134	.0973	.0962	**.053**
2	.1750	.1750	.1333	.1333	.1158	.1146	**.067**
1	.2660	.2660	.1901	.1901	.1562	.1534	**.087**
0	.3039	.3039	.2223	.2223	.1855	.1825	**.109**
00	.3505	.3505	.2624	.2624	.2223	.2190	**.137**
000	.4072	.4072	.3117	.3117	.2679	.2642	**.173**
0000	.4757	.4757	.3718	.3718	.3237	.3197	**.219**

❑ **Table 5 – THW Example**

What is the cross-sectional area (square inch) for one No. 14 THW conductor, Fig. 5–8?

(a) 0.0206 square inch (b) 0.0172 square inch

(c) 0.0209 square inch (d) 0.0278 square inch

- Answer: (c) 0.0209 square inch

❑ **Table 5 – RHW (With Outer Cover) Example**
What is the cross-sectional area (square inch) for one No. 12 RHW conductor (with outer cover)?

(a) 0.0206 square inch
(b) 0.0172 square inch
(c) 0.0353 square inch
(d) 0.0278 square inch

- Answer: (c) 0.0353 square inch

❑ **Table 5 – RHH (Without Outer Cover) Example**
What is the cross-sectional area (square inch) for one No. 10 RHH (without an outer cover)?

(a) 0.0117 square inch
(b) 0.0333 square inch
(c) 0.0252 square inch
(d) 0.0278 square inch

- Answer: (b) 0.0333 square inch

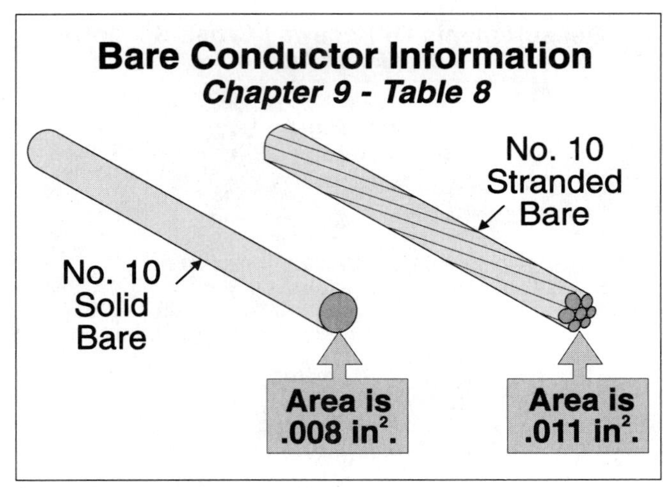

Figure 5-9

Bare Conductor Information

Chapter 9, Table 5 – Compact Aluminum Building Wire Nominal Dimensions And Areas
Tables 5A Chapter 9 list the cross-sectional area for compact aluminum building wires. We will not use these tables for this unit.

Chapter 9, Table 6 – None

Chapter 9, Table 7 – None

Chapter 9, Table 8 – Conductor Properties
Table 8 contains conductor properties such as: cross-sectional area in circular mils, number of strands per conductor, cross-sectional area in square inch for bare conductors, and the conductors resistance at 75°C for direct current for both copper and aluminum wire.

❑ **Bare Conductors Cross-Sectional Area – Example**
What is the cross-sectional area in square inches for one No. 10 bare conductor, Fig. 5–9?

(a) 0.008 Solid (b) 0.011 Stranded (c) 0.038 (d) a or b

- Answer: (d) 0.008 square inch for solid and 0.011 square inch for stranded

Chapter 9, Table 9 – AC Resistance For Conductors In Conduit Or Tubing
Table 9 contains the alternating current resistance for copper and aluminum conductors.

❑ **Alternating Current Resistance For Conductors Example**
What is the alternating current resistance of No. 1/0 copper installed in a metal conduit? Conductor length 1,000 feet.

(a) 0.12 ohm (b) 0.13 ohm (c) 0.14 ohm (d) 0.15 ohm

- Answer: (a) 0.12 ohm

Chapter 9, Table 10 – Expansion Characteristics Of PVC Rigid Nonmetallic Conduit
This table contains the expansion characteristics of PVC.

❑ **Expansion Characteristics Of PVC Rigid Nonmetallic Conduit Example**
What is the expansion characteristic (in inches) for 50 feet of PVC, when the ambient temperature change is 100°F?

(a) 1 inch (b) 2 inches (c) 3 inches (d) 4 inches

- Answer: (b) 2 inches

5–2 *RACEWAY AND NIPPLE CALCULATION*

Appendix C – Tables 1 through 13 cannot be used to determine raceway sizing when conductors of different sizes (or types of insulation) are installed in the same raceway. The following steps can be used to determine the raceway size and nipple size.

Step 1: → Determine the cross-sectional area (square inch) for each conductor from Table 5 of Chapter 9 for insulated conductors, and Table 8 of Chapter 9 for bare conductors.

Step 2: → Determine the total cross-sectional area for all conductors.

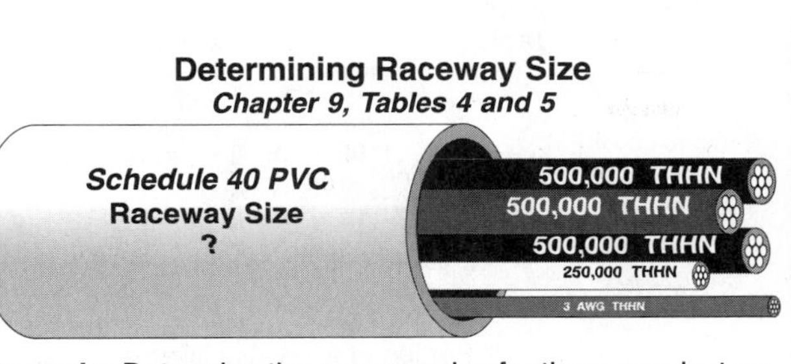

Example: Determine the raceway size for these conductors.

Determine the area in square inches of the conductors, *Chapter 9, Table 5.*

1- 500 kcmil THHN = .7073 in² x 3 conductors = 2.1219 in²
1- 250 kcmil THHN = .3970 in² x 1 conductor = .3970 in²
1- No. 3 THHN = .0973 in² x 1 conductor = .0973 in²
Total area of the conductors = 2.6162 in²

Chapter 9, Table 4, assume 40% fill (*Chapter 9, Table 1).*
3 inch raceway required at 2.907 square inches.

Figure 5-10

Determining Raceway Size

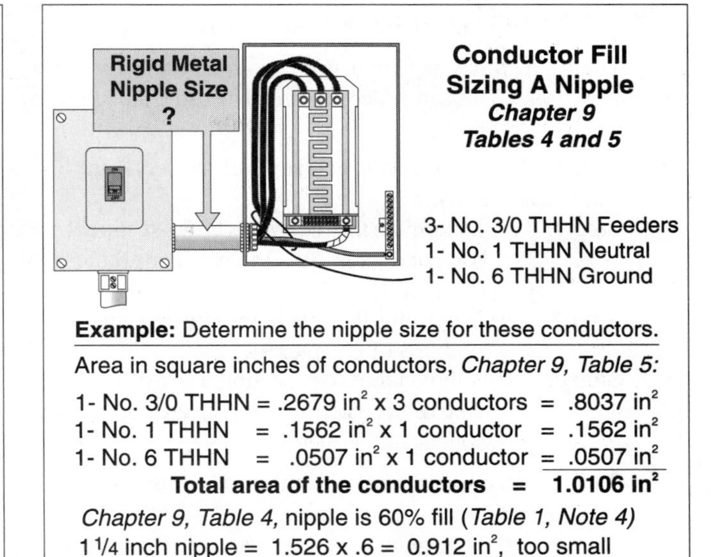

Example: Determine the nipple size for these conductors.

Area in square inches of conductors, *Chapter 9, Table 5:*

1- No. 3/0 THHN = .2679 in² x 3 conductors = .8037 in²
1- No. 1 THHN = .1562 in² x 1 conductor = .1562 in²
1- No. 6 THHN = .0507 in² x 1 conductor = .0507 in²
Total area of the conductors = 1.0106 in²

Chapter 9, Table 4, nipple is 60% fill (*Table 1, Note 4)*
1 1/4 inch nipple = 1.526 x .6 = 0.912 in², too small
1 1/2 inch nipple = 2.071 x .6 = 1.242 in², just right
2 inch nipple = 3.408 x .6 = 2.0448 in², too big

Figure 5-11

Conductor Fill – Sizing A Nipple

Step 3: → Size the raceway according to Table 4–4 using the percent fill as listed in Table 1 of Chapter 9:
40% for three or more conductors
60% for raceways 24 inches or less in length (nipples).

❑ **Raceway Size Example**

A 400-ampere feeder is installed in schedule 40 rigid nonmetallic conduit. This raceway contains three 500-kcmil THHN conductors, one 250-kcmil THHN conductor, and one No. 3 THHN conductor. What size raceway is required for these conductors, Fig. 5–10?

(a) 2 inch (b) 2½ inch (c) 3 inch (d) 3½ inch

• Answer: (c) 3 inch

Step 1: → Cross- sectional area of the conductors, Table 5 of Chapter 9.
500 kcmil THHN = 0.7073 square inch × 3 wires = 2.1219 square inches
250 kcmil THHN = 0.3970 square inch × 1 wire = 0.3970 square inch
No. 3 THHN = 0.0973 square inch × 1 wire = 0.0973 square inch

Step 2: → Total cross-sectional area of all conductors = 2.6162 square inches

Step 3: → Size the conduit at 40% fill [Chapter 9, Table 1] using Table 4,
3 inch schedule 40 PVC has a cross-sectional area of 2.907 square inches permitted for conductor fill.

❑ **Nipple Size Example**

What size rigid metal nipple is required for three No. 3/0 THHN conductors, one No. 1 THHN conductor and one No. 6 THHN conductor, Fig. 5–11?

(a) 2 inch (b) 1½ inch (c) 1¼ inch (d) None of these

• Answer: (b) 1½ inch

Step 1: → Cross-sectional area of the conductors, Table 5 of Chapter 9.
No. 3/0 THHN = 0.2679 square inch × 3 wires = 0.8037 square inch
No. 1 THHN = 0.1562 square inch × wire = 0.1562 square inch
No. 6 THHN = 0.0507 square inch × 1 = 0.0507 square inch

Step 2: → Total cross-sectional area of the conductors = 1.0106 square inches

Step 3: → Size the conduit at 60% fill [Table 1, Note 4 of Chapter 9] using Table 4.
1¼ inch nipple = 1.526 × 0.6 = 0.912 square inch, too small
1½ inch nipple = 2.07 × 0.6 = 1.242 square inches, *just right*
2 inch nipple = 3.4 08 × 0.6 = 2.0448 square inches, too big

5–3 EXISTING RACEWAY CALCULATION

There are times that you need to add conductors to an existing raceway. This can be accomplished by using the following steps:

Part 1 – Determine the raceway's cross-sectional spare space area.

Step 1: → Determine the raceway's cross-sectional area for conductor fill [Table 1 and Table 4 of Chapter 9].

Step 2: → Determine the area of the existing conductors [Table 5 of Chapter 9].

Step 3: → Subtract the cross-sectional area of the existing conductors (Step 2) from the area of permitted conductor fill (Step 1).

Part 2 – To determine the number of conductors permitted in spare space area:

Step 4: → Determine the cross-sectional area of the conductors to be added [Table 5 of Chapter 9 for insulated conductors and Table 8 of Chapter 9 for bare conductors].

Step 5: → Divide the spare space area (Step 3 – Part 1) by the conductors cross-sectional area (Step 4 – Part 2).

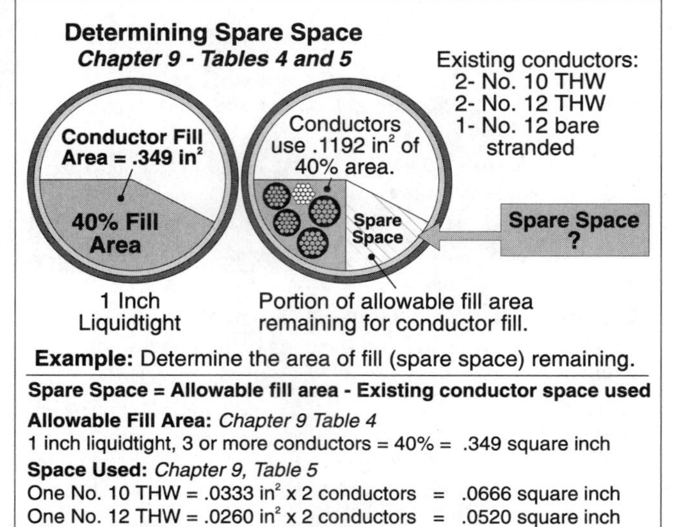

Figure 5-12

Determining Raceway Spare Space Area

❑ **Spare Space Area Example**

An existing one inch liquidtight flexible metallic conduit contains two No. 12 THW conductors, two No. 10 THW conductors, and one No. 12 bare (stranded). What is the area remaining for additional conductors, Fig. 5–12?

(a) 0.2264 square inch (if the raceway is more than 24 inches long)

(b) 0.3992 square inch (if the raceway is less than 24 inches long)

(c) There is no spare space.

(d) a and b are correct

- Answer: (d) a and b

Step 1: → Cross-sectional area permitted for conductor fill [Table 1, Note 4 and Table 4 of Chapter 9]
At 40% = 0.864 × 0.4 = 0.3456 square inch
Nipple at 60% = 0.864 × 0.6 = 0.5184 square inch

Step 2: → Cross-sectional area of existing conductors.
No. 12 THW 0.0260 square inch × 2 wires = 0.0520 square inch
No. 10 THW 0.0333 square inch × 2 wires = 0.0666 square inch
No. 12 bare 0.006 square inch × 1 wire = 0.006 square inch
(Ground wires must be counted for raceway fill – Table 1, Note 3 of Chapter 9.)
Total cross-sectional area of existing conductors = 0.1192 square inch

Step 3: → Subtract the area of the existing conductors from the permitted area of conductor fill.
Raceway more than 24 inches long = 0.3456 – 0.1192 = 0.2264 square inch
Nipple (less than 24 inches long) = 0.5184 – 0.1192 = 0.3992 square inch

❑ **Number Of Conductors Permitted In Spare Space Area Example**

An existing one-inch liquidtight flexible metallic conduit contains two No. 12 THW conductors, two No. 10 THW conductors, and one No. 12 bare (stranded). How many additional No. 10 THHN conductors can be added to this raceway, Fig. 5–13?

(a) 10 conductors if the raceway is more than 24 inches long

(b) 18 conductors if the raceway is less than 24 inches long

(c) 15 conductors regardless of the raceway length

(d) a and b are correct

- Answer: (d) both a and b are correct

Step 1: → Cross-sectional area permitted for conductor fill [Table 1, Note 4 and Table 4 of Chapter 9]
At 40%) = .864 × 0.4 = 0.3456 square inch
Nipple at 60% = 0.864 × 0.6 = .5184 square inch

Step 2: → Cross-sectional area of existing conductors.
No. 12 THW 0.0260 square inch × 2 wires = 0.0520 square inch
No. 10 THW 0.0333 square inch × 2 wires = 0.0666 square inch
No. 12 bare 0.006 square inch × 1 wire = 0.006 square inch
(Ground wires must be counted for raceway fill – Table 1, Note 3 of Chapter 9.)
Total cross-sectional area of existing conductors = 0.1192 square inch

Step 3: → Subtract the area of the existing conductors from the permitted area of conductor fill.
Raceway more than 24 inches long = 0.3456 – 0.1192 = 0.2264 square inch
Nipple (less than 24 inches long) = 0.5184 – 0.1192 = 0.3992 square inch

Step 3: → Cross-sectional area of the conductors to be installed [Table 5 of Chapter 9]. No. 10 THHN = 0.0211 square inch.

Step 3: → Divide the spare space area (Step 3) by the conductor area.
Raceway at 40% = 0.2264 square inch/0.0211 = 10.7 or 10 conductors
Nipple at 60% = 0.3992 square inch/0.0211 = 18.92 or 18 conductors. We must round down to 18 conductors because Note 8, Table 1 of Chapter 9 only applies if all of the conductors are the same size and same insulation.

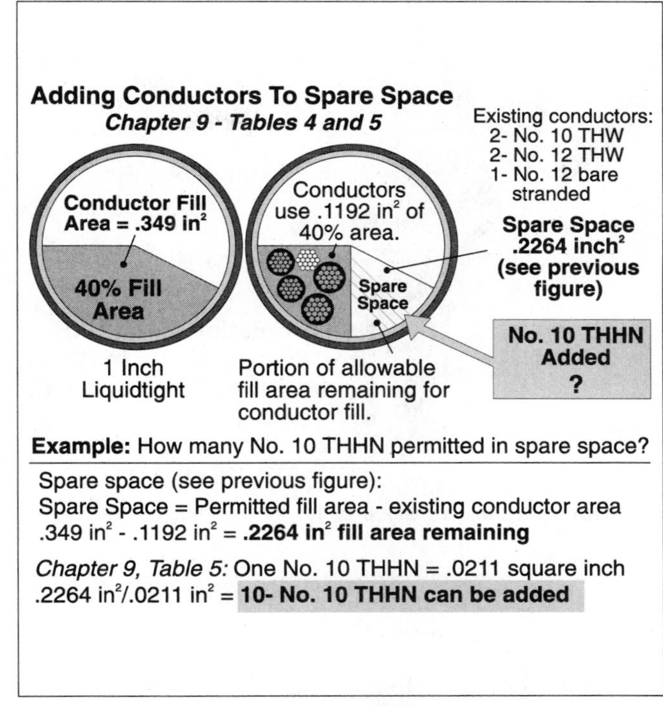

Figure 5-13

Adding Conductors to Spare Space

5–4 *TIPS FOR RACEWAY CALCULATIONS*

Tip 1: → Take your time
Tip 2: → Use a ruler or straight-edge when using tables
Tip 3: → Watch out for the different types of raceways and conductor insulations, particularly RHH/RHW with or without outer cover.

PART B – *OUTLET BOX FILL CALCULATIONS*

INTRODUCTION [370-16]

Boxes shall be of sufficient size to provide free space for all conductors. An outlet box is generally used for the attachment of devices and fixtures and has a specific amount of space (volume) for conductors, devices, and fittings. The volume taken up by conductors, devices, and fittings in a box must not exceed the box fill capacity. The volume of a box is the total volume of its assembled parts, including plaster rings, industrial raised covers, and extension rings. The total volume includes only those fittings that are marked with their volume in cubic inches [370-16(a)].

5–5 *SIZING BOX – CONDUCTORS ALL THE SAME SIZE [Table 370–16(a)]*

When all of the conductors in an *outlet box* are the same size (insulation doesn't matter), Table 370–16(a) of the National Electrical Code can be used to:

(1) Determine the number of conductors permitted in the outlet box, or
(2) Determine the size outlet box required for the given number of conductors.

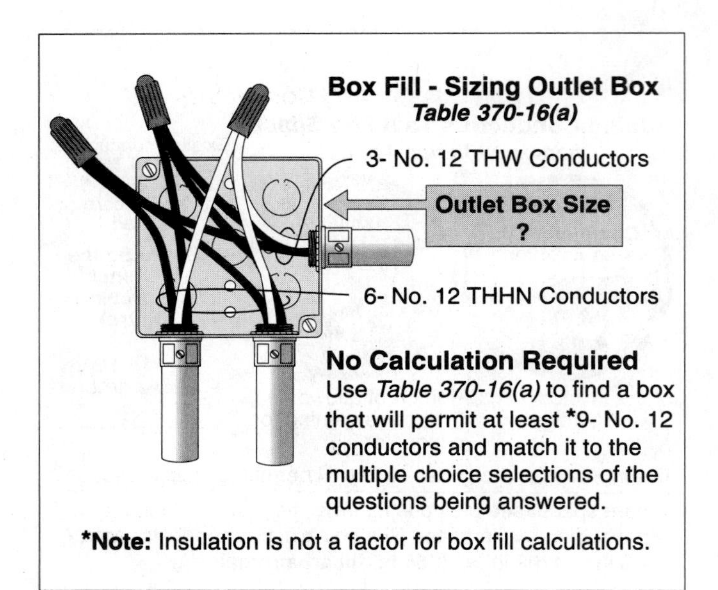

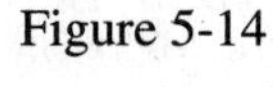

Figure 5-14

Sizing Outlet Box

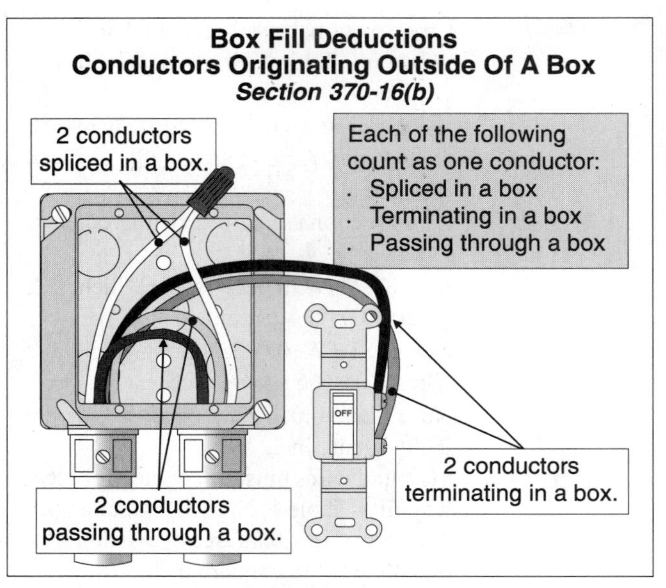

Figure 5-15

Box Fill Deductions

Note: Table 370–16(a) applies only if the outlet box contains no switches, receptacles, fixture studs, fixture hickeys, manufactured cable clamps, or grounding conductors (not likely).

❑ **Outlet Box Size Example**

What size outlet box is required for six No. 12 THHN conductors, and three No. 12 THW conductors, Fig. 5–14?

(a) 4 inches × 1¼ inches square (b) 4 inches × 1½ inches square

(c) 4 inches × 1¼ inches round (d) 4 inches × 1½ inches round

- Answer: (b) 4 inches × 1½ inches square,

  Table 370–16(a) permits nine No. 12 conductors, insulation is not a factor.

❑ **Number of Conductors in Outlet Box Example**

Using Table 370-16(a), how many No. 14 THHN conductors are permitted in a 4 inch × 1½ inch round box?

(a) 7 conductors (b) 9 conductors

(c) 10 conductors (d) 11 conductors

- Answer: (c) 7 conductors

5–6 *CONDUCTOR EQUIVALENTS [370–16(b)]*

Table 370–16(a) does not take into consideration the fill requirements of clamps, support fittings, devices or equipment grounding conductors within the outlet box. In no case can the volume of the box and its assembled sections be less than the fill calculation as listed below:

(1) Conductor Fill – Conductors Running Through. Each conductor that originates outside the box and terminates or is spliced within the box is considered as one conductor, Fig. 5–15.

Conductor Running Through The Box. Each conductor that runs through the box is considered as one conductor, Fig. 5–15. Conductors, no part of which leaves the box, shall not be counted, this includes equipment bonding jumpers and pigtails, Fig. 5–16.

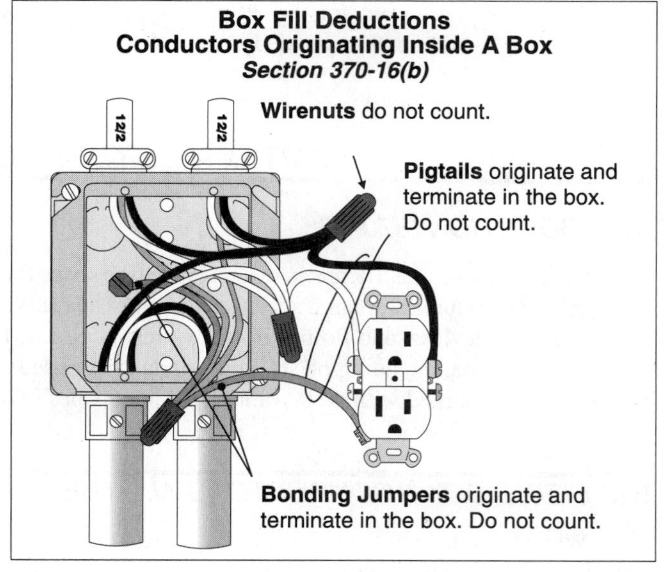

Figure 5-16

Box Fill Deductions

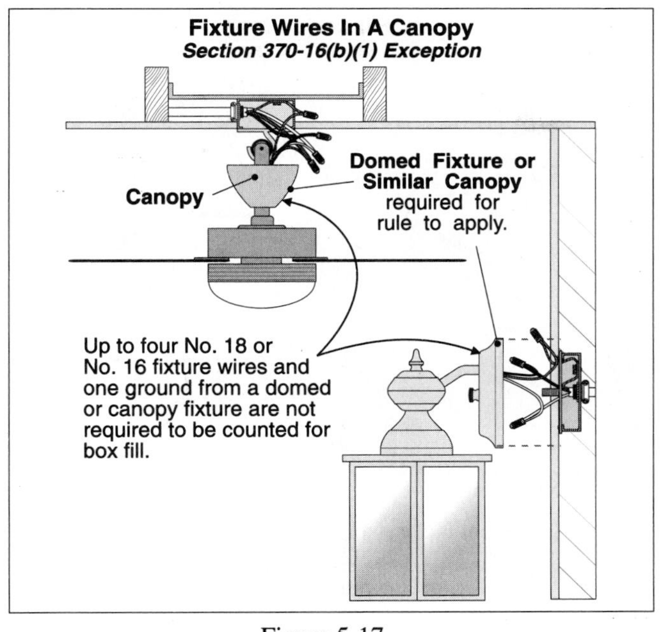

Figure 5-17

Fixture Wires In A Canopy

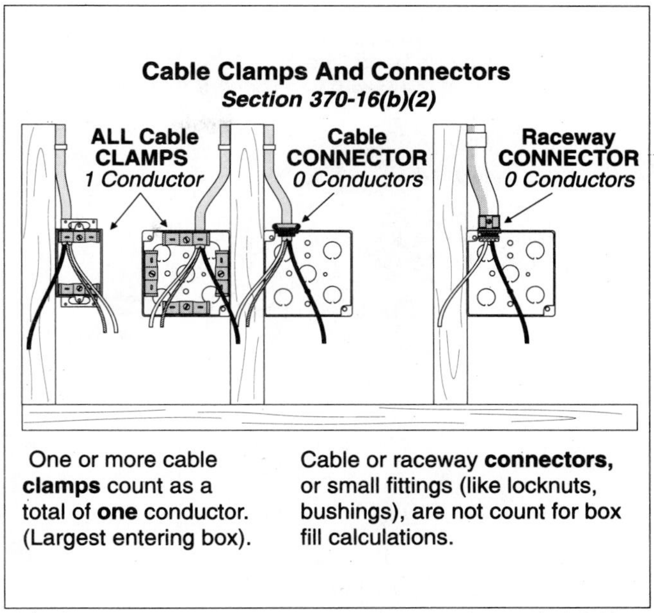

Cable Clamps And Connectors

Section 370-16(b)(2)

ALL Cable CLAMPS *1 Conductor*

Cable CONNECTOR *0 Conductors*

Raceway CONNECTOR *0 Conductors*

One or more cable **clamps** count as a total of **one** conductor. (Largest entering box).

Cable or raceway **connectors,** or small fittings (like locknuts, bushings), are not count for box fill calculations.

Figure 5-18

Cable Clamps And Connectors

Exception. Fixture wires smaller than No. 14 from a domed fixture or similar canopy are not counted, Fig. 5–17.

(2) Cable Fill. One or more internal cable clamps in the box are considered as one conductor volume in accordance with the volume listed in Table 370-16(b), based on the largest conductor that enters the outlet box, Fig. 5–18.

Note: Small fittings such as locknut and bushings are not counted [370-16(b)].

(3) Support Fittings Fill. One or more fixture studs or hickeys within the box are considered as one conductor volume in accordance with the volume listed in Table 370–16(b), based on the largest conductor that enters the outlet box, Fig. 5–19.

(4) Device or Equipment Fill. Each yoke of strap containing one or more devices or equipment is considered as two conductors volume in accordance with the volume listed in Table 370-16(b), based on the largest conductor that terminates on the yoke, Fig. 5–20.

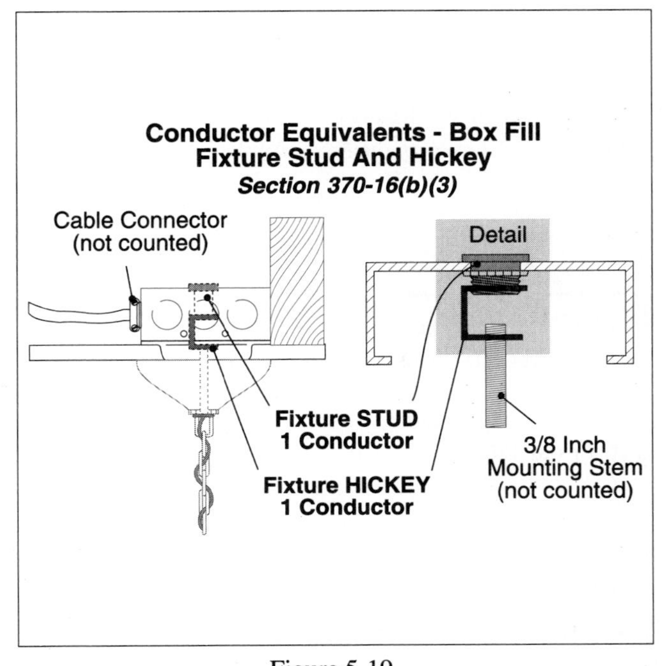

Figure 5-19

Box Fill – Fixture Stud And Hickey

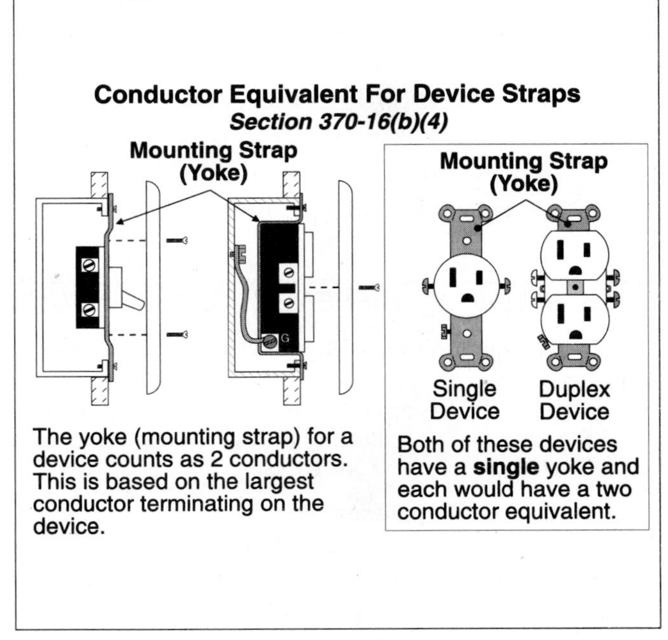

Figure 5-20

Conductor Equivalent For Device Straps

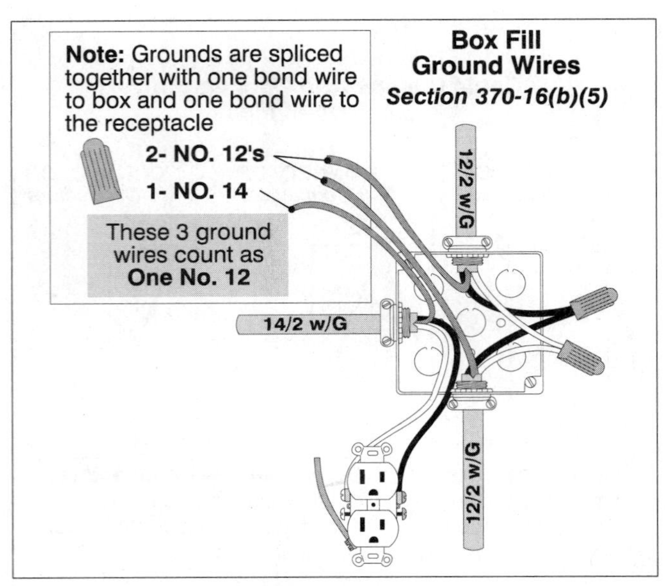

Figure 5-21

Box Fill – Ground Wires

Box Fill Deductions
Section 370-16(a)

14/2 & 14/3 Counts as 5 CONDUCTORS.
Box comes with four cable clamps. Counts as 1 CONDUCTOR.
1- Strap (Yoke) Counts as 2 CONDUCTORS.
1- Strap (Yoke) Counts as 2 CONDUCTORS.
2- Ground Wires Counts as 1 CONDUCTOR.
Wirenut & Pigtail Counts as 0 CONDUCTOR.
Box has the equivalent of 11- No. 14 conductors.

Figure 5-22

Box Fill Deductions

(5) Grounding Conductors. One or more grounding conductors are considered as one conductor volume in accordance with the volume listed in Table 370–16(b), based on the largest grounding conductor that enters the outlet box, Fig. 5–21.

Note: Fixture ground wires smaller than No. 14 from a domed fixture or similar canopy are not counted [370-16(b)(1) Exception].

What's Not Counted

Wirenuts, cable connectors, raceway fittings, and conductors that originate and terminate within the outlet box (such as equipment bonding jumpers and pigtails) are not counted for box fill calculations [370–16(a)].

❑ **Number of Conductors Example**

What is the total number of conductors used for box fill calculations in Figure 5–22?

(a) 5 conductors (b) 7 conductors

(c) 9 conductors (d) 11 conductors

- Answer: (d) 11 conductors

  Switch – Five No. 14 conductors, two conductors for device and three conductors terminating.

  Receptacle – Four No. 12 conductors, two conductors for the device and two conductors terminating.

  Ground Wire – One conductor.

  Cable Clamps – One conductor.

5–7 *SIZING BOX – DIFFERENT SIZE CONDUCTORS [370–16(b)]*

To determine the size of the outlet box when the conductors are of different sizes (insulation is not a factor), the following steps can be used:

Step 1: → Determine the number and size of the conductors' equivalents in the box.
Step 2: → Determine the volume of the conductors' equivalents from Table 370–16(b).
Step 3: → Size the box by using Table 370–16(a).

❑ **Outlet Box Sizing Example**

What size outlet box is required for 14/3 romex (with ground) that terminates on a switch, with 14/2 romex that terminates on a receptacle, if the box has internal cable clamps factory installed, Fig. 5–22?

(a) 4 inches × 1¼ inch square (b) 4 inches × 1½ inch square

(c) 4 inches × 2⅛ inch square (d) any of these

- Answer: (c) 4 inches × 2⅛ inch square

Step 1: → Determine the number and size of conductors.
14/3 romex – three No. 14's
14/2 romex – two No. 14's
Cable Clamps – one No. 14
Switch – two No. 14's
Receptacles – two No. 14's
Ground wires – one No. 14
Total – eleven No. 14's

Step 2: → Determine the volume of the conductors [Table 370–16(b)].
No. 14 = 2 cubic inches
Eleven No. 14 conductors = 11 wires × 2 cubic inches = 22 cubic inches

Step 3: → Select the outlet box from Table 370–16(a).
4 inches × 1¼ inch square, 18 cubic inches, too small
4 inches × 1½ inch square, 21 cubic inches, *just right*
4 inches × 2⅛ inch square, 30.3 cubic inches, larger than necessary

❑ **Domed Fixture Canopy Example [370–16(a), Exception].**

A round 3 inch × ½ inch box has a total volume of 6 cubic inches and has factory internal cable clamps. Can this pancake box be used with a lighting fixture that has a domed canopy? The branch circuit wiring is 14/2 nonmetallic sheath cable and the paddle fan has three fixture wires and one ground wire all smaller than No. 14, Fig. 5–23.

(a) Yes (b) No

- Answer: (a) No,

The box is limited to 6 cubic inches, and the conductors total 8 cubic inches [370–16(a)].

Step 1: → Determine the number and size of conductors within the box.
14/2 romex – two No. 14's
Cable clamps – one No. 14
Ground wire – one No. 14
Total – four No. 14's

Step 2: → Determine the volume of the conductors [Table 370–16(b)].
No. 14 = 2 cubic inches
Four No. 14 conductors = 4 wires × 2 cubic inches = 8 cubic inches.

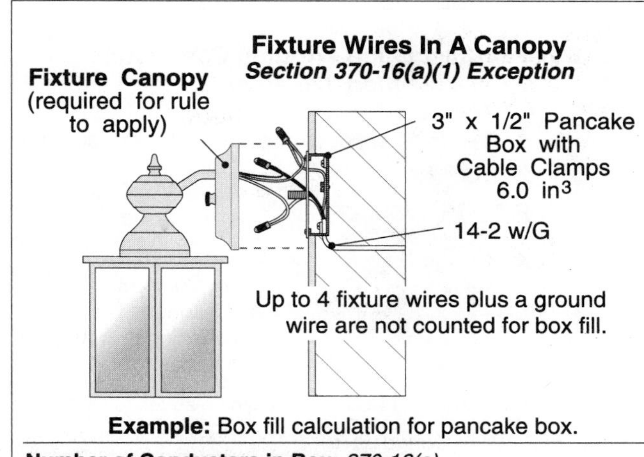

Example: Box fill calculation for pancake box.

Number of Conductors in Box: *370-16(a)*
14-2 NM Cable = 2 conductors
14-2 Ground = 1 conductor
Cable Clamps = 1 conductor
Fixture Wires = 0 conductors, *(370-16(a)(1) Exception)*
Total = 4- No. 14 conductors

Volume of Conductors: *Table 370-16(b)*
1- No. 14 = 2 cubic inches x 4 conductors = **8 cubic inches**

Volume of box is **6 cubic** inches (too small).
Installation is a **VIOLATION.**

Figure 5-23

Fixture Wires In A Canopy

❑ **Conductors Added To Existing Box Example**

How many No. 14 THHN conductors can be pulled through a 4 inch square × 2⅛-inch deep box that has a plaster ring of 3.6 cubic inches? The box already contains two receptacles, five No. 12 THHN conductors, and one No. 12 bare grounding conductor, Fig. 5–24.

(a) 4 conductors (b) 5 conductors (c) 6 conductors (d) 7 conductors

- Answer: (b) 5 conductors

Step 1: → Determine the number and size of the existing conductors.
Two Receptacles – 2 yokes × 2 conductors each = four No. 12 conductors
Five No. 12's – five No. 12 conductors
One ground – one No. 12 conductor
Total conductors – ten No. 12 conductors

Step 2: → Determine the volume of the existing conductors [Table 370–16(b)].
No. 12 conductor = 2.25 cubic inches, 10 wires × 2.25 cubic inches = 22.5 cubic inches

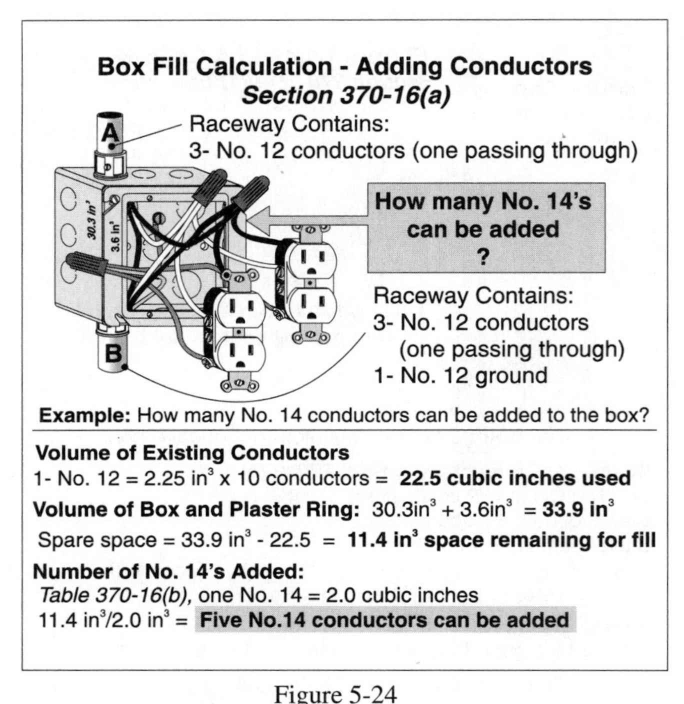

Figure 5-24

Box Fill Calculation – Adding Conductors

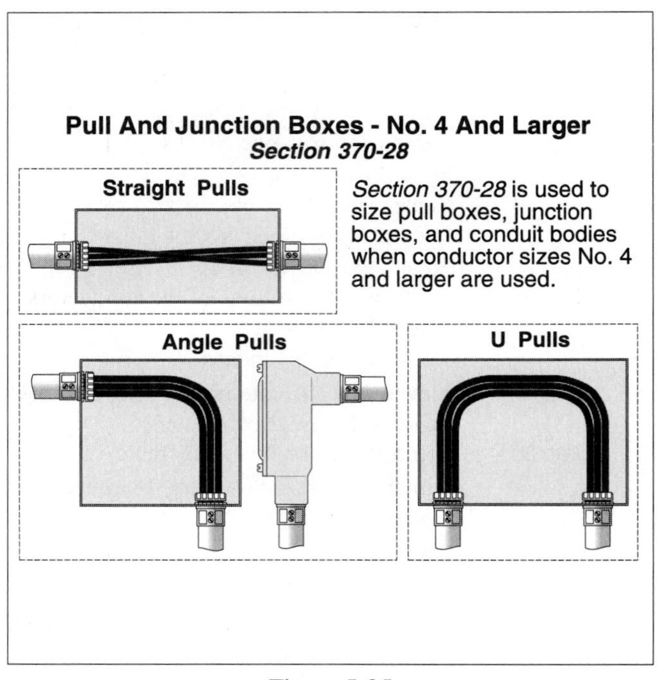

Figure 5-25

Pull And Junction Boxes – No. 4 And Larger

Step 3: → Determine the space remaining for the additional No. 14 conductors.
Remaining space = Total space less existing conductors
Total space = 30.3 cubic inches (box) [Table 370–16(a)] + 3.6 cubic inches (ring) = 33.9 cubic inches
Remaining space = 33.9 cubic inches – 22.5 cubic inches (ten No. 12 conductors)
Remaining space = 11.4 cubic inches

Step 4: → Determine the number of No. 14 conductors permitted in the spare space.
Conductors added = Remaining space/added conductors volume
Conductors added = 11.4 cubic inches/2 cubic inches [Table 370–16(b)]
Conductors added = five No. 14 conductors.

PART C – *PULL, JUNCTION BOXES, AND CONDUIT BODIES*

INTRODUCTION

Pull boxes, *junction boxes*, and *conduit bodies* must be sized to permit conductors to be installed so that the conductor insulation will not be damaged. For conductors No. 4 and larger, we must size pull boxes, junction boxes, and conduit bodies according to the requirements of Section 370–28 of the National Electrical Code, Fig. 5–25.

5–8 *PULL AND JUNCTION BOX SIZE CALCULATIONS*

Straight Pull Calculation [370–28(a)(1)]

A straight pull calculation applies when conductors enter one side of a box and leave through the opposite wall of the box. The minimum distance from where the raceway enters to the opposite wall must not be less than eight times the trade size of the largest raceway, Fig. 5–26.

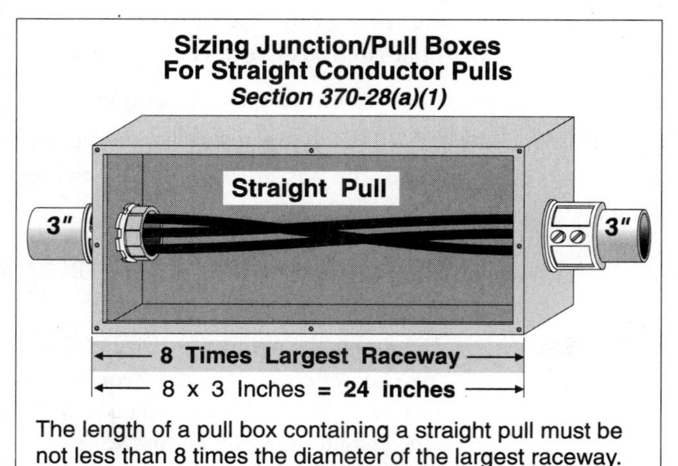

Figure 5-26

Straight Pull Sizing – 370–28(a)(1)

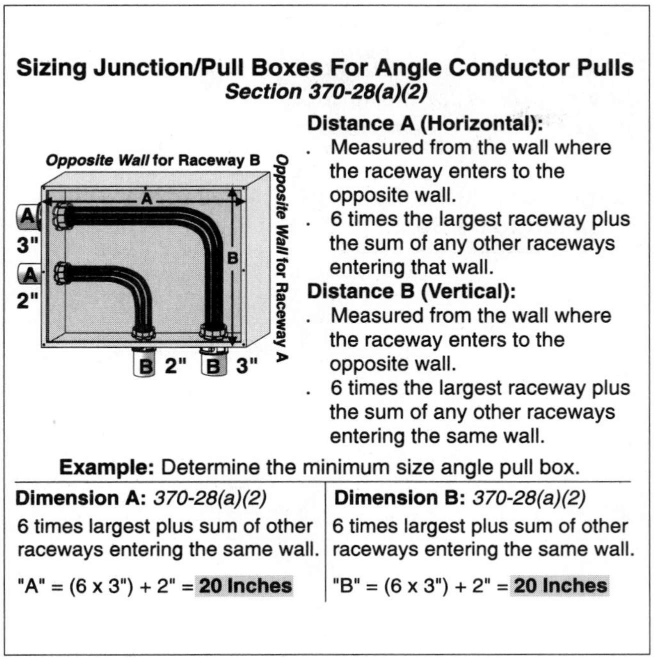

Sizing Junction/Pull Boxes For Angle Conductor Pulls
Section 370-28(a)(2)

Opposite Wall for Raceway B
Opposite Wall for Raceway A
A 3"
A 2"
B 2" B 3"

Distance A (Horizontal):
- Measured from the wall where the raceway enters to the opposite wall.
- 6 times the largest raceway plus the sum of any other raceways entering that wall.

Distance B (Vertical):
- Measured from the wall where the raceway enters to the opposite wall.
- 6 times the largest raceway plus the sum of any other raceways entering the same wall.

Example: Determine the minimum size angle pull box.

Dimension A: *370-28(a)(2)*	**Dimension B:** *370-28(a)(2)*
6 times largest plus sum of other raceways entering the same wall.	6 times largest plus sum of other raceways entering the same wall.
"A" = (6 x 3") + 2" = **20 Inches**	"B" = (6 x 3") + 2" = **20 Inches**

Figure 5-27

Angle Pull Sizing – 270–28(a)(2)

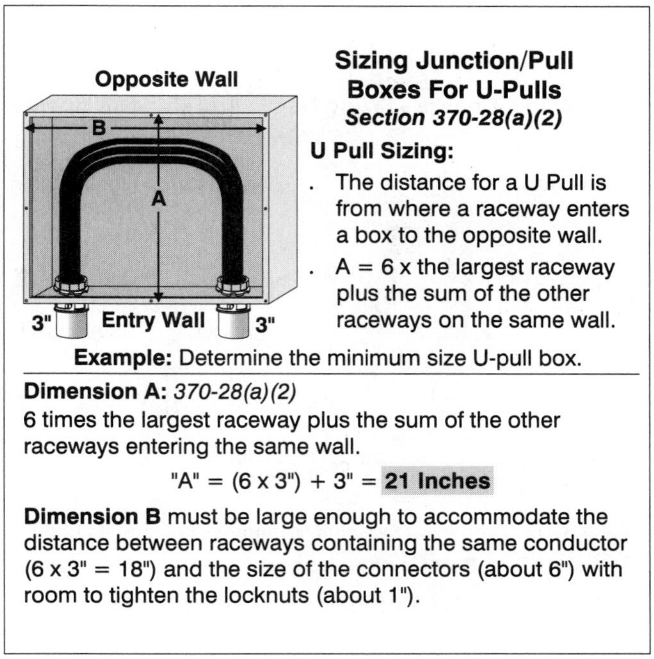

Sizing Junction/Pull Boxes For U-Pulls
Section 370-28(a)(2)

Opposite Wall
B
A
3" Entry Wall 3"

U Pull Sizing:
- The distance for a U Pull is from where a raceway enters a box to the opposite wall.
- A = 6 x the largest raceway plus the sum of the other raceways on the same wall.

Example: Determine the minimum size U-pull box.

Dimension A: *370-28(a)(2)*
6 times the largest raceway plus the sum of the other raceways entering the same wall.

"A" = (6 x 3") + 3" = **21 Inches**

Dimension B must be large enough to accommodate the distance between raceways containing the same conductor (6 x 3" = 18") and the size of the connectors (about 6") with room to tighten the locknuts (about 1").

Figure 5-28

Sizing Junction/Pull Boxes For U-Pulls

Angle Pull Calculation [370–28(a)(2)]

An angle pull calculation applies when conductors enter one wall and leave the enclosure not opposite the wall of the conductor entry. The distance for angle pull calculations from where the raceway enters to the opposite wall must not be less than six times the trade diameter of the largest raceway, plus the sum of the diameters of the remaining raceways on the same wall *and row*, Fig. 5–27. When there is more than one *row*, each row shall be calculated separately and the row with the largest calculation shall be considered the minimum angle pull dimension.

U–Pull Calculations [370–28(a)(2)]

A U–pull calculation applies when the conductors enter and leave from the same wall. The distance from where the raceways enter to the opposite wall must not be less than six times the trade diameter of the largest raceway, plus the sum of the diameters of the remaining raceways on the same wall, Fig. 5–28.

Distance Between Raceways Containing The Same Conductor Calculation [370–28(a)(2)]

After sizing the pull box, the raceways must be installed so that the distance between raceways enclosing the same conductors shall not be less than six times the trade diameter of the largest raceway. This distance is measured from the nearest edge of one raceway to the nearest edge of the other raceway, Figure 5–29.

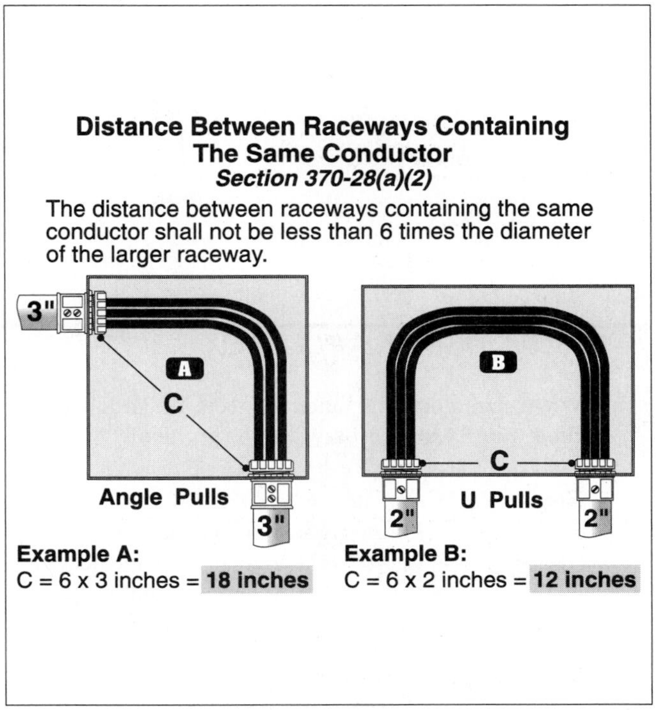

Distance Between Raceways Containing The Same Conductor
Section 370-28(a)(2)

The distance between raceways containing the same conductor shall not be less than 6 times the diameter of the larger raceway.

3" A C Angle Pulls 3"

B C U Pulls 2" 2"

Example A:
C = 6 x 3 inches = **18 inches**

Example B:
C = 6 x 2 inches = **12 inches**

Figure 5-29

Distance Between Raceways Containing The Same Conductor

5–9 *DEPTH OF BOX AND CONDUIT BODY SIZING [370–28(a)(2), Exception]*

When conductors enter an enclosure opposite a *removable cover,* such as the back of a pull box or conduit body, the distance from where the conductors enter to the removable cover shall not be less than the distances listed on Table 373–6(a); one wire per terminal, Fig. 5–30.

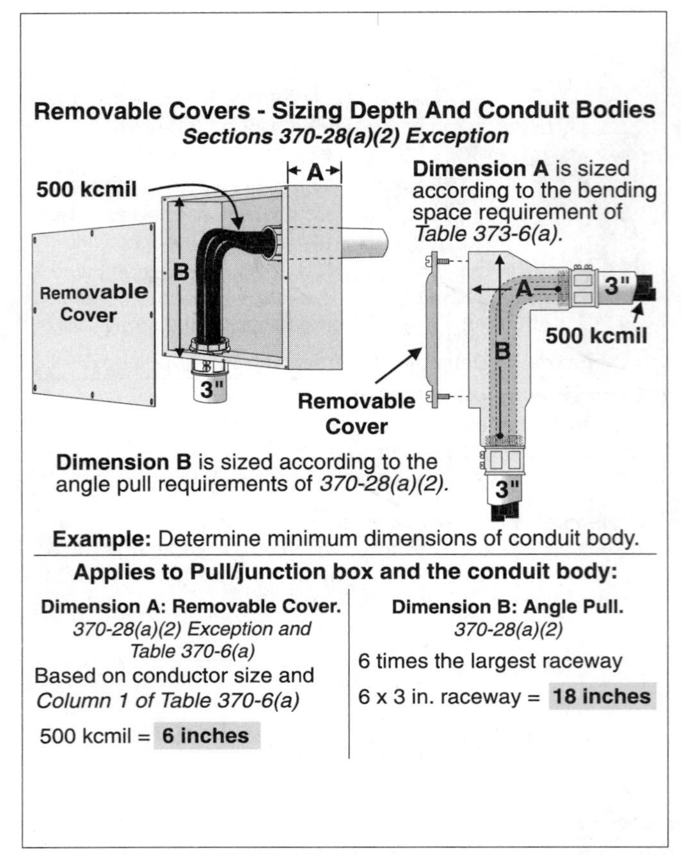

Figure. 5-30

Removable Covers – Sizing Depth And Conduit Bodies

TABLE 373-6(a) - Depth of Box or Conduit Body Distance measured from where conductors enter to removable cover

Largest Conductor Required To Be Bent	Minimum Depth
No. 4 through No. 2/0	Less than 4 inches
No. 3/0 and No. 4/0	4 inches
250 kcmil	4½ inches
300 - 350 kcmil	5 inches
400 - 500 kcmil	6 inches
600 - 900 kcmil	8 inches
1,000 - 1,250 kcmil	10 inches
1,500 - 2,000 kcmil	12 inches

❑ Depth Of Pull Or Junction Box Example

A 24 inch × 24 inch pull box has two 2-inch conduits that enter the back of the box with 4/0 conductors. What is the minimum depth of the box?

(a) 4 inches (b) 6 inches (c) 8 inches (d) 10 inches

- Answer: (a) 4 inches, Table 373-6(a)

5–10 *JUNCTION AND PULL BOX SIZING TIPS*

When sizing pull and junction boxes, the following suggestions should be helpful.

Step 1: → Always draw out the problem.

Step 2: → Calculate the HORIZONTAL distance(s):
- Left to right straight calculation
- Left to right angle or U-pull calculation
- Right to left straight calculation
- Right to left angle or U-pull calculation

Step 3: → Calculate the VERTICAL distance(s):
- Top to bottom straight calculation
- Top to bottom angle or U-pull calculation
- Bottom to top straight calculation
- Bottom to top angle or U-pull calculation

5–11 *PULL BOX EXAMPLES*

❑ Pull Box Sizing Example No. 1

A junction box contains two 3-inch raceways on the left side and one 3-inch raceway on the right side. The conductors from one of the 3-inch raceways on the left wall are pulled through the 3-inch raceway on the right wall. The other 3-inch raceway's conductors are pulled through a raceway at the bottom of the pull box, Fig. 5–31.

➛ Horizontal Dimension Question A

What is the horizontal dimension of this box?

(a) 18 inches (b) 21 inches (c) 24 inches (d) none of these

- Answer: (c) 24 inches, Section 370-28, Figure 31, Part A

  Left wall to the right wall angle pull = $(6 \times 3$ inch$) + 3$ inch $= 21$ inches

  Left wall to the right wall straight pull = 8×3 inch $= 24$ inches

  Right wall to left wall angle pull = No calculation

  Right wall to the left wall straight pull = 8×3 inch $= 24$ inches

➛ Vertical Dimension Question B

What is the vertical dimension of this box?

(a) 18 inches (b) 21 inches (c) 24 inches (d) none of these

- Answer: (a) 18 inches, Section 370-28, Figure 31, Part B

  Top to bottom wall angle pull = No calculation

  Top to bottom wall straight pull = No calculation

  Bottom to top wall angle pull = $(6 \times 3) = 18$ inches

  Bottom to top wall straight pull = No calculation

➛ Distance Between Raceways Question C

What is the minimum distance between the two 3-inch raceways that contain the same conductors?

(a) 18 inches (b) 21 inches

(c) 24 inches (d) None of these

- Answer: (a) 18 inches, $(6 \times 3$ inches$) = 18$ inches, Section 370-28, Fig. 5–31, Part C

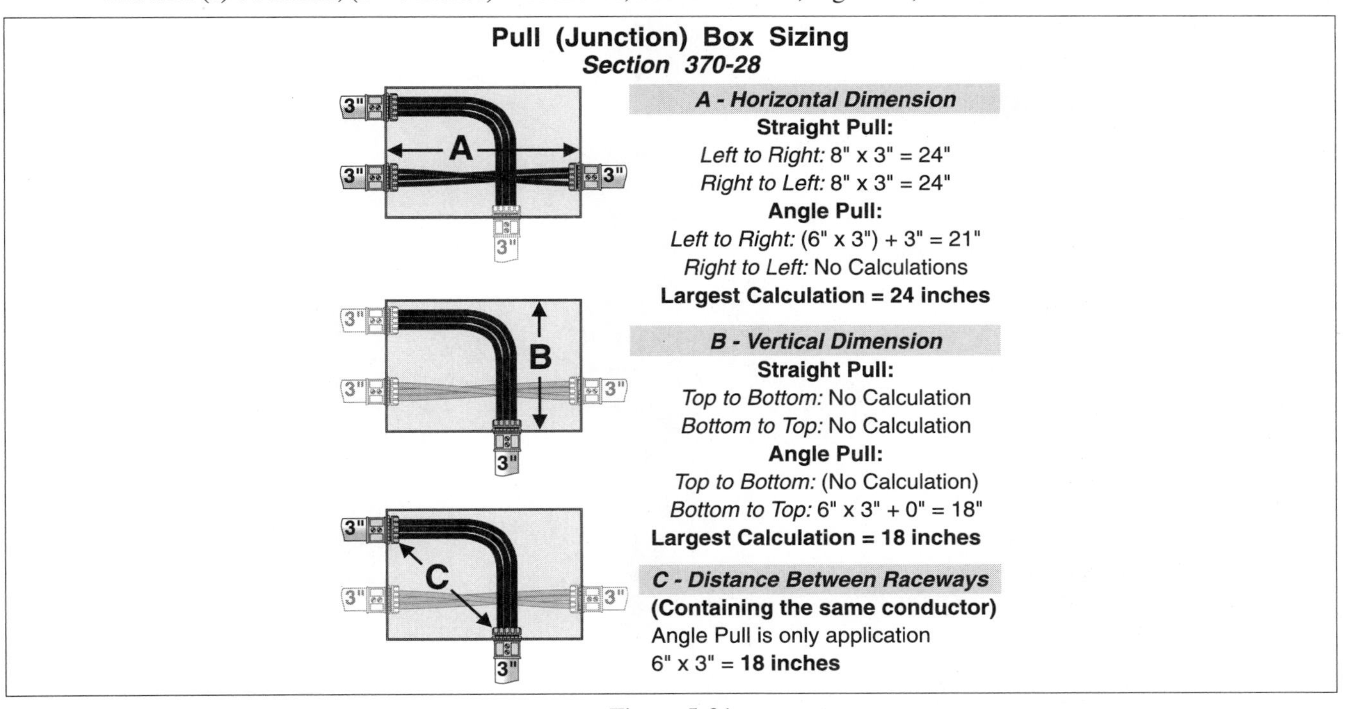

Figure 5-31

Pull (Junction) Box Sizing Example

❑ **Pull Box Sizing Example No. 2**

A pull box contains two 4-inch raceways on the left side and two 2-inch raceways on the top.

➤ **Horizontal Dimension Question A**

What is the horizontal dimension of the box?

(a) 28 inches (b) 21 inches (c) 24 inches (d) none of these

- Answer: (a) 28 inches, Section 370-28(a)(2)

  Left wall to the right wall angle pull = (6 × 4 inches) + 4 inches = 28 inches
  Left wall to the right wall straight pull = No calculation
  Right wall to left wall angle pull = No calculation
  Right wall to the left wall straight pull = No calculation

➤ **Vertical Dimension Question B**

What is the vertical dimension of the box?

(a) 18 inches (b) 21 inches (c) 24 inches (d) 14 inches

- Answer: (d) 14 inches, Section 370-28(a)(2)

  Top to bottom wall angle pull = (6 × 2 inches) + 2 inches = 14 inches
  Top to bottom wall - straight pull = No calculation
  Bottom to top wall angle pull = No calculation
  Bottom to top wall straight pull = No calculation

➤ **Distance Between Raceways Question C**

What is the minimum distance between the two 4 inch raceways that contain the same conductors?

(a) 18 inches

(b) 21 inches

(c) 24 inches

(d) None of these

- Answer: (c) 24 inches, (6 × 4 inches) = 24 inches, Section 370-28(a)(2)

Unit 5 – Raceway Fill, Box Fill, Junction Boxes, And Conduit Bodies Summary Questions

Part A – Raceway Fill

5–1 Understanding Chapter 9 Tables

1. When all the conductors are the same size and insulation, the number of conductors permitted in a raceway can be determined simply by looking at the Tables listed in ________.
(a) Chapter 9 (b) Appendix B (c) Appendix C (d) none of these

2. When equipment grounding conductors are installed in a raceway, the actual area of the conductor must be used when calculating raceway fill.
(a) True (b) False

3. When a raceway does not exceed 24 inches, the raceway is permitted to be filled to _________ percent of its cross–section area.
(a) 53 (b) 31 (c) 40 (d) 60

4. How many No. 16 TFFN conductors can be installed in a ¾-inch electrical metallic tubing?
(a) 40 (b) 26 (c) 30 (d) 29

5. How many No. 6 RHH (without outer cover) can be installed in a 1¼-inch nonmetallic tubing?
(a) 25 (b) 16 (c) 13 (d) 7

6. How many No. 1/0 XHHW can be installed in a 2-inch flexible metal conduit?
(a) 7 (b) 6 (c) 16 (d) 13

7. How many No. 12 RHH (with outer cover) can be installed in a 1-inch raceway?
(a) 7 (b) 11 (c) 5 (d) 4

8. If we have a 2½-inch rigid metal conduit and we want to install three THHN compact conductors, what is the largest conductor permitted to be installed?
(a) No. 4/0 (b) 250 kcmil (c) 300 kcmil (d) 350 kcmil

9. The actual area of conductor fill is dependent on the raceway size and the number of conductors installed. If there are three or more conductors installed in a raceway the total area of conductor fill is limited to _______ percent.
(a) 53 (b) 31 (c) 40 (d) 60

10. • What is the area in square inches for a No. 10 THW?
(a) .0333 (b) .0172 (c) .0252 (d) .0278

11. What is the area in square inches for a No. 14 RHW (without cover)?
(a) .0209 (b) .0172 (c) .0252 (d) .0278

12. What is the area in square inches for a No. 10 THHN?
(a) .0117 (b) .0172 (c) .0252 (d) .0211

13. What is the area in square inches for a No. 12 RHH (with outer cover)?
(a) .0117 (b) .0353 (c) .0252 (d) .0327

14. • What is the area in square inches of a No. 8 bare solid?
(a) .013 (b) .027 (c) .038 (d) .045

5–2 Raceway And Nipple Calculations

15. The number of conductors permitted in a raceway is dependent on _________.
(a) the area (square-inch) of the raceway
(b) the percent area fill as listed in Chapter 9, Table 1
(c) the area (square-inch) of the conductors as listed in Chapter 9, Tables 5 and 8
(d) all of these

16. A 200-ampere feeder installed in rigid nonmetallic conduit schedule 80, has three – No. 3/0 THHN, one – No. 2 THHN, and one – No. 6 THHN. What size raceway is required?
(a) 2-inch (b) 2½-inch (c) 3-inch (d) 3½ -inch

17. What size rigid metal nipple is required for three – No. 4/0 THHN, one – No. 1/0 THHN and No. 4 THHN?
(a) 1½-inch (b) 2-inch (c) 2½-inch (d) none of these

5–3 Existing Raceway

18. • An existing rigid metal nipple contains 4 – No. 10 THHN and 1 – No. 10 (bare stranded) ground wire in a ¾-inch raceway. How many additional No. 10 THHN can be installed?
(a) 5 (b) 7 (c) 9 (d) 11

Part B – Outlet Box Fill Calculations

5–4 Conductors All The Same Size [Table 370–16]

19. What size box is required for six No. 14 THHN and three No. 14 THW?
(a) 4 × 1¼ inch square (b) 4 × 1½ inch round
(c) 4 × 1⅛ inch round (d) None of these

20. How many No. 10 THHN are permitted in a 4 × 1½ square box?
(a) 8 conductors (b) 9 conductors
(c) 10 conductors (d) 11 conductors

5–5 Conductor Equivalents [370–16]

21. Table 370–16 does not take into consideration the volume of ______.
(a) switches and receptacles
(b) fixture studs and hickeys
(c) manufactured cable clamps
(d) all of these

22. When determining the number of conductors for box fill calculations, which of the following statements are true?
(a) A fixture stud or hickey is considered as one conductor for each type, based on the largest conductor that enters the outlet box.
(b) Internal factory cable clamps are considered as one conductor for one or more cable clamps, based on the largest conductor that enters outlet box.
(c) The device yoke is considered as two conductors, based on the largest conductor that terminates on the strap (device mounting fitting).
(d) All of these.

23. • When determining the number of conductors for box fill calculations, which of the following statements are true?
(a) Each conductor that runs through the box without loop (without splice) is considered as one conductor.
(b) Each conductor that originates outside the box and terminates in the box is considered as one conductor.
(c) Wirenuts, cable connectors, raceway fittings, and conductors that originate and terminate within the outlet box (equipment bonding jumpers and pigtails) are not counted for box fill calculations.
(d) All of these.

24. It is permitted to omit one equipment grounding conductor and not more than _______ that enter a box from a fixture canopy.
(a) four fixture wires (b) four No. 16 fixture wires
(c) four No. 18 fixture wires (d) b and c

25. Can a round 4 inch × $\frac{1}{2}$ inch (pancake) box marked as 8 cubic inches with manufactured cable clamps supplied with 14/2 romex be used with a fixture that has two No. 18 TFN and a canopy cover?
(a) Yes (b) No

5–7 Different Size Conductors

26. What size outlet box is required for, 12/2 romex that terminates on a switch, 12/3 romex that terminates on a receptacle, and the box has manufactured cable clamps.
(a) 4 × $1\frac{1}{4}$ inch square (b) 4 × $1\frac{1}{2}$ inch square
(c) 4 × $2\frac{1}{8}$ inch square (d) any of these

27. • How many No. 14 THHN conductors can be pulled through a 4 inch square × $1\frac{1}{2}$ inch deep box with a plaster ring of 3.6 cubic inches. The box contains two duplex receptacles, five No. 14 THHN and two grounding conductors.
(a) One (b) Two (c) Three (d) Four

Part C – Pull, Junction Boxes, And Conduit Bodies

5–8 Pull And Junction Box Size Calculations

28. When conductors No. 4 and larger are installed in boxes and conduit bodies, we must size the enclosure according to which of the following requirements?
(a) The minimum distance for straight pull calculations from where the conductors enter to the opposite wall must not be less than eight times the trade size of the largest raceway.
(b) The distance for angle pull calculations from the raceway entry to the opposite wall must not be less than six times the trade diameter of the largest raceway, plus the sum of the diameters of the remaining raceways on the same wall and row.
(c) The distance for U–pull calculations from where the raceways enter to the opposite wall must not be less than six times the trade diameter of the largest raceway, plus the sum of the diameters of the remaining raceways on the same wall and row.
(d) The distance between raceways enclosing the same conductors shall not be less than six times the trade diameter of the largest raceway.
(e) All of the above are correct.

29. When conductors enter an enclosure opposite a removable cover, the distance from where the conductors enter to the removable cover shall not be less than _______.
(a) six times the largest raceway
(b) eight times the largest raceway
(c) a or b
(d) none of these

The following information applies to Questions 30 and 31.
A junction box contains two $2\frac{1}{2}$-inch raceways on the left side and one $2\frac{1}{2}$-inch raceway on the right side. The conductors from one 2 1/2-inch raceway (left wall) are pulled through the raceway on the right wall. The other $2\frac{1}{2}$-inch raceway conductors are pulled through a raceway at the bottom of the pull box.

30. What is the distance from the left wall to the right wall?
(a) 18 inches (b) 21 inches (c) 24 inches (d) 20 inches

31. • What is the distance from the bottom wall to the top wall?
(a) 18 inches (b) 21 inches (c) 24 inches (d) 15 inches

32. What is the distance from between the raceways that contain the same conductors?
(a) 18 inches (b) 21 inches (c) 24 inches (d) 15 inches

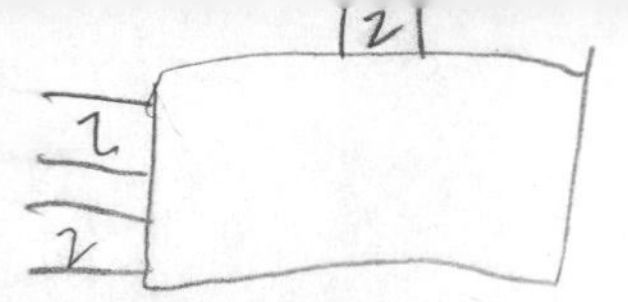

The following information applies to Questions 33, 34, and 35.
A junction box contains two 2-inch raceways on the left side and two 2-inch raceway on the top.

33. What is the distance from the left wall to the right wall?
(a) 28 inches (b) 21 inches (c) 24 inches (d) 14 inches

34. What is the distance from the bottom wall to the top wall?
(a) 18 inches (b) 21 inches (c) 24 inches (d) 14 inches

35. What is the distance from between the 2-inch raceways that contain the same conductors?
(a) 18 inches (b) 21 inches (c) 24 inches (d) 12 inches

Challenge Questions

Part A – Raceway Calculations

36. • A 3-inch schedule 40 PVC raceway contains seven No. 1 RHW without cover. How many number No. 2 THW conductors may be installed in this raceway with the existing conductors?
(a) 11 (b) 15 (c) 20 (d) 25

Part B – Box Fill Calculations

37. • Determine the minimum cubic inches required for two No. 10 TW passing through a box, four No. 14 THHN terminating, two No. 12 TW terminating to a receptacle, and one No 12 equipment bonding jumper from the receptacle to the box.
(a) 18.5 cubic inches (b) 22 cubic inches (c) 20 cubic inches (d) 21.75 cubic inches

38. • When determining the number of conductors in a box fill two No. 18 fixture wires, one 14/3 nonmetallic sheathed cable with ground, one duplex switch, and two cable clamps would count as ______.
(a) 9 conductors (b) 8 conductors (c) 10 conductors (d) 6 conductors

Part C – Pull And Junction Box Calculations

The Figure 5–32 applies to Questions 39, 40, 41, and 42.

39. The minimum horizontal dimension for the junction box shown in the diagram is ______ inches, Fig. 5–32?
(a) 21 (b) 18
(c) 24 (d) 20

40. • The minimum vertical dimension for the junction box is ______ inches, Fig. 5–32?
(a) 16 (b) 18
(c) 20 (d) 24½

41. The minimum distance between the two 2-inch raceways that contains the same conductor (C) would be ______ inches, Fig. 5–32?
(a) 12 (b) 18
(c) 24 (d) 30

Figure 5-32

42. • If a 3-inch raceway entry (250-kcmils) is in the wall opposite to a removable cover, the distance from that wall to the cover must not be less than ______ inches, Fig. 5–32?
(a) 4 (b) 4½
(c) 5 (d) 5½

Unit 6

Conductor Sizing And Protection Calculations

OBJECTIVES

After reading this unit, the student should be able to briefly explain the following concepts:

- Conductor allowable ampacity
- Conductor ampacity
- Conductor bunching derating factor, Note 8(a)
- Conductor sizing summary
- Conductor insulation property
- Conductor size – voltage drop
- Conductors in parallel
- Current-carrying conductors
- Equipment conductors size and protection examples
- Minimum conductor size
- Overcurrent protection of equipment conductors
- Overcurrent protection
- Overcurrent protection of conductors
- Terminal ratings

After reading this unit, the student should be able to briefly explain the following terms:

- Ambient temperature
- American wire gauge
- Ampacity
- Ampacity derating factors
- Conductor bundled
- Conductor properties
- Continuous load
- Current-carrying conductors
- Fault current
- Interrupting rating
- Overcurrent protection device
- Overcurrent protection
- Parallel conductors
- Temperature correction factors
- Terminal ratings
- Voltage drop

6–1 *CONDUCTOR INSULATION PROPERTY [Table 310–13]*

Table 310–13 of the NEC provides information on conductor properties such as, permitted use, maximum operating temperature, and other insulation details, Fig. 6–1. See table 310–13 on the next page.

The following aberrations and explanations should be helpful in understanding Table 310–13 as well as Table 310–16.

-2 - Conductor is permitted to be used at a continuous 90°C operation temperature
F - Fixture wire (solid or 7 strand) [Table 402–3]
FF - Flexible Fixture wire (19 strands) [Table 402–3] 60°C Insulation rating
H - 75°C Insulation rating
HH - 90°C Insulation
N - Nylon outer cover
T - Thermoplastic insulation
W - Wet or Damp

For details on fixture wires, see Article 402, Table 402–3 and Table 402–5.

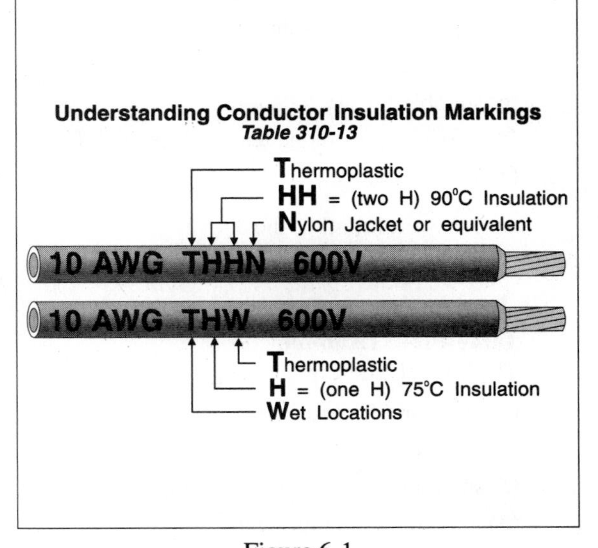

Figure 6-1

Conductor Insulation Markings

TABLE 310–13 CONDUCTOR INFORMATION					
	Column 2	Column 3	Column 4	Column 5	Column 6
Type Letter	Trade Name	Maximum Operating Temperature	Applications Provisions	Sizes Available	Outer Covering
THHN	Heat resistant thermoplastic	90ºC	Dry and damp locations	14 – 1000	Nylon jacket or equivalent
THHW	Moisture- & heat-resistant thermoplastic	75ºC 90ºC	Wet locations Dry and damp locations	14 – 1000	None
THW	Moisture- & heat-resistant thermoplastic	75ºC 90ºC	Dry, damp, and wet locations Within electrical discharge lighting equipment *See Section 410–31*	14 – 2000	None
THWN	Moisture- & heat-resistant thermoplastic	75ºC	Dry and wet locations	14 – 1000	Nylon jacket or equivalent
TW	Moisture- resistant thermoplastic	60ºC	Dry and wet locations	14 – 2000	None
XHHW	Moisture- & heat-resistant cross-linked synthetic polymer	90ºC 75ºC	Dry and damp locations Wet locations	14 – 2000	None

❑ **Table 310–13 Example**

TW can be described as __________ Fig. 6–2?

(a) thermoplastic insulation
(b) suitable for wet locations
(c) suitable for dry locations
(d) maximum operating temperature of 60ºC
(e) all of the above

- Answer: (e) all of the above

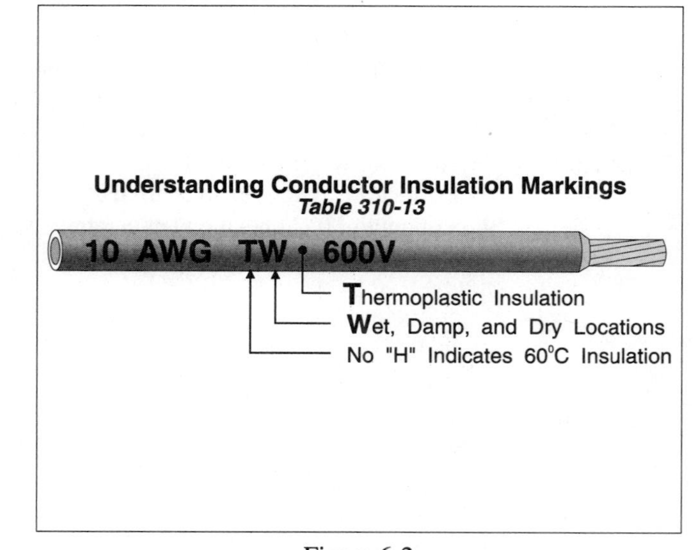

Figure 6-2

Conductor Insulation Marking Example

❑ **Table 402–3 Example**

TFFN can be described as ___________?

(a) 19-strand fixture wire
(b) thermoplastic insulation with a nylon outer cover
(c) suitable for dry and wet locations
(d) maximum operating temperature of 90ºC
(e) all of the above except (c)

- Answer: (e) all of the above except (c)

6–2 *CONDUCTOR ALLOWABLE AMPACITY [310–15]*

The ampacity of a conductor is the current the conductors can carry continuously under the specific condition of use [Article 100 definition]. The ampacity of a conductor is listed in Table 310-16 under the condition of no more than three current carrying conductors bundled together in an ambient temperature of 86°C. The ampacity of a conductor changes if the ambient temperature is not 86°F or if more than three current-carrying conductors are bundled together for more than two feet.

Section 310–15 lists 2 ways of determining conductor ampacity.

Ampacity Method 1: Simply use the values listed in Tables 310–16 through 310–19 of the NEC according to the specific conditions of use.

Table 310-16 is based on an ambient temperature of 86°F and 3 current-carrying conductors in a raceway or cable.

Figure 6-3

Conductor Ampacity Table 310–16

Ampacity Method 2: Engineer supervised formula: $I = \sqrt{\frac{TC - (TA + DELTA\ TD)}{RDC\ (1 + YC)\ RCA}}$

For all practical purposes, the electrical trade uses the ampacities as listed in Table 310–16 of the National Electrical Code.

TABLE 310–16. ALLOWABLE AMPACITIES OF INSULATED CONDUCTORS
Based On Not More Than Three Current-Carrying Conductors And Ambient Temperature of 30°C (86°F)

Size	Temperature Rating of Conductor, See Table 310–13						Size
	60°C (40°F)	75°C (167°F)	90°C (194°F)	60°C (40°F)	75°C (167°F)	90°C (194°F)	
AWG kcmil	†TW	†THHN †THW †THWN †XHHW Wet Location	†THHN †THHW †XHHW Dry Location	†TW	†THHN †THW †THWN †XHHW Wet Location	†THHN †THHW †XHHW Dry Location	AWG kcmil
	COPPER			ALUMINUM/COPPER-CLAD ALUMINUM			
14	**20†**	20†	25†				12
12	**25†**	25†	30†	20†	20†	25†	10
10	**30**	35†	40†	25	30†	35†	8
8	**40**	50	55	30	40	45	8
6	**55**	65	75	40	50	60	6
4	**70**	85	95	55	65	75	4
3	85	**100**	110	65	75	85	3
2	95	**115**	130	75	90	100	2
1	110	**130**	150	85	100	115	1
1/0	125	**150**	170	100	120	135	1/0
2/0	145	**175**	195	115	135	150	2/0
3/0	165	**200**	225	130	155	175	3/0
4/0	195	**230**	260	150	180	205	4/0
250	215	**255**	290	170	205	230	250
300	240	**285**	320	190	230	255	300
350	260	**310**	350	210	250	280	350
400	280	**335**	380	225	270	305	400
500	320	**380**	430	260	310	350	500

† The maximum *overcurrent protection device* permitted for copper No. 14 is 15 amperes, No. 12 is 20 amperes, and No. 10 is 30 amperes. The maximum overcurrent protection device permitted for aluminum No. 12 is 15 amperes and No. 10 is 25 amperes. This † note does not establish the conductor ampacity, but limits the maximum size protection device that may be placed on the conductor, Fig. 6–4.

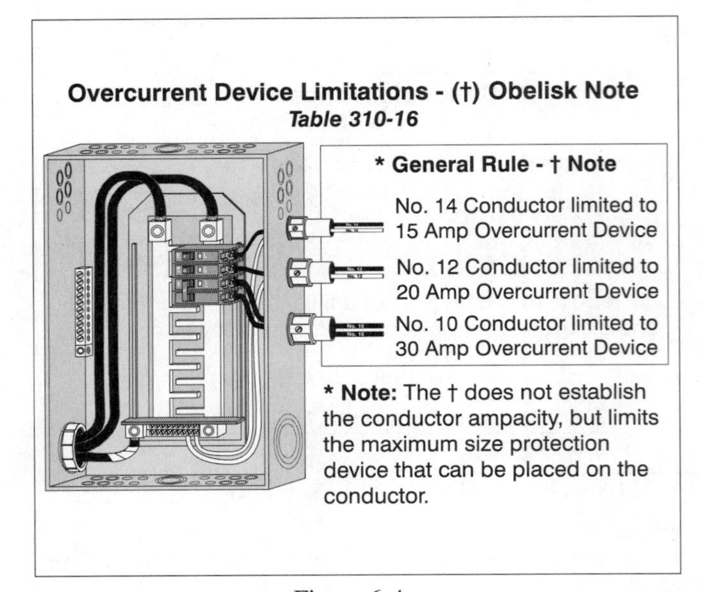

Figure 6-4

† Note of Table 310–16

Cross-Sections And Trade Sizes Of NEC Conductors
Tables 310-13 and 310-16

AWG Sizes 18 - 4/0
18 16 14 12 10 8 6 4 3 2 1

kcmil Sizes 250 - 2000

1/0 (0) 2/0 (00) 3/0 (000) 4/0 (0000) 250 kcmil 300 kcmil 350 kcmil

Sizes 18 through 4/0 are American Wire Gage (AWG). Sizes 250 kcmil plus are in circular mils (cmil) with "k" meaning 1,000. See *Section 110-6.*

Figure 6-5

Conductor Size

CAUTION: The overcurrent protection device limitations of the † Note to Table 310–16 does not apply to the following: See Section 240–3(d) through (m).
Air conditioning conductor Section 440–22
Capacitor conductors Article 460
Class 1 remote control conductors Article 725
Cooking equipment taps 210–19(b) Exception No. 1
Feeder taps Section 240–21
Fixture wires and taps 210–19(c) Exception
Motor branch circuit conductors Section 430–52
Motor feeder conductor Section 430–62
Motor control conductors Section 430–72
Motor taps branch circuits Section 430–53(d)
Motor taps feeder conductors Section 430–28
Power loss hazard conductors Section 240–3(a)
Two-wire transformers Section 240–3(i)
Transformer Tap Conductor Section 240–21(b)(d) and (m)
Welders Article 630

6–3 *CONDUCTOR SIZING [110–6]*

Conductors are sized according to the American Wire Gage (AWG) from Number 40 through Number 0000. The smaller the AWG size, the larger the conductor. Conductors larger than 0000 (4/0) are identified according to their circular mil area, such as 250,000, 300,000, 500,000. The circular mil size is often expressed in kcmil, such as 250 kcmil, 300 kcmil, 500 kcmil, etc, Fig. 6–5.

Smallest Conductor Size

The smallest size conductor permitted by the National Electrical Code for branch circuits, feeders, or services is No. 14 copper or No. 12 aluminum [Table 310–5]. Some local codes require a minimum No. 12 for commercial and industrial installations. Conductors smaller than No. 14 are permitted for:

Class 1 circuits [402–11, Exception, and 725–27]
Cords [Table 400–5(a)]
Fixture wire [402–5 and 410–24]
Flexible cords [400–12]
Motor control circuits [430–72]
Nonpower limited fire alarm circuits [760–27]
Power limited fire alarm circuits [760–54(d)].

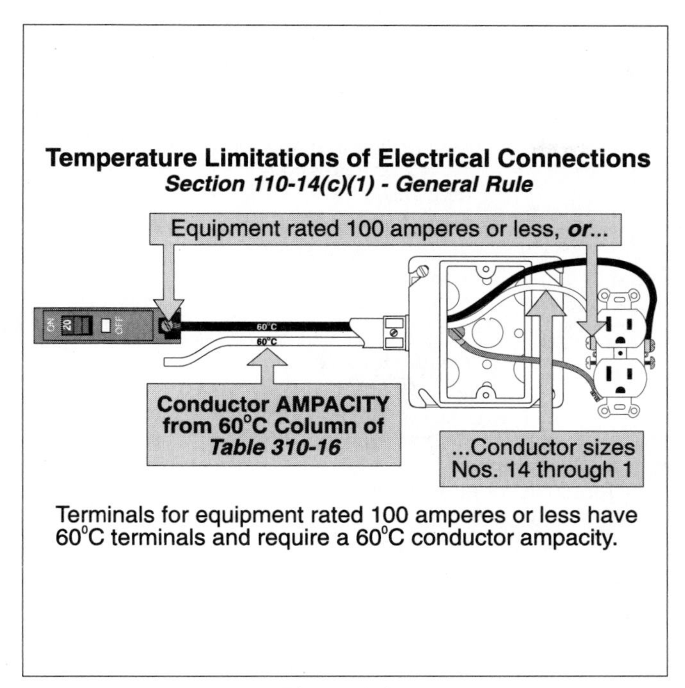

Figure 6-6

Temperature Limitations of Electrical Connections

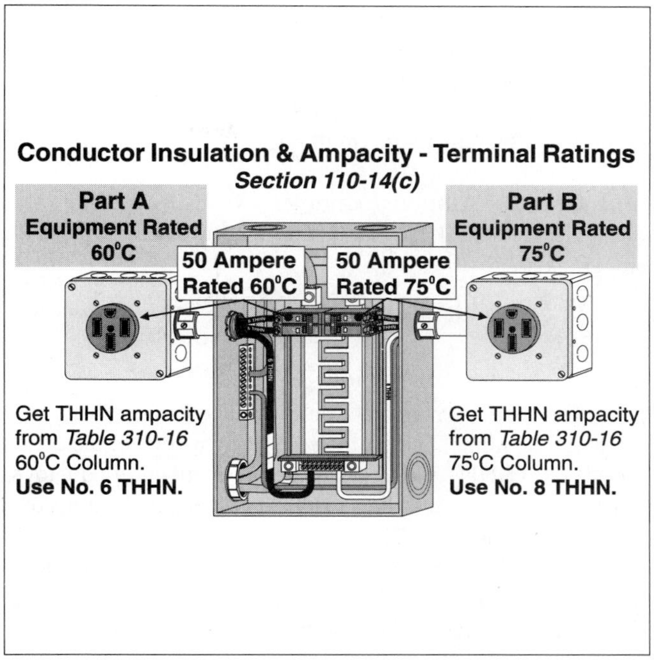

Figure 6-7

Conductor Insulation and Ampacity – Terminal Ratings

6–4 *TERMINAL RATINGS [110–14(c)]*

When selecting a conductor for a circuit, we must always select the conductors not smaller than the terminal ratings of the equipment.

Circuits Rated 100 Amperes And Less

Equipment terminals rated 100 amperes or less (and pressure connector terminals for No. 14 through No. 1 conductors), shall have the conductor sized no smaller than the 60°C temperature rating listed in Table 310–16, unless the terminals are marked otherwise, Fig. 6–6.

❑ **Terminal Rated 60°C Example**

What size THHN conductor is required for a 50-ampere circuit if the equipment is listed for use at 60°C, Fig. 6–7 Part A?

(a) No. 10 (b) No. 8
(c) No. 6 (d) Any of these

- Answer: (c) No. 6

Conductors must be sized to the lowest temperature rating of either the equipment (60°C) or the conductor (90°C). THHN insulation can be used, but the conductor size must be selected based on the 60°C terminal rating of the equipment, not the 90°C rating of the insulation. Using the 60°C column of Table 310–16, this 50 ampere circuit requires a No. 6 THHN conductor (rated 55 amperes at 60°C).

❑ **Terminal Rated 75°C Example**

What size THHN conductor is required for a 50-ampere circuit if the equipment is listed for use at 75°C, Fig. 6–7 Part B?

(a) No. 10
(b) No. 8
(c) No. 6
(d) Any of these

- Answer: (b) No. 8

Conductors must be sized according to the lowest temperature rating of either the equipment (75°C) or the conductor (90°C). THHN conductors can be used, but the conductor size must be selected according to the 75°C terminal rating of the equipment, not the 90°C rating of the insulation. Using the 75°C column of Table 310–16, this installation would permit No. 8 THHN (rated 50 amperes 75°C) to supply the 50-ampere load.

Circuits Over 100 Amperes

Terminals for equipment rated over 100 amperes and pressure connector terminals for conductors larger than No. 1, shall have the conductor sized according to the 75°C temperature rating listed in Table 310–16, Fig. 6–8.

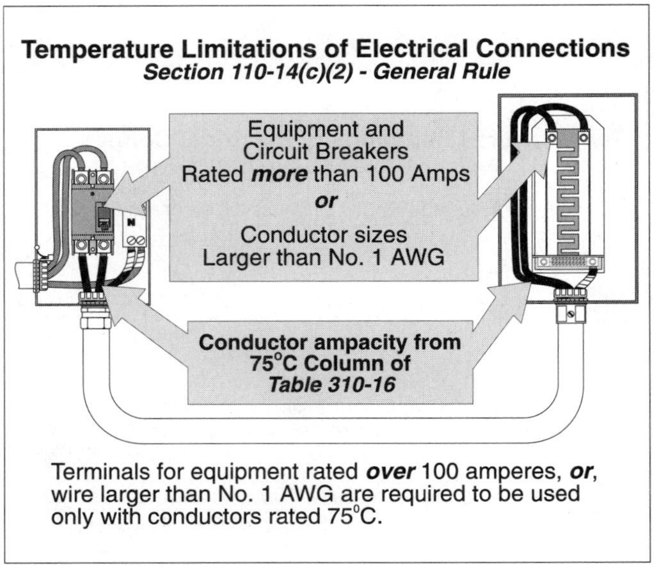

Figure 6-8

Temperature Limitations Of Electrical Connections

❑ Over 100 Ampere Example

What size THHN conductor is required to supply a 225-ampere feeder?

(a) No. 1/0 (b) No. 2/0

(c) No. 3/0 (d) No. 4/0

- Answer: (d) No. 4/0

The conductors in this example must be sized to the lowest temperature rating of either the equipment (75°C) or the conductor (90°C). THHN conductors can be used, but the conductor size must be selected according to the 75°C terminal rating of the equipment. Using the 75°C column of Table 310–16, this would require a No. 4/0 THHN (rated 230 amperes at 75°C) to supply the 225 ampere load. No. 3/0 THHN is rated 225 amperes at 90°C, but we must size the conductor to the terminal rating at 75°C.

Minimum Conductor Size Table

When sizing conductors the following table must always be used to determine the minimum size conductor.

CAUTION: When sizing conductors, we must consider conductor voltage drop, ambient temperature correction and conductor bunching derating factors. These subjects are covered later in this book.

Terminal Size And Matching Copper Conductor		
Terminal ampacity	**60°C Terminals Wire Size**	**75°C Terminals Wire Size**
15	14	14
20	12	12
30	10	10
40	8	8
50	6	8
60	4	6
70	4	4
100	1	3
125	1/0	1
150	–	1/0
200	–	3/0
225	–	4/0
250	–	250 kcmil
300	–	350 kcmil
400	–	2 – 3/0
500	–	2 – 250 kcmil

What Is The Purpose Of THHN [110-14(c)]?

In general, 90°C rated conductor ampacities cannot be used for sizing a circuit conductor. However, THHN offers the opportunity of having a greater conductor ampacity for conductor ampacity derating. The higher ampacity of THHN can permit a conductor to be used without having to increase its size because of conductor ampacity derating. Remember, the advantage of THHN is not to permit a smaller conductor, but to prevent you from having to install a larger conductor when ampacity derating factors are applied, Fig. 6–9.

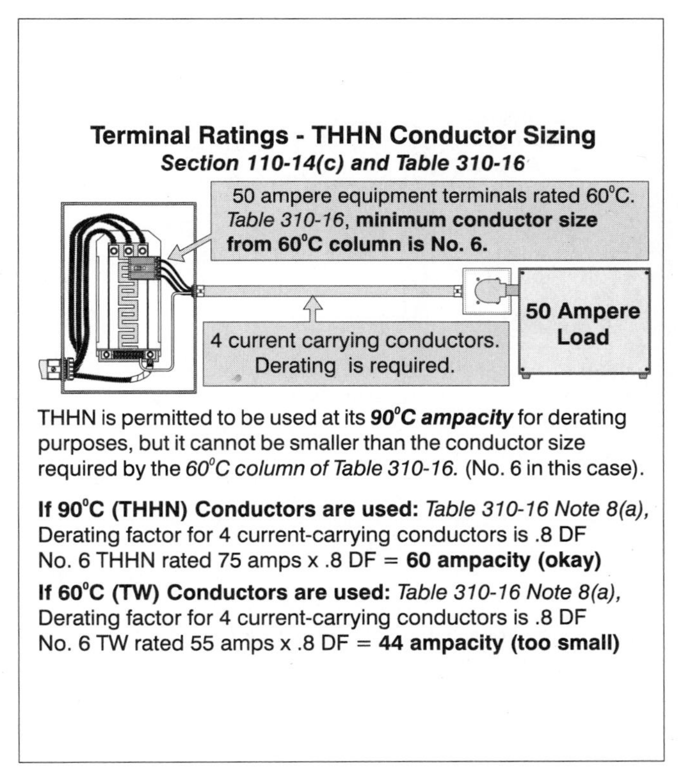

Figure 6-9

Purpose of THHN

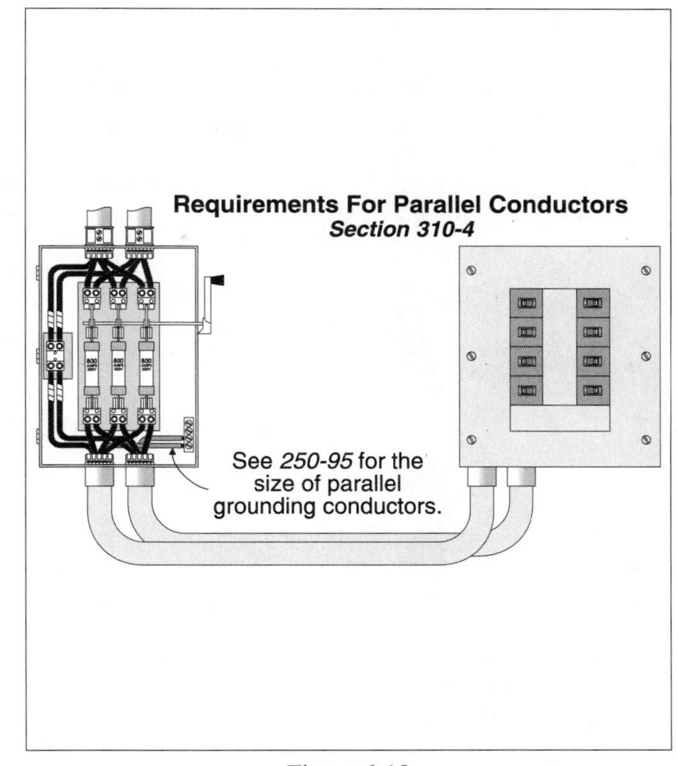

Figure 6-10

Requirements for Conductors in Parallel

6–5 *CONDUCTORS IN PARALLEL [310–4]*

As the conductor size increases, the cross-sectional area required to carry one ampere increases. Parallel conductors (electrically joined at both ends) permit a reduced cross-sectional area per ampere which can result in significant cost saving, Fig. 6–10.

Conductor Size	Circular Mils Chapter 9, Table 8	Ampacity 75°C	Circular Mils Per Amperes
No. 1/0	105,600 cm	150 amperes	704 cm/per amp
No. 3/0	167,600 cm	200 amperes	838 cm/per amp
250 kcmil	250,000 cm	255 amperes	980 cm/per amp
500 kcmil	500,000 cm	380 amperes	1,316 cm/per amp
750 kcmil	750,000 cm	475 amperes	1,579 cm/per amp

❑ **Conductors In Parallel Example**

What size 75°C conductor is required for a 600-ampere service that has a calculated demand load of 550 amperes, Fig. 6–11?

(a) 500 kcmil (b) 750 kcmil (c) 1,000 kcmil (d) 1,250 kcmil

- Answer: (d) 1,250 kcmil rated 590 amperes [Table 310–16]

❑ **Parallel Conductor Example**

What size conductor would be required in each of two raceways for a 600 ampere service? The calculated demand load is 550 amperes, Fig. 6–12?

(a) two – 300 kcmil (b) two – 250 kcmil (c) two – 500 kcmil (d) two– 750 kcmil

- Answer: (a) 2– 300 kcmil each rated 285 amperes.

  285 amperes × 2 conductors = 570 amperes [Table 310–16]

  When we can install two 300-kcmil conductors (600,000 circular mils total) instead of one 1,250-kcmil conductor, see previous example.

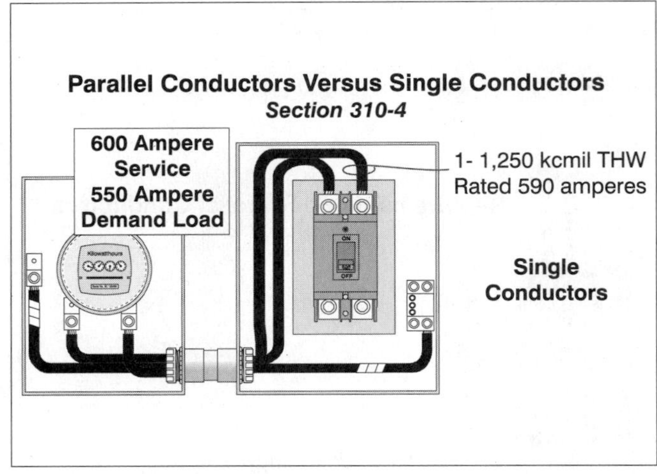

Figure 6-11

Non-parallel Example

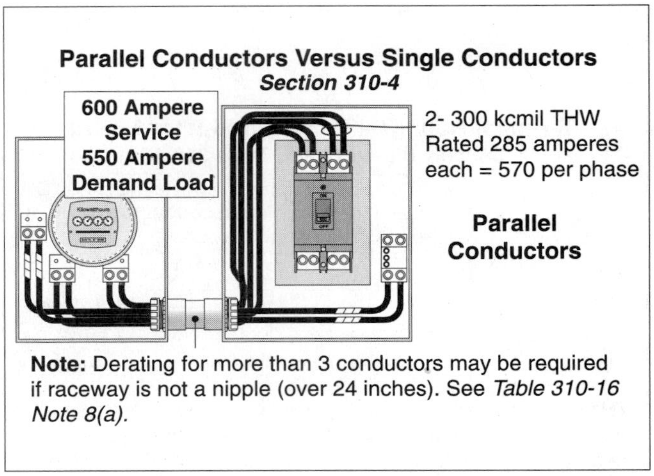

Parallel Conductors Versus Single Conductors
Section 310-4

600 Ampere Service
550 Ampere Demand Load

2- 300 kcmil THW Rated 285 amperes each = 570 per phase

Parallel Conductors

Note: Derating for more than 3 conductors may be required if raceway is not a nipple (over 24 inches). See *Table 310-16 Note 8(a).*

Figure 6-12

Parallel Example

Grounding Conductors In Parallel

When equipment grounding conductors are installed with circuit conductors that are run in parallel, each raceway must have an equipment grounding conductor. The equipment grounding conductor installed in each raceway must be sized according to the overcurrent protection device rating that protects the circuit [250–95]. Parallel equipment grounding conductors are not required to be a minimum No. 1/0 [310–4], Fig. 6–13.

❑ **Grounding Conductors In Parallel Example**

What size equipment grounding conductor is required in each of two raceways for a 600-ampere feeder, Fig. 6–14?

(a) No. 3 (b) No. 2 (c) No. 1 (d) No. 1/0

- Answer: (c) No. 1, Section and Table 250–95

  There must be a total of two No. 1 equipment grounding conductors, one No. 1 in each raceway.

6–6 *CONDUCTOR SIZE – VOLTAGE DROP [210–19(a), FPN No. 4, And 215–2(b), FPN No. 2]*

Conductor voltage drop is the result of the conductor's resistance opposing the current flow, $E_{vd} = R \times I$. The NEC® does not require that we limit the voltage drop on conductors, but the Code does suggest that we consider its effects. The subject of conductor voltage drop is covered in detail in Unit 8.

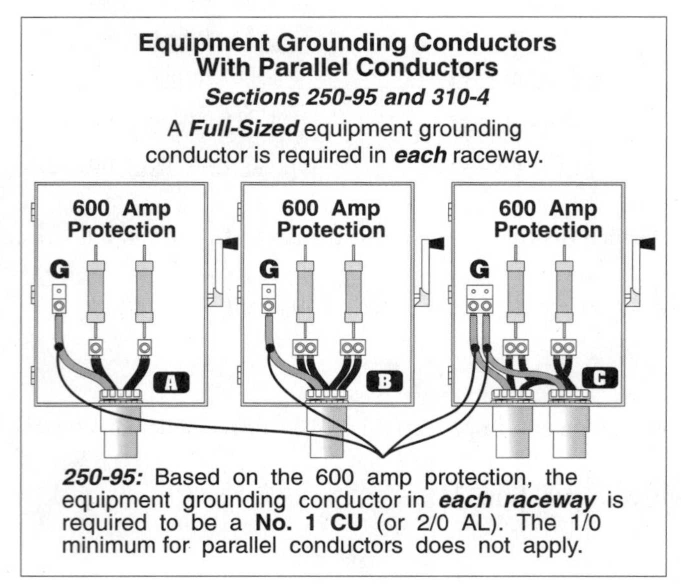

Figure 6-13

Parallel Grounding Conductor

Sizing Parallel Equipment Grounding Conductors
Section 250-95

The equipment grounding conductor must be ***full size*** to ***each*** parallel raceway.

Protection Device: 600 Amperes

Feeder Conductors: 300 kcmil for each phase in each parallel raceway.

Equipment Grounding Conductor: *Table 250-95*, based on the 600 ampere ovrcurrent device.

Use No. *1 AWG in *each* raceway

600A Device

***Note:** 1/0 minimum parallel conductor required by *Section 310-4* does not apply to equipment grounding conductors.

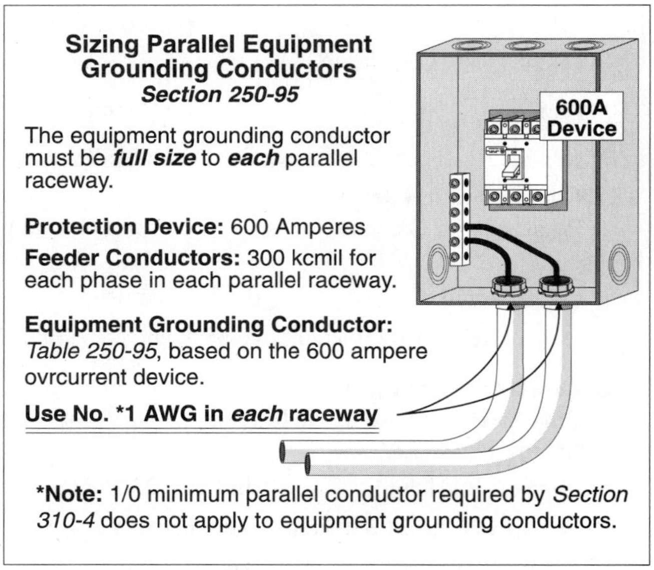

Figure 6-14

Parallel Grounding Conductor Example

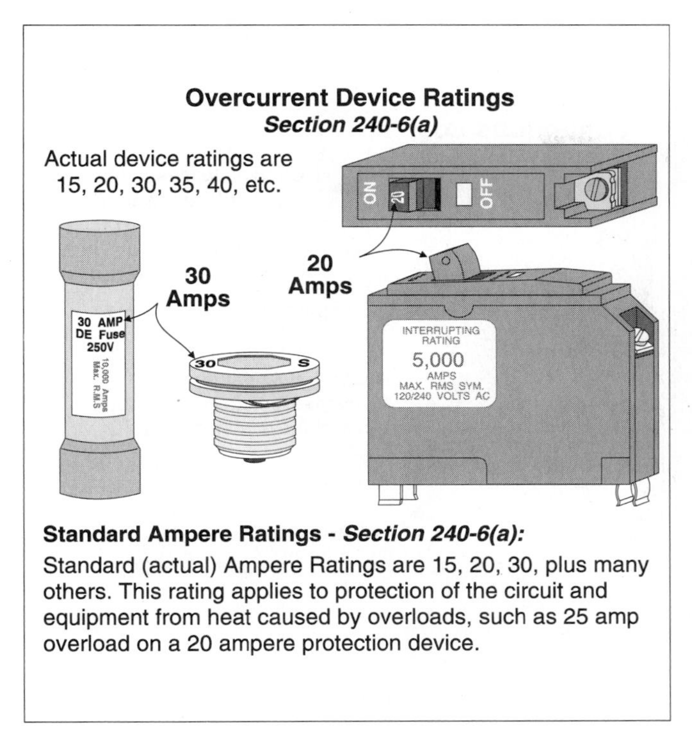

Figure 6-15

Overcurrent Protection Device Rating

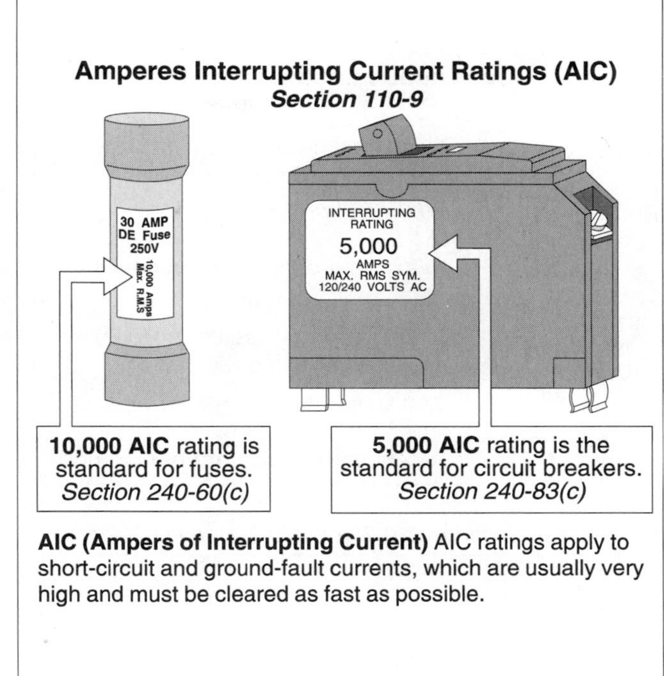

Figure 6-16

Interrupting Rating

6–7 *OVERCURRENT PROTECTION [Article 240]*

One purpose of an overcurrent protection device is to open a circuit to prevent damage to persons or property due to excessive or dangerous heat. Overcurrent protection devices have two ratings, overcurrent and interrupting (AIC).

Overcurrent Rating

This is the actual ampere rating of the protection device, such as 15, 20, or 30 amperes. If the current flowing through the protection device exceeds the device setting for a specific period of time, the protection device will open to remove the danger of excess heat, Fig. 6–15.

Interrupting Rating (Short-circuit) [110–9]

The overcurrent protection device is designed to clear fault-current and it must have a short-circuit interrupting rating sufficient for the available fault current. The minimum interruption rating for circuit breakers is 5,000 amperes [240–83(c)] and 10,000 amperes for fuses [240–60(c)]. The overcurrent protection device must clear fault-current and prevent the increase of damage to other equipment or other parts of the circuit, Fig. 6–16.

If the overcurrent protection device does not have an interrupting rating above the available fault current, it could explode while attempting to clear the fault. In addition, downstream equipment must have a withstand rating so that it will not suffer serious damage during a ground-fault [110–10].

Standard Size Protection Devices [240–6(a)]

The National Electrical Code lists standard sized overcurrent protection devices: 15, 20, 25, 30, 35, 40, 45, 50, 60, 70, 80, 90, 100, 110, 125, 150, 175, 200, 225, 250, 300, 350, 400, 450, 500, 600, 700, 800, 1000, 1200, 1600, 2000, 2500, 3000, 4000, 5000, and 6000 amperes. *Exception:* Additional standard ratings for fuses include : 1, 3, 6, 10, and 601 amperes.

Continuous Load [384–16(c)]

Overcurrent protection devices are sized no less than 125 percent of the continuous load, plus 100 percent of the noncontinuous load. Conductors for continuous loads are sized based on the ampacities as listed on Table 310-16 "before any ampacity derating factors are applied" [110-14(c), 210–22(c), 220–3(a), 220–10(b), and 384–16(c)], Fig. 6–17.

❑ **Continuous Load Example**

What size protection device is required for a 100-ampere continuous load, Fig. 6–18?

(a) 150 ampere (b) 100 ampere (c) 125 ampere (d) any of these

- Answer: (c) 125 ampere, 100 amperes × 1.25 = 125 amperes [240–6(a)]

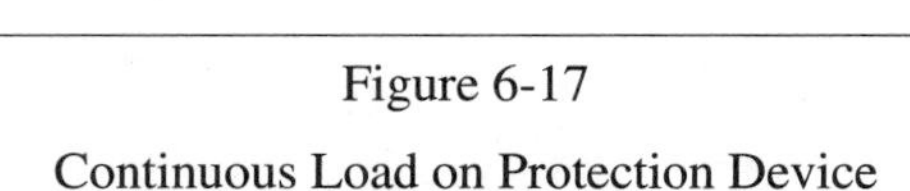

Figure 6-17

Continuous Load on Protection Device

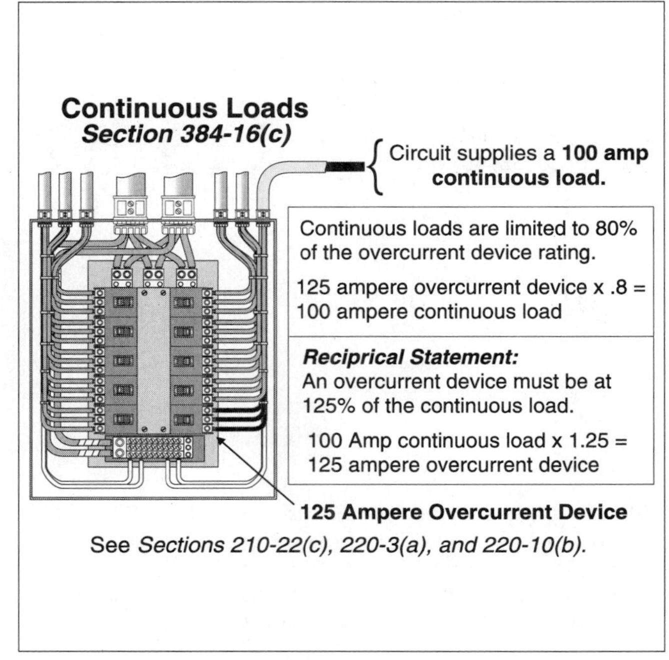

Figure 6-18

Continuous Load Example

6–8 *OVERCURRENT PROTECTION OF CONDUCTORS – GENERAL REQUIREMENTS [240–3]*

There are many different rules for sizing and protecting conductors. It is not simply a 20-ampere breaker on No. 12 wire. The general rule for conductor overcurrent protection is that a conductor must be protected against overcurrent in accordance with its ampacity [240–3].

However, if the ampacity of a conductor does not correspond with the standard ampere rating of a fuse or circuit breaker as listed in 240–6(a), the next larger overcurrent protection device (not over 800 amperes) is permitted [240–3(b)]. The next size up rule only applies if the conductor does not supply multioutlet receptacles for cord-and-plug connected loads and does not exceed 800 amperes, Fig. 6–19.

❑ **Overcurrent Protection Of Conductors Example**

What size conductor is required for a 130-ampere load that is protected with a 150-ampere breaker, Fig. 6–20?

(a) No. 1/0 (b) No. 1

(c) No. 2 (d) any of these

- Answer: (b) No. 1

The conductor must be sized according to the 75°C column of Table 310–16, which is No. 1 [110-14(c)(2)]. This No. 1 conductor is rated 130 amperes at 75°C column of Table 310–16 and is permitted to be protected by a 150-ampere protection device. [240–3(b)]

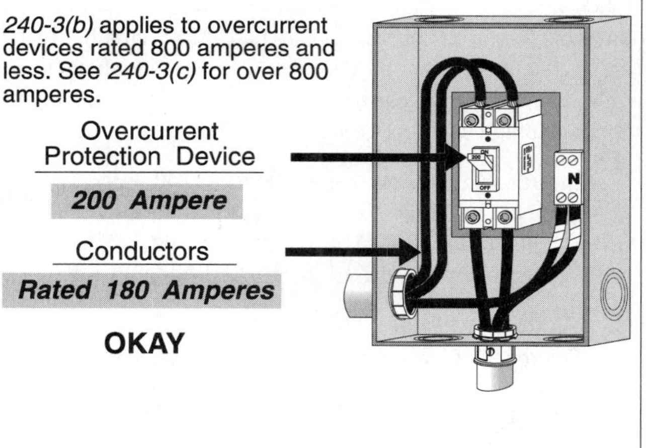

Figure 6-19

Next Size Larger Protection

6–9 *OVERCURRENT PROTECTION OF CONDUCTORS – SPECIFIC REQUIREMENTS*

When sizing and protecting conductors for equipment, be sure to apply the specific NEC® requirement.

Equipment

Air Conditioning [440–22, and 440–32]

Appliances [422–4, 422–5 and 422–28]

Cooking Appliances [210–19(b), 210–20, Exception No. 1, Note 4 of Table 220–19]
Electric Heating Equipment [424–3(b)]
Fire Protective Signaling Circuits [760–23]
Motors:
- Branch-Circuits [430–22(a), and 430–52]
- Feeders [430–24, and 430–62]
- Remote Control [430–72]

Panelboard [384–16(a)]
Transformers [240–21 and 450–3]

Feeders And Services
Dwelling Unit Feeders and Neutral [215–2 and Note 3 of Table 310–16]
Feeder Conductor [215–2 and 215–3]
Not Over 800 Amperes [240–3(b)]
Feeders Over 800 Amperes [240–3(c)]
Service Conductors [230–90(a)]
Temporary Conductors [305–4]

Grounded Conductor
Neutral Calculations [220–22]
Grounded Service Size [230–23(b)]

Tap Conductors
Tap – Ten foot [240–21(b)]
Tap – Twenty-five foot [240–21]
Tap – One hundred foot [240–21(e)]
Tap – Outside Feeder [240–21(m)]

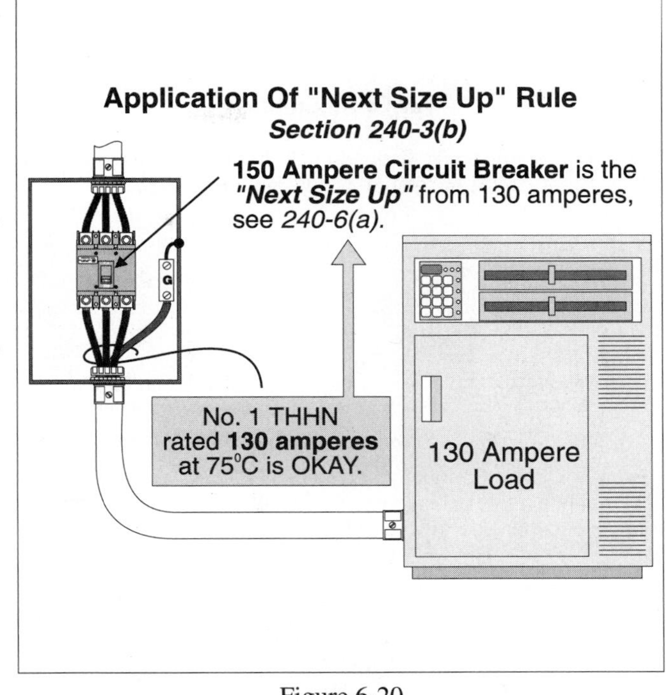

Figure 6-20
Next Size Larger Protection Example

6–10 *EQUIPMENT CONDUCTORS SIZE AND PROTECTION EXAMPLES*

❑ **Air Conditioning Example**

An air conditioner nameplate indicates the minimum circuit ampacity of 53 amperes and maximum fuse size of 60 amperes. What is the minimum size branch circuit conductor and the maximum size overcurrent protection device, Fig. 6–21?

(a) No. 10 with a 60-ampere fuse
(b) No. 8 with a 60-ampere fuse
(c) No. 8 with a 50-ampere fuse
(d) No. 10 with a 30-ampere fuse

- Answer:(a) No. 6 with a 60 ampere Fuse

  Conductor: The conductors must be sized based on the 60ºC column of Table 310–16, which is No. 10.

  Overcurrent Protection: The protection device must not be greater than a 45-ampere fuse, preferably a dual-element fuse because of high inrush starting current.

Note: The obelisk (†) note on the bottom of Table 310-16 limiting the overcurrent protection device to 30 amperes for No. 10 wire does not apply to air conditioning equipment [240-3(h)].

❑ **Water Heater Example**

What size conductor and protection device is required for a 4,500 VA water heater, Fig. 6–22?

(a) No. 10 wire with 20-ampere protection
(b) No. 12 wire with 20-ampere protection
(c) No. 10 wire with 25-ampere protection
(d) No. 10 wire with 30-ampere protection
(e) c or d

- Answer: (e) c or d

  I = VA/E, I = 4,500 VA/240 volts, I = 18.75 amperes

Conductor: The conductor is sized at 125 percent of the water heater rating [422–14(b)],
Minimum conductor = 18.75 amperes × 1.25 = 23.4 amperes
Conductor is sized according to the 60ºC column of Table 310–16, = No. 10 rated 30 amperes

Overcurrent Protection: The overcurrent protection device is sized no more than 150 percent of the appliance rating, [422–28(e)], 18.75 amperes × 1.50 = 28.1 amperes, next size up = 30 amperes. Please note that the exception to 422–28(e) permits the next size up.

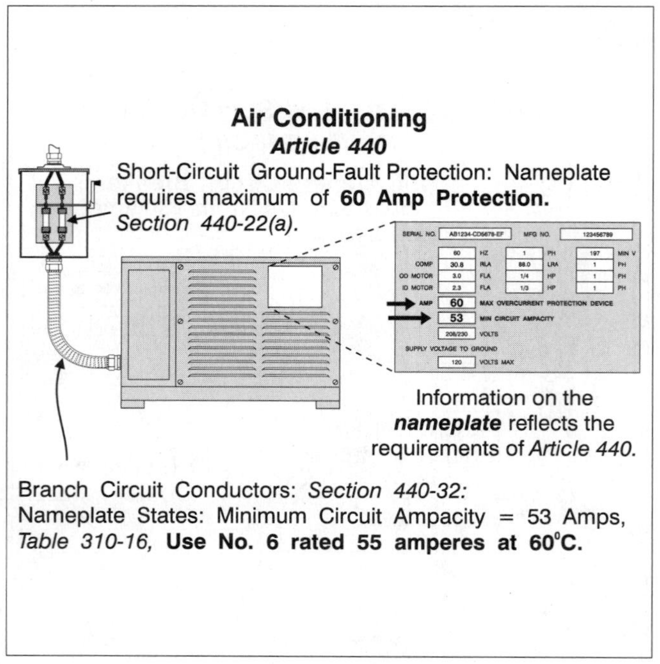

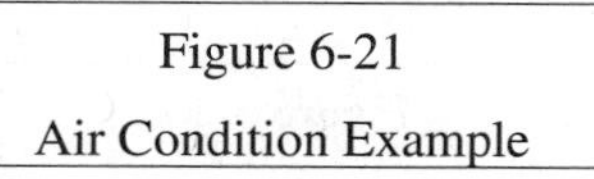

Figure 6-21

Air Condition Example

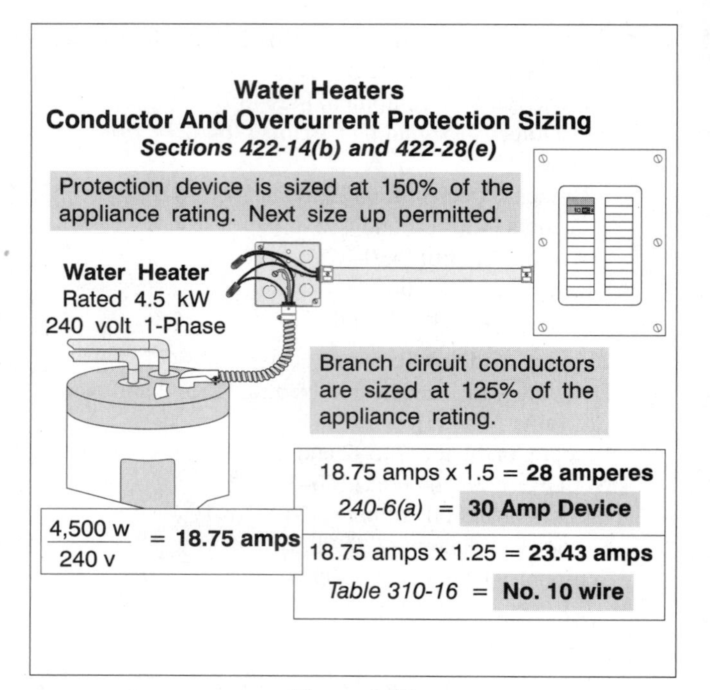

Figure 6-22

Water Heater Example

❑ **Motor Example**

What size branch circuit conductor and short-circuit protection (circuit breaker) is required for a 2-horsepower motor rated 230 volts, Fig. 6–23?

(a) No. 14 with a 15-ampere breaker

(b) No. 12 with a 20-ampere breaker

(c) No. 12 with a 30-ampere breaker

(d) No. 14 with a 30-ampere breaker

- Answer: (d) No. 14 with a 30-ampere protection device

Conductors: Conductors are sized no less than 125 percent of the motor full load current [430–22(a)].
12 amperes × 1.25 = 15 amperes, Table 310–16, No. 14 is rated 20 amperes 60°C .

Protection: The short-circuit protection (circuit breaker) is sized at 250 percent of the motor full load current. 12 amperes × 2.5= 30 amperes [240–6(a) and 430–52(c)(1) Exception No. 1].

Note: The obelisk (†) note on the bottom of Table 310-16 limiting the overcurrent protection device to 15 amperes for No. 14 wire does not apply to motors [240-3(f)]. This concept is explained in detail in Unit 4 on Motors and Transformers.

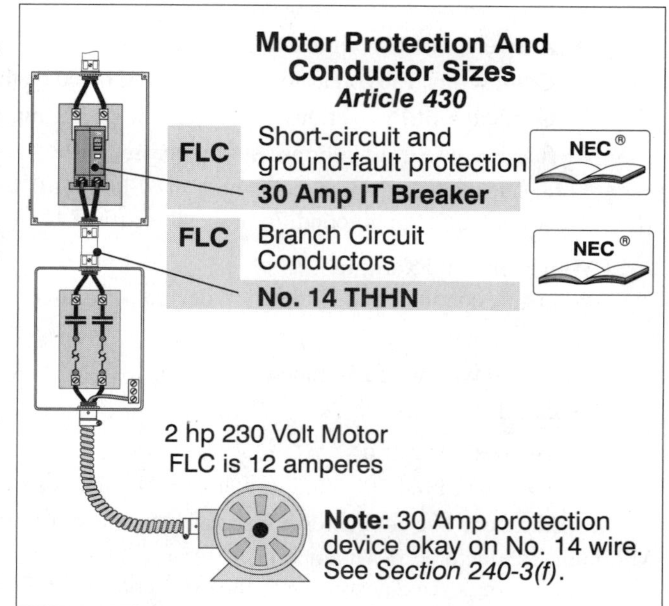

Figure 6-23

Motor Example

6–11 *CONDUCTOR AMPACITY [310–10]*

The insulation temperature rating of a conductor is limited to an operating temperature that prevents serious heat damage to the conductor's insulation. If the conductor carries excessive current, the I^2R heating within the conductor can destroy the conductor insulation. To limit elevated conductor operation temperatures, we must limit the current flow (ampacity) on the conductors.

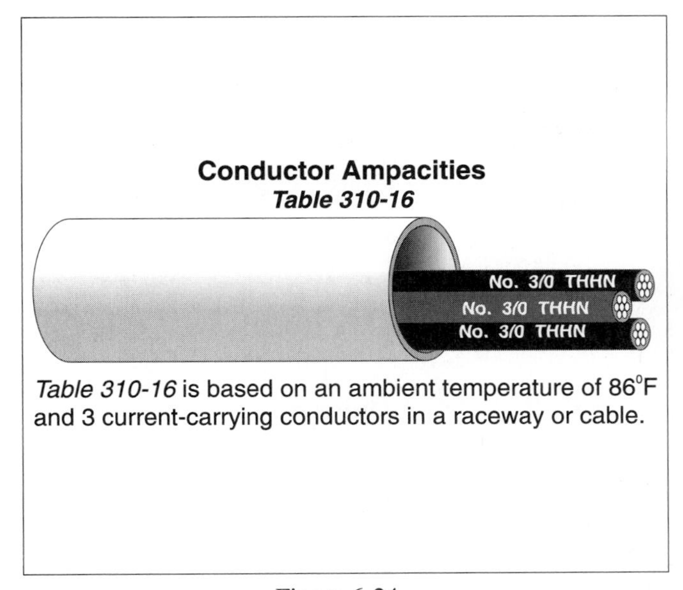

Figure 6-24

Conductor Ampacity, Table 310–16

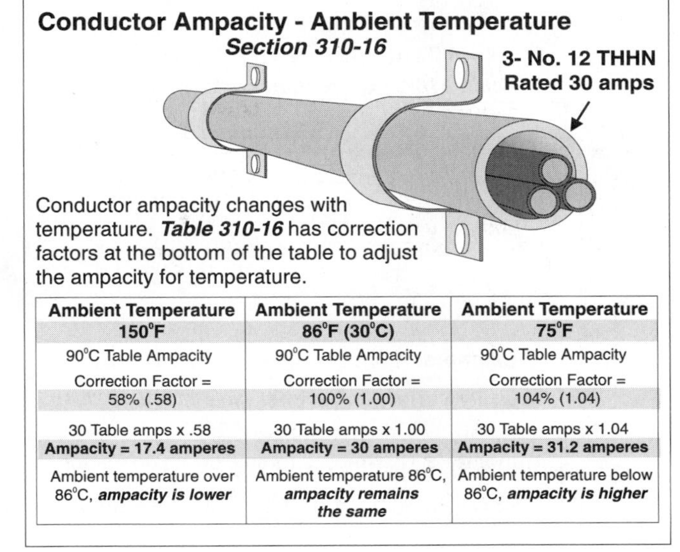

Ambient Temperature 150°F	Ambient Temperature 86°F (30°C)	Ambient Temperature 75°F
90°C Table Ampacity	90°C Table Ampacity	90°C Table Ampacity
Correction Factor = 58% (.58)	Correction Factor = 100% (1.00)	Correction Factor = 104% (1.04)
30 Table amps x .58 **Ampacity = 17.4 amperes**	30 Table amps x 1.00 **Ampacity = 30 amperes**	30 Table amps x 1.04 **Ampacity = 31.2 amperes**
Ambient temperature over 86°C, ***ampacity is lower***	Ambient temperature 86°C, ***ampacity remains the same***	Ambient temperature below 86°C, ***ampacity is higher***

Figure 6-25

Conductor Ampacity Affected by Temperature

Allowable Ampacities

The ampacity of a conductor is the current the conductors can carry *continuously* under the specific condition of use [Article 100 definition]. The ampacity of a conductor is listed in Table 310-16 under the condition of no more than three current-carrying conductors bundled together in an ambient temperature of 86°C. The ampacity of a conductor changes if the ambient temperature is not 86°F or if more than three current-carrying conductors are bundled together in any way, Fig. 6-24.

6–12 *AMBIENT TEMPERATURE DERATING FACTOR [Table 310–16]*

The ampacity of a conductor as listed on Table 310–16 is based on the conductor operating at an ambient temperature of 86°F (30°C). When the ambient temperature is different than 86°F (30°C) for a prolonged period of time, the conductor ampacity as listed in Table 310–16 must be adjusted to a new ampacity, Fig. 6–25.

In general, 90°C rated conductor ampacities cannot be used for sizing a circuit conductor. However higher insulation temperature rating offers the opportunity of having a greater conductor ampacity for conductor ampacity derating and a reduced temperature correction factor. The temperature correction factors used to determine the new conductor ampacity are listed at the bottom of Table 310–16. The following formula can be used to determine the conductor's new ampacity when the ambient temperature is not 86°F (30°C).

Ampacity = Allowable Ampacity × Temperature Correction Factor

Note: If the length of the conductor in the different ambient is 10 feet or less and does not exceed 10 percent of the total length of the conductor, then the ambient temperature correction factors of Table 310–16 do not apply [310–15(c) Exception].

❑ **Ambient Temperature Below 86°F (30°C) Example**

What is the ampacity of No. 12 THHN when installed in a walk-in cooler that has an ambient temperature of 50°F, Fig. 6–26?

(a) 31 amperes (b) 35 amperes (c) 30 amperes (d) 20 amperes

- Answer: (a) 31 amperes

  New Ampacity = Table 310–16 Ampacity × Temperature Correction Factor

  Table 310–16 ampacity for No. 12 THHN is 30 amperes at 90°C.

  Temperature Correction Factor, 90°C conductor rating installed at 50°C is 1.04

  New Ampacity = 30 amperes × 1.04 = 31.2 amperes

  Ampacity increases when the ambient temperature is less than 86°F (30°C).

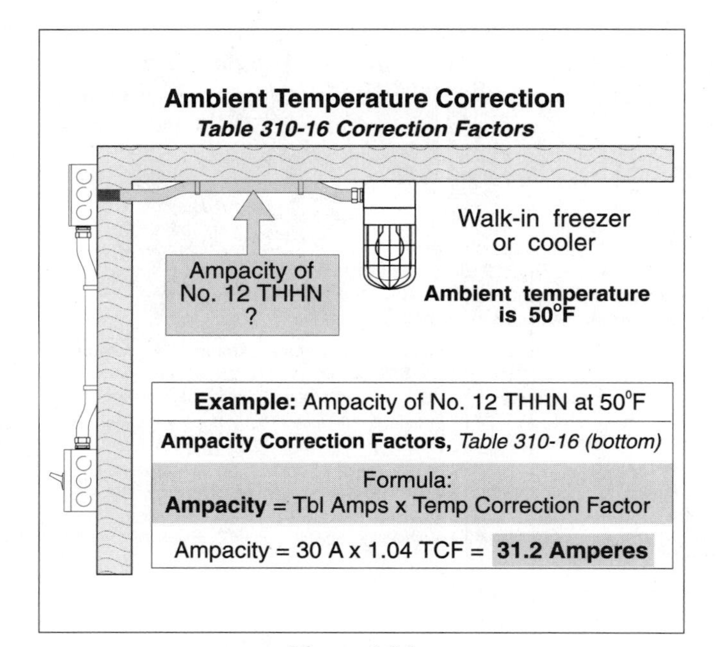

Figure 6-26

Ambient Temperature Correction – Table 310–16

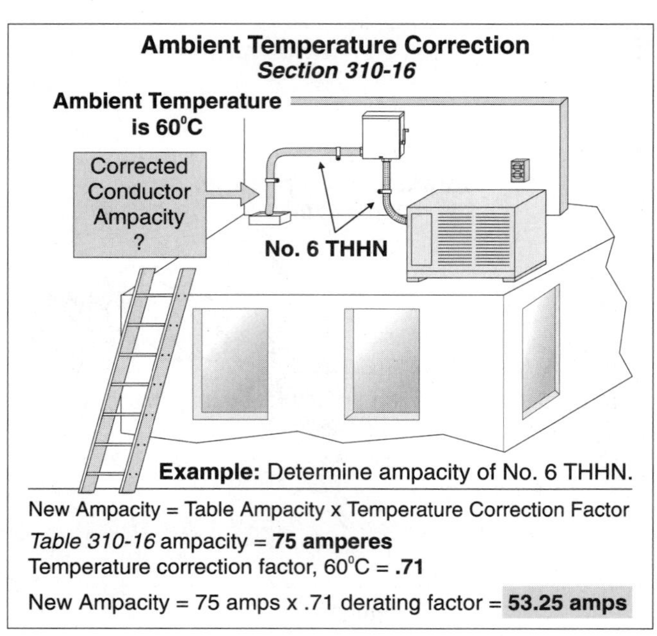

Figure 6-27

Conductor Ampacity – Temperature Below 86ºF

AMBIENT TEMPERATURE CORRECTION FACTORS

When the ambient temperature is other than 86ºF (30ºC), multiply the ampacities of 310–16 by the appropriate factor shown below

Ambient Temp °F Fahrenheit	60°C Insulation	75°C Insulation	90°C Insulation	Ambient Temp°C Centigrade
below 78	1.08	1.05	1.04	below 26
78–86	1.00	1.00	1.00	26–30
87–95	.91	.94	.96	31–35
96–104	.82	.88	.91	36–40
105–113	.71	.82	.87	41–45
114–122	.58	.75	.82	46–50
123–131	.41	.67	.76	51–55
132–140	*	.58	.71	56–60
141–158	*	.33	.58	61–70
159–176	*	*	.41	71–80

* Conductor with this insulation can not be installed in this ambient temperature location.

❑ **Ambient Temperature Above 86ºF (30ºC) Example**

What is the ampacity of No. 6 THHN when installed on a roof that has an ambient temperature of 60ºC, Fig. 6–27?

(a) 53 amperes (b) 35 amperes (c) 75 amperes (d) 60 amperes

- Answer: (a) 53 amperes

  New Ampacity = Table 310–16 Ampacity × Ambient Temperature Correction Factor

  Table 310–16 ampacity for No. 6 THHN is 75 amperes at 90ºC.

  Temperature Correction Factor, 90ºC conductor rating installed at 60ºC is .71

  New Ampacity = 75 amperes × 0.71 = 53.25 amperes

  Ampacity decreases when the ambient temperature is more than 86ºF (30ºC).

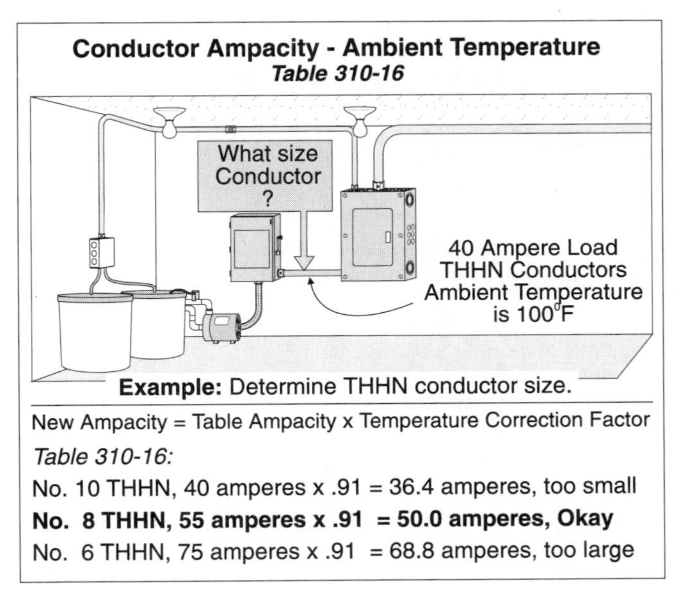

Figure 6-28

Conductor Ampacity, Ambient Temperature Above 86°F

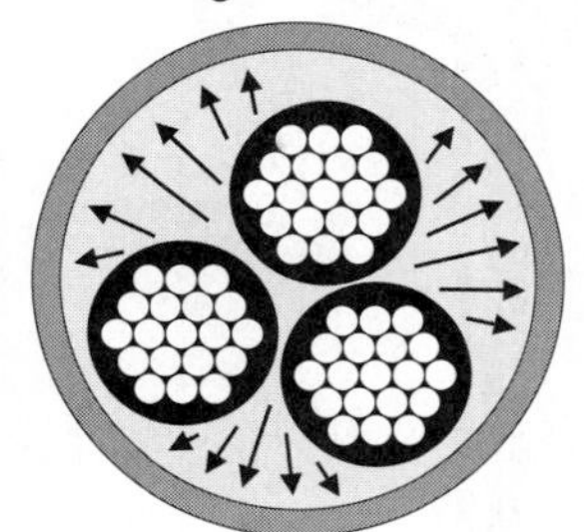
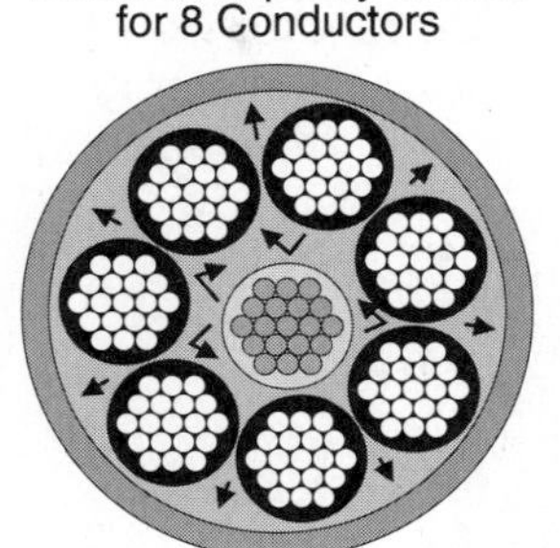

Figure 6-29

Conductor Size Example

❑ **Conductor Size Example**

What size conductor is required to supply a 40 ampere load? The conductors pass through a room where the ambient temperature is 100ºF, Fig. 6–28.

(a) No. 10 THHN (b) No. 8 THHN (c) No. 6 THHN (d) any of these

- Answer: (b) No. 8 THHN

  The conductor to the load must have an ampacity of 40 amperes after applying the ambient temperature derating factor.

  New Ampacity = Table 310–16 Amperes × Ambient Temperature Correction.

  No. 10 THHN, 40 amperes × 0.91 = 36.4 amperes

  No. 8 THHN, 55 amperes × .91 = 50.0 amperes

  No. 6 THHN, 75 amperes × .91 = 68.0 amperes

6–13 *CONDUCTOR BUNCHING DERATING FACTOR, NOTE 8(a) OF TABLE 310–16*

When many conductors are bundled together, the ability of the conductors to dissipate heat is reduced. The National Electrical Code requires that the ampacity of a conductor be reduced whenever four or more current-carrying conductors are bundled together, Fig. 6–29. In general, 90ºC rated conductor ampacities cannot be used for sizing a circuit conductor. However higher insulation temperature rating offers the opportunity of having a greater conductor ampacity for conductor ampacity derating.

The ampacity derating factors used to determine the new ampacity is listed in Note 8(a) of Table 310–16. The following formula can be used to determine the new conductor ampacity when more than three current-carrying conductors are bundled together.

New Ampacity = Table 310–16 Ampacity × Note 8(a) Correction Factor

Note: Note 8(a) derating factors do not apply when bundling conductor in a nipple (not exceeding 24 inches) [See Exception No. 3 to Note 8(a)], Fig. 6–30.

❑ **Conductor Ampacity Example**

What is the ampacity of four current carrying No. 10 THHN conductors installed in a raceway or cable, Fig. 6–31?

(a) 20 amperes (b) 24 amperes (c) 32 amperes (d) None of these

- Answer: (c) 32 amperes.

  New Ampacity = Table 310–16 Ampacity × Note 8(a) Derating Factor

  Table 310–16 ampacity for No. 10 THHN is 40 amperes at 90ºC

  Note 8(a) derating factor for four current-carrying conductors is 0.8

  New Ampacity = 40 amperes × 0.8 = 32 amperes

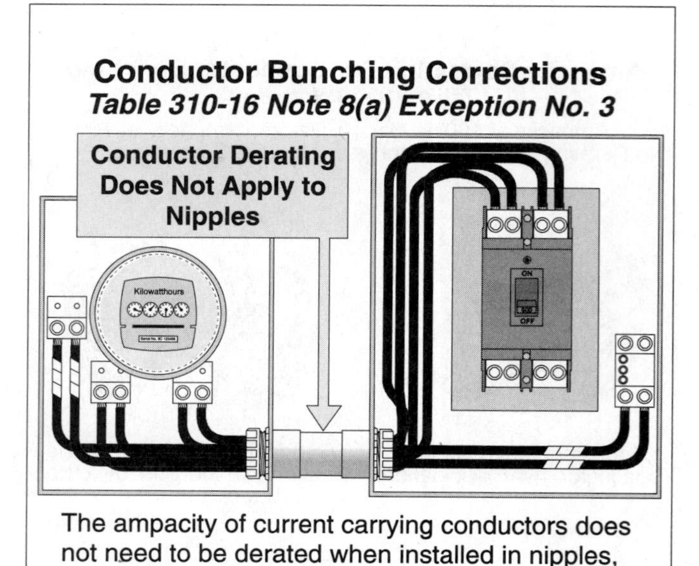

Figure 6-30

Conductor Bunching

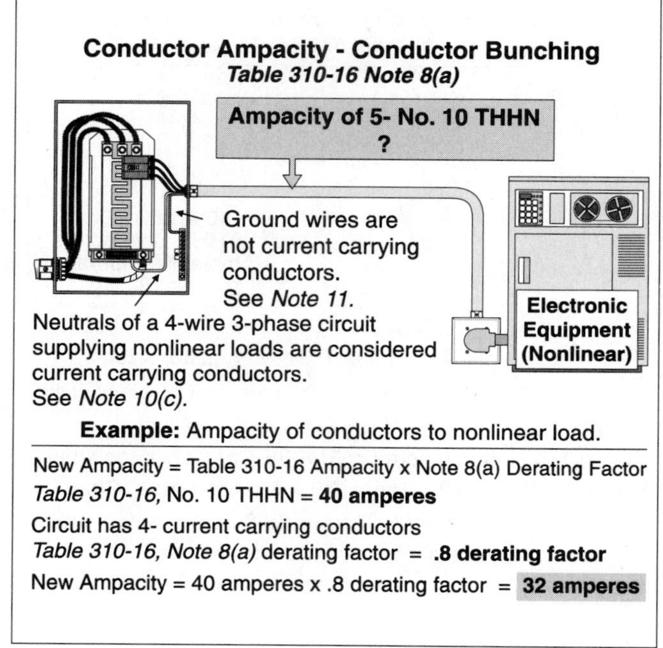

Figure 6-31

Note 8(a) Derating Factors Do Not Apply to Nipples

Conductor Bundling Derating Factors Note 8(a) of Table 310–16	
Number of Current Carrying Conductors	**Conductor Ampacity Derating Factor**
4 –6	80% or .80
7 – 9	70% or .70
10 – 20	050% or .5
21 – 30	45% or .45
31 – 40	40% or .40
41 – 60	35% or .35

❑ **Conductor Size Example**

A raceway contains four current-carrying conductors. What size conductor is required to supply a 40-ampere load, Fig. 6–32?

(a) No. 10 THHN (b) No. 8 THHN
(c) No. 6 THHN (d) None of these

- Answer: (b) No. 8 THHW

  The conductor must have an ampacity of 40 amperes after applying Note 8 derating factor.

  Ampacity = 310–16 Ampacity × Note 8 Derating Factor.

  No. 10 THHN, 40 amperes × 0.8 = 32 amperes

  No. 8 THHN, 55 amperes × 0.8 = 44 amperes

  No. 6 THHN, 75 amperes 0.8 = 60 amperes

6–14 *AMBIENT TEMPERATURE AND CONDUCTOR BUNDLING DERATING FACTORS*

If the ambient temperature is different than 86ºF (30ºC) and there are more than three current-carrying conductors bundled together, then the ampacity listed in Table 310–16 must be adjusted for both conditions, Fig. 6–33.

The following formula can be used to determine the new conductor ampacity when both ambient temperature and Note 8(a) derating factors apply:

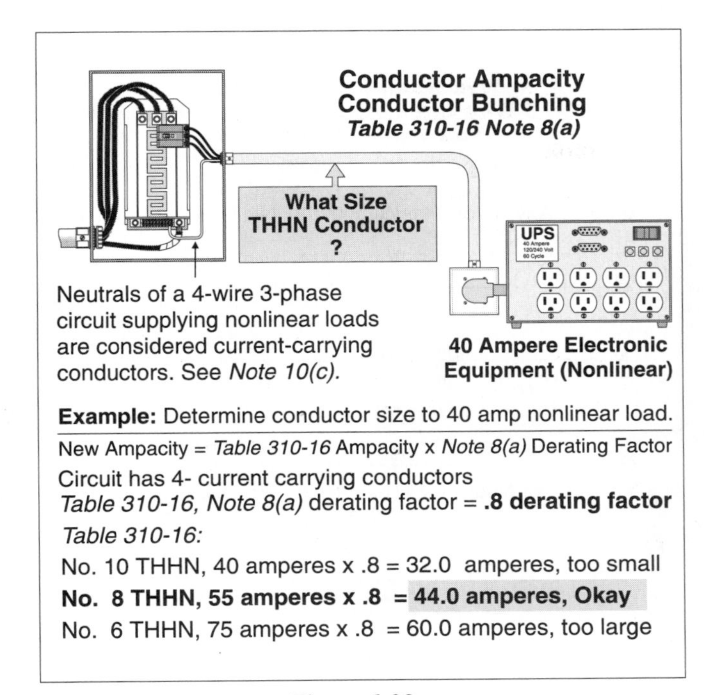

Figure 6-32

Conductor Bundle Ampacity Example

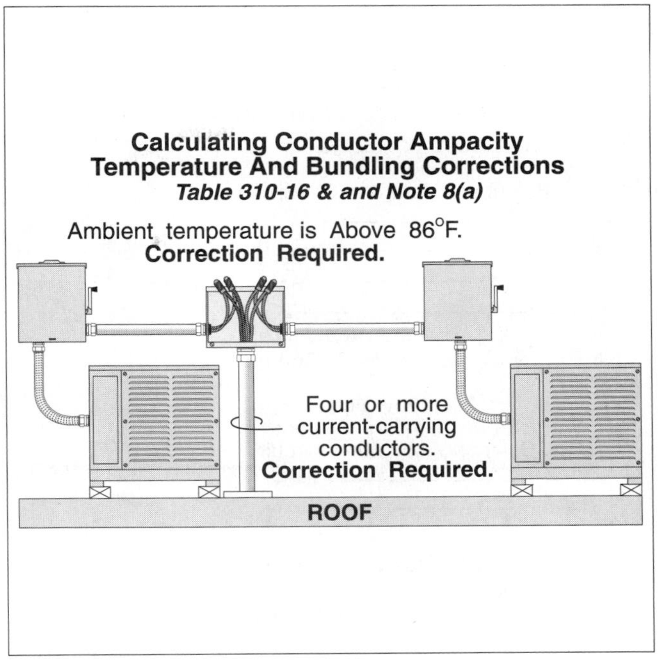

Calculating Conductor Ampacity
Temperature And Bundling Corrections
Table 310-16 & and Note 8(a)

Ambient temperature is Above 86°F.
Correction Required.

Four or more current-carrying conductors.
Correction Required.

ROOF

Figure 6-33

Calculating Conductor Ampacity

Ampacity = Table 310–16 Ampacity × Temperature Correction Factor × Note 8(a) Correction Factor

Note: If the length of conductor bunching or ambient temperature is 10 feet or less and does not exceed 10 percent of the conductor length, then neither ampacity correction factor applies [310–15(c), Exception].

❑ **Conductor Ampacity Example**

What is the ampacity of four current-carrying No. 8 THHN conductors installed in ambient temperature of 100ºF, Fig. 6–34?

(a) 25 amperes (b) 40 amperes (c) 55 amperes (d) 60 amperes

- Answer: (b) 40 amperes

New Ampacity = Table 310–16 Ampacity × Temperature Factor × Note 8(a) Factor

Table 310–16 ampacity of No. 8 THHN is 55 amperes at 90ºC.

Temperature correction factor for 90ºC conductor insulation at 100ºF is 0.91.

Note 8(a) derating factor for four conductors is 0.8.

New Ampacity = 55 amperes × 0.91 × 0.8

New Ampacity = 40 amperes

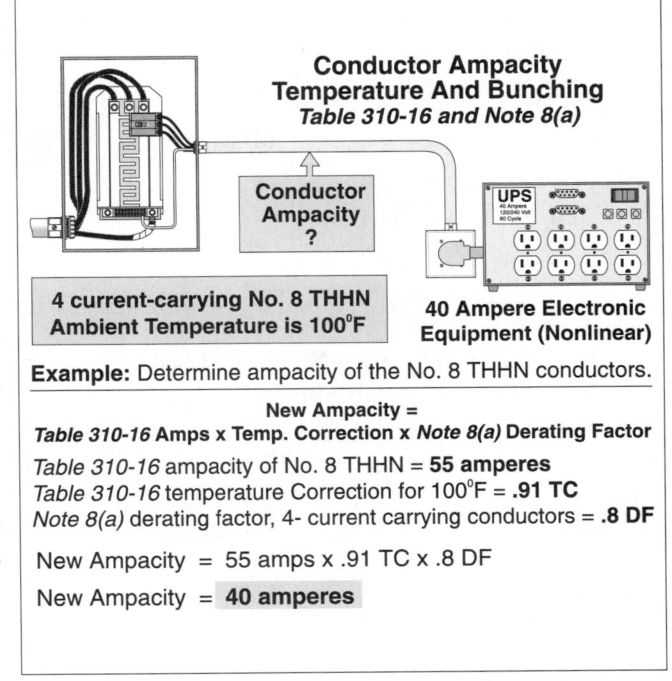

Figure 6-34

Ampacity Correction, Both Temperature and Bundling

6–15 *CURRENT-CARRYING CONDUCTORS*

Note 8(a) of Table 310–16 derating factors only apply when there is more than three current-carrying conductors bundled. Naturally all phase conductors are considered current-carrying and the following should be helpful to determine which other conductors are considered current carrying.

Grounded Conductor – Balanced Circuits, Note 10(a) of Table 310–16

The grounded (neutral) conductor of a balanced 3-wire circuit, or a balanced 4-wire wye circuit is not considered a current-carrying conductor, Fig. 6–35.

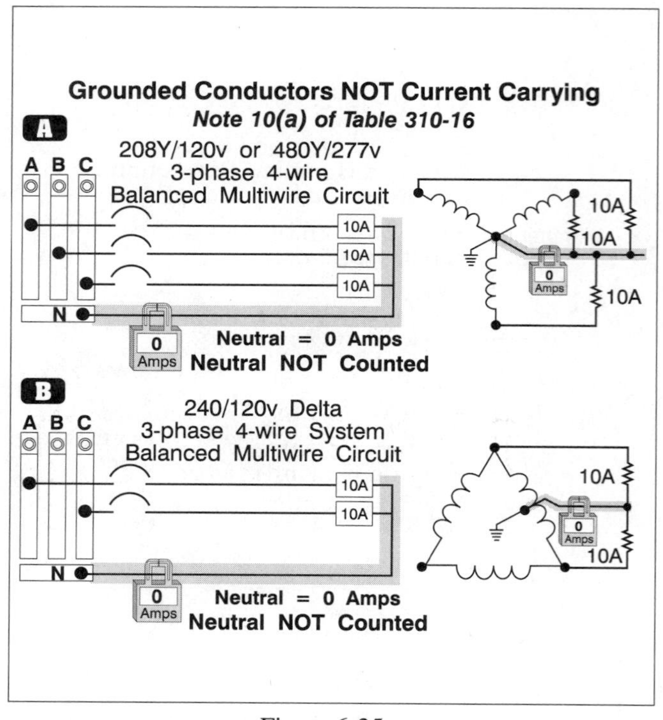

Figure 6-35

Conductor Ampacity Correction Example

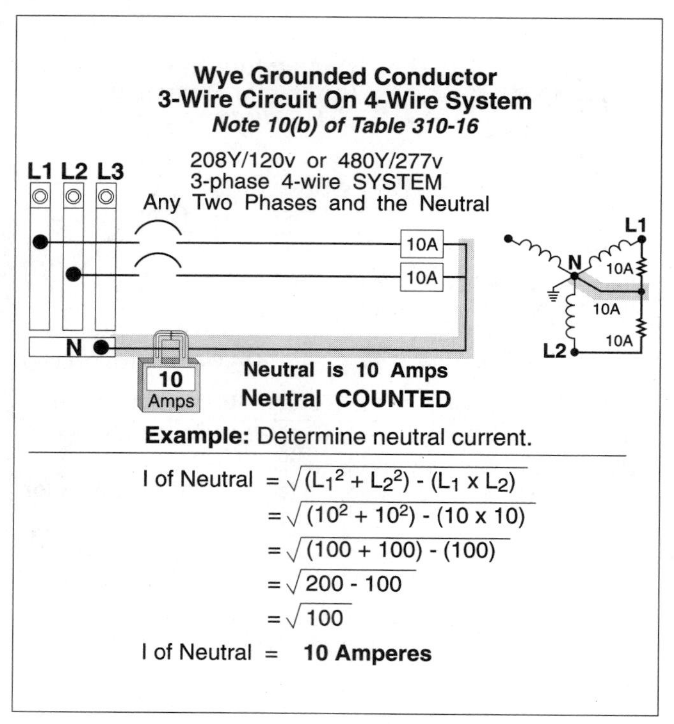

Figure 6-36

Balanced Circuits

Grounded Conductor – Unbalanced 3-wire Wye Circuit, Note 10(b) of Table 310–16

The grounded (neutral) conductor of balanced 3-wire wye circuit is considered a current-carrying conductor, Fig. 6–36. This can be proven with the following formula:

$I_n = \sqrt{(L_1 + L_2) - (L_1 \times L_2)}$ **L_1 = Current of one phase** **L_2 = Current of the other phase**

❑ **Grounded Conductor Example**

What is the neutral current for a balanced 16-ampere, 3-wire, 208Y/120-volt branch circuit that supplies fluorescent lighting, Fig. 6–37?

(a) 8 amperes (b) 16 amperes

(c) 32 amperes (d) none of these

- Answer: (b) 16 amperes

$I_n = \sqrt{(L_1 + L_2) - (L_1 \times l_2)}$

$I_n = \sqrt{(16^2 + 16^2) - (16 \times 16)}$

$I_n = \sqrt{512 - 256}$ $I_n = \sqrt{256}$ I_n = 16 amperes

Grounded Conductor – Nonlinear Loads, Note 10(c) of Table 310–16

The grounded (neutral) conductor of a balanced 4-wire wye circuit that is at least 50 percent loaded with nonlinear loads (computers, electric discharge lighting, etc.) is considered a current-carrying conductor, Fig. 6–38.

CAUTION: Because of nonlinear loads, triplen harmonic currents add on the neutral conductor. The actual current on the neutral can be almost 200 percent of the ungrounded conductor's current, Fig. 6–39.

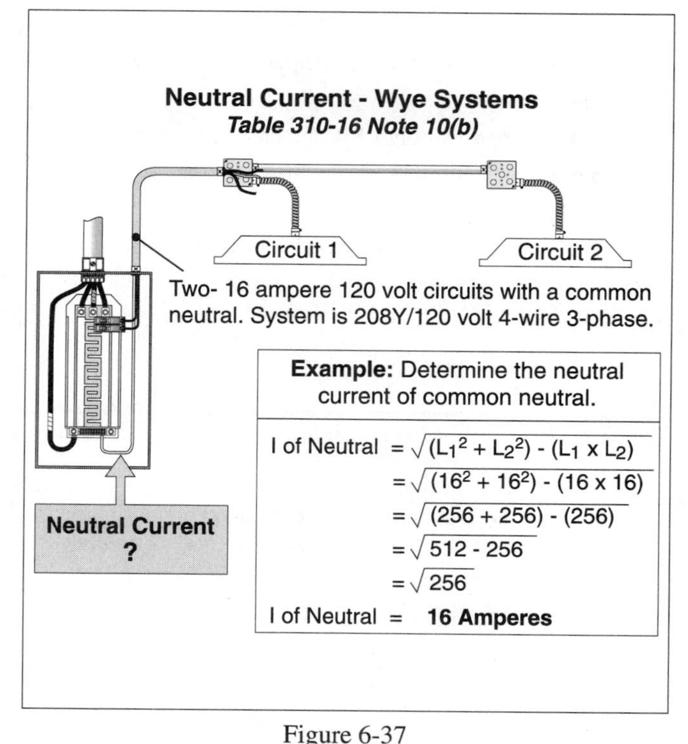

Figure 6-37

Wye 3-wire Unbalanced Circuit

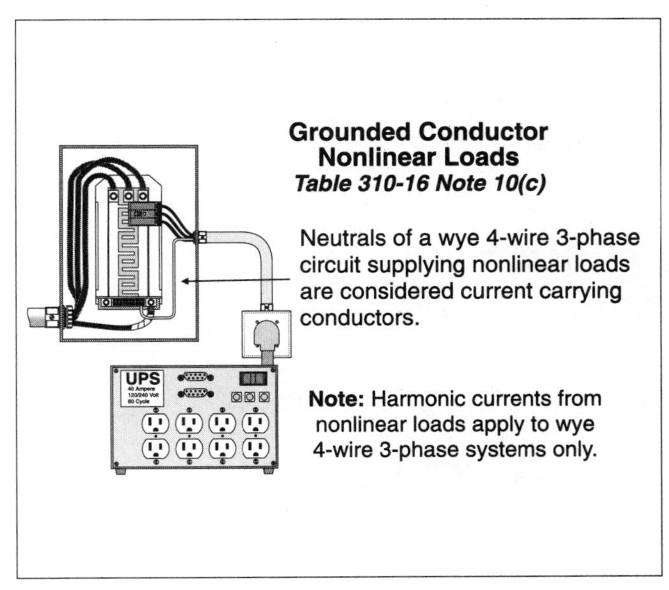

Figure 6-38

Unbalanced Wye Example

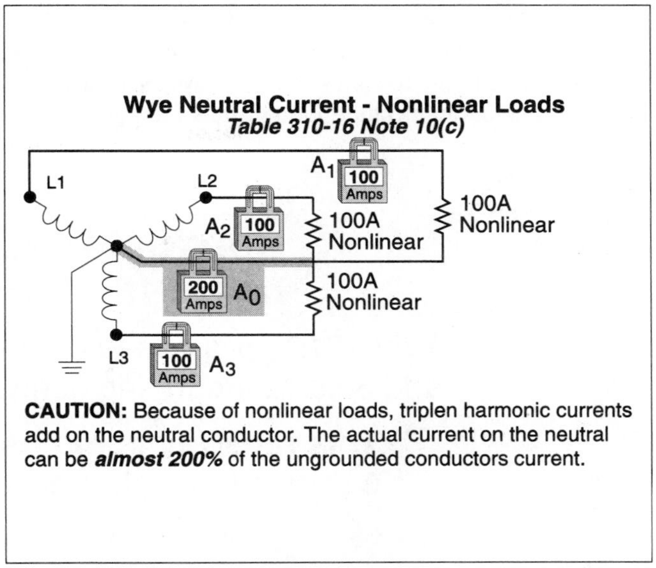

Wye Neutral Current - Nonlinear Loads
Table 310-16 Note 10(c)

L1
L2
L3
A_1 100 Amps
A_2 100 Amps
A_0 200 Amps
A_3 100 Amps
100A Nonlinear
100A Nonlinear
100A Nonlinear

CAUTION: Because of nonlinear loads, triplen harmonic currents add on the neutral conductor. The actual current on the neutral can be ***almost 200%*** of the ungrounded conductors current.

Figure 6-39

Nonlinear Loads

Two-wire Circuits

Both the grounded and ungrounded conductor of a two-wire circuit carries current and both are considered current-carrying, Fig. 6–40.

Grounding and Bonding Conductors – Note 11 of Table 310–16

Grounding and bonding conductors do not normally carry current and are not considered current-carrying , Fig. 6–41.

6–16 *CONDUCTOR SIZING SUMMARY*

Conductor Ampacity, Section 310–10

The ampacity of a conductor is dynamic and changes with changing conditions. The factors that affect conductor ampacity are, Fig. 6–42:

1) The allowable ampacity as listed in Table 310–16.
2) The ambient temperature correction factors, if the ambient temperature is not 86°F.
3) Conductor bunching derating factors, if more than three current-carrying conductors are bundled together.

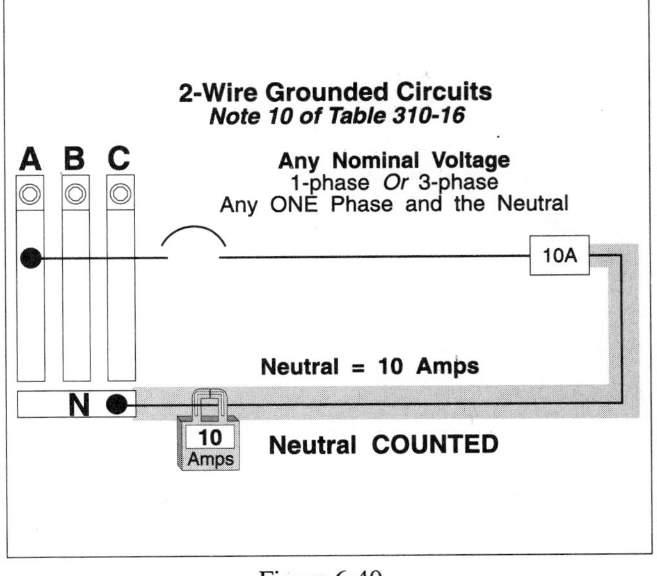

Figure 6-40

Neutral Overload

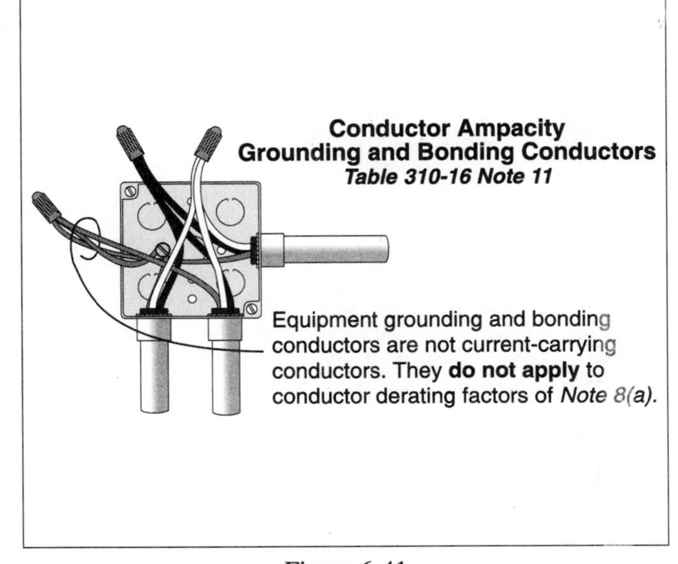

Figure 6-41

Two Wire Circuits

Terminal Ratings, Section 110–14(c)

Equipment rated 100 amperes or less must have the conductor sized no smaller than the 60ºC column of Table 310–16. Equipment rated over 100 amperes must have the conductors sized no smaller than the 75ºC column of Table 310–16. However higher insulation temperature rating offers the opportunity of having a greater conductor ampacity for conductor ampacity derating.

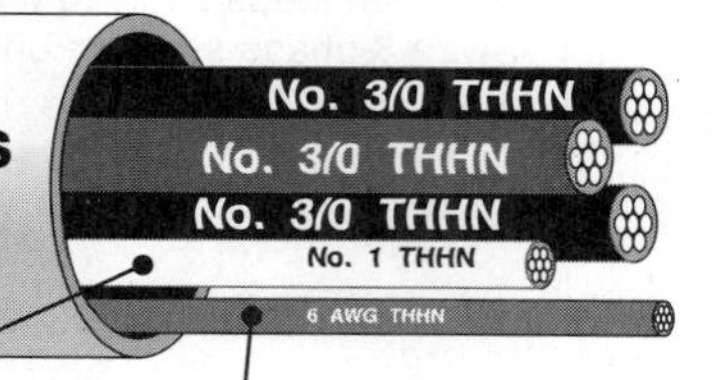

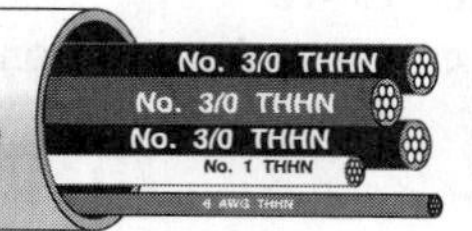

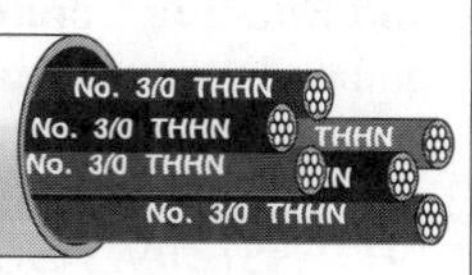

Figure 6-42

Conductor Sizing Summary

Unit 6 – Conductor Sizing And Protection Summary Questions

6–1 Conductor Insulation Property [Table 310-13]

1. THHN can be described as ________?
   (a) thermoplastic insulation with an nylon outer cover
   (b) suitable for dry and wet locations
   (c) maximum operating temperature of 90°C
   (d) a and c

6–2 Conductor Allowable Ampacity [310–15]

2. • The maximum overcurrent protection device size for No. 14 is 15 amperes, No. 12 is 20 amperes, and No. 10 is 30 amperes. This is a general rule, but it does not apply to motors or air conditioners according to Section 240-3.
   (a) True (b) False

6–3 Conductor Sizing

3. • Conductor sizes are expressed in American Wire Gage (AWG) from No. 40 through 0000. Conductors larger than ________ are expressed in circular mils.
   (a) No. 1/0 (b) No. 1
   (c) No. 3/0 (d) No. 4/0

4. The smallest size conductor permitted for branch circuits, feeders, and services for residential, commercial, and industrial locations is ________.
   (a) No. 14 copper (b) No. 12 Aluminun
   (c) No. 12 copper (d) a and b

6–4 Terminal Rating [110-14(c)]

5. Equipment terminals rated 100 amperes or less (circuit breakers, fuses, etc.) and pressure connector terminals for No. 14 through No. 1 conductors, shall have the conductor sized according to 60°C temperature rating as listed in Table 310–16.
   (a) True (b) False

6. • What is the minimum size THHN conductor that is permitted to terminate on a 70-ampere circuit breaker of fuse? Be sure to comply with the requirements of Section 110-14(c)(1).
   (a) No. 8 (b) No. 6
   (c) No. 4 (d) none of these

7. • What size THHN conductor is required for a 70-ampere branch-circuit if the circuit breaker and equipment is listed for 75°C terminals and the load does not exceed 65 amperes?
   (a) No. 10 (b) No. 8 (c) No. 6 (d) No. 4

8. Terminals for equipment rated over 100 amperes and pressure connector terminals for conductors larger than No. 1, shall have the conductor sized according to a 75°C temperature rating as listed in Table 310–16.
   (a) True (b) False

9. • What size THHN conductor is required for an air-conditioning unit, if the nameplate requires a conductor ampacity of 34 amperes? Terminals of all the equipment and circuit breakers are rated 75°C.
   (a) No. 12 (b) No. 10
   (c) No. 8 (d) No. 14

10. What is the minumum size THHN conductor is required for a 150-ampere circuit breaker or fuse? Be sure to comply with the requirements of Section 110-14(c)(2).
    (a) No. 1/0 (b) No. 2/0
    (c) No. 3/0 (d) No. 4/0

11. In general, THHN (90°C) conductor ampacities cannot be used when sizing conductor. However, when more than three current-conductors conductors are bundled together, or if the ambient temperature is greater than 86°F, the allowable conductor ampacity must be decreased. THHN offers the opportunity of having a greater ampacity for conductor derating purposes, thereby permitting the same conductor to be used without having to increase the conductor size.
(a) True (b) False

12. • What size conductor is required to supply a 190-ampere load in a dry location? Terminals are rated 75°C.
(a) 300 kcmils THHN (b) No. 4/0 XHHW
(c) No. 3/0 THWN (d) none of these

6–5 Conductors In Parallel [310–4]

13. • Phase and grounded (neutral) conductors sized No. 1 AWG and larger are permitted to be connected in parallel.
(a) True (b) False

14. To insure that the currents are evenly distributed between the parallel conductors, each conductor within a parallel set must be installed in the same type of raceway (metal or nonmetallic) and must be the same length, material, mils area, insulation type, and must terminate in the same method.
(a) True (b) False

15. • Paralleling of conductors is done by sets. One phase or neutral is not required to be paralleled the same as those of another phase or neutral.
(a) True (b) False

16. When an electric relay (coil) is energized, the initial current can be very high causing significant voltage drop. The reduced voltage at the coil (because of voltage drop) can cause the coil contacts to chatter (open and close like a buzzer) or not close at all. Paralleling of control wiring conductors is often necessary to reduce the effects of voltage drop for long control runs.
(a) True (b) False

17. When equipment grounding conductors are installed in parallel, each raceway must have a full-size equipment grounding conductor sized according to the overcurrent protection device rating of that circuit.
(a) True (b) False

18. All parallel equipment grounding conductors must be the same length, material, circular mils, insulation, terminate in the same manner, and are required to be a minimum No. 1/0.
(a) True (b) False

19. What size equipment grounding conductor is required in each raceway for an 800-ampere, 500-kcmil feeder parallel in two raceways?
(a) No. 3 (b) No. 2 (c) No. 1 (d) No. 1/0

20. If we have an 800-ampere service, with a calculated demand load of 750 amperes, what size 75°C conductors would be required if parallel in two raceways?
(a) two – No. 4/0 (b) two – 250 kcmil (c) two – 500 kcmil (d) two – 750 kcmil

21. • What are the circular mils required for a 250-ampere feeder paralleled in two raceways?
(a) two – No. 3 (b) two – No. 2 (c) two – No. 2/0 (d) two – No. 1/0

6–6 Conductor Size – Voltage Drop

22. There is no mandatory rule in the NEC® to limit the voltage drop on conductors, but the NEC® recommends that you consider it effects.
(a) True (b) False

6–7 Overcurrent Protection [Article 240]

23. One of the purposes of conductor overcurrent protection is to protect the conductors against excessive or dangerous heat.
(a) True (b) False

24. The overcurrent device must be designed and rated to clear fault current and must have a short-circuit interrupting rating sufficient for the available fault levels. The minimum interruption rating for circuit breakers is ________ amperes and ________ amperes for fuses.
(a) 10,000; 10,000 (b) 5,000; 5,000 (c) 5,000; 10,000 (d) 10,000; 5,000

25. The following are the standard sized circuit breakers and fuses _______ ampere.
(a) 25 (b) 90
(c) 350 (d) any of these

26. The maximum continuous load permitted on an overcurrent protection device is limited to ______ percent of the device rating.
(a) 80 (b) 100 (c) 125 (d) 150

6–8 Conductor Sizing And Protection [240-3]

27. If the ampacity of a conductor does not correspond with the standard ampere rating of a fuse or circuit breaker, the next size up protection device is permitted. This applies only if the conductors supply multioutlet receptacles for portable cord- and plug-connected loads.
(a) True (b) False

28. What size conductor (75°C) is required for a 70-ampere breaker that supplies a 70-ampere load?
(a) No. 8 (b) No. 6 (c) No. 4 (d) Any of these

6–11 Conductor Ampacity [310-10 And 310-15]

29. The temperature rating of a conductor is the maximum operating temperature the conductor insulation can withstand (without serious damage) over a prolonged period of time. The ___________ provide guidance for adjusting the conductors ampacities for the different conditions.
(a) conductor allowable ampacities
(b) ambient temperature correction factors
(c) Notes correction factors
(d) continuous load factor
(e) a, b and c

6–12 Ambient Temperature Correction Factor [Table 310–16]

30. The ampacities listed in Table 310–16 apply only when the ambient temperature is 40°C and there are no more than two current-carrying conductors bundled together. If the ambient temperature is not 40°C, or there are more than two current-carrying conductors in a raceway, the allowable ampacities must be adjusted to reflect the ampacity under the condition of use.
(a) True (b) False

31. • What is the ampacity of No. 8 THHN conductors when installed in a walk-in cooler if the ambient temperature is 50°F?
(a) 40 amperes (b) 50 amperes (c) 55 amperes (d) 57 amperes

32. • What size conductor is required to feed a 16-ampere load, when the conductors (in a raceway) pass over a roof, ambient temperature of 100°F? The circuit is protected with a 20-ampere protection device.
(a) No. 14 THHN (b) No. 12 THHN (c) No.10 THHN (d) No. 8 THHN

6–13 Conductor Bunching Derating Factor, Note 8(a) Of Table 310–16

33. When four or more current-carrying conductors are bundled together for more than _____ inches, the conductor allowable ampacity must be reduced according to the factors listed in Note 8(a) of Table 310–16.
(a) 12 (b) 24 (c) 36 (d) 48

34. What is the ampacity of four No. 1/0 THHN conductors?
(a) 110 amperes (b) 135 amperes (c) 155 amperes (d) 175 amperes

35. • A raceway contains eight current-carrying conductors. What size conductor is required to feed a 21 ampere noncontinuous lighting load? Overcurrent protection device rated 30 amperes.
(a) No. 14 THHN (b) No. 12 THHN (c) No. 10 THHN (d) Any of these

6–14 Ambient Temperature And Conductor Bunching Derating Factors

36. What is the ampacity of eight current-carrying No. 10 THHN conductors installed in ambient temperature of 100ºF?
(a) 21 amperes (b) 26 amperes (c) 32 amperes (d) 40 amperes

6–15 Current-Carrying Conductors

37. • The neutral conductor of a balanced three-wire, delta circuit, or four-wire, wye circuit is considered a current-carrying conductor for the purpose of applying Note 8 derating factors.
(a) True (b) False

38. The neutral conductor of a balanced four-wire wye circuit that is at least 50 percent loaded with nonlinear loads (electric-discharge lighting, electronic ballast, dimmers, controls, computers, laboratory test equipment, medical test equipment, recording studio equipment, etc.) is not considered a current-carrying conductor for the purpose of applying Note 8 derating factors.
(a) True (b) False

39. • The neutral conductor of balanced three-wire wye circuit is not considered a current-carrying conductor for the purpose of applying Note 8(a) derating factors.
(a) True (b) False

6–16 Conductor Sizing Summary

40. • The ampacity of a conductor can be different along the length of the conductor. The higher ampacity is permitted to be used for the lower ampacity, if the length of the lower ampacity is no more than 10 feet or no more than 10 percent of the length of the circuit conductors.
(a) True (b) False

41. Most terminals are rated 60ºC for equipment 100 amperes and less and 75ºC for equipment terminals rated over 100 amperes. Despite the conductor ampacity, conductors must be sized no smaller than the terminal temperature rating.
(a) True (b) False

Challenge Questions

6–7 Overcurrent Protection - Continuous Load [220–3(a), 220–10(b), 384–16(a)]

42. • A continuous load of 27 amperes requires the circuit overcurrent device to be sized at ______ amperes.
(a) 20 (b) 30 (c) 40 (d) 35

43. • What size overcurrent protection device is required for a 45 ampere continuous load? The circuit is in a raceway with 14 current-carrying conductors.
(a) 45 amperes (b) 50 amperes (c) 60 amperes (d) 70 amperes

44. • A 65 ampere continuous load requires a ______ ampere overcurrent protection device.
(a) 60 (b) 70 (c) 75 (d) 90

45. • A department store (continuous load) feeder supplies a lighting load of 103 amperes. The minimum size overcurrent protection device permitted for this feeder is ______ amperes.
(a) 110 (b) 125 (c) 150 (d) 175

6–12 Ambient Temperature Derating Factor [Table 310–16]

46. • A No. 2 TW conductor is installed in a location where the ambient temperature is expected to be 102ºF. The temperature correction factor for conductor ampacity in this location is ______.
(a) .96 (b) .88 (c) .82 (d) .71

47. • If the ambient temperature is 71ºC, the minimum insulation that a conductor must have and still have the capacity to carry current is ______.
(a) 60ºC (b) 105ºC (c) 90ºC (d) any of these

6–13 Conductor Bunching Derating Factor, Note 8(a) Of Table 310–16

48. • The ampacity of six current-carrying No. 4/0 XHHW aluminum conductors installed in a ground floor slab (wet location) is ______ amperes.
(a) 135 (b) 185 (c) 144 (d) 210

6–14 Ambient Temperature And Conductor Bundling Derating Factors

49. • The ampacity of fifteen current-carrying No. 10 RHW aluminum conductors, in an ambient temperature of 75°F, would be ______ amperes.
(a) 30 (b) 22 (c) 16 (d) 12

50. • A No. ______ THHN conductor is required for a 19.7-ampere load if the ambient temperature is 75°F, and there are nine current-carrying conductors in the raceway.
(a) 14 (b) 12 (c) 10 (d) 8

51. • The ampacity of the nine current-carrying No. 10 THW conductors installed in a 20-inch-long raceway is ______ amperes.
(a) 25 (b) 30 (c) 35 (d) none of these

52. • The ampacity of ten current-carrying No. 6 THHW conductors installed in an 18-inch conduit, with an ambient temperature of 39°C is ______ amperes.
(a) 47 (b) 68 (c) 66 (d) 75

6–15 Current-Carrying Conductors

53. A raceway contains the following: One four-wire multiwire circuit that supplies a balanced incandescent 120 volt lighting load; one four-wire multiwire circuit that supplies a balanced 120 volt fluorescent lighting load; two conductors that supply a receptacle; and there is one equipment grounding conductor. The system is three phase 208Y/120 volts. How many of these conductors are considered current-carrying?
(a) 7 conductors (b) 8 conductors (c) 9 conductors (d) 11 conductors

54. • There are a total of nine No. 10 THW conductors in a raceway. The system voltage is three-phase, 208Y/120 volts. One of the conductors is an equipment grounding conductor; four conductors supply a four-wire multiwire 120-volt branch circuit for balanced fluorescent lighting, and the remaining conductors supply a four-wire multiwire 120-volt branch circuit for balanced incandescent lighting. Taking all of these factors into consideration, how many of these conductors are considered current-carrying?
(a) 6 conductors (b) 7 conductors (c) 9 conductors (d) 10 conductors

Unit 7

Motor Calculations

OBJECTIVES

After reading this unit, the student should be able to briefly explain the following concepts:

- Branch circuit short-circuit
- Ground-fault protection
- Feeder-protection
- Feeder conductor size
- Highest rated motor
- Motor overcurrent protection
- Motor branch circuit conductors
- Motor VA calculations
- Motor calculation steps
- Overload protection

After reading this unit, the student should be able to briefly explain the following terms:

- Code letter
- Dual-element fuse
- Ground-fault
- Inverse time breaker
- Heaters
- Motor full-load current
- Motor nameplate current rating
- Next smaller device size
- One-time fuse
- Overcurrent protection
- Overcurrent
- Overload
- Service factor (SF)
- Short-circuit
- Temperature rise

INTRODUCTION

When sizing conductors and *overcurrent protection* for motors, we must comply with the requirements of the National Electrical Code, specifically Article 430 [240–3(f)].

7–1 *MOTOR BRANCH CIRCUIT CONDUCTORS [430–22(a)]*

Branch circuit conductors to a single motor must have an ampacity of not less than 125 percent of the motor's full-load current as listed in Tables 430–147 through 430–150 [430–6(a)]. The actual conductor size must be selected from Table 310–16 according to the terminal temperature rating (60° or 75°C) of the equipment [110–14(c)], Fig. 7–1.

❑ **Motor Branch Circuit Conductors Example**

What size THHN conductor is required for a 2-horsepower, 240-volt single-phase motor, Fig. 7–2?

(a) No. 14 (b) No. 12
(c) No. 10 (d) No. 8

- Answer: (a) No. 14

Motor Full-Load Current – Table 430–148: 2-horsepower, 240-volts single-phase = 12 amperes.

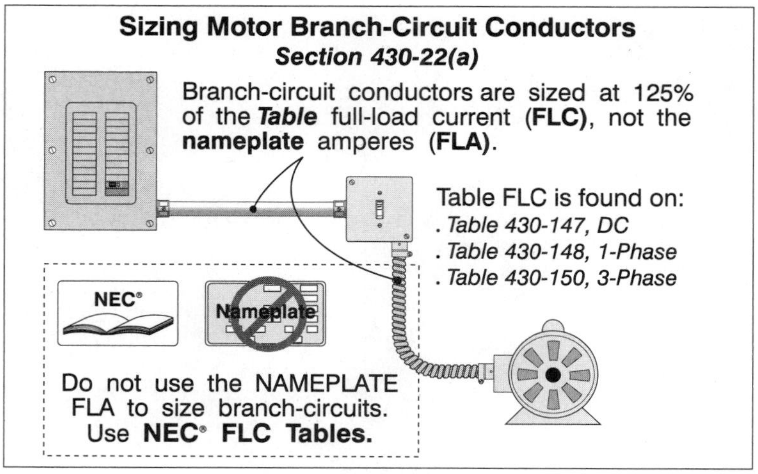

Figure 7-1

Branch Circuit Conductor Size

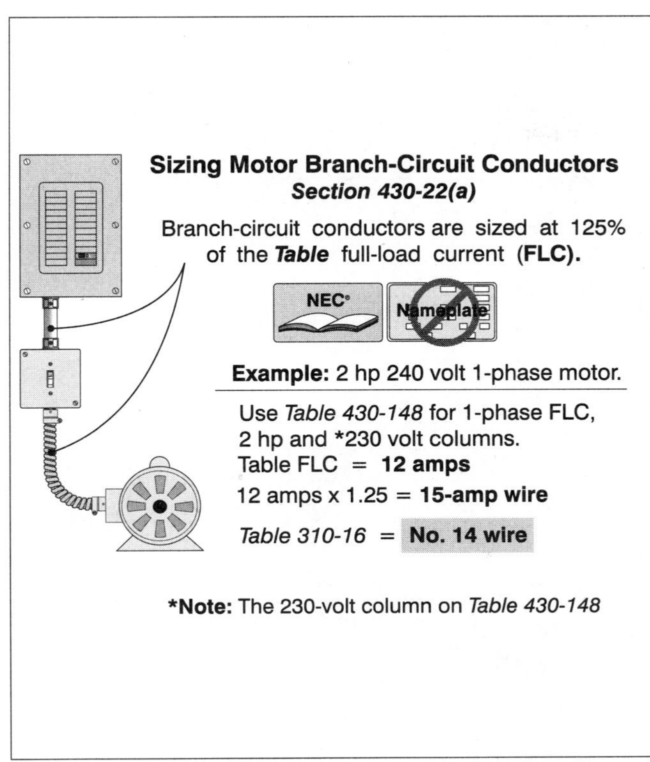

Figure 7-2

Conductor Size Example

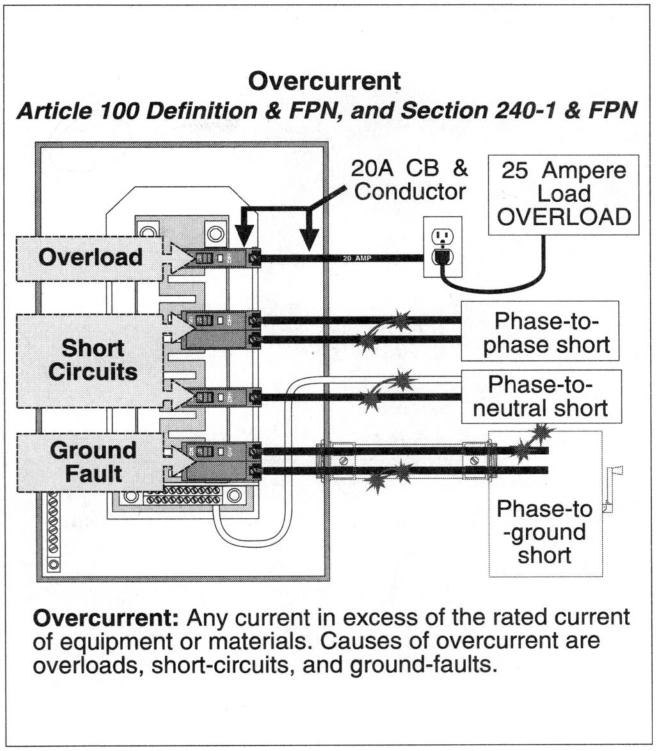

Overcurrent

Article 100 Definition & FPN, and Section 240-1 & FPN

Overload

Short Circuits

Ground Fault

20A CB & Conductor

25 Ampere Load OVERLOAD

Phase-to-phase short

Phase-to-neutral short

Phase-to -ground short

Overcurrent: Any current in excess of the rated current of equipment or materials. Causes of overcurrent are overloads, short-circuits, and ground-faults.

Figure 7-3

Overcurrent–Definition

Conductor Sized at 125 percent Of Motor Full-Load Current:
12 amperes × 1.25 = 15 amperes, Table 310–16, No. 14 THHN rated 20 amperes at 60°C.

Note: The minimum size conductor permitted for building wiring is No. 14 [310–5], however local codes and many industrial facilities contain requirements that No. 12 be used as the smallest building wire.

7–2 *MOTOR OVERCURRENT PROTECTION*

Motors and their associated equipment must be protected against overcurrent (overload, short-circuit or ground-fault) [Article 100], Fig. 7–3. Due to the special characteristics of induction motors, overcurrent protection is generally accomplished by having the overload protection separated from the *short-circuit* and *ground-fault protection* device, Fig. 7–4. Article 430, Part C contains the requirements for motor overload protection and Part D of Article 430 contains the requirements for motor short-circuit and ground-fault protection.

Overload Protection

Overload is the condition in which current exceeds the equipment ampere rating, which can result in equipment damage due to dangerous overheating [Article 100]. Overload protection devices, sometimes called heaters, are intended to protect the motor, the motor-control equipment, and the branch-circuit conductors from excessive heating due to motor overload [430–31]. Overload protection is not intended to protect against short-circuits or ground-fault currents, Fig. 7–5.

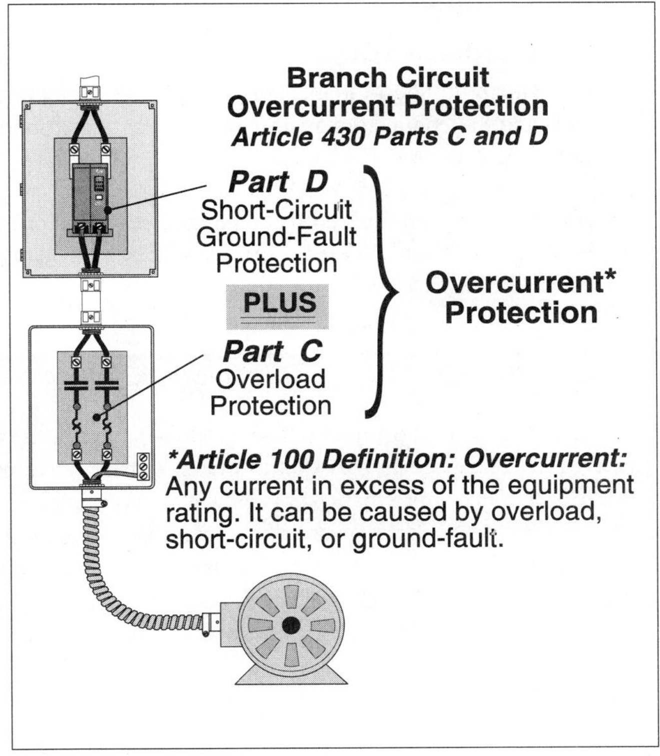

Figure 7-4

Motor Overcurrent Protection

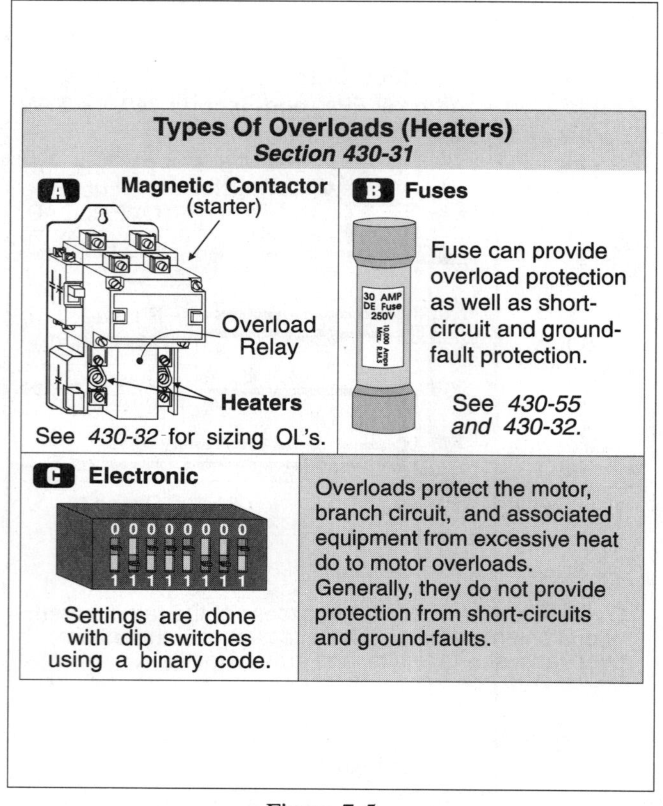

Figure 7-5

Types of Overloads

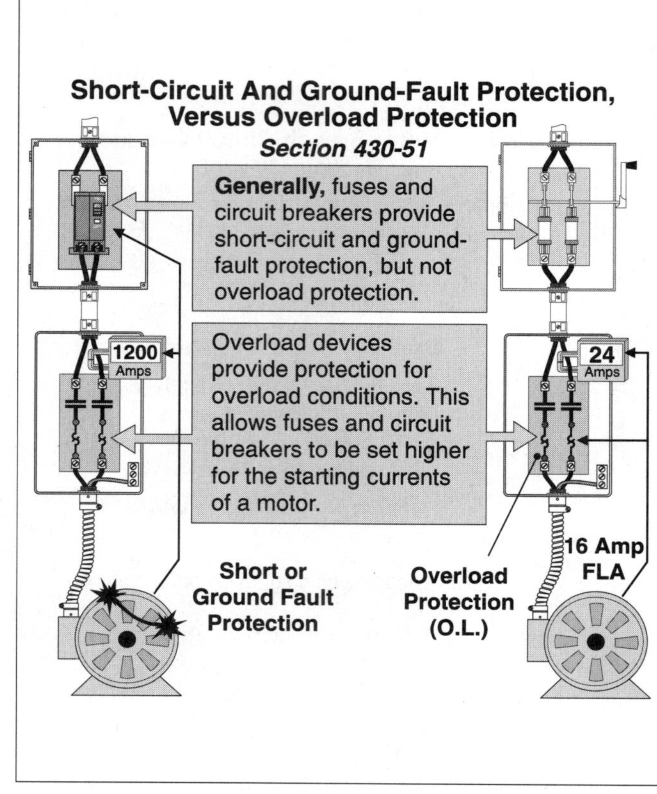

Figure 7-6

Short Circuit and Ground-Fault Protection

If overload protection is to be accomplished by the use of fuses, a fuse must be installed to protect each ungrounded conductor [430–36 and 430–55].

Short-Circuit And Ground-Fault Protection

Branch-circuit short-circuit and ground-fault protection devices are intended to protect the motor, the motor control apparatus, and the conductors against short-circuit or ground-faults, but they are not intended to protect against an overload [430–51], Fig. 7–6.

Note: The ground-fault protection device required for motor circuits is not the type required for personnel [210–8], feeders [215–9 and 240–13], services [230–95], or temporary wiring for receptacles [305–6].

7–3 *OVERLOAD PROTECTION [430–32(a)]*

In addition to short-circuit and ground-fault protection, motors must be protected against overload. Generally, the motor overload device is part of the motor starter; however, a separate overload device like a dual-element fuse can be used. Motors rated more than 1-horsepower without integral thermal protection, and motors 1-horsepower or less (automatically started) [430–32(c)], must have an overload device sized in response to the motor nameplate current rating [430–6(a)]. The overload device must be sized no larger than the requirements of Section 430–32. However, if the overload is sized according to Section 430–32 and it is not capable of carrying the motor starting or running current, the next larger size overload can be used [430–34]. The sizing of the overload device is dependent on the following factors.

Service Factor

Motors with a nameplate *service factor* (SF) rating of 1.15 or more, shall have the overload protection device sized no more than 125 percent of the motor nameplate current rating. If the overload is sized at 125 percent and it is not capable of carrying the motor starting or running current, the next size larger protection device can be used. But when we use the next size protection, it cannot exceed 140 percent of the motor nameplate current rating [430–34].

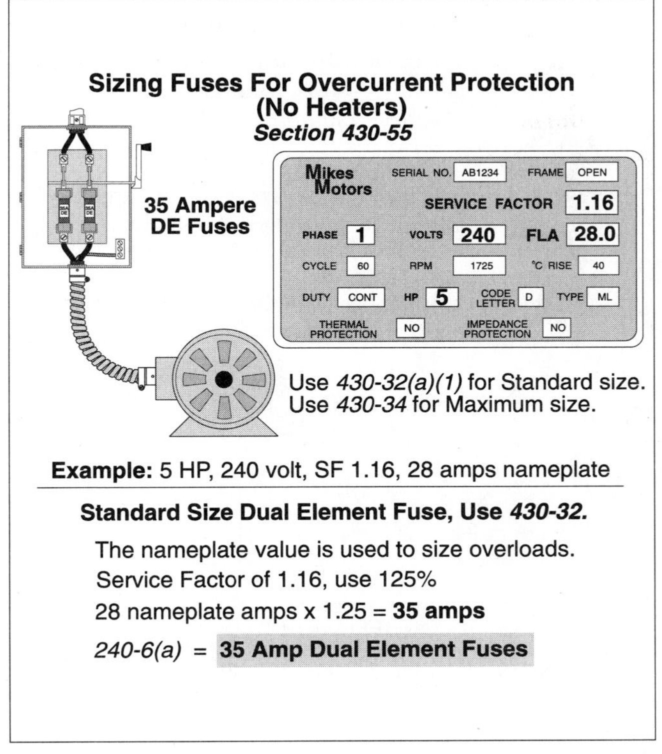

Figure 7-7

Overload Size with Service Factor

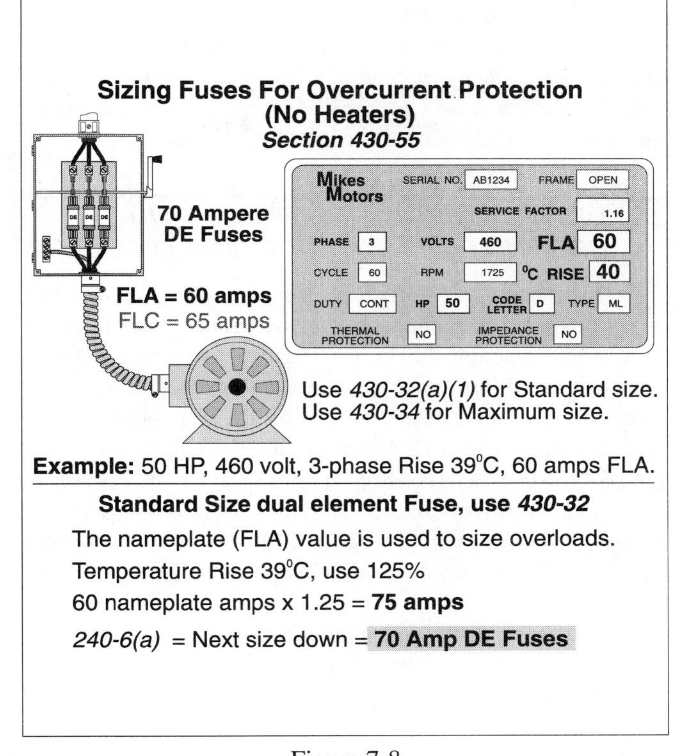

Figure 7-8

Overload Size–Temperature Factor

❑ Service Factor Example

If a dual-element fuse is used for overload protection, what size fuse is required for a 5-horsepower, 230-volt, single-phase motor, service factor 1.16 if the motor nameplate current rating is 28 amperes, Fig. 7–7?

(a) 25 amperes (b) 30 amperes (c) 35 amperes (d) 40 amperes

- Answer: (c) 35-ampere dual-element fuse

  Overload protection is sized to motor nameplate current rating [430–6(a) and 430–32(a)(1)]. Standard Size, 28 amperes × 1.25 = 35 amperes [240–6(a)].

Temperature Rise

Motors with a nameplate temperature rise rating not over 40ºC, shall have the overload protection device sized no more than 125 percent of the motor nameplate current rating. If the overload is sized at 125 percent and it is not capable of carrying the motor starting or running current, then the next size larger protection device can be used. When we use the next size protection, it cannot exceed 140 percent of the motor nameplate current rating [430–34].

❑ Temperature Rise Example

If a dual-element fuse is used for the overload protection, what size fuse is required for a 50-horsepower, 460-volt, three-phase motor, Temperature Rise 39ºC, motor nameplate current rating of 60 amperes (FLA), Fig. 7–8?

(a) 40 amperes (b) 50 amperes (c) 60 amperes (d) 70 amperes

- Answer: (d) 70 amperes

Overloads are sized according to the motor nameplate current rating of 60 amperes, not the 65 amperes Full Load Current rating listed in Table 430-150. 60 amperes × 1.25 = 75 amperes, 70 ampere [240–6(a) and 430–32(a)(1)]. If the 70 ampere fuse blows, the maximum size overload is limited to 140 percent of the motor nameplate current rating. 60 amperes × 1.4 = 84 ampere, next size down protection is 80 amperes [430–34].

All Other Motors

Motors that do not have a service factor rating of 1.15 and up, or a temperature rise rating of 40ºC and less, must have the overload protection device sized at not more than 115 percent of the motor nameplate ampere rating. If the overload is sized at 115 percent and it is not capable of carrying the motor starting or running current, the next larger protection device can be used. But when we use the next size protection, it cannot exceed 130 percent of the motor nameplate current rating [430–34].

Number of Overloads [430–37)

An overload protection device must be installed in each *ungrounded conductor* according to the requirements of Table 430–37. If a fuse is used for overload protection, a fuse must be installed in each ungrounded conductor [430–36].

7–4 *BRANCH CIRCUIT SHORT-CIRCUIT GROUND-FAULT PROTECTION [430–52(c)(1)]*

In addition to overload protection each motor and its accessories requires short-circuit and ground-fault protection. NEC® Section 430–52(c) requires the motor branch circuit short-circuit and ground-fault protection (except torque motors) to be sized no greater than the percentages listed in Table 430–152. When the short-circuit ground-fault protection device value determined from Table 430–152 does not correspond with the standard rating or setting of overcurrent protection devices as listed in Section 240–6(a), the next higher protection device size may be used [430-52(c)(1) Exception No. 1], Fig. 7–9.

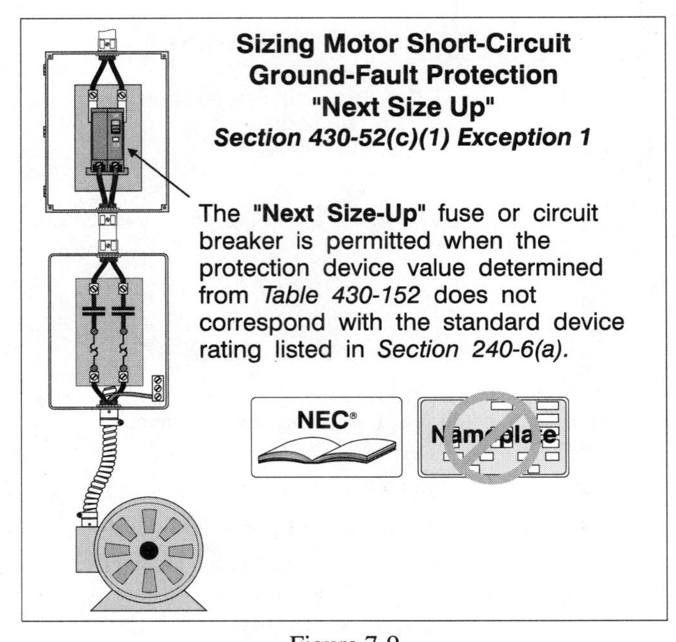

Figure 7-9

Short-Circuit and Ground-Fault Protection

To determine the percentage from Table 430–152 to be used to size the motor branch circuit short-circuit ground-fault protection device, the following steps should be helpful.

Step 1:

→ Locate the motor type on Table 430–152: Wound rotor, direct-current or all other motors.

Step 3:

→ Select the percentage from Table 430–152 according to the type of protection device, such as nontime delay (one-time) fuse, dual element fuse, or inverse time circuit breaker.

Note: Where the protection device rating determined by Table 430-152 does not correspond with the standard size or rating of fuses or circuit breakers listed in Section 240-6(a), the next higher standard size rating shall be permitted [430-52(c)(1) Exception No. 1].

NEC® Table 430–152			
	Percent of Full-Load Current (FLC) Tables 430–147, 148 and 150		
Type of Motor	One-Time Fuse	Dual-Element Fuse	Circuit Breaker
Direct Current and Wound-Rotor Motors	150	150	150
All Other Motors	300	175	250

❑ **Branch Circuit Example**

Which of the following statements are true?

(a) The branch circuit short-circuit protection (nontime delay fuse) for a 3-horsepower, 115-volt motor, single-phase shall not exceed 110 amperes.

(b) The branch circuit short-circuit protection (dual element fuse) for a 5-horsepower, Design E, 230-volt motor, single-phase shall not exceed 50 amperes.

(c) The branch circuit short-circuit protection (inverse time breaker) for a 25-horsepower, synchronous, 460-volt motor, three-phase shall not exceed 70 amperes.

(d) All of these are correct.

- Answer: (d) All of these are correct.

Short-circuit and ground-fault protection, 430-53(c)(1) Exception No. 1 and Table 430-152:

Table 430-148 – 34 amperes × 3.00 = 102 amperes, next size up permitted, 110 amperes.

Table 430-148 – 28 amperes × 1.75 = 49 amperes, next size up permitted, 50 amperes.

Table 430-150 – 26 amperes × 2.50 = 65 amperes, next size up permitted, 70 amperes.

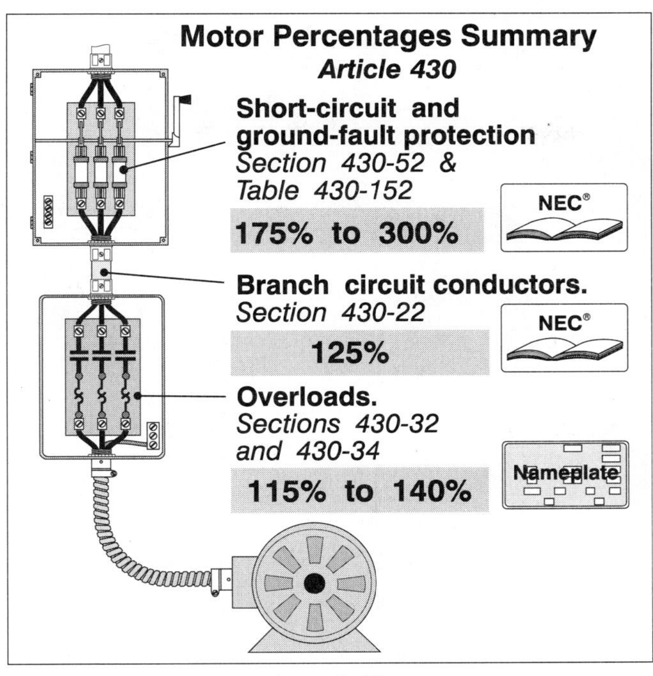

Figure 7-10

Motor Percentages Summary

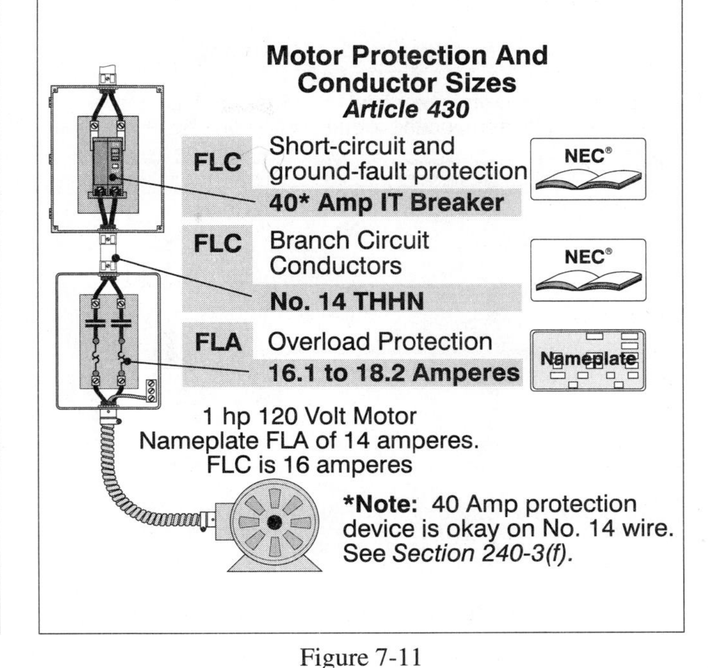

Figure 7-11

Motor Protection And Conductor Sizes

WARNING: Conductors are sized at 125 percent of the motor full-load current [430-22(a)], overloads are sized from 115 percent to 140 percent of the motor *nameplate* current rating [430-32(a)(1)], and the short-circuit ground-fault protection device is sized from 175 percent to 300 percent of the *motor full-load* current [Table 430-152]. As you can see, there is no relationship between the branch circuit conductor ampacity (125 percent) and the short-circuit ground-fault protection device (175 percent up to 300 percent), Fig. 7–10.

❑ **Branch Circuit Example**

Which of the following statements are true for a 1-horsepower, 120-volt motor, nameplate current rating of 14 amperes, Fig. 7–11?

(a) The branch circuit conductors can be No. 14 THHN.

(b) Overload protection is from 16.1 amperes to 18.2 amperes.

(c) Short-circuit and ground-fault protection is permitted to be a 40 ampere circuit breaker.

(d) All of these are correct.

- Answer: (d) All of these are correct.

  Conductor Size [430–22(a)]: 16 amperes × 1.25 = 20 amperes = No. 14 at 60°C, Table 310–16.

  Overload Protection Size: Standard – 14 amperes (nameplate) × 1.15 = 16.1 amperes [430–32(a)(1)].

  Maximum – 14 amperes × 1.3 = 18.2 amperes [430–34].

  Short-Circuit And Ground-Fault Protection: 16 amperes (FLC) × 2.50 = 40 ampere circuit breaker, [430–52(c)(1), Tables 430–152 and 240–6].

This bothers many electrical people, but you need to realize that the No. 14 THHN conductors and motor are protected against overcurrent by the 16-ampere overload device and the 40-ampere short-circuit protection device.

7–5 *FEEDER CONDUCTOR SIZE [430–24]*

Motors Only

Conductors that supply several motors must have an ampacity of not less than:

(1) 125 percent of the highest rated motor full-load current (FLC) [430–17], plus

(2) the sum of the full-load currents of the other motors (on the same phase) [430–6(a)].

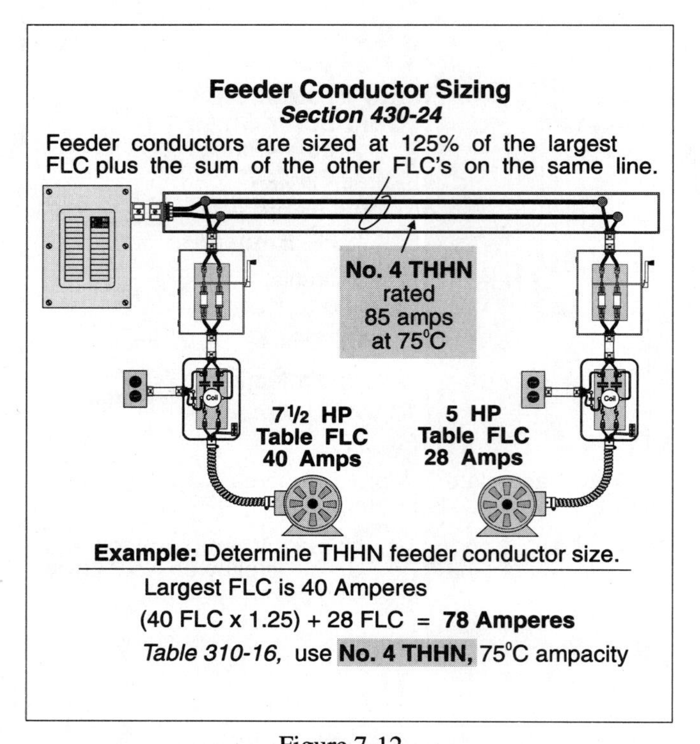

Figure 7-12

Feeder Conductor Sizing

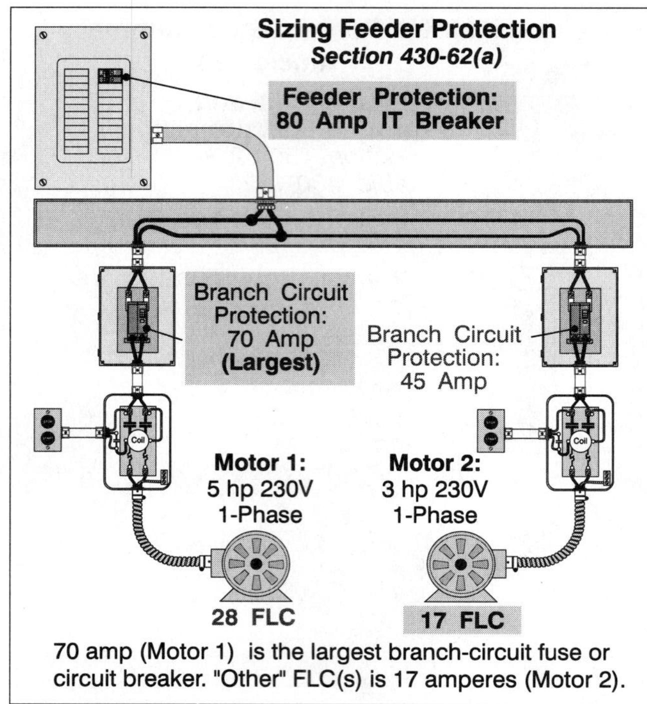

Figure 7-13

Sizing Feeder Protection

❑ **Feeder Conductor Size – Motors Only Example**

What size feeder conductor (in amperes) is required for two motors: Motor 1 – 7½-horsepower, single-phase 230-volts (40 amperes) and motor 2 – 5-horsepower, single-phase 230-volts (28 amperes)? Terminals rated for 75°C, Fig. 7–12.

(a) 50 amperes (b) 60 amperes (c) 70 amperes (d) 80 amperes

- Answer: (d) 80 amperes, (40 amperes × 1.25) + 28 amperes = 78 amperes
  No. 4 conductor at 75°C, Table 310-16 is rated for 85 amperes.

7–6 *FEEDER PROTECTION [430–62(a)]*

Motors Only

Motor *feeder* conductors must have protection against short-circuits and ground-faults but not overload. The protection device must be sized not greater than: The largest branch-circuit short-circuit ground-fault protection device [430–52(c)] of any motor of the group, plus the sum of the full-load currents of the other motors (on the same time and phase).

❑ **Feeder Protection – Motors Only Example**

What size feeder protection (inverse time breaker) is required for a 5-horsepower, 230-volts single-phase motor and a 3-horsepower, 230-volt single-phase motor, Fig. 7–13?

(a) 30 ampere breaker (b) 40 ampere breaker

(c) 50 ampere breaker (d) 60 ampere breaker

- Answer: (d) 60-ampere breaker

Motor Full Load Current [Table 430–148]
5-horsepower, motor FLC = 28 amperes [Table 430–148]
3-horsepower, motor FLC = 17 amperes [Table 430–148]
Branch Circuit Protection [430–52(c)(1), Table 430–152 and 240–6(a)]
5-horsepower: 28 amperes × 2.5 = 70 amperes
3 horsepower: 17 amperes × 2.5 = 42.5 amperes; Next size up, 45 amperes

Feeder Conductor [430–24(a)]
28 amperes × 1.25 + 17 amperes = 52 amperes, No. 6 rated 55 amperes at 60°C, Table 310-16

Feeder Protection [430–62]
Not greater than 70-amperes protection, plus 17 amperes = 87 amperes; Next size down, 80 amperes

7–7 HIGHEST RATED MOTOR [430–17]

When selecting the feeder conductors or feeder short-circuit ground-fault protection device, the highest rated motor shall be the highest rated motor full-load current, not the highest-rated horsepower [430–17].

❑ Highest Rated Motor Example

Which is the highest-rated motor of the following, Fig. 7–14.

(a) 10-horsepower, three-phase, 208-volt

(b) 5-horsepower, single-phase, 208-volt

(c) 3-horsepower, single-phase, 120-volt

(d) any of these

- Answer: (c) 3-horsepower, single-phase, 120-volt, Full Load Current of 34 amperes

  10-horsepower = 30.8 amperes [Table 430–150]

  5-horsepower = 30.8 amperes [Table 430–148]

  3-horsepower = 34.0 amperes [Table 430–148]

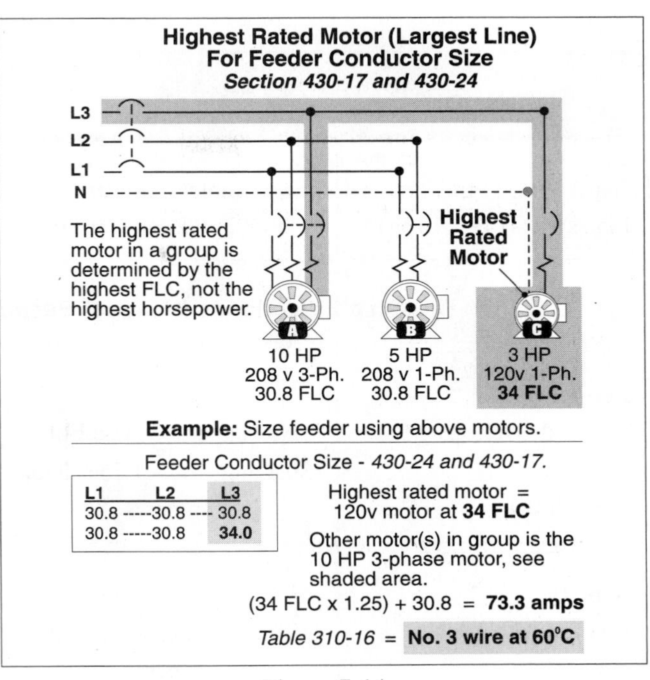

Figure 7-14

Highest Rated Motor For Feeder Conductor Size

7–8 MOTOR CALCULATION STEPS

Steps & NEC Rules	M1	M2	M3
Step 1 Motor FLC Tables 430–147, 148 & 150	________FLC	________FLC	________FLC
Step 2 Overloads (Heaters) Based On Motor Nameplate Current Rating			
Standard 430–32(a)(1)	____ × 1.____ = ____	____ × 1.____ = ____	____ × 1.____ = ____
Maximum 430–34	____ × 1.____ = ____	____ × 1.____ = ____	____ × 1.____ = ____
Step 3 Branch Circuit Conductor. 430–22(a), Table 310–16	____ × 1.25 = ____	____ × 1.25 = ____	____ × 1.25 = ____
Step 4 Branch Circuit Protection. Tables 430–152, 430–52(c), 240–6(a)	____ × ____ = ____ Next size UP	____ × ____ = ____ Next size Up	____ × ____ = ____ Next size UP
Step 5 Feeder Conductor 430–24 and Table 310–16	____ × 1.25 + ____ + ____ + ____ = ____ Table 310–16, Use No. ____ (60 or 75°C?)		
Step 6 Feeder Protection 430–62, Tables 430–152, and 240–6(a)	____ + ____ + ____ + ____ = ____ Next size down		

❑ Motor Calculation Steps Example

Given: One 10-horsepower, 208-volt, three-phase motor and three – 1-horsepower, 120-volt motors. Determine the standard and maximum overload sizes, branch circuit conductor (THHN), branch circuit short-circuit ground-fault protection device (Inverse Time Breaker), for all motors; and then determine the feeder conductor and protection size, Fig. 7–15.

SOLUTION	M1	M2	M3
Step 1 Motor FLC Tables 430–147, 148 or 150	Table 430–150 30.8 Full Load Current	Table 430–147 16 Full Load Current	
Step 2 Overloads (Heaters) Based On Motor Nameplate Current Rating			
Step 2A Standard Overload Nameplate 430–32(a)(1)	No Nameplate, Use FLC 30.8 amperes × 1.15 = 35.4 amperes	No Nameplate, Use FLC 16 amperes × 1.15 = 18.4 amperes	
Step 2B Maximum Overload Nameplate 430–34	30.8 amperes × 1.3 = 40 amperes	16 amperes × 1.3 = 20.8 amperes	
Step 3 Branch Circuit Conductors 125 percent of FLC 110–14(c), 430–22(a), Table 310–16	30.8 amperes × 1.25 = 38.5 amperes Table 310–16 No. 8 THHN 60°C	16 amperes × 1.25 = 20 amperes Table 310–16 No. 14 THHN 60°C	
Step 4 Branch Circuit Protection FLC × Table 430–152 percent 430–52(c), 240–6(a)	30.8 amperes × 2.5 = 77 amperes 240–6(a) = 80 amps	16 amperes × 2.5 = 40 amperes 240–6(a) = 40 amps	
Step 5 Feeder Conductor FLC × 125 percent + FLC others 430–24, Table 310–16	(30.8 amperes × 1.25) + 16 amperes = 54.5 amperes Table 310–16, Use No. 6 THHN, rated 55 amperes 60°C		
Step 6 Feeder Protection Largest Branch protection + FLCs of other motors 430–62, Table 430–152, 240–6(a)	Inverse Time Breaker 80 amperes + 16 amperes + = 96 amperes, next size down, 90 amps		

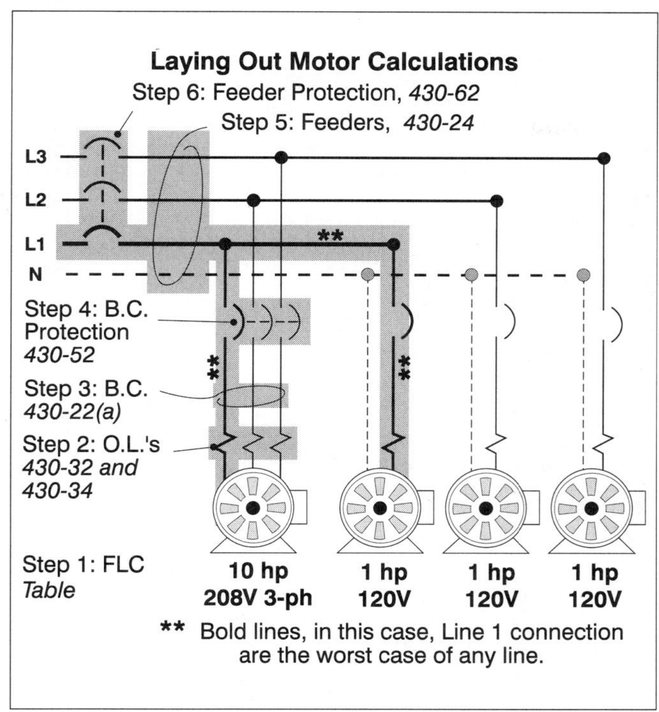

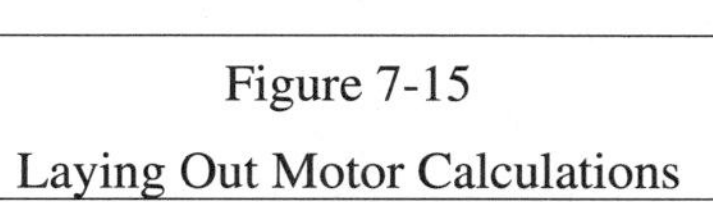

Figure 7-15

Laying Out Motor Calculations

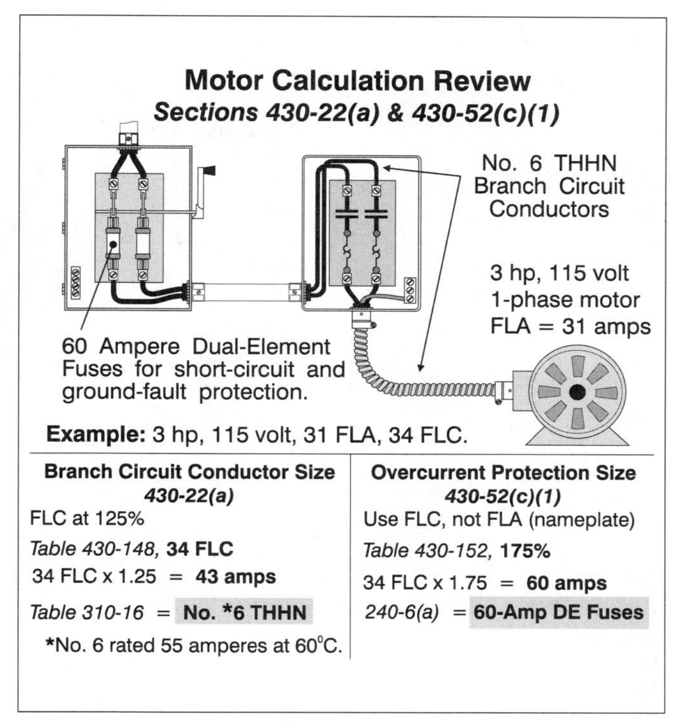

Figure 7-16

Motor Calculation Review

7–9 *MOTOR CALCULATION REVIEW*

❑ Branch Circuit Review Example

Size the branch circuit conductors (THHN) and short-circuit ground-fault protection device for a 3-horsepower, 115-volt, single-phase motor. The motor nameplate full-load amperes is 31 amperes and dual-element fuses to be used for short-circuit and ground-fault protection, Fig. 7–16.

Branch Circuit Conductors [430–22(a)]: Branch circuit conductors to a single motor must have an ampacity of not less than 125 percent of the motor full-load *current* as listed in Tables 430–147 through 430–150 [430–6(a)]. Three-horsepower, full-load current is 34 amperes, 34 amperes × 125 percent = 43 amperes. Table 310–16, at 60°C terminals, the conductor must be a No. 6 THHN rated 55 amperes [110–14(c)(1)].

Branch Circuit Short-Circuit Protection [430–52(c)(1)]: The branch-circuit short-circuit and ground-fault protection device protects the motor, the motor control apparatus, and the conductors against overcurrent due to short-circuits or ground-faults, but not overload [430–51]. The branch-circuit short-circuit and ground-fault protection device is sized by considering the type of motor and the type of protection device according to the motor full-load current listed in Table 430–152. When the protection device values determined from Table 430–152 do not correspond with the standard rating of overcurrent protection devices as listed in Section 240–6(a), the next higher overcurrent protection device must be installed. The branch-circuit short-circuit protection is sized at 34 amperes × 175 percent = 60 amperes dual-element fuse [240–6(a)]. See Example No. 8 in the back of the NEC® Code Book.

❑ Feeder and Branch Circuit Review Example

Size the feeder conductor (THHN) and protection device (inverse time breakers 75°C terminal rating) for the following motors: Three 1-horsepower, 115-volt single-phase motors, and three 5-horsepower, 208-volt, single-phase motors, and one 15-horsepower, wound-rotor, 208-volt, three-phase motor, Fig. 7–17.

Branch Circuit Conductors [110–14(c) 75°C terminals, Table 310–16 and 430–22(a)]:

15 horsepower: 46.2 amperes [Table 430–150] × 125 percent = 58 amperes, No. 6 THHN, rated 65 amperes

5 horsepower: 30.8 amperes [Table 430–148] × 125 percent = 39 amperes, No. 8 THHN, rated 50 amperes

1 horsepower: 16 amperes [Table 430–148] × 125 percent = 20 amperes, No. 14 THHN, rated 20 amperes

Branch circuit short-circuit protection [240–6(a), 430–52(c)(1), and Table 430–152]:

15 horsepower: 46.2 amperes × 150 percent (wound-rotor) = 69 amperes; Next size up = 70 amperes.

5 horsepower: 30.8 amperes × 250 percent = 77 amperes; Next size up = 80 amperes

1 horsepower: 16 amperes × 250 percent = 40 amperes

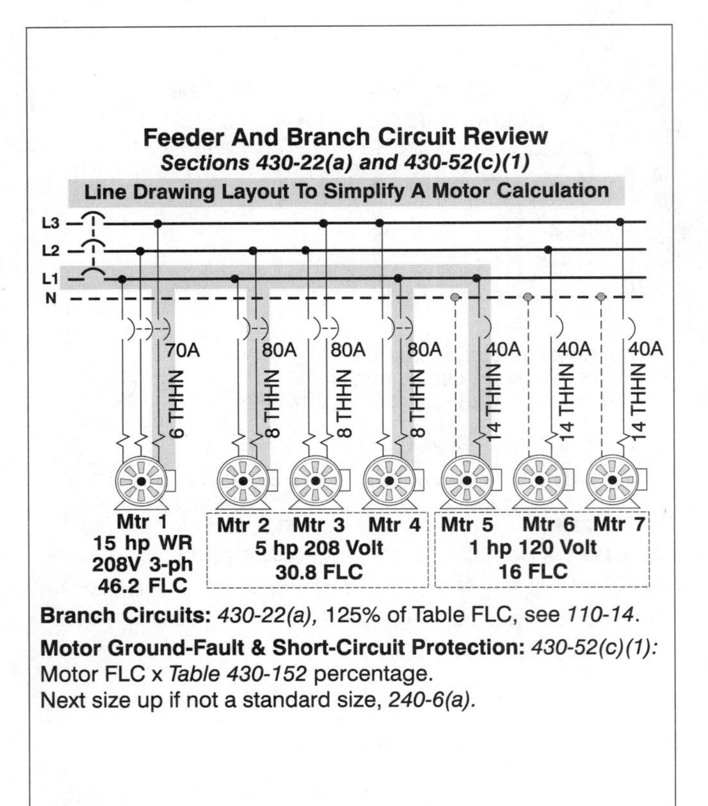

Figure 7-17

Branch Circuit Review

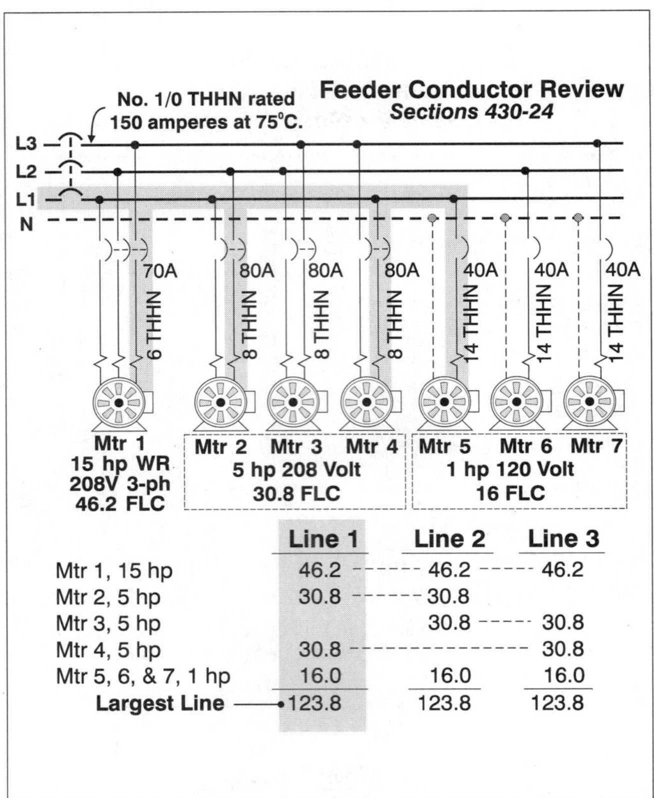

Figure 7-18

Feeder Conductor Review

Feeder Conductor [430–24]: Conductors that supply several motors must have an ampacity of not less than 125 percent of the highest rated motor full-load current [430–17], plus the sum of the other motor full-load currents [430–6(a)]. These conductors must be protected against short-circuits and ground-faults according to Section 430–62. The feeder conductor is sized at (46.2 amperes × 1.25) + 30.8 amperes + 30.8 amperes + 16 amperes = 136 amperes, [Table 310–16, No. 1/0 THHN], rated 150 amperes, Fig. 7–18

Note: When sizing the feeder conductor, be sure to only include the motors that are on the same phase. For that reason, only four motors are used for the feeder conductor size.

Feeder Protection [430–62]: Feeder conductors must be protected against short-circuits and ground-faults and sized no greater than the maximum branch-circuit short-circuit ground-fault protection device [430–52(c)(1)] plus the sum of the full-load currents of the other motors (on the same phase). The protection device must not be greater than: (80 amperes + 30.8 amperes + 46.2 amperes + 16 amperes) = 173 amperes; Next size down, 150 ampere circuit breaker [240–6(a)]. See example No. 8 in the back of the NEC® Code Book, Fig. 7–19.

Note: When sizing the feeder protection, be sure to only include the motors that are on the same phase. For that reason, only four motors are used for the feeder conductor size.

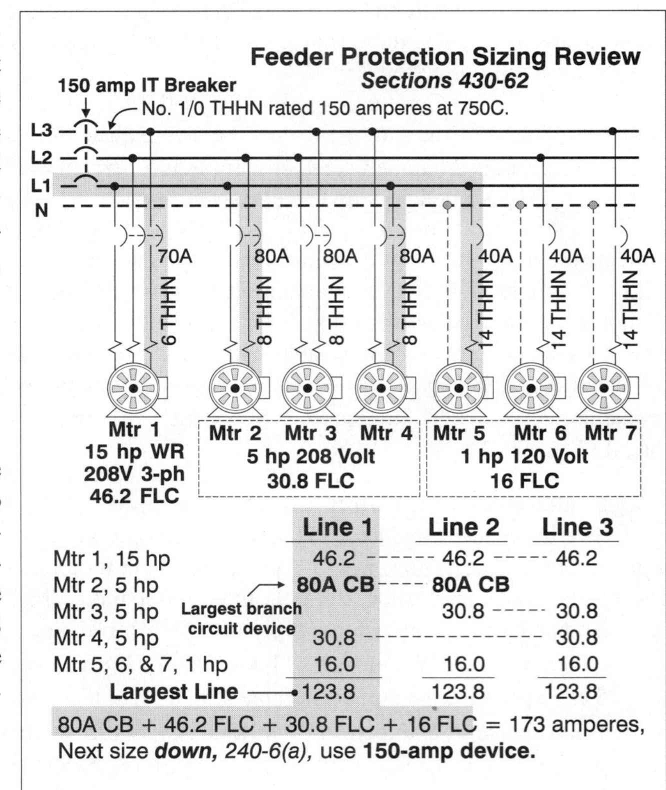

Figure 7-19

Feeder Protection Sizing Review

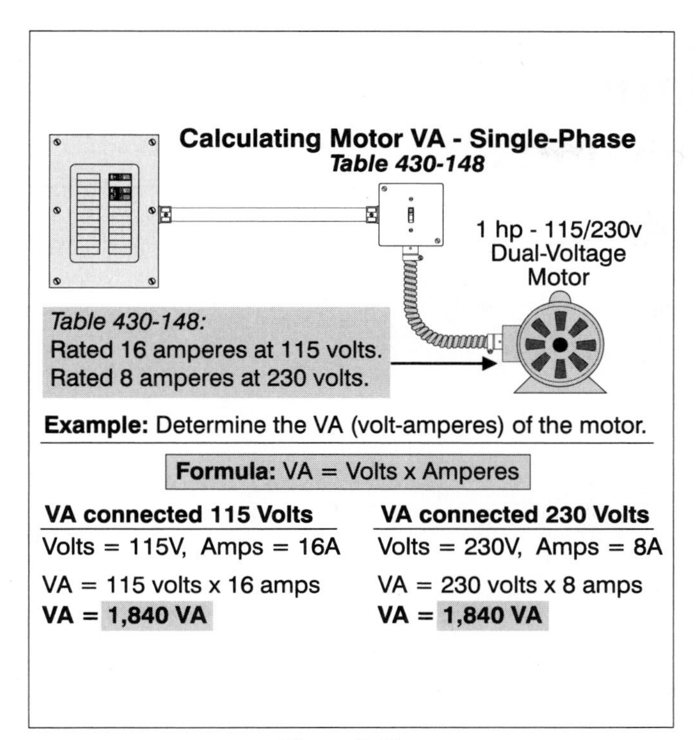

Figure 7-20

Calculating Motor VA – Single-Phase

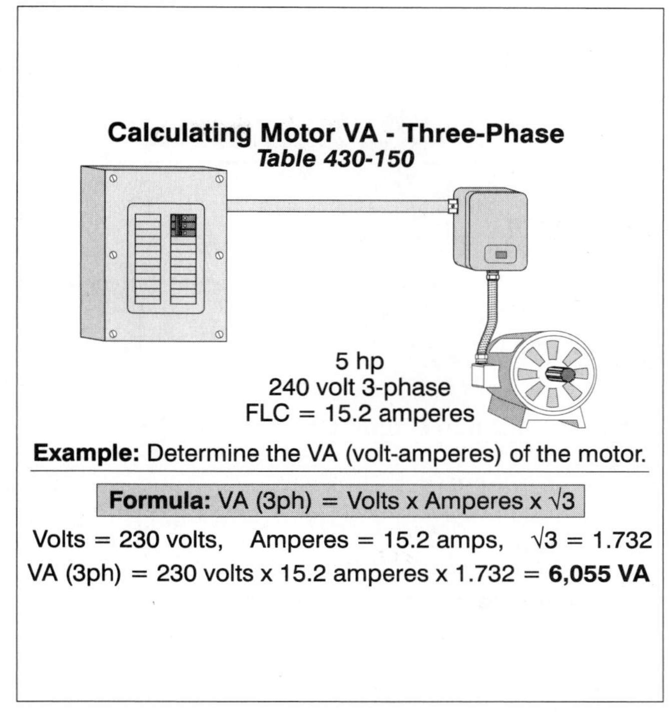

Figure 7-21

Calculating Motor VA – Three-Phase

7–10 *MOTOR VA CALCULATIONS*

The input VA of a motor is determined by multiplying the motor volts times the motor amperes. To determine the *motor VA rating*, the following formulas can be used:

Single-Phase Motor VA = Volts × Motor Amperes
Three-Phase motor VA = Volts × Motor Amperes × $\sqrt{3}$

Note: Many people believe that a 230-volt motor will consume less power than a 115-volt motor. A motor will consume approximately the same amount of power whether it's wired at 115 volts or 230 volts.

❑ **Single-Phase Motor VA Example**

What is the motor input VA of a 1-horespower, motor rated 115/230-volts? The system voltage is 115/230-volts, Fig. 7–20.

(a) 1,840 VA at 115 volts (b) 1,840 VA at 230 volts (c) a and b (d) none of these

- Answer: (c) a and b

  Motor VA = Volts × Full Load Current

  VA of 1-horsepower 230 volts = 230 volts × 8 amperes = 1,840 VA

  VA of 1-horsepower 115 volts = 115 volts × 16 amperes = 1,840 VA

CAUTION: Some exam testing agencies use the table amperes times the given voltage. Under this condition, the answer would be 120 volts × 16 amperes or 240 volts × 8 amperes = 1,920 VA

❑ **Three-Phase Motor VA Example**

What is the input VA of a 5-horsepower, 240-volt, three-phase motor, Fig. 7–21?

(a) 6,055 VA (b) 3,730 VA (c) 6,440 VA (d) 8,050 VA

- Answer: (a) 6,055 VA Table 430–150

  Motor VA = Volts × Table Full-Load Current × $\sqrt{3}$

  Table 430–150 Full-Load Current = 15.2 amperes

  Motor VA = 230 volts × 15.2 amperes × 1.732 = 6,055 VA or

  240 volts × 15.2 amperes = 6,318 VA

Unit 7 – Motor Calculations Summary Questions

Introduction

1. When sizing conductors and overcurrent protection for motors, we must comply with the requirements of NEC® Article 430, not Article 240.
(a) True (b) False

7–1 Motor Branch Circuit Conductors [430–22(a)]

2. What size THHN conductor is required for a 5-horsepower, 230-volt, single-phase motor? Terminals are rated 75ºC.
(a) No. 14 (b) No. 12 (c) No. 10 (d) No. 8

7–2 Motor Overcurrent Protection

3. Motors and their associated equipment must be protected against overcurrent (overload, short-circuit or ground-fault), but because of the special characteristics of induction motors, overcurrent protection is generally accomplished by having the overload protection separate from the short circuit and ground fault protection.
(a) True (b) False

4. • Which Parts of Article 430 contain the requirements for motor overcurrent protection?
(a) Overload protection – Part C
(b) Short-circuit ground-fault protection Part D
(c) a and b
(d) None of these

5. • Overload is the condition where current is greater than the equipment ampacity rating resulting in equipment damage due to dangerous overheating [Article 100]. Overload protection devices, sometimes called heaters, are intended to protect the _________ from dangerous overheating.
(a) motor (b) motor-control equipment
(c) branch-circuit conductors (d) all of these

6. The branch-circuit short-circuit and ground-fault protection device is intended to protect the motor, the motor control apparatus and the conductors against overcurrent due to _________.
(a) short-circuits (b) ground-faults
(c) overloads (d) a and b

7–3 Overload Protection

7. • The NEC® requires motor overloads to be sized according to the motor full-load current rating as listed in Tables 430–147, 148, or 150.
(a) True (b) False

8. The standard size overload must be sized according to the requirements of Section 430–32 of the NEC®. If the overload sized according to Section 430–32 is not capable of carrying the motor starting and running current, the next size up overload can be used if sized according to the requirements of Section 430–34.
(a) True (b) False

9. Motors with a nameplate service factor (SF) rating of 1.15 or more must have the overload protection device sized at no more than _____ percent of the motor nameplate current rating. If the overload is not capable of carrying the motor starting or running current, the next size up protection device is permitted, but it cannot exceed _____ percent of the motor nameplate current rating.
(a) 100, 125 (b) 115, 130 (c) 115, 140 (d) 125, 140

10. Motors with a nameplate temperature rise rating not over 40°C, must have the overload protection device sized at no more than ______ percent of the motor nameplate current rating. If the overload is not capable of carrying the motor starting or running current of the motor, the next size up is permitted, but it cannot exceed ______ percent of the motor nameplate current rating.
(a) 100, 125 (b) 115, 130 (c) 115, 140 (d) 125, 140

11. Motors that do not have a service factor of 1.15 and up or a temperature rise rating not over 40°C, must have the overload protection device sized at not more than ______ percent of the motor nameplate ampere rating. If the overload is not capable of carrying the motor starting or running current, the next size up device is permitted, but it cannot exceed ______ percent of the motor nameplate current rating.
(a) 100, 125 (b) 115, 130 (c)115, 140 (d) 125, 140

12. If a dual-element fuse is used for overload protection, what size fuse is required for a 5-horsepower, 208-volt, three-phase motor, service factor 1.16, motor nameplate current rating of 16 amperes (FLA)?
(a) 20 amperes (b) 25 amperes (c) 30 amperes (d) 35 amperes

13. If a dual-element fuse is used for the overload protection, what size fuse is required for a 30-horsepower, 460-volt, three-phase synchronous motor, temperature rise 39°C?
(a) 20 amperes (b) 25 amperes
(c) 30 amperes (d) 40 amperes

7–4 Branch Circuit Short-Circuit Ground-Fault Protection [430–52(a)]

14. In addition to overload protection, each motor and its accessories requires short-circuit and ground-fault protection according to the requirements of Section 430–52. When sizing the branch circuit protection device, we must consider which of the following factors?
(a) the motor type, such as induction, synchronous, wound-rotor, etc.
(b) the motor Code letter starting characteristics
(c) the type of protection device to be used, fuse or breaker
(d) all of these

15. The NEC® requires motor branch circuit short-circuit and ground-fault protection to be sized not greater than the percentages listed in Table 430–152. When the short-circuit ground-fault protection device value determined from Table 430–152 does not correspond with the standard rating of overcurrent protection devices as listed in Section 240–6(a), the next ________ device size must be used.
(a) smaller (b) larger (c) a or b (d) none of these

16. To determine the percentage of the motor FLC from Table 430–152 that is to be used to size the motor branch circuit short-circuit ground-fault protection device, the following steps should be used:
(a) Locate the motor type on Table 430–152, such as direct current, wound rotor, high-reactance, autotransformer start, or all other motors.
(b) Locate the motor starting conditions, such as Code letter or no Code letter.
(c) Select the percentage from Table 430–152 according to the type of protection device, such as one-time fuse, dual-element fuse, or circuit breaker.
(d) all of these

17. If the branch circuit short-circuit ground-fault protection dual-element fuse selected (sized not greater than the percentages listed in Table 430–152,) is not capable of carrying the load, the next larger size dual-element fuse can be used. The next size dual-element fuse cannot exceed _____ percent of the motor full-load current rating.
(a) 125 (b) 150 (c) 175 (d) 225

18. Conductors are sized at _____ percent of the motor full-load currents [430–6 and 430–22], overloads from _____ percent, and the motor short-circuit ground-fault protection device (inverse time circuit breaker) is sized up to _____. As you can see, there is no relationship between the branch circuit conductor ampacity and the short-circuit ground-fault protection device!
(a) 125, 115, 250 (b) 100, 125, 150
(c) 125, 125, 125 (d) 100, 100, 100

19. Which of the following statements are true for a 10-horsepower, 208-volt, three-phase motor, nameplate current 29 amperes?
(a) The branch circuit conductors can be No. 8 THHN.
(b) Overload protection is from 33 amperes to 38 amperes.
(c) Short-circuit and ground-fault protection is an 80-ampere circuit breaker.
(d) All of these are correct.

FLA = 29 × 1.15 = 33.35
29 × 1.30 = 37.7

7–5 Feeder Conductor Size [430–24]

20. Feeder conductors that supply several motors must have an ampacity of not less than:
I. 125 percent of the highest rated motor FLC
II. The sum of the full-load currents of the other motors on the same phase [430–6(a)].
(a) I only
(b) II
(c) I or II
(d) I and II

21. Motor feeder conductors (sized according to 430–24) must have a feeder protection device to protect against short-circuits and ground-faults (not overloads), sized not greater than:
I. The largest branch-circuit short-circuit ground-fault protection device [430–52] of any motor of the group
II. The sum of full-load currents of the other motors on the same time phase
(a) I only
(b) II
(c) I or II
(d) I and II

7–6 Feeder – Protection [430–62(a)]

22. • Which of the following statements about a 30-horsepower, 460-volt, three-phase synchronous motor and a 10-horsepower, 460-volt, three-phase motor are true?
(a) The 30-horsepower motor has No. 8 THHN with a 70-ampere breaker.
(b) The 10-horsepower motor has No. 14 THHN with a 35-ampere breaker.
(c) The feeder conductors must have a No. 6 THHN with a 80-ampere breaker.
(d) All of these.

7–7 Highest Rated Motor [430–17]

23. When selecting the feeder conductors and short-circuit ground-fault protection device, the highest rated motor shall be the highest rated ________.
(a) horsepower
(b) full-load current
(c) nameplate current
(d) any of these

24. • Which is the highest rated motor of the following?
I. 25-horsepower synchronous, three-phase, 460-volt
II. 20-horsepower, three-phase, 460-volt
III. 15-horsepower, three- phase, 460-volt
IV. 3-horsepower, 120-volt
(a) I only
(b) II only
(c) III only
(d) IV only

7–10 Motor VA Calculations

25. What is the VA input of a dual-voltage 5-horsepower, three-phase motor rated 460/230 volts?
(a) 3,027 VA at 460 volts
(b) 6,055 VA at 230 volts
(c) 6,055 VA at 460 volts
(d) b and c

26. What is the input VA of a 3-horsepower, 208-volt, single-phase motor?
(a) 3,890 VA
(b) 6,440 VA
(c) 6,720 VA
(d) none of these

Challenge Questions

7–1 Motor Branch Circuit Conductors [430–22(a)]

27. • The branch circuit conductors of a 5-horsepower, 230-volt, motor (with a nameplate rating of 25 amperes), shall have an ampacity of not less than ______ amperes. Note: The motor is used for intermittent duty and cannot run for more than 5 minutes at any one time due to the nature of the apparatus it drives.
(a) 33 (b) 37 (c) 21 (d) 23

7–3 Motor Overload [430–32(a)(1) And 430–34]

28. The standard overload protective device for a 2-horsepower, 115-volt motor that has a full-load current rating of 24 amperes, and a nameplate rating of 21.5 amperes shall not exceed ______ amperes.
(a) 20.6 (b) 24.7 (c) 29.9 (d) 33.8

29. • The maximum overload protective device for a 2-horsepower, 115-volt motor with a nameplate rating of 22 amperes is ______ amperes. The Service Factor is 1.2.
(a) 30.8 (b) 33.8 (c) 33.6 (d) 22.6

Ultimate Trip Setting [430–32(a)(2)]

30. The ultimate trip overload device of a thermally protected 1½-horsepower, 120-volt motor would be rated no more than ______ amperes.
(a) 31.2 (b) 26 (c) 28 (d) 23

7–4 Branch Circuit Protection [430–52 And Table 430–152]

31. A 2-horsepower, 120-volt wound rotor motor would require a ______ branch circuit short circuit protection device.
(a) 15 ampere (b) 20 ampere (c) 25 ampere (d) 40 ampere

32. • The branch circuit short circuit protection device for a 10-horsepower, 230-volt, motor shall not exceed ______ amperes. *Note:* Use an inverse time breaker for protection.
(a) 60 (b) 70 (c) 75 (d) 80

33. The branch circuit protection for a 125-horsepower, 240-volt direct current motor is ______ amperes.
(a) 400 (b) 600 (c) 700 (d) 800

7–5 Motor Feeder Conductor Size [430–24]

34. • The motor controller (below) would require a No. ______ THHN for the feeder, if the motor terminals are rated for 75°C.

15-horsepower, 208v 3-Phase	15-horsepower, 208v 3-Phase	15-horsepower, 208v 3-Phase	Motor Controller 208/120-volt, 3-Phase
1-horsepower, 120v 1-Phase	1-horsepower, 120v 1-Phase	1-horsepower, 120v 1-Phase	
3-horsepower, 208v 1-Phase	3-horsepower, 208v 1-Phase	3-horsepower, 208v 1-Phase	

(a) 2 (b) 3/0 (c) 4/0 (d) 250 kcmils

7–6 Motor Feeder Protection [430–62(a)]

35. • There are three motors: Motor 1 is 5-horsepower, 230-volt, with service factor of 1.2; Motors 2 and 3 are 1½-horsepower, 120-volt. This group is fed with a 3-wire feeder. Using an inverse time breaker, the feeder conductor protection device after balancing all three motors would be ______ amperes.
(a) 60 (b) 70 (c) 80 (d) 90

36. • If an inverse time breaker is used for the feeder short circuit protection, what size protection is required for the following three-phase motors?
Motor 1 = 40-horsepower, 52 FLC
Motor 2 = 20-horsepower, 27 FLC
Motor 3 = 10-horsepower, 14 FLC
Motor 4 = 5-horsepower, 7.6 FLC
(a) 225 amperes (b) 200 amperes (c) 125 amperes (d) 175 amperes

37. • If dual-element fuses are used to protect a three wire 115/230-volt feeder conductor for twenty-two ½-horsepower, single-phase, 115-volt motors, the fuse size selected should be sized not greater than ______ amperes.
(a) 125 (b) 90 (c) 100 (d) 110

38. • The feeder protection for a 25-horsepower, 208-volt, 3-phase and three 3-horsepower, 120-volt motors would be ______ amperes, after balancing. *Note:* Use inverse time breakers (ITB).
(a) 225 (b) 200 (c) 300 (d) 250

Unit 8

Voltage Drop Calculations

OBJECTIVES

After reading this unit, the student should be able to briefly explain the following concepts:

- Alternating current resistance as compared to direct current
- Conductor resistance
- Conductor resistance – alternating current circuits
- Conductor resistance – direct current circuits
- Determining circuit voltage drop
- Extending circuits
- Limiting current to limit voltage drop
- Limiting conductor length to limit voltage drop
- Resistance – alternating current
- Sizing conductors to prevent excessive voltage drop
- Voltage drop considerations
- Voltage drop recommendations

After reading this unit, the student should be able to briefly explain the following terms:

- American wire gauge
- CM = circular mils
- Conductor
- Cross-sectional area
- D = distance
- Eddy currents
- E_{VD} = Conductor voltage drop expressed in volts
- I = load in amperes at 100 percent
- K = direct current constant
- Ohm's law method
- Q = alternating current adjustment factor
- R = resistance
- Skin effect
- Temperature coefficient
- VD = volts dropped

8–1 *CONDUCTOR RESISTANCE*

Metals that carry electric current are called conductors or wires and oppose the flow of electrons. Conductors can be solid, stranded, copper or aluminum. The conductor's opposition to the flow of current (resistance) is determined by the material type (copper/aluminum), the cross-sectional area (wire size) and the conductor's length and operating temperature. The *resistance* of any property is measured in ohms.

Material

Silver is the best conductor because it has the lowest resistance, but its high cost limits its use to special applications. *Aluminum* is often used when weight or cost are important considerations, but *copper* is the most common type of metal used for electrical conductors, Fig. 8–1.

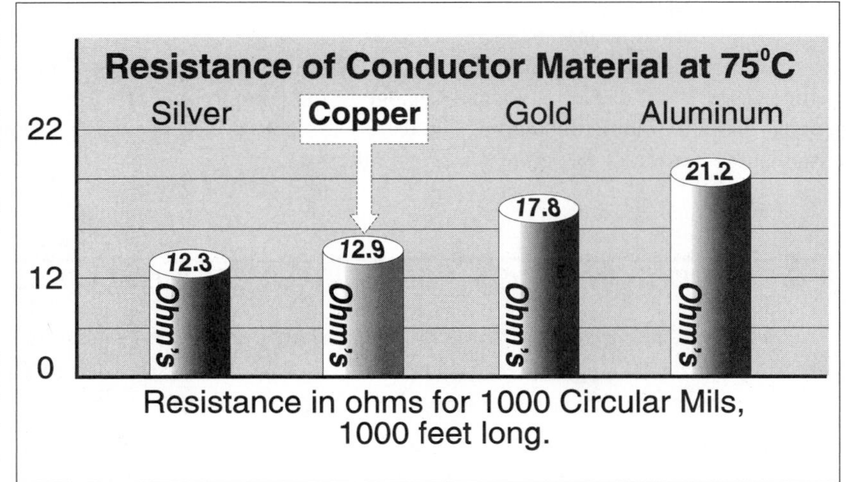

Figure 8-1

Resistance of Conductor Materials

Cross-sectional Area

The *cross-sectional area* of a conductor is the conductor's surface area expressed in circular mils. The greater the conductor cross-sectional area, the greater the number of available electron paths, and the lower the conductor resistance. Conductors are sized according to the *American Wire Gauge* which ranges from a small of No. 40 to the largest of No. 0000 (4/0). The conductor's resistance varies inversely with the conductor's diameter. That is, the smaller the wire size the greater the resistance, and the larger the wire size the lower the resistance, Fig. 8–2.

Conductor Length

The resistance of a conductor is directly proportional to the conductor length. The following table provides examples of conductor resistance and *area circular mils* for conductor lengths of 1,000 feet. Naturally, longer or shorter lengths will result in different conductor resistances.

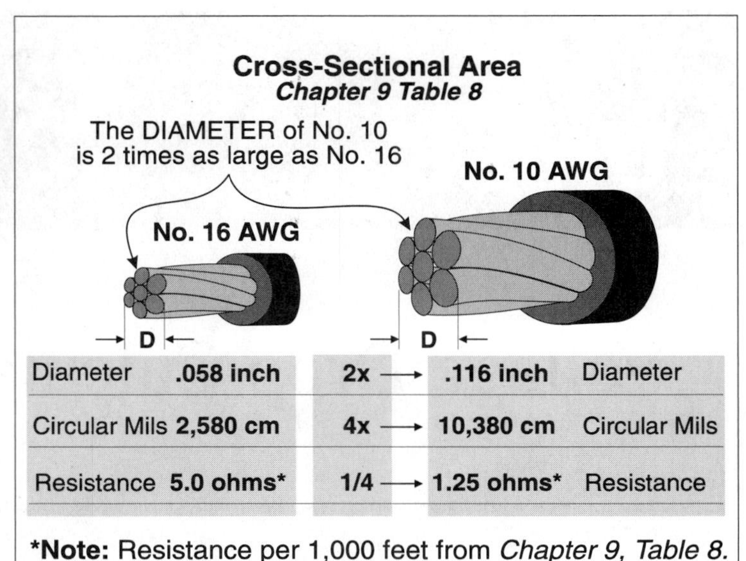

Figure 8-2

Conductor Cross-Sectional Area

Conductor Properties – NEC Chapter 9, Table 8			
Conductor Size American Wire Gauge	**Conductor Resistance Per 1,000 Feet At 75°C**	**Conductor Diameter In Mils**	**Conductor Area Circular Mils**
No. 14	3.1 ohms	64.084	4,107
No. 12	2.0 ohms	80.808	6,530
No. 10	1.2 ohms	101.89	10,380
No. 8	.78 ohm	128.49	16,510
No. 6	.49 ohm	162.02	26,240

Temperature

The resistance of a conductor changes with changing temperature. This is called *temperature coefficient.* Temperature coefficient describes the effect that temperature has on the resistance of a conductor. Positive temperature coefficient indicates that as the temperature rises, the conductor resistance will also rise. Examples of conductors that have a positive temperature coefficient are silver, copper, gold, and aluminum conductors. Negative temperate coefficient means that as the temperature increases, the conductor resistance decreases.

The conductor resistances listed in the National Electrical Code Table 8 and 9 of Chapter 9, are based on an operating temperature of 75°C. A three-degree change in temperature will result in a 1 percent change in conductor resistance for both copper and aluminum conductors. The formula to determine the change in conductor resistance with changing temperature is listed at the bottom of Table 8, Chapter 9 in the NEC®.

8–2 *CONDUCTOR RESISTANCE – DIRECT CURRENT CIRCUITS, [Chapter 9, Table 8 of the NEC®]*

The National Electric Code lists the resistance and area circular mils for both direct current and alternating current circuit conductors. Direct current circuit conductor resistances are listed in Chapter 9, Table 8, and alternating current circuit conductor resistances are listed in Chapter 9, Table 9 of the NEC.

The *direct current conductor resistances* listed in Chapter 9, Table 8 apply to conductor lengths of 1,000 feet. The following formula can be used to determine the conductor resistance for conductor lengths other than 1,000 feet:

$$\textbf{Direct current Conductor Resistance} = \frac{\textbf{Conductor Resistance Ohms}}{\textbf{1,000 Feet}} \times \textbf{Conductor's Length}$$

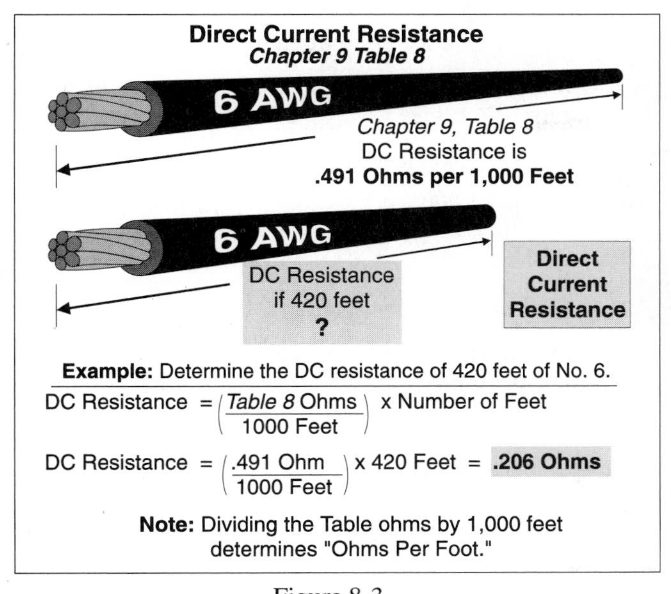

Figure 8-3

Direct Current Conductor Resistance

Skin Effect of Alternating Current
Chapter 9 Table 9

Eddy currents are stronger towards the center of the conductor. This forces current to flow toward the outside of the conductor.

Cross-Section of Conductor

Skin effect is directly proportional to frequency. 60 hertz is the standard frequency in the United States and does not have a major effect when stranded wire is used.

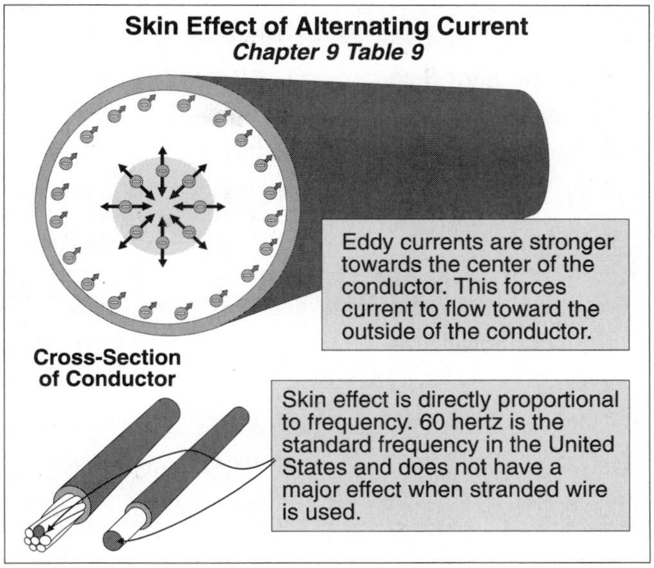

Figure 8-4

Skin Effect

❑ Direct Current Conductor Resistance Examples

What is the resistance of 420 feet of No. 6 copper, Fig. 8–3?

- Answer: .206 ohms

The resistance of No. 6 copper 1,000 feet long is .491 ohms, Chapter 9, Table 8. The resistance for 420 feet is:

(.491 ohms /1,000 feet) × 420 feet = .206 ohms

What is the resistance of 1,490 feet of No. 3 aluminum?

- Answer: .600 ohms

The resistance of No. 3 aluminum 1,000 feet long is .403 ohms, Chapter 9, Table 8. The resistance for 1,490 feet is:

(.403 ohms/1,000 feet) × 1,490 feet = .600 ohms

8–3 *CONDUCTOR RESISTANCE – ALTERNATING CURRENT CIRCUITS*

In direct current circuits, the only property that opposes the flow of electrons is resistance. In alternating current circuits, the expanding and collapsing magnetic field within the conductor induces an electromotive force that opposes the flow of alternating current. This opposition to the flow of alternating current is called *inductive reactance* which is measured in ohms.

In addition, alternating current flowing through a conductor generates small erratic independent currents called *eddy currents*. Eddy currents are greatest in the center of the conductors and repel the flowing electrons toward the conductor surface; this is known as *skin effect*. Because of skin effect, the effective cross-sectional area of an alternating current conductor is reduced, which results in an increase of the conductor resistance. The total opposition to the flow of alternating current (resistance and inductive reactance) is called *impedance* and is measured in ohms, Fig. 8–4.

8–4 *ALTERNATING CURRENT RESISTANCE AS COMPARED TO DIRECT CURRENT*

The opposition to current flow is greater for alternating current as compared to direct current circuits, because of inductive reactance, eddy currents and skin effect. The following two tables give examples of the difference between alternating current circuits as compared to direct current circuits, Fig. 8–5.

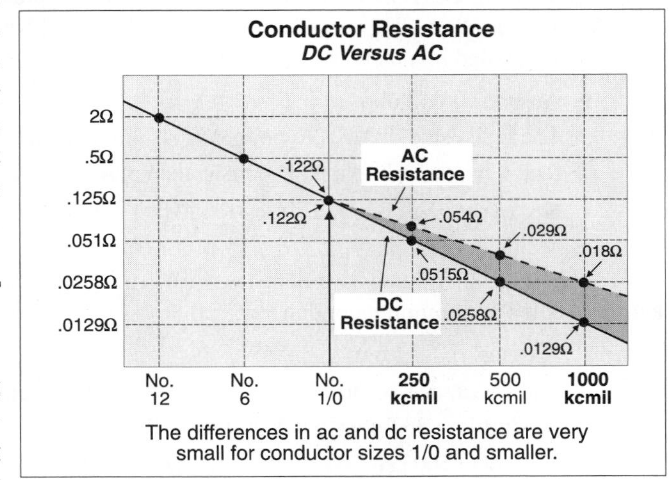

Figure 8-5

Conductor Resistance AC versus DC

COPPER – Alternating Current versus Direct Current Resistance at 75°C			
Conductor Size	**Alternating Current Chapter 9, Table 9**	**Direct Current Chapter 9, Table 8**	**AC resistance greater than DC resistance by %**
250,000	0.054 ohm per 1,000 feet	0.0515 ohm per 1,000 feet	4.85%
500,000	0.029 ohm per 1,000 feet	0.0258 ohm per 1,000 feet	12.40%
1,000,000	0.018 ohm per 1,000 feet	0.0129 ohm per 1,000 feet	39.50%

ALUMINUM – Alternating Current versus Direct Current Resistance at 75°C			
Conductor Size	**Alternating Current Chapter 9, Table 9**	**Direct Current Chapter 9, Table 8**	**AC resistance greater than DC resistance by %**
250,000	0.086 ohm per k feet	0.0847 ohm per k feet	1.5%
500,000	0.045 ohm per k feet	0.0424 ohm per k feet	6.13%
1,000,000	0.025 ohm per k feet	0.0212 ohm per k feet	17.12%

8–5 *RESISTANCE ALTERNATING CURRENT [Chapter 9, Table 9 Of The NEC®]*

Alternating current conductor resistances are listed in Chapter 9, Table 9 of the NEC®. The alternating current resistance of a conductor is dependent on the conductors material (copper or aluminum) and on the magnetic property of the raceway.

❑ **Chapter 9, Table 9 Alternating Current Resistance Example**

- Answer: What is the alternating current resistance for a 250,000 circular mils conductor 1,000 feet long?
  Copper conductor in nonmetallic raceway = 0.052 ohm per 1,000 feet
  Copper conductor in aluminum raceway = 0.057 ohm per 1,000 feet
  Copper conductor in steel raceway = 0.054 ohm per 1,000 feet
  Aluminum conductor in nonmetallic raceway = 0.085 ohm per 1,000 feet
  Aluminum conductors in aluminum raceway = 0.090 ohm per 1,000 feet
  Aluminum conductors in steel raceway = 0.086 ohm per 1,000 feet

Alternating Current Conductor Resistance Formula
The following formula can be used to determine the resistance of different lengths of conductors.

Alternating Current Resistance =

$$\frac{\textbf{Conductor Resistance Ohms}}{\textbf{1,000 Feet}} \times \textbf{Conductor Length}$$

❑ **Alternating Current Conductor Resistance Examples**
What is the alternating current resistance of 420 feet of No. 2/0 copper installed in a metal raceway, Fig. 8–6?

- Answer: 0.042 ohm
  The resistance of No. 2/0 copper is .1 ohm per 1,000 feet, Chapter 9, Table 9. The resistance of 420 feet of No. 2/0 is: (0.1 ohm/1,000 feet) × 420 feet = 0.042 ohm

What is the alternating current resistance of 169 feet of 500 kcmil installed in aluminum conduit?

- Answer: 0.0054 ohms
  The resistance of 500 kcmils installed in aluminum conduit is 0.032 ohm per 1,000 feet. The resistance of 169 feet of 500 kcmils in aluminum conduit is:
  (0.032 ohm/1,000 feet) × 169 feet = 0.0054 ohm

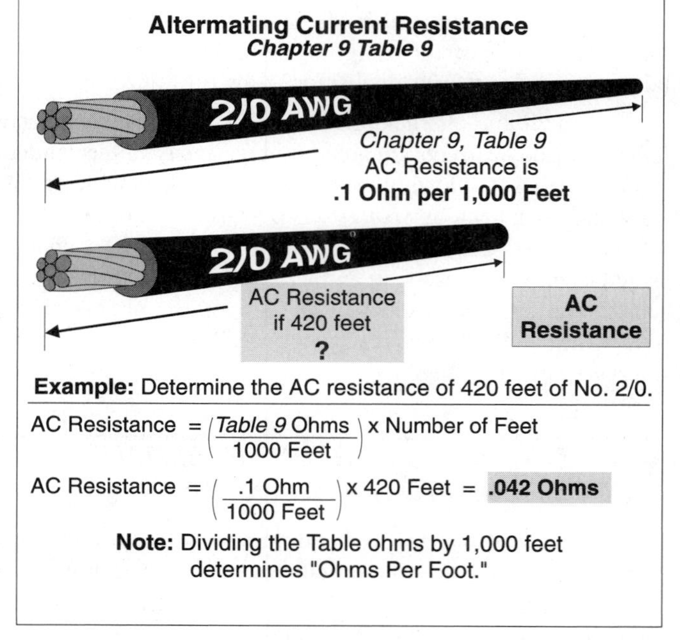

Figure 8-6
Alternating Current Resistance

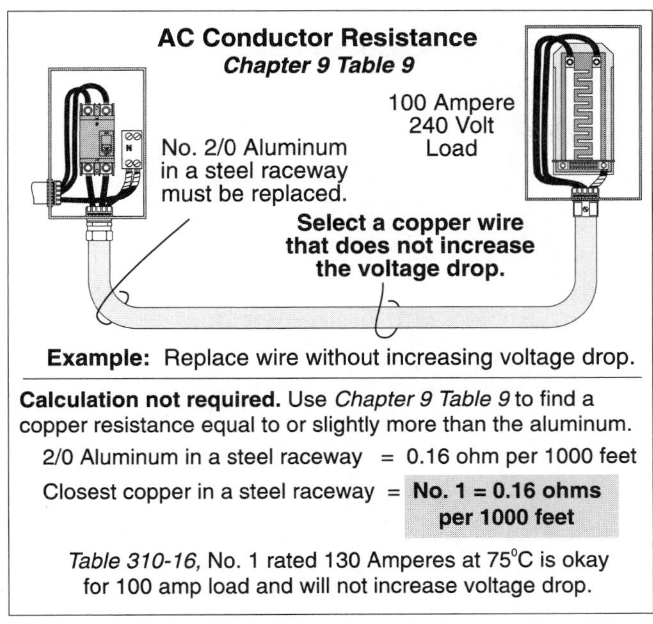

Figure 8-7

Alternating Current Conductor Resistance Example

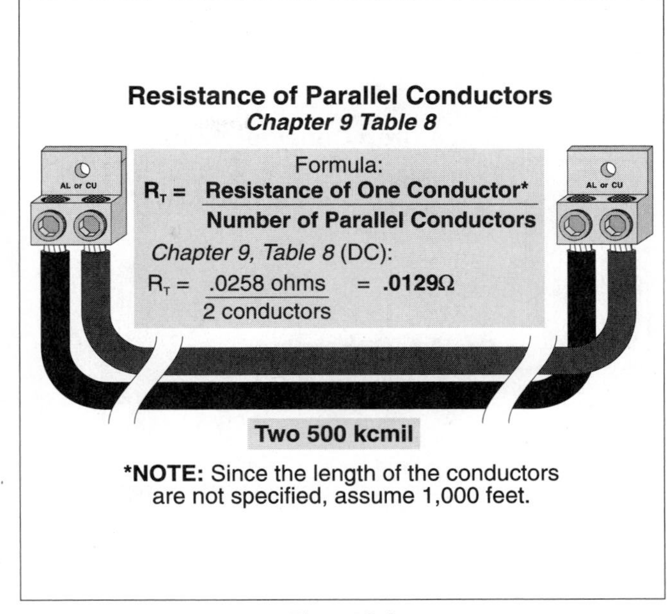

Figure 8-8

Resistance Parallel Conductors Example

Converting Copper To Aluminum Or Aluminum To Copper

When requested to determine the replacement conductor for copper or aluminum the following steps should be helpful.

Step 1: → Determine the resistance of the existing conductor using Table 9, Chapter 9 for 1,00 0 feet.

Step 2: → Using Table 9, Chapter 9 to locate a replacement conductor that has a resistance of not more than the existing conductors.

Step 2: → Verify that the replacement conductor has an ampacity [Table 310–16] sufficient for the load.

❑ **Converting Copper To Aluminum Or Aluminum To Copper Example**

A 100 ampere, 240-volt, single-phase load is wired with No. 2/0 aluminum conductors in a steel raceway. What size copper wire can we use to replace the aluminum wires and not have a greater voltage drop. Note: The wire selected must have an ampacity of at least 100 amperes, Fig. 8–7?

(a) No. 1/0 (b) No. 1 (c) No. 2 (d) No. 3

- Answer: (b) No. 1 copper

The resistance of No. 2/0 aluminum in a steel raceway is 0.16 ohm per 1,000 feet. No. 1 copper in a steel raceway has a resistance of 0.16 ohms and it has an ampacity of 130 amperes at 75°C [Table 310–16 and 110–14(c)].

Determining The Resistance Of Parallel Conductors

The resistance total in a parallel circuit is always less than the smallest resistor. The equal resistors formula can be used to determine the resistance total of parallel conductors:

$$\textbf{Resistance Total} = \frac{\textbf{Resistance of One Conductor*}}{\textbf{Number of Parallel Conductors}}$$

*Resistance according to Chapter 9 Table 8 or Table 9 of the NEC®, assuming 1,000 feet unless specified otherwise.

❑ **Resistance of Parallel Conductors Example**

What is the direct current resistance for two 500 kcmil conductors in parallel, Fig. 8–8?

(a) 0.0129 ohm (b) 0.0258 ohm (c) 0.0518 ohm (d) 0.0347 ohm

- Answer: (a) 0.0129 ohms

$$\text{Resistance Total} = \frac{\text{Resistance of One Conductor}}{\text{Number of Parallel Conductors}}$$

$$\text{Resistance Total} = \frac{0.0258 \text{ ohm}}{2 \text{ conductors}} = 0.0129 \text{ ohm}$$

Note: You could also look up the resistance of 1,000 kcmils in Chapter 9, Table 8, which is 0.0129 ohm.

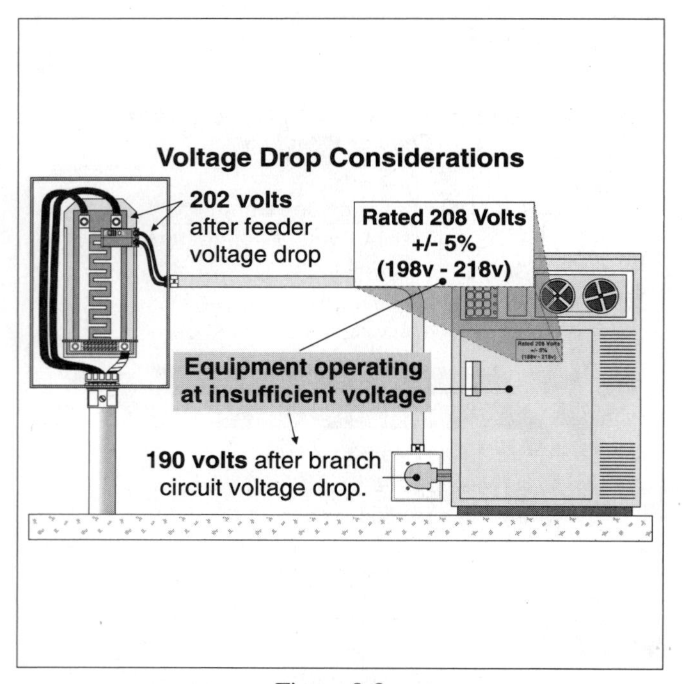

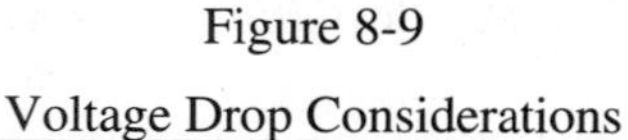

Figure 8-9

Voltage Drop Considerations

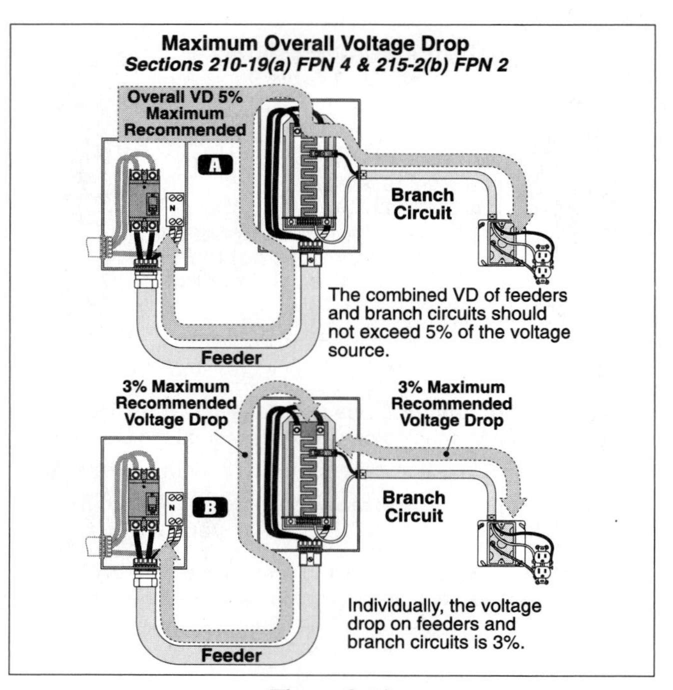

Figure 8-10

Maximum Recommended Voltage Drop

8–6 *VOLTAGE DROP CONSIDERATIONS*

The resistance of a conductor's opposition to current flow results in conductor *voltage drop*; E (voltage drop) = I (amperes) × R (ohms). Because of the conductor's voltage drop, the operating voltage of the equipment will be reduced. Inductive loads, such as motors and ballasts, can overheat at reduced *operating voltage*, which can result in reduced equipment operating life as well as inconvenience for the customer. Electronic equipment such as computers, laser printers, copy machines, etc., can suddenly power down resulting in data loss if the operating voltage is too low. Resistive equipment simply will not provide the power output at reduced operating voltage, Fig. 8–9.

8–7 *NEC® VOLTAGE DROP RECOMMENDATIONS*

Contrary to many beliefs, the NEC® does not contain any requirements for sizing ungrounded conductors for voltage drop. It does recommend in many areas of the NEC® that we consider the effects of conductor voltage drop when sizing conductors. See some of these recommendations in the Fine Print Notes to Sections 210–19(a), 215–2(b), 230–31(c) and 310–15. Please be aware that Fine Print Notes in the NEC® are recommendations, they are not requirements [90–5]. The NEC® recommends that the maximum combined voltage drop for both the feeder and branch circuit should not exceed 5 percent, and the maximum on the feeder *or* branch circuit should not exceed 3 percent, Figure 8–10.

❑ **NEC® Voltage Drop Recommendation Example**

What are the minimum NEC® recommended operating volts for a 115-volt rated load that is connected to a 120/240-volt source, Fig. 8–11?

(a) 120 volts (b) 115 volt (c) 114 volts (d) 116 volts

- Answer: (c) 114 volts

The maximum conductor voltage drop recommended for both the feeder and branch circuit is 5 percent of the voltage source (120 volts). The total conductor voltage drop (feeder and branch circuit) should not exceed: 120 volts × .05 = 6 volts. The operating voltage at the load is calculated by subtracting the conductor's voltage drop from the voltage source: 120 volts – 6 volts drop = 114 volts.

8–8 *DETERMINING CIRCUIT CONDUCTORS VOLTAGE DROP*

When the circuit conductors have already been installed, the *voltage drop* of the conductors can be determined by the Ohm's Law Method or by the formula method.

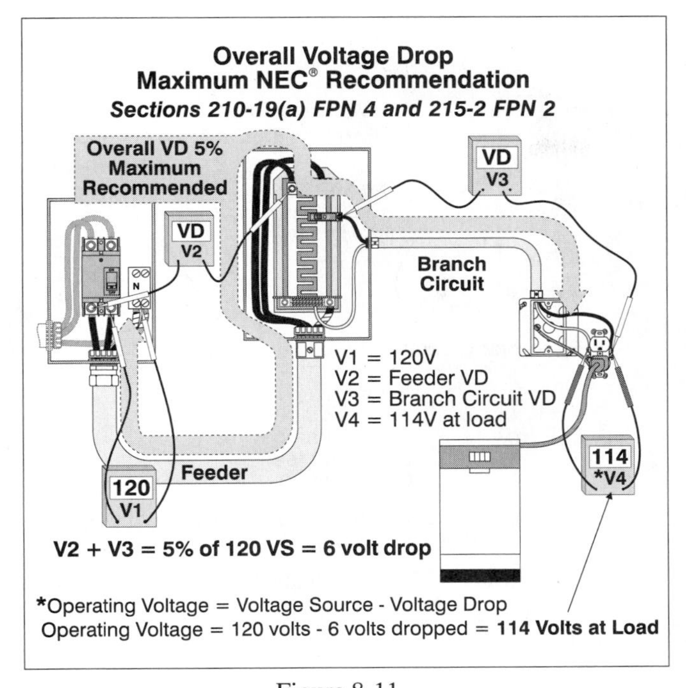

Figure 8-11

Overall Voltage Drop Example

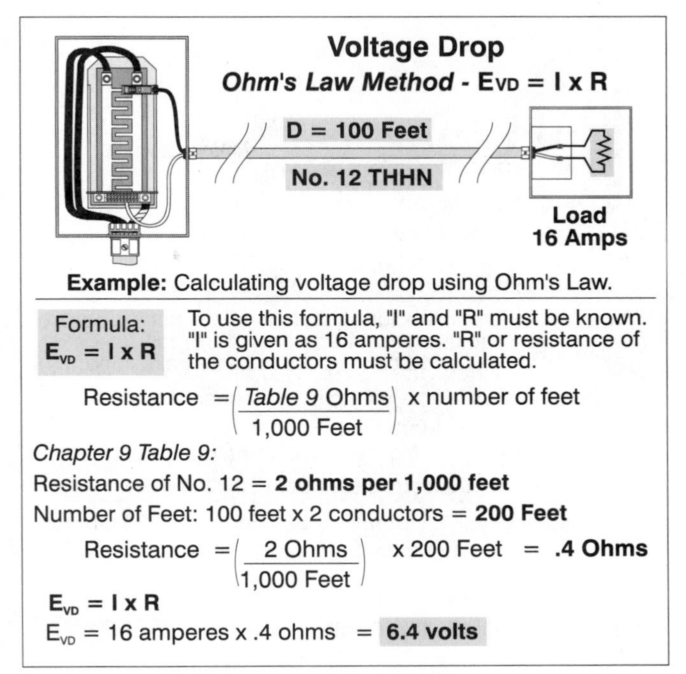

Figure 8-12

Voltage Drop – Ohm's Law, 120 Volt Example

Ohm's Law Method – Single-Phase Only

$E_{vd} = I \times R$

E_{vd} = Conductor voltage drop expressed in volts

I = The load in amperes at 100 percent, not at 125 percent for motors or continuous loads

R* = Conductor Resistance, Chapter 9, Table 8 for direct current circuits, or Chapter 9, Table 9 for alternating current circuits

*For conductors No. 1/0 and smaller, the difference in resistance between direct current and alternating current circuits is so little that it can be ignored. In addition, we can ignore the small difference in resistance between stranded or solid wires.

❑ **Ohm's Law Conductor Voltage Drop 120 Volt Example**

What is the voltage drop of two No. 12 THHN conductors that supply a 16-ampere, 120 volt continuous load located 100 feet from the power supply, Fig. 8–12?

(a) 3.2 volts (b) 6.4 volts (c) 9.6 volts (d) 12.8 volts

- Answer: (b) 6.4 volts

  $E_{vd} = I \times R$

  I = Amperes = 16 amperes

  R = 2 ohms per 1,000 feet, Chapter 9, Table 9. The resistance No. 12 wire, 200 feet long is:

  (2 ohms/1,000 feet) × 200 feet = 0.4 ohm

  E_{vd} = 16 amperes × 0.4 ohms E_{vd} = 6.4 volts

❑ **Ohm's Law Conductor Voltage Drop 240 Volt Example**

A single-phase, 24-ampere, 240 volt load is located 160 feet from the panelboard and is wired with No. 10 THHN. What is the voltage drop of the circuit conductors, Fig. 8–13?

(a) 4.5 volts (b) 9.25 volts (c) 3.6 volts (d) 5.5 volts

- Answer: (b) 9.25 volts is the closest answer

  Evd = I × R

  I = 24 amperes

  R = 1.2 ohms per 1,000 feet, Chapter 9, Table 9. The resistance for 320 feet of No. 10 is:

  (1.2 ohms/1,000 feet) × 320 feet = 0.384 ohm

  E_{vd} = 24 amperes × 0.384 ohm E_{vd} = 9.216 volts drop

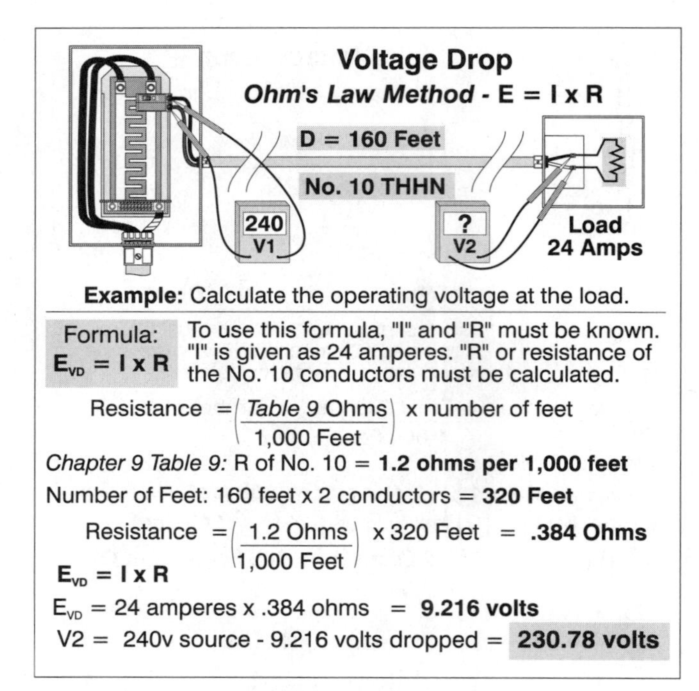

Figure 8-13

Voltage Drop – Ohm's Law, 240 Volt Example

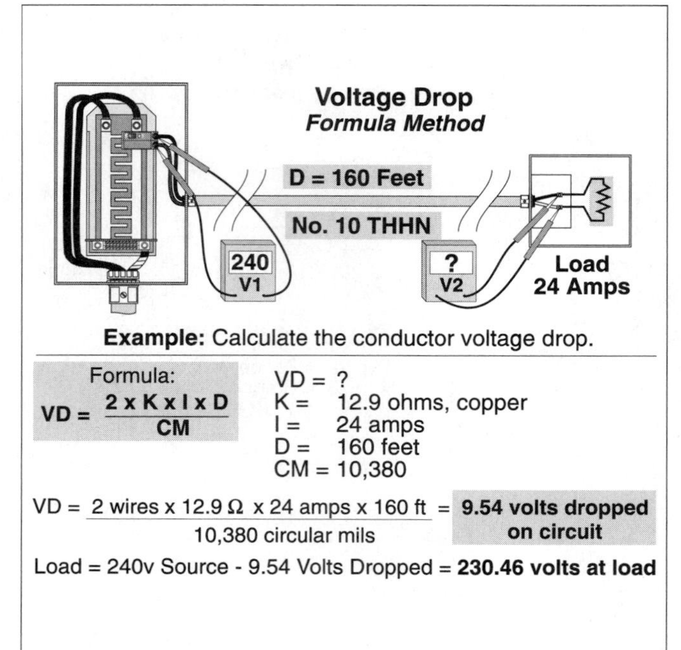

Figure 8-14

Voltage Drop – Formula Method, Single-Phase Example

Formula Method

In addition to the Ohm's Law Method, the following formula can be used to determine the conductor voltage drop.

Single–Phase $E_{VD} = \dfrac{2 \times (K \times Q) \times I \times D}{CM}$ **Three–Phase** $E_{VD} = \dfrac{\sqrt{3} \times (K \times Q) \times I \times D}{CM}$

VD = Volts Dropped: The voltage drop of the circuit expressed in volts. The NEC® recommends a maximum 3 percent voltage drop for either the branch circuit or feeder.

K = Direct Current Constant: This constant *K* represents the direct current resistance for a one thousand circular mils conductor that is one thousand feet long, at an operating temperature of 75°C. The constant *K* value is 12.9 ohms for copper, and 21.2 ohms for aluminum.

I = Amperes: The load in amperes at 100 percent, not at 125 percent for motors or continuous loads!

Q = Alternating Current Adjustment Factor: For alternating current circuits with conductors No. 2/0 and larger, the direct current resistance constant *K* must be adjusted for the effects of self-induction (eddy currents). The Q-Adjustment Factor is calculated by dividing the alternating current resistance listed in Chapter 9, Table 9, by the direct current resistance listed in Chapter 9, Table 8 in the NEC®. For all practical exam purposes, this resistance adjustment factor can be ignored because exams rarely give alternating current voltage drop questions with conductors larger than No. 1/0.

D = Distance: The distance the load is from the power supply.

CM = Circular-mils: The circular mils of the circuit conductor as listed in Chapter 9, Table 8 NEC®.

❑ Voltage Drop Formula Single-Phase Example

A 24-ampere, 240-volt load is located 160 feet from a panelboard and is wired with No. 10 THHN. What is the approximate voltage drop of the branch circuit conductors, Fig. 8–14?

(a) 4.25 volts (b) 9.5 volts (c) 3 percent (d) 5 percent

- Answer: (b) 9.5 volts is the closest

$$VD = \frac{2 \times K \times I \times D}{CM}$$

K = 12.9 ohms, Copper I = 24 amperes

D = 160 feet CM = No. 10, 10,380 circular mils, Chapter 9, Table 8

$$VD = \frac{2 \text{ wires} \times 12.9 \text{ ohms} \times 24 \text{ amps} \times 160 \text{ feet}}{10{,}380 \text{ circular mils}}$$

VD = 9.54 volts dropped

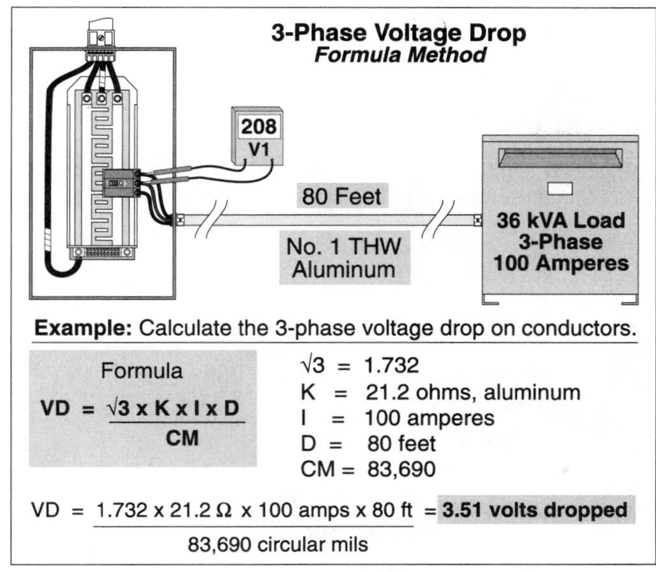

Figure 8-15

Voltage Drop – Formula Method, Three-Phase Example

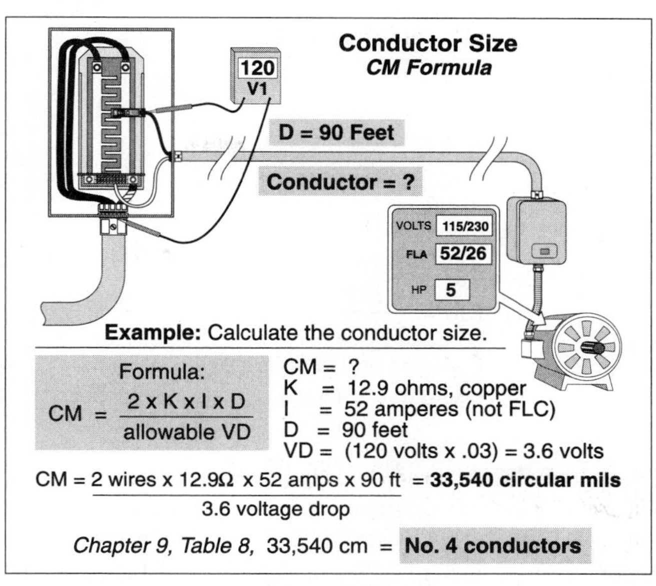

Figure 8-16

Conductor Size Example

❑ **Voltage Drop Formulas Three-Phase Example**

A three-phase, 36-kVA load rated 208-volts is located 80 feet from the panelboard and is wired with No. 1 THHN aluminum. What is the approximate voltage drop of the feeder circuit conductors, Fig. 8–15?

(a) 3.5 volts (b) 7 volts (c) 3 percent (d) 5 percent

- Answer: (a) 3.5 volts is the closest

$$VD = \frac{\sqrt{3} \times K \times I \times D}{CM}$$

$\sqrt{3} = 1.732$

K = 21.2 ohms, aluminum

$I = 100 \text{ amperes}, I = \frac{VA}{(E \times \sqrt{3})}, = \frac{36{,}000 \text{ VA}}{(208 \text{ volts} \times 1.732)}$

D = 80 feet

CM = No. 1, 83,690 circular mils, Chapter 9, Table 8

$$VD = \frac{1.732 \times 21.2 \text{ ohms} \times 100 \text{ amps} \times 80 \text{ feet}}{83{,}690 \text{ circular mils}}$$

VD = 3.51 volts dropped

8–9 *SIZING CONDUCTORS TO PREVENT EXCESSIVE VOLTAGE DROP*

The size of a conductor (actually its resistance) affects the conductor voltage drop. If we want to decrease the voltage drop of a circuit, we can increase the size of the conductor (reduce its resistance). When sizing conductors to prevent excessive voltage drop, the following formulas can be used:

$$\textbf{Single–Phase CM} = \frac{2 \times K \times I \times D}{\text{Voltage Dropped}} \qquad \textbf{Three–Phase CM} = \frac{\sqrt{3} \times K \times I \times D}{\text{Voltage Dropped}}$$

❑ **Size Conductor Single-Phase Example**

A 5 horsepower motor is located 90 feet from a 120/240-volt panelboard. What size conductor should we use if the motor nameplate indicates 52 amperes at 115 volts? Terminals rated for 75°C, Fig. 8–16.

(a) No. 10 THHN (b) No. 8 THHN
(c) No. 6 THHN (d) No. 4 THHN

- Answer: (d) No. 4 THHN

$$CM = \frac{2 \times K \times I \times D}{VD}$$

K = 12.9 ohms, copper

I = 52 amperes at 115 volts

D = 90 feet

VD = 3.6 volts, = 120 volts × 0.03

$$CM = \frac{2 \text{ wires} \times 12.9 \text{ ohms} \times 52 \text{ amperes} \times 90 \text{ feet}}{3.6 \text{ volts}}$$

CM = 33,540 circular mils

CM = No. 4, Chapter 9, Table 8

Note:

The NEC Section 430–22(a), requires that the motor conductors be sized not less than 125 percent of the motor full load currents as listed in Table 430–148. The motor full-load current is 56 amperes, and the conductor must be sized at; 56 amperes × 1.25 = 70 amperes. The No. 4 THHN required for voltage drop is rated for 75 amperes at 75°C according to Table 310–16 and 110–14(c).

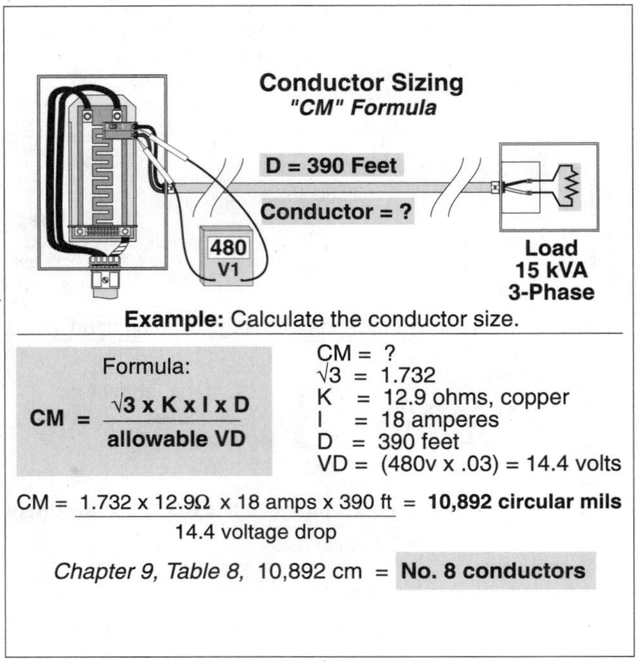

Conductor Sizing
"CM" Formula
D = 390 Feet
Conductor = ?
480
V1
Load
15 kVA
3-Phase
Example: Calculate the conductor size.
Formula:
$CM = \frac{\sqrt{3} \times K \times I \times D}{\text{allowable VD}}$
CM = ?
√3 = 1.732
K = 12.9 ohms, copper
I = 18 amperes
D = 390 feet
VD = (480v x .03) = 14.4 volts
$CM = \frac{1.732 \times 12.9\Omega \times 18 \text{ amps} \times 390 \text{ ft}}{14.4 \text{ voltage drop}}$ = **10,892 circular mils**
Chapter 9, Table 8, 10,892 cm = **No. 8 conductors**

Figure 8-17

Conductor Size, Three-Phase Example

❑ Size Conductor Three-Phase Example

A three-phase, 15-kVA load rated 480 volts is located 390 feet from the panelboard. What size conductor is required to prevent the voltage drop from exceeding 3 percent, Fig. 8–17?

(a) No. 10 THHN (b) No. 8 THHN (c) No. 6 THHN (d) No. 4 THHN

- Answer: (b) No. 8 THHN

$$CM = \frac{\sqrt{3} \times K \times I \times D}{VD}$$

K = 12.9 ohms, copper

$$I = 18 \text{ amperes}, \quad \frac{VA}{(E \times 1.732)} = \frac{15{,}000 \text{ VA}}{(480 \text{ volts} \times 1.732)} = 18 \text{ amperes}$$

D = 390 feet

VD = 14.4 volts, = 480 volts × 0.03

$$CM \quad \frac{1.732 \times 12.9 \text{ ohms} \times 18 \text{ amperes} \times 390 \text{ feet}}{14.4 \text{ volts}}$$

CM = 10,892 = No. 8, Chapter 9, Table 8

8–10 *LIMITING CONDUCTOR LENGTH TO LIMIT VOLTAGE DROP*

Voltage drop can also be reduced by limiting the length of the conductors. The following formulas can be used to help determine the maximum conductor length to limit the voltage drop to NEC® suggestions.

$$\textbf{Single–Phase} = \mathbf{D} = \frac{\mathbf{CM \times VD}}{\mathbf{2 \times K \times I}}$$

$$\textbf{Three–Phase} = \mathbf{D} = \frac{\mathbf{CM \times VD}}{\mathbf{\sqrt{3} \times K \times I}}$$

❑ Limit Distance Single-Phase Example

What is the maximum distance a single-phase, 10-kVA, 240-volt load can be located from the panelboard so the voltage drop does not exceed 3 percent? The load is wired with No. 8 THHN, Fig 8–18.

(a) 55 feet (b) 110 feet (c) 165 feet (d) 220 feet

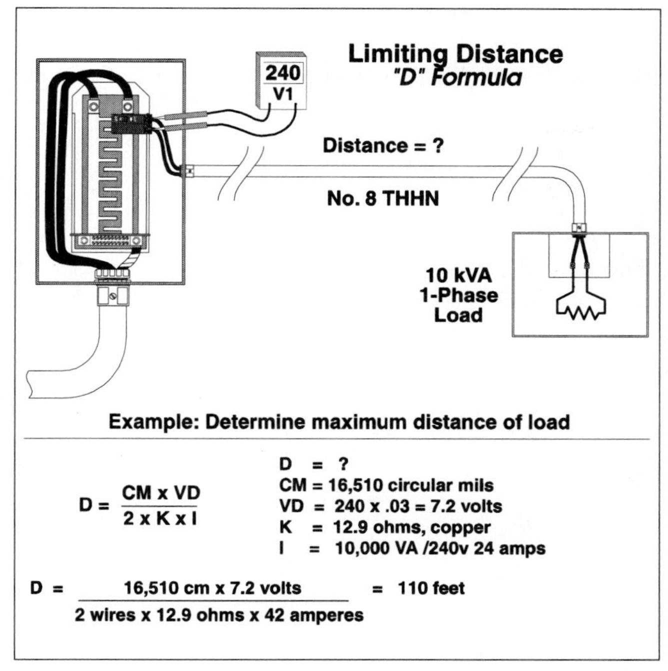

Figure 8-18

Maximum Distance, Single-Phase Example

480
V1

Maximum Distance
"D" Formula

Distance = ?

No. 6 THHN

37.5 kVA
480 v 3-Phase
45 Amperes

Example: Calculate distance of conductors.

Formula:

$$D = \frac{CM \times VD}{\sqrt{3} \times K \times I}$$

D = ?
CM = 26,240
VD = 480v x .03 = 14.4 volts
√3 = 1.732
K = 12.9 ohms, copper
I = 45 amperes

$$D = \frac{26{,}240 \text{ cm} \times 14.4 \text{ volts}}{1.732 \times 12.9 \text{ ohms} \times 45 \text{ amps}} = \textbf{376 feet}$$

Figure 8-19

Maximum Distance,Three-Phase Example

- Answer: (b) 110 feet

$$D = \frac{CM \times VD}{2 \times K \times I}$$

CM = No. 8, 16,510 circular mils, Chapter 9, Table 8

VD = 7.2 volts, = 240 volts × 0.03

K = 12.9 ohms, copper

I = 42 amperes, I = VA/E, I = 10,000 VA/240 volts

$$D = \frac{16{,}510 \text{ circular mils} \times 7.2 \text{ volts}}{2 \text{ wires} \times 12.9 \text{ ohms} \times 42 \text{ amperes}}$$

D = 110 feet

❑ Limit Distance Three-Phase Example

What is the maximum distance a three-phase, 37.5-kVA, 480-volt transformer, wired with No. 6 THHN, can be located from the panelboard so that the voltage drop does not exceed 3 percent, Fig. 8–19?

(a) 275 feet (b) 325 feet (c) 375 feet (d) 425 feet

- Answer: (c) 375 feet

$$D = \frac{CM \times VD}{\sqrt{3} \times K \times I}$$

CM = No. 6, 26,240 circular mils, Chapter 9, Table 8

K = 12.9 ohms, copper

$\sqrt{3} = 1.732$

VD = 14.4 volts, = 480 volts × 0.03

$$I = 45 \text{ amperes}, = \frac{VA}{(\text{Volts} \times 1.732)} = \frac{37{,}500 \text{ VA}}{(480 \text{ volts} \times 1.732)}$$

$$D = \frac{26{,}240 \text{ circular mils} \times 14.4 \text{ volts}}{1.732 \times 12.9 \text{ ohms} \times 45 \text{ amperes}}$$

D = 376 feet

8–11 *LIMITING CURRENT TO LIMIT VOLTAGE DROP*

Sometimes the only method of limiting the circuit voltage drop is to limit the load on the conductors. The following formulas can be used to determine the maximum load.

Single–Phase $= I = \dfrac{CM \times VD}{2 \times K \times D}$ **Three–Phase** $= I = \dfrac{CM \times VD}{\sqrt{3} \times K \times D}$

❑ **Maximum Load Single-Phase Example**

An existing installation contains No. 1/0 THHN aluminum conductors in a nonmetallic raceway to a panelboard located 220 feet from a 230-volt power source. What is the maximum load that can be placed on the panelboard so that the NEC® recommendations for voltage drop are not exceeded, Fig. 8–20?

(a) 50 amperes (b) 75 amperes (c) 100 amperes (d) 150 amperes

- Answer: (b) 75 amperes is the closest answer

$I = \dfrac{CM \times VD}{2 \times K \times D}$

CM = No. 1/0, 105,600 circular mils, Chapter 9, Table 8

VD = 6.9 volts, = 230 volts × 0.03

K = 21.2 ohms, aluminum

D = 220 feet

$I = \dfrac{105{,}600 \text{ circular mils} \times 6.9 \text{ volts drop}}{2 \text{ wires} \times 21.2 \text{ ohms} \times 220 \text{ feet}} =$

I = 78 amperes

Note: The maximum load permitted on 1/0 THHN Aluminum at 75°C is 120 amperes; Table 310–16.

❑ **Maximum Load Three-Phase Example**

An existing installation contains No. 1 THHN conductors in an aluminum raceway to a panelboard located 300 feet from a three-phase 460/230 volt power source. What is the maximum load the conductors can carry so that the NEC® recommendation for voltage drop is not exceeded, Fig. 8–21?

(a) 170 amperes (b) 190 amperes (c) 210 amperes (c) 240 amperes

- Answer: (a) 170 amperes

$I = \dfrac{CM \times VD}{\sqrt{3} \times 2 \times K \times D}$

CM = No. 1, 83,690 circular mils, Chapter 9, Table 8

VD = 13.8 volts, = 460 volts × 0.03

$\sqrt{3} = 1.732$

K = 12.9 ohms, copper

D - 220 feet

$I = \dfrac{83{,}690 \text{ circular mils} \times 13.8 \text{ volts}}{1.732 \times 12.9 \text{ ohms} \times 300 \text{ feet}}$

I = 172 amperes

Note: The maximum load permitted on No. 1 THHN at 75°C is 130 amperes; Table 310–16 and 110–14(c).

8–12 *EXTENDING BRANCH CIRCUITS*

If you want to extend an existing circuit and you want to limit the voltage drop, follow these steps.

Step 1: → Determine the voltage drop of the existing conductors.

Single–Phase VD $= \dfrac{2 \times K \times I \times D}{CM}$

Three–Phase VD $= \dfrac{\sqrt{3} \times K \times I \times D}{CM}$

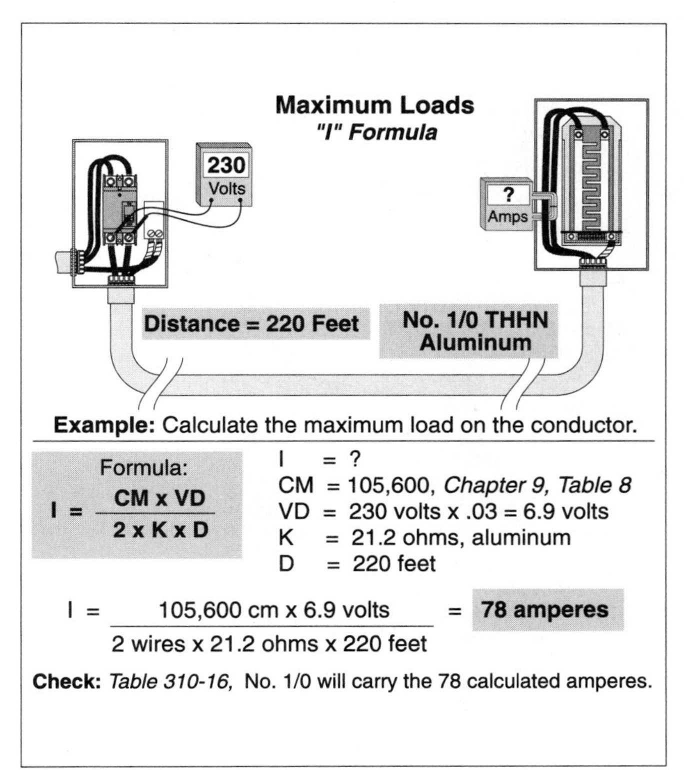

Figure 8-20

Maximum Load, Single-Phase Example

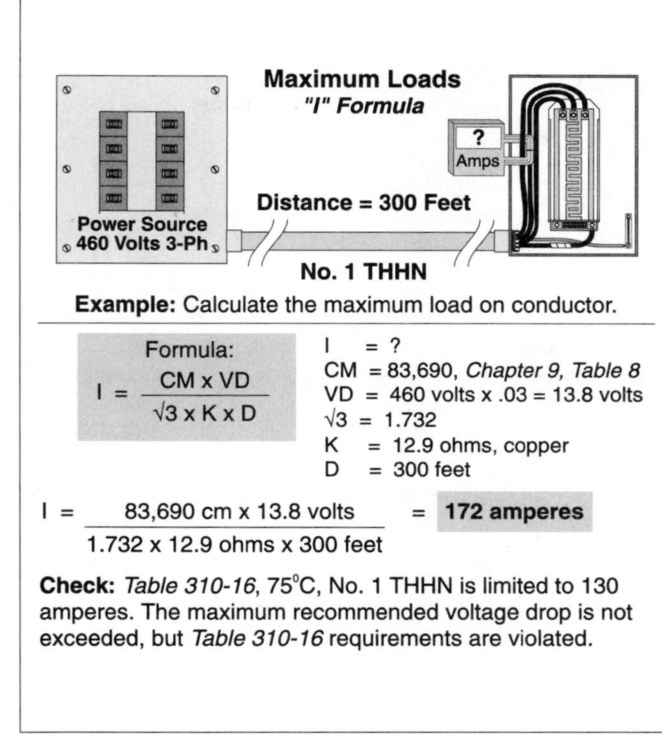

Figure 8-21

Maximum Distance, Three-Phase Example

Step 2: → Determine the voltage drop permitted for the extension by subtracting the voltage drop of the existing conductors from the permitted voltage drop.

Step 3: → Determine the extended conductor size

$$\text{Single–Phase CM} = \frac{2 \times K \times I \times D}{VD}$$

$$\text{Three–Phase CM} = \frac{\sqrt{3} \times K \times I \times D}{VD}$$

❑ Single-Phase Example

An existing junction box is located 55 feet from the panelboard and contains No. 4 THW aluminum. We want to extend this circuit 65 feet and supply a 50 ampere, 240-volt load. What size copper conductors can we use for the extension, Fig. 8–22?

(a) No. 8 THHN (b) No. 6 THHN (c) No. 4 THHN (d) No. 5 THHN

- Answer: (b) No. 6 THHN

Step 1: → Determine the voltage drop of the existing conductors:

$$\text{Single Phase VD} = \frac{2 \times K \times I \times D}{CM}$$

K = 21.2 ohms, aluminum
D = 55 feet
I = 50 amperes
CM = No. 4, 41,740 circular mils, Chapter 9, Table 8

$$VD = \frac{2 \text{ wires} \times 21.2 \text{ ohms} \times 50 \text{ amps} \times 55 \text{ feet}}{41{,}740 \text{ circular mils}}$$

VD = 2.79 volts

Step 2: → Determine the voltage drop permitted for the extension by subtracting the voltage drop of the existing conductors from the total permitted voltage drop. Total permitted voltage drop = 240 volts × 0.03 = 7.2 volts less existing conductor voltage drop of 2.79 volts = 4.41 volts.

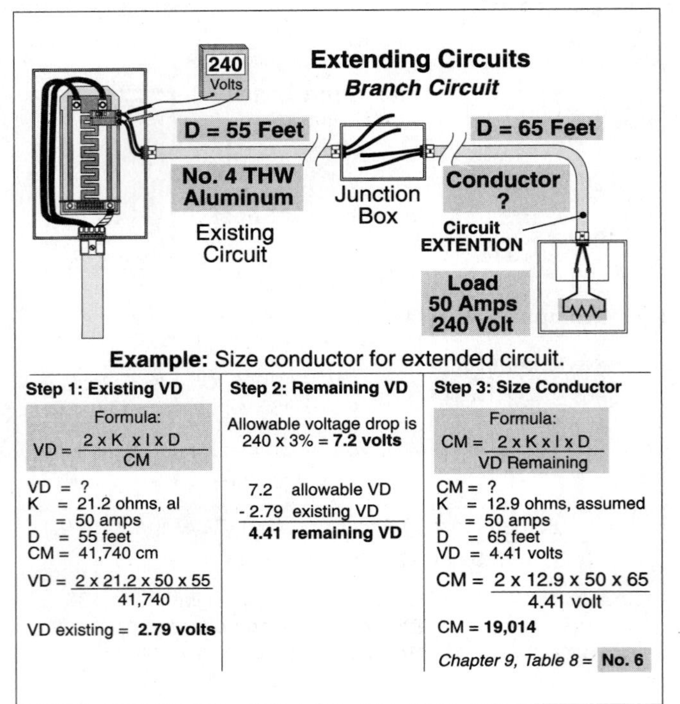

Figure. 8-22

Extending Circuit Example

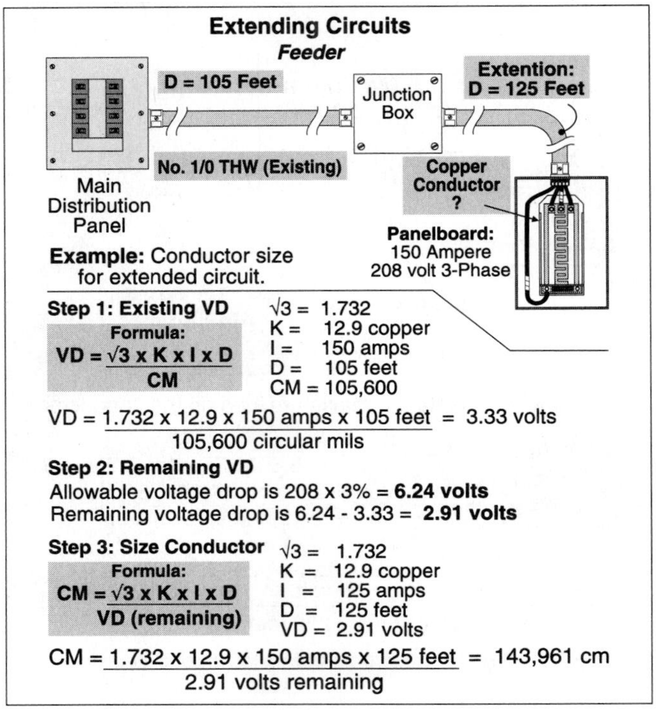

Extending Circuits

Feeder

D = 105 Feet

Junction Box

Extention: D = 125 Feet

No. 1/0 THW (Existing)

Main Distribution Panel

Copper Conductor ?

Panelboard: 150 Ampere 208 volt 3-Phase

Example: Conductor size for extended circuit.

Step 1: Existing VD

Formula: $VD = \frac{\sqrt{3} \times K \times I \times D}{CM}$

√3 = 1.732
K = 12.9 copper
I = 150 amps
D = 105 feet
CM = 105,600

$VD = \frac{1.732 \times 12.9 \times 150 \text{ amps} \times 105 \text{ feet}}{105{,}600 \text{ circular mils}} = 3.33 \text{ volts}$

Step 2: Remaining VD

Allowable voltage drop is 208 x 3% = **6.24 volts**
Remaining voltage drop is 6.24 - 3.33 = **2.91 volts**

Step 3: Size Conductor

Formula: $CM = \frac{\sqrt{3} \times K \times I \times D}{VD \text{ (remaining)}}$

√3 = 1.732
K = 12.9 copper
I = 125 amps
D = 125 feet
VD = 2.91 volts

$CM = \frac{1.732 \times 12.9 \times 150 \text{ amps} \times 125 \text{ feet}}{2.91 \text{ volts remaining}} = 143{,}961 \text{ cm}$

Figure. 8-23

Extending Circuits – Feeder

Step 3: → Determine the extended conductor size.

$$CM = \frac{2 \times K \times I \times D}{VD}$$

K = 12.9 ohms, copper
I = 50 amperes
D = 65 feet
VD = 7.2 volts total less 2.79 volts = 4.41 volts

$$CM = \frac{2 \text{ wires} \times 12.9 \text{ ohms} \times 50 \text{ amps} \times 65 \text{ feet}}{4.41 \text{ volts drop}}$$

CM = 19,014 circular mils
CM = No. 6 conductor, Chapter 9, Table 8

❑ Three-Phase Example

An existing junction box is located 105 feet from the Main Distribution Panel (MDP) and contains No. 1/0 THW. We want to extend this feeder 125 feet to supply a 150 ampere, 208 volt three-phase panelboard. What size copper conductors can we use for the 125 foot extension, Fig. 8–23?

(a) 3/0 THHN (b) 250 kcmil THHN (c) 300 kcmil THHN (d) 500 kcmil THHN

- Answer: (b) 250 kcmil THHN

Step 1: → Determine the voltage drop of the existing conductors:

$$VD = \frac{\sqrt{3} \times K \times I \times D}{CM}$$

K = 12.9 Copper
I = 150 amperes
D = 105 feet
CM = 105,600

$$VD = \frac{1.732 \times 12.9 \text{ ohms} \times 150 \text{ amps} \times 105 \text{ feet}}{105{,}600 \text{ circular mils}}$$

VD = 3.33 volts

Step 2 → Determine the voltage drop permitted for the extension by subtracting the voltage drop of the existing conductors from the permitted voltage drop. The voltage drop permitted for the extension is 2.91 volts. (208 volts × .03 = 6.24 volts - 3.33 volts drop = 2.91 volts)

Step 3 → Determine the extended conductor size:

$$CM = \frac{\sqrt{3} \times K \times I \times D}{VD}$$

K = 12.9 ohms, copper
I = 150 amperes
D= 125 feet
VD = 2.91 (6.24 - 3.33)

$$CM = \frac{1.732 \times 12.9 \text{ ohms} \times 150 \text{ amps} \times 125 \text{ feet}}{2.91 \text{ volts drop}}$$

CM = 143,961 circular mils = No. 3/0 (Chapter 9, Table 8)

Note: No. 3/0 THHN is rated for 200 amperes [Table 310-16], which is sufficient to carry the 150 ampere load.

8–13 *SIZING CONDUCTORS WHEN LOADS ARE AT DIFFERENT DISTANCES*

When a circuit contains more than one load at different distances, the conductors can be sized according to the following steps:

❑ **Example A** Fig. 8–24.

A jogging track has 4 lighting fixtures located 250 feet apart from each other. The first fixture is located 150 feet from the panelboard. Each fixture is rated 150 watts at 120 volts (+ 5, or - 10%) and has an ampere rating of 1.5 amperes. The calculated voltage at the panelboard is 114 volts when the jogging lights are on. What size conductor is required to each fixture?

Step 1: → Determine the CM to each fixture assuming each fixture is wired independent of all other fixtures.

Fixture 1 = 968 circular mils
Fixture 2 = 2,580 circular mils
Fixture 3 = 4,193 circular mils
Fixture 4 = 5,805 circular mils

- Answer: The actual conductor to each fixture is determined by adding the circular mils of the other fixtures.
  Fixture 1 = No. 8 - 13,546 circular mils (968 CM + 2,580 CM + 4,193 CM + 5,805 CM)
  Fixture 2 = No. 8 - 12,578 circular mils (2,580 CM + 4,193 CM + 5,805 CM)
  Fixture 3 = No. 10 - 9,998 circular mils (4,193 CM + 5,805 CM)
  Fixture 4 = No. 12 - 5,805 circular mils

➛ **Fixture No. 1 Calculation**

$$CM = \frac{2 \times K \times I \times D}{VD}$$

$$CM = \frac{2 \times 12.9 \text{ ohms} \times 1.5 \text{ amps} \times 150 \text{ feet}}{6 \text{ volts drop}}$$

K = 12.9 ohms, copper
I = 1.5 amperes
D = 150 feet
VD = 6 volts (114 volts - 108 volts*)
CM = 968

* The minimum voltage at the load is 120 volts × .90 = 108 volts. The question gave a voltage tolerance of (+ 5%, or - 10%). 120 volts less 10 percent = 108 volts as the minimum voltage at the load to maintain the tolerance level. 120 - 10 percent is the same as 120 volts × 90 percent or 120 × .9.

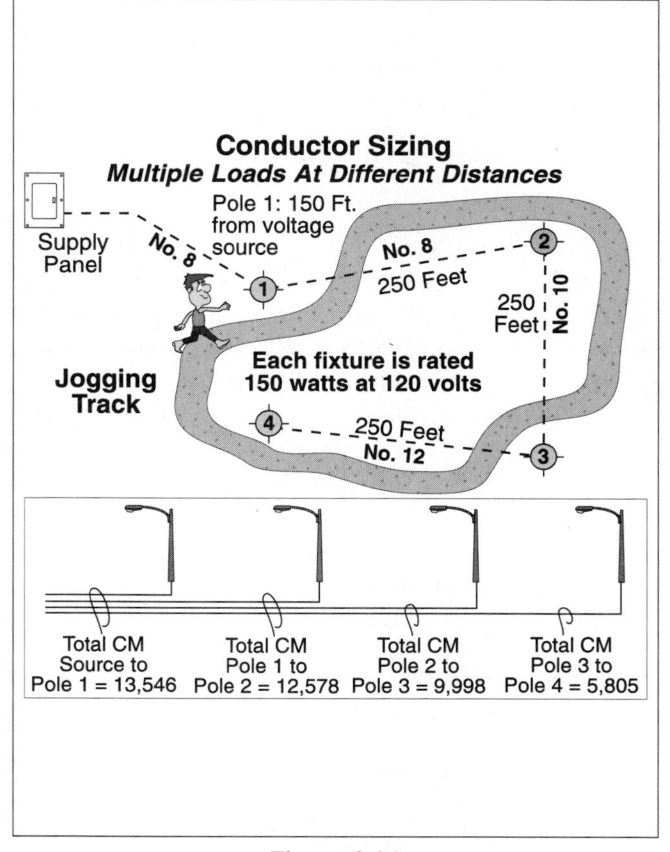

Figure. 8-24
Conductor Sizing – Multiple Loads At Different Distances

➛ **Fixture No. 2 Calculation**

$$CM = \frac{2 \times K \times I \times D}{VD}$$

$$CM = \frac{2 \times 12.9 \text{ ohms} \times 1.5 \text{ amps} \times 400 \text{ feet}}{6 \text{ volts drop}}$$

K = 12.9 ohms. copper
I = 1.5 amperes
D = 400 feet (150 feet + 250 feet)
VD = 6 volts (114 volts - 108 volts*)
CM = 2,580

➛ **Fixture No. 3 Calculation**

$$CM = \frac{2 \times K \times I \times D}{VD}$$

K = 12.9 ohms, copper
I = 1.5 amperes
D = 650 feet (150 feet + 250 feet + 250 feet)
VD = 6 volts (114 volts - 108 volts*)

$$CM = \frac{2 \times 12.9 \text{ ohms} \times 1.5 \text{ amps} \times 650 \text{ feet}}{6 \text{ volts drop}}$$

CM = 4,193

➛ **Fixture No. 4 Calculation**

$$CM = \frac{2 \times K \times I \times D}{VD}$$

K = 12.9 ohms, copper
I = 1.5 amperes
D = 900 feet (150 feet + 250 feet + 250 feet + 250 feet)
VD = 6 volts (114 volts - 108 volts*)

$$CM = \frac{2 \times 12.9 \text{ ohms} \times 1.5 \text{ amps} \times 900 \text{ feet}}{6 \text{ volts drop}}$$

CM = 5,805

❑ **Example B**

A parking lot contains three poles with fixtures. Pole No. 1 is located 575 feet from the power supply and contains one single-phase 480 volt, 2.5 amperes high-pressure sodium fixture. Pole No. 2 is located 600 feet from Pole No. 1 and it contains two fixtures. Pole No. 3 is located 550 feet from Pole No. 2 and it contains one fixture. What size conductor is required to each pole?

- Answer: The conductor size to each pole is:

  Pole No. 1 = No. 6 - 20,829 circular mils (2,576 CM + 10,526 CM + 7,727 CM)

  Pole No. 2 = No. 6 - 18,253 circular mils (10,526 CM + 7,727 CM)

  Pole No. 3 = No. 10 - 5,805 circular mils

➛ **Pole No. 1 Calculation**

$$CM = \frac{2 \times K \times I \times D}{VD}$$

K = 12.9 ohms, copper
I = 2.5 amperes
D = 575 feet
VD = 14.4 volts (480 volts × .03*)

$$CM = \frac{2 \times 12.9 \text{ ohms} \times 2.5 \text{ amps} \times 575 \text{ feet}}{14.4 \text{ volts drop}}$$

CM = 2,576

* Since we have no specification, we will use the NEC® recommendation of 3 percent.

➛ **Pole No. 2 Calculation**

$$CM = \frac{2 \times K \times I \times D}{VD}$$

K = 12.9 ohms, copper

I = 5 amperes

D = 1,175 feet (575 + 600)

VD = 14.4 volts (480 volts × .03*)

$$CM = \frac{2 \times 12.9 \text{ ohms} \times 5 \text{ amps} \times 1{,}175 \text{ feet}}{14.4 \text{ volts drop}}$$

CM = 10,526

➛ **Pole No. 3 Calculation**

$$CM = \frac{2 \times K \times I \times D}{VD}$$

K = 12.9 ohms, copper

I = 2.5 amperes

D = 1,725 feet (575 feet + 600 feet + 550 feet)

VD = 14.4 volts (480 volts × .03)

$$CM = \frac{2 \times 12.9 \text{ ohms} \times 2.5 \text{ amps} \times 1{,}725 \text{ feet}}{14.4 \text{ volts drop}}$$

CM = 7,727

8–14 *FEEDER AND BRANCH CIRCUIT COMBINED CALCULATIONS*

When adding branch circuits to an existing panelboard, the branch circuit conductors must be sized so the voltage drop does not exceed the equipment specification. When sizing the branch circuit conductors, the following steps should be used.

Step 1: → Determine the feeder voltage drop.

$$\text{Single Phase} = \frac{2 \times I \times D}{CM}$$

$$\text{Three Phase} = \frac{\sqrt{3} \times I \times D}{CM}$$

Step 2: → Determine the maximum voltage drop for the equipment (check specifications). If no specifications are given, use the NEC® recommendations of 5 percent for both the feeder and the branch circuit.

Step 3: → Determine the voltage drop permitted for the branch circuit by subtracting the feeder voltage from the maximum overall voltage drop. If no specifications are given, the NEC® recommends a maximum of 3 percent.

Step 4: → Size the branch circuit conductors.

$$\text{Single Phase CM} = \frac{2 \times K \times I \times D}{VD}$$

$$\text{Three Phase CM} + \frac{\sqrt{3} \times K \times I \times D}{VD}$$

❑ **Single-Phase Example**

A 200 ampere, single-phase, 3-wire 120/240 volt feeder supplies a panelboard located 180 feet from the power supply. The feeder conductors are No. 4/0 aluminum and the load on the feeder is 160 amperes (including the 5 kVA load). What size branch circuit conductor is required to supply a 5 kVA, 240 volt computer load located 120 feet from the panelboard, Fig. 8–25?

(a) No. 12 THHN (b) No. 10 THHN

(c) No. 8 THHN (d) No. 6 THHN

• Answer: (b) No. 10 THHN

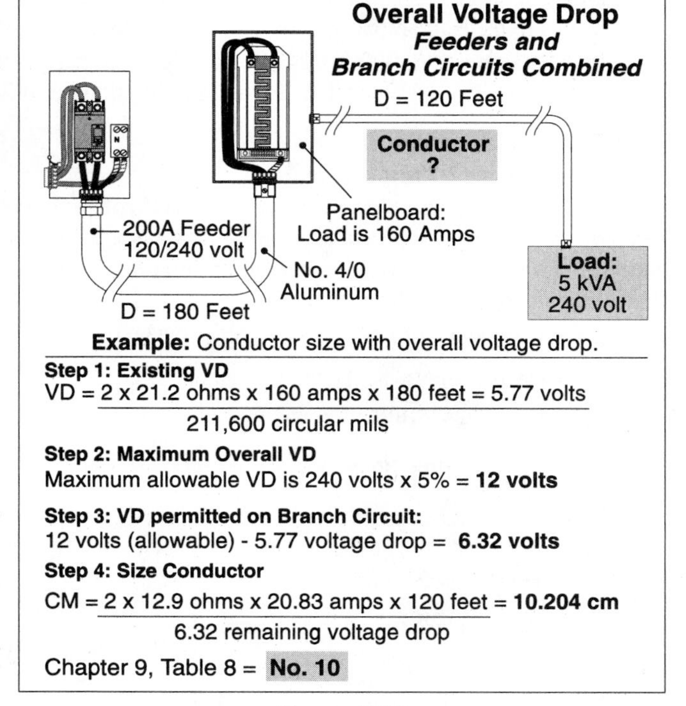

Figure. 8-25

Overall Voltage Drop – Feeders And Branch Circuits Combined

Step 1: → Determine the feeder voltage drop for No. 4/0 aluminum conductors.

$$\text{Single Phase VD} = \frac{2 \times K \times I \times D}{CM}$$

K = 21.2 ohms, aluminum
I = 160 amperes
D = 180 feet
CM = 211,600 = No. 4/0 - Chapter 9, Table 8

$$VD = \frac{2 \times 21.2 \text{ ohms} \times 160 \text{ amps} \times 180 \text{ feet}}{211{,}600 \text{ circular mils}}$$

VD = 5.77 volts

Step 2: → Determine the voltage drop suggested for both the feeder and branch circuit. Since no specifications are given, use the NEC® recommendation, 5 percent.
Feeder and Branch Voltage Drop = 240 volts × .05 = 12 volts

Step 3: → Determine the voltage drop permitted for the branch circuit (subtract the feeder voltage drop from the maximum overall voltage drop). Since no specifications are given, be sure the voltage drop does not exceed the NEC® recommendation, 3 percent.
Branch Volt Drop = 12 volts - 5.77 volts = 6.32 volts, this is less than the maximum recommended by the NEC® (240 volts × .03 = 7.2 volts)

Step 4: → Size the branch circuit conductors

$$\text{Single Phase CM} = \frac{2 \times K \times \times I \times D}{VD}$$

K = 12.9 ohms, copper
I = 20.83 amperes (5,000 VA/240 volts)
D = 120 feet
VD = 6.32 volts (12 volts - 5.77 volts)

$$CM = \frac{2 \times 12.9 \text{ ohms} \times 20.83 \text{ amps} \times 120 \text{ feet}}{6.32 \text{ volts drop}}$$

CM = 10,204 = No. 10 Copper (Chapter 9, Table 8)

Note: Section 645-5(a) requires branch circuit conductors to data processing equipment located in data processing rooms to have an ampacity of not less than 125 percent of the load rating. The branch circuit ampacity shall not be less than 26 amperes (5,000 VA/240 volts = 20.83 amperes × 1.25). According to Table 310-16, No. 10 conductor of any insulation (TW, THW or THHN) has a rating of at least 26 amperes.

❑ Three-Phase Example

A 100 ampere, three-phase, 4-wire 208/120 volt wye feeder supplies a panelboard rated 100 amperes. The panelboard is located 110 feet from source and is supplied by No. 3 THHN. The existing load on the panelboard is 70 amperes. What size aluminum branch circuit conductors are required to supply 7.5 kVA, three-phase 208 volt heating equipment, located 95 feet from the panelboard?

(a) No. 12 THHN (b) No. 10 THHN (c) No. 8 THHN (d) No. 6 THHN

- Answer: (c) No. 8 THHN

Step 1: → Determine the feeder voltage drop,

$$\text{Three Phase VD} = \frac{\sqrt{3} \times K \times I \times D}{CM}$$

$\sqrt{3} = 1.732$
K = 12.9 ohms, copper
I = 91 amperes

$$I = P/(E \times \sqrt{3}) = [70 \text{ amps} + \frac{7{,}500 \text{ VA}}{(208 \text{ volts} \times 1.732)}]$$

D = 110 feet
CM = 52,620, Chapter 9, Table 8

$$VD = \frac{1.732 \times 12.9 \text{ ohms} \times 91 \text{ amps} \times 110 \text{ feet}}{52{,}620 \text{ circular mils}}$$

VD = 4.25 volts

Step 2 → Determine the maximum voltage drop for both the feeder and branch circuit combined. Since no specifications are given, the NEC® 5 percent recommendation should be followed.
Feeder and Branch Circuit Voltage Drop = 208 volts × .05 = 10.4 volts

Step 3 → Determine the voltage drop permitted for the branch circuit by subtracting the feeder voltage drop from the maximum overall voltage drop. Since no specifications are given, be sure the voltage drop does not exceed the NEC® recommendation of 3 percent.
Branch Circuit Voltage Drop = 10.4 volts - 4.25 volts = 6.15 volts. The NEC® recommends that the voltage drop should not exceed 3 percent,
208 volts × .03 = 6.24 volts

Step 4 → Size the aluminum branch circuit conductors

$$\text{Three Phase CM} = \frac{\sqrt{3} \times K \times I \times D}{VD}$$

$\sqrt{3} = 1.732$
K = 21.2 ohms, aluminum

$$I = 21 \text{ amperes } I = P/(E \times \sqrt{3}) = \frac{7{,}500 \text{ VA}}{208 \text{ volts} \times 1.732}$$

D = 55 feet
VD = 6.15 volts

$$CM = \frac{1.732 \times 21.2 \text{ ohms} \times 21 \text{ amps} \times 95 \text{ feet}}{6.15 \text{ volts drop}}$$

CM = 11,911 = No. 8 Aluminum (Chapter 9, Table 8)

Unit 8 – Voltage Drop Summary Questions

8–1 Conductor Resistance

1. The ________ the number of free electrons, the better the conductivity of the conductor.
(a) greater (b) fewer

2. Conductor resistance is determined by the ________.
(a) material type
(b) cross-sectional area
(c) conductor length
(d) conductor operating temperature
(e) all of these

3. ________ is the best conductor, better than gold, but the high cost limits its use to special applications, such as fuse elements and some switch contacts.
(a) Silver (b) Copper (c) Aluminum (d) None of these

4. ________ conductors are often used when weight or cost are important consideration.
(a) Silver (b) Copper
(c) Aluminum (d) None of these

5. • Conductor cross-sectional area is the surface area expressed in ________,
(a) square inches (b) mils (c) circular mils (d) none of these

6. The resistance of a conductor is directly proportional to the conductor length.
(a) True (b) False

7. The resistance of a conductor changes with temperature. Temperature coefficient describes the effect that temperature has on the resistance of a conductor. Conductors with a ________ temperature coefficient have an increase in resistance with an increase in temperature.
(a) positive (b) negative (c) neutral (d) none of these

8–2 Conductor Resistance – Direct Current [Chapter 9, Table 8 Of The NEC®]

8. The National Electric Code lists the resistance and circular mils for both direct current and alternating current conductors. Direct current conductor resistances are listed in Chapter 9, Table ________ and alternating current conductor resistances are listed in Chapter 9, Table ________.
(a) 1, 5 (b) 3, 4 (c) 9, 8 (d) 8, 9

9. What is the direct current resistance of 400 feet of No. 6?
(a) .2 ohm (b) .3 ohm (c) .4 ohm (d) .5 ohm

10. What is the direct current resistance of 150 feet of No. 1?
(a) .023 ohm (b) .031 ohm (c) .042 ohm (d) .056 ohm

11. What is the direct current resistance of 1,400 feet of No. 3 AL?
(a) .23 ohm (b) .31 ohm (c) .42 ohm (d) .56 ohm

12. What is the direct current resistance of 800 feet of No. 1/0 AL?
(a) .23 ohm (b) .16 ohm (c) .08 ohm (d) .56 ohm

13. What is the direct current resistance of 120 feet of No. 1?
(a) .23 ohm (b) .16 ohm (c) .02 ohm (d) .56 ohm

14. What is the direct current resistance of 1,249 feet of No. 3 AL?
(a) .23 ohm (b) .16 ohm (c) .02 ohm (d) .5 ohm

8–3 Conductor Resistance – Alternating Current Circuits

15. The intensity of the magnetic field is dependent on the intensity of alternating current circuits. The greater the current flow, the greater the overall magnetic field.
(a) True (b) False

16. • The expanding and collapsing magnetic field within the conductor, exerts a force on the moving electrons. This force is called counter-electromotive-force.
(a) True (b) False

17. ________ currents are small independent currents that are produced as a result of the expanding and collapsing magnetic field. They flow erratically within the conductor opposing current flow and consuming power.
(a) Lentz (b) Ohms (c) Eddy (d) Kerchoffs

18. The expanding and collapsing magnetic field induces a counter voltage within the conductors which repels the flowing electrons towards the conductor surface. This is known as ________ effect.
(a) inductive (b) skin (c) surface (d) watt

8–4 Alternating Current Resistance As Compared To Direct Current

19. The opposition to current flow is greater for alternating current circuits because of ________ than for direct current circuits.
(a) eddy currents (b) skin effect (c) cemf (d) all of these

8–5 Conductor Resistance – Alternating Current [Chapter 9, Table 9 Of The NEC®]

20. The alternating current conductor resistances listed in Chapter 9, Table 9 of the NEC® are different for copper and aluminum and for nonmagnetic and magnetic raceways.
(a) True (b) False

21. What is the alternating current resistance of 320 feet of No. 2/0?
(a) .03 ohm (b) .04 ohm (c) .05 ohm (d) .06 ohm

22. What is the alternating current resistance of 1,000 feet of 500 kcmil when installed in an aluminum raceway?
(a) .032 ohm (b) .027 ohm (c) .029 ohm (d) .030 ohm

23. What is the alternating current resistance of 220 feet of No. 2/0?
(a) .012 ohm (b) .022 ohm (c) .33 ohm (d) .43 ohm

24. What is the alternating current resistance of 369 feet of 500 kcmil installed in nonmetallic conduit?
(a) .01 ohm (b) .02 ohm (c) .03 ohm (d) .04 ohm

25. A 36 ampere load is located 80 feet from the panelboard and is wired with No. 1 THHN aluminum. What is the total resistance of the circuit conductors?
(a) .04 ohm (b) .25 ohm (c) .5 ohm (d) all of these

26. What size copper conductors can be used to replace No. 1/0 aluminum that supplies a 110 ampere load. We do not want to increase the circuit voltage drop.
(a) No. 3 (b) No. 2 (c) No. 1 (d) No. 1/0

27. What is the direct current resistance in ohms for three 300 kcmil conductors in parallel?
(a) .014 ohm (b) .026 ohm (c) .052 ohm (d) .047 ohm

28. What is the ac resistance in ohms for three 1/0 THHN aluminum conductors in parallel?
(a) .1 ohms (b) .2 ohms (c) .4 ohms (d) .067 ohms

8–6 Voltage Drop Consideration

29. Because of the great demand for electricity, utilities sometimes are required to reduce their output voltage. In addition, the utility and customers' transformers, services, feeders and branch circuit conductors oppose the flow of current. The opposition to current flow results in voltage drop. All circuits have voltage drop, simply because all conductors have resistance.
(a) True (b) False

30. When sizing conductors for feeders and branch circuits, the NEC® ________ that we take voltage drop into consideration [210–19(a) FPN No. 4, 215–2 FPN No. 2 and 310–15 FPN].
(a) permits (b) suggests (c) requires (d) demands

31. ________ equipment such as motors and electromagnetic ballast can overheat at reduced voltage. This results in reduced equipment operating life and inconvenience to the customer.
(a) Inductive (b) Electronic (c) Resistive (d) All of these

32. • ________ equipment such as computers, laser printers, copy machines, etc., can suddenly power down because of reduced voltage, resulting in data losses.
(a) Inductive (b) Electronic (c) Resistive (d) All of these

33. Resistive loads such as incandescent lighting and electric space heating can have their power output decreased by the square of the voltage.
(a) True (b) False

34. What is the power consumed of a 4.5 kW 230 volt water heater operating at 200 volts?
(a) 2,700 watts (b) 3,400 watts (c) 4,500 watts (d) 5,500 watts

35. How can conductor voltage drop be reduced?
(a) reduce the conductor resistance
(b) increase conductor size
(c) decrease conductor length
(d) all of these

8–7 NEC® Voltage Drop Recommendations

36. If the branch circuit supply voltage is 208 volts, the maximum recommended voltage drop of the circuit should not be more than ________ volts.
(a) 3.6 (b) 6.24 (c) 6.9 (d) 7.2

37. If the feeder supply voltage is 240 volts, the maximum recommended voltage drop of the feeder should not be more than ________ volts.
(a) 3.6 (b) 6.24 (c) 6.9 (d) 7.2

8–8 Determining Circuit Conductors Voltage Drop

38. What is the voltage drop of two No. 12 THHN conductors supplying a 12-ampere continuous load. The continuous load is located 100 feet from the power supply.
(a) 3.2 volts (b) 4.75 volts (c) 6.4 volts (d) 12.8 volts

39. A single-phase, 24-amperes, 240 volt-load is located 160 feet from the panelboard. The load is wired with No. 10 THHN. What is the approximate voltage drop of the branch circuit conductors?
(a) 4.25 volts (b) 9.5 volts (c) 3 percent (d) 5 percent

40. A three-phase, 36-kVA load, rated 208-volts is located 100 feet from the panelboard and is wired with No. 1 THHN aluminum. What is the approximate voltage drop of the circuit conductors?
(a) 3.5 volts (b) 5 volts (c) 3 volts (d) 4.4 volts

8–9 Sizing Conductors To Prevent Excessive Voltage Drop

41. A single-phase, 5-horsepower motor is located 110 feet from the panelboard. The nameplate indicates that the voltage is 115/230 and the FLA is 52/26 amperes. What size conductor is required if the motor winding is connected in parallel and operates at 115 volts?
(a) No. 10 THHN (b) No. 8 THHN (c) No. 6 THHN (d) No. 3 THHN

42. A single-phase, 5-horsepower motor is located 110 feet from the panelboard. The nameplate indicates that the voltage is 115/230 and the FLA is 52/26 amperes. What size conductor is required if the motor winding is connected in series and operates at 230 volts?
(a) No. 10 THHN (b) No. 8 THHN (c) No. 6 THHN (d) No. 4 THHN

43. A three-phase, 15-kW, 480-volt load is located 300 feet from the panelboard. What size conductor is required to prevent the voltage drop from exceeding 3 percent?
(a) No. 10 THHN (b) No. 8 THHN (c) No. 6 THHN (d) No. 4 THHN

8–10 Limiting Conductor Length To Limit Voltage Drop

44. What is the approximate distance a single phase 7.5-kVA, 240-volt load can be located from the panelboard so the voltage drop does not exceed 3 percent? The load is wired with No. 8 THHN.
(a) 55 feet (b) 110 feet (c) 145 feet (d) 220 feet

45. What is the approximate distance a three phase 37.5-kVA, 460-volt transformer, wired with No. 6 THHN, can be located from the panelboard so the voltage drop does not exceed 3 percent?
(a) 250 feet (b) 300 feet (c) 345 feet (d) 400 feet

8–11 Limiting Current To Limit Voltage Drop

46. An existing installation contains No. 1/0 THHN aluminum conductors in a nonmetallic raceway to a panelboard located 190 feet from a 240 volt power source. What is the maximum load that can be placed on the panelboard so that the NEC® recommendations for voltage drop are not exceeded?
(a) 90 amperes (b) 110 amperes (c) 70 amperes (d) 175 amperes

47. An existing installation contains No. 2 THHN conductors in an aluminum raceway to a panelboard located 275 feet from a three-phase 460/230 volt power source. What is the maximum load the conductors can carry without exceeding the NEC® recommendation for voltage drop?
(a) 110 amperes (b) 175 amperes (c) 190 amperes (d) 149 amperes

8–12 Extending Circuits

48. An existing junction box is located 65 feet from the panelboard and contains No. 4 THHN Aluminum. We want to extend this circuit 85 feet and supply a 50-ampere, 208-volt load. What size copper conductors can we use for the extension?
(a) No. 8 THHN (b) No. 6 THHN (c) No. 4 THHN (d) No. 5 THHN

8–13 Miscellaneous

49. What is the circuit voltage if the conductor voltage drop is 3.3 volts? Assume 3 percent voltage drop.
(a) 110 volts (b) 115 volts (c) 120 volts (d) none of these

Challenge Questions

8–1 Conductor Resistance

50. The resistance of a conductor is affected by temperature change. This is called the ______.
(a) temperature correction factor
(b) temperature coefficient
(c) ambient temperature factor
(d) none of these

8–3 Conductor Resistance – Alternating Current Circuits

51. The total opposition to current flow in an alternating current circuit is expressed in ohms and is called ______.
(a) impedance (b) conductance (c) reluctance (d) resistance

8–5 Alternating Current Conductor Resistance [Chapter 9, Table 9 Of The NEC®]

52. A 40-ampere, 240-volt, single-phase load is located 150 feet from an existing junction box. The junction box is located 50 feet from the panelboard and is wired with No. 4 THHN aluminum wire. The total resistance of the two No. 4 conductors from the panelboard to the junction box is approximately ______ ohm.
(a) .03 (b) .09 (c) .05 (d) .04

53. A load is located 100 feet from a 230-volt power supply and is wired with No. 4 THHN aluminum conductors. What size copper conductor can be used to replace the aluminum conductors and not increase the conductor voltage drop?
(a) No. 6 (b) No. 8 (c) No. 1 (d) No. 1/0

8–7 NEC® Voltage Drop Recommendations

54. A 40-ampere, 240-volt rated, single-phase load is wired 150 feet from a junction box, and the junction box is located 50 feet from a panelboard (for a total of 200 feet). If the voltage at the panelboard is 240 volts, what is the minimum voltage recommended by the NEC® at the 40-ampere load?
(a) 228.2 volts (b) 232.8 volts (c) 236.2 volts (d) 117.7 volts

8–8 Determining Circuit Conductors Voltage Drop

55. What is the voltage drop of two No. 4 aluminum conductors that supply a 5-horsepower, single-phase, 208-volt motor that has a nameplate rating of 55 amperes? The motor is located 95 feet from the power supply.
(a) 3.25 volts (b) 5.31 volts (c) 6.24 volts (d) 7.26 volts

8–9 Sizing Conductors To Prevent Excessive Voltage Drop

56. A 40-ampere, 240-volt, single-phase load is located 150 feet from an existing junction box. The junction box is located 50 feet from the panelboard. When the 40-ampere load is on, the voltage at the junction box would be calculated to be 236 volts. The NEC® recommends that voltage drop for this branch circuit does not exceed 3 percent of the 240-voltage source, which is 7.2 volts. What size conductor could be installed from the junction box to the load and still meet the NEC recommendations?
(a) No. 3 (b) No. 1 (c) No. 1/0 (d) No. 6

8–10 Limiting Conductor Length To Limit Voltage Drop

57. How far can a 50-ampere, three-phase, 230-volt load be located from the panel, if fed with No. 3 THHN, and still meet the NEC® recommendations for voltage drop?
(a) 275 feet (b) 300 feet (c) 325 feet (d) 350 feet

8–11 Limiting Current To Limit Voltage Drop

58. Two No. 8 THHN supply a 120-volt load that is located 225 feet from the panelboard. What is the maximum load, in amperes, that can be applied to these conductors without exceeding the NEC® recommendation on conductor voltage drop?
(a) 0 amperes (b) 5 amperes (c) 10 amperes (d) 15 amperes

8–13 Miscelleous Questions

59. An 8-ohm resistor is connected to a 120-volt power supply. Using a voltmeter, we measure 112 volts across the resistor. What is the current of the 8 ohm resistor in amperes?
(a) 14.00 amperes (b) 13.37 amperes (c) 15.73 amperes (d) 19.41 amperes

60. A 480-volt, single-phase feeder carries 400 amperes and has a 7.2 voltage drop. What is the total resistance of the conductors in this circuit?
(a) .1880 ohm (b) .1108 ohm (c) .0190 ohm (d) .0180 ohm

61. An 8 ohm resistor operates at 112 volts and is connected to a 115 volt power supply. The percent of voltage drop for the circuit is ______ percent.
(a) 2.6 (b) 3.0 (c) 5.0 (d) 7.0

Unit 9

One-Family Dwelling-Unit Load Calculations

OBJECTIVES

After reading this unit, the student should be able to briefly explain the following concepts:

- Air conditioning versus heat
- Appliance (small) circuits
- Appliance demand load
- Clothes dryer demand load
- Cooking equipment calculations
- Dwelling-unit calculations – optional method
- Dwelling-unit calculations – standard method
- Laundry circuit
- Lighting and receptacles

After reading this unit, the student should be able to briefly explain the following terms:

- General lighting
- General–use receptacles
- Optional method
- Rounding
- Standard method
- Unbalanced demand load
- Voltages

PART A - *GENERAL REQUIREMENTS*

9–1 *GENERAL REQUIREMENTS*

Article 220 provides the requirement for residential branch circuits, feeders, and service calculations. Other applicable articles are Branch Circuits – 210, Feeders – 215, Services – 230, Overcurrent Protection – 240, Wiring Methods – 300, Conductors – 310, Appliances– 422, Electric Space Heating Equipment – 424, Motors – 430, and Air Conditioning – 440.

9–2 *VOLTAGES [220–2]*

Unless other voltages are specified, branch-circuit, feeder, and service loads shall be computed at nominal system voltages of 120, 120/240, 208Y/120, or 240, Fig. 9–1.

9–3 *FRACTION OF AN AMPERE*

There are no specific NEC® rules for rounding when a calculation results in a fraction of an ampere, but Chapter 9, Part B Examples contains the following note: "except where the computations result in a major fraction of an ampere (.5 or larger), such fractions may be dropped."

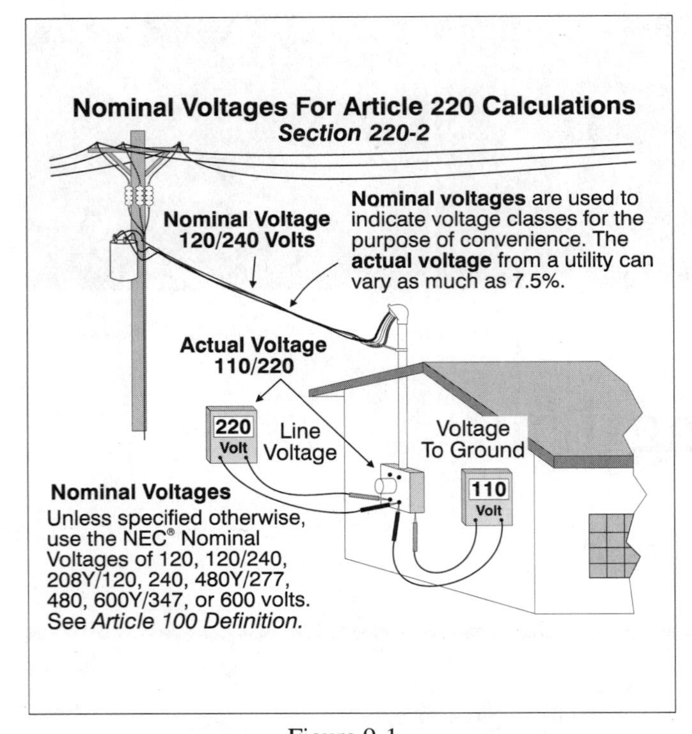

Figure 9-1

Nominal Voltages

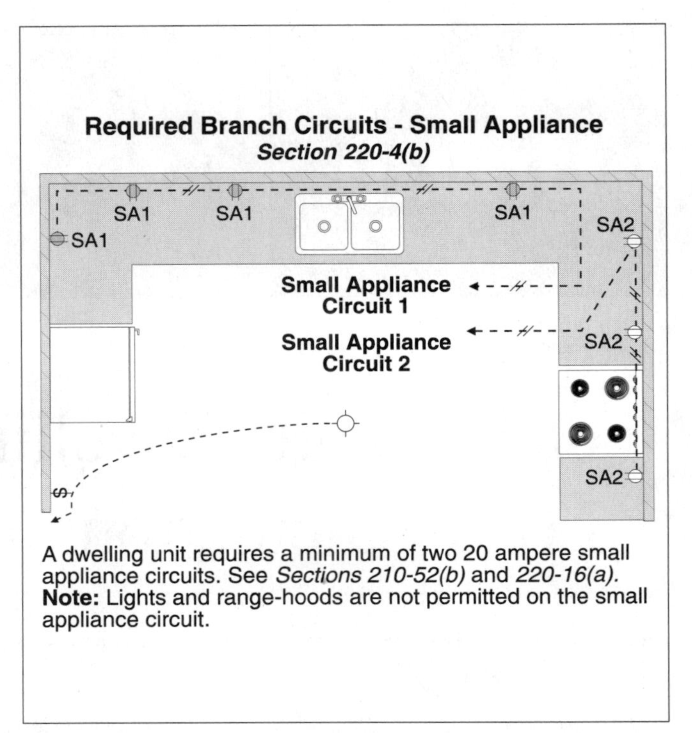

Figure 9-2

Required Branch Circuits – Small Appliances

9–4 *APPLIANCE (SMALL) CIRCUITS [220–4(b)]*

A minimum of two 20-ampere *small appliance branch circuits* is required for receptacle outlets in the kitchen, dining room, breakfast room, pantry, or similar dining areas [220–4(b)]. In general, no other receptacle or lighting outlet can be connected to these 20-ampere small appliance branch circuits [210–52(b) Exceptions], Fig. 9–2.

Feeder And Service

When sizing the feeder or service, each dwelling-unit shall have two 20-ampere small appliance branch circuits with a feeder load of 1,500 volt-amperes for each circuit [220–16(a)].

Other Related Code Sections

Areas that the small appliance circuits supply [210–52(b)] and
Receptacle outlets required for kitchen countertops [210–52(c)].
Fifteen- or 20-ampere rated receptacles can be used on 20-ampere circuit [210-21(b)(3) and 220–16(a)].

9–5 *COOKING EQUIPMENT – BRANCH CIRCUIT [Table 220–19, Note 4]*

To determine the branch circuit demand load for cooking equipment, we must comply with the requirements of Table 220–19, Note 4. The branch circuit demand load for one range shall be according to the demand loads listed in Table 220–19, but a minimum 40-ampere circuit is required for 8.75 kW and larger ranges [210–19(b)].

❑ **Less than 12 kW, Column *A* Example**

What is the branch circuit demand load (in amperes) for one 9-kW range, Fig. 9–3?

(a) 21 amperes (b) 27 amperes (c) 38 amperes (d) 33 amperes

- Answer: (d) 33 amperes, 8 kW

  The demand load for one range in Table 220–19 Column A is 8 kW. This can be converted to amperes by dividing the power by the voltage, $I = \frac{P}{E}$

  $$I = \frac{8\text{ kW} \times 1{,}000}{240\text{ volts}^*}$$

  * Assume 120/240 volts for all calculations unless the question gives a specific voltage and system.

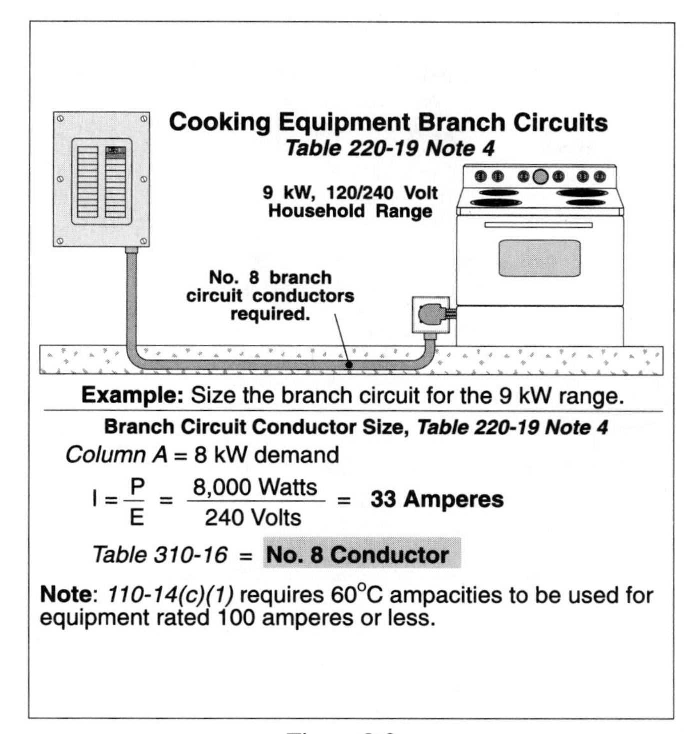

Figure 9-3

Branch Circuit Column "A" Example

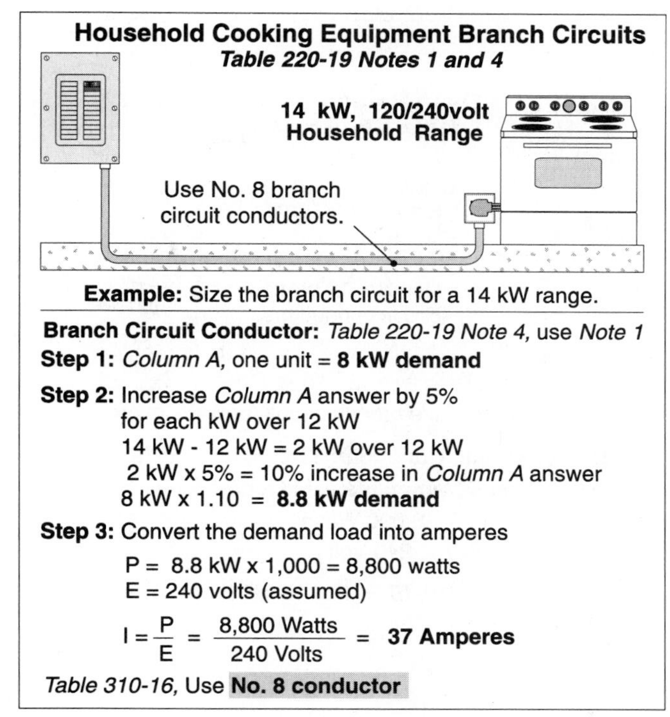

Figure 9-4

Branch Circuit Note 1 Example

❑ **More Than 12 kW, Note 1 Example**

What is the branch circuit load (in amperes) for one 14-kW range, Fig. 9–4?

(a) 33 amperes (b) 37 amperes (c) 50 amperes (d) 58 amperes

- Answer: (b) 37 amperes

Step 1: → Since the range exceeds 12 kW, we must comply with Note 1 of Table 220–19. The first step is to determine the demand load as listed in Column A of Table 220–19 for one unit, = 8 kW

Step 2: → We must increase the Column A value (8 kW) by 5 percent for each kW that the average range (in this case 14 kW) exceeds 12 kW. This results is an increase of the Column A value (8 kW) by 10 percent.

8 kW × 1.1 = 8.8 kW or (8,000 watts + 800 watts)

Step 3: → Convert the demand load in kW to amperes.

$$I = \frac{P}{E},\ I = \frac{8{,}800\ \text{VA}}{240\ \text{volts}} = 36.7\ \text{amperes}$$

One Wall-Mounted Oven Or One Counter-Mounted Cooking Unit [220–19 Note 4]

The branch circuit demand load for one wall-mounted oven or one counter-mounted cooking unit shall be the nameplate rating.

❑ **One Oven Example**

What is the branch circuit load (in amperes) for one 6-kW wall-mounted oven, Fig. 9–5?

(a) 20 amperes (b) 25 amperes (c) 30 amperes (d) 40 amperes

- Answer: (b) 25 amperes

Nameplate = 6,000 VA $I = \frac{VA}{E},\ I = \frac{6{,}000\ \text{VA}}{240\ \text{volts}} = 25\ \text{amperes}$

❑ **One Counter-Mounted Cooking Unit Example**

What is the branch circuit load (in amperes) for one 4.8-kW counter-mounted cooking unit?

(a) 15 amperes (b) 20 amperes (c) 25 amperes (c) 30 amperes

- Answer: (b) 20 amperes,

Nameplate = 4,800 VA, $I = \frac{VA}{E},\ I = \frac{4{,}800\ \text{VA}}{240\ \text{volts}} = 20\ \text{amperes}$

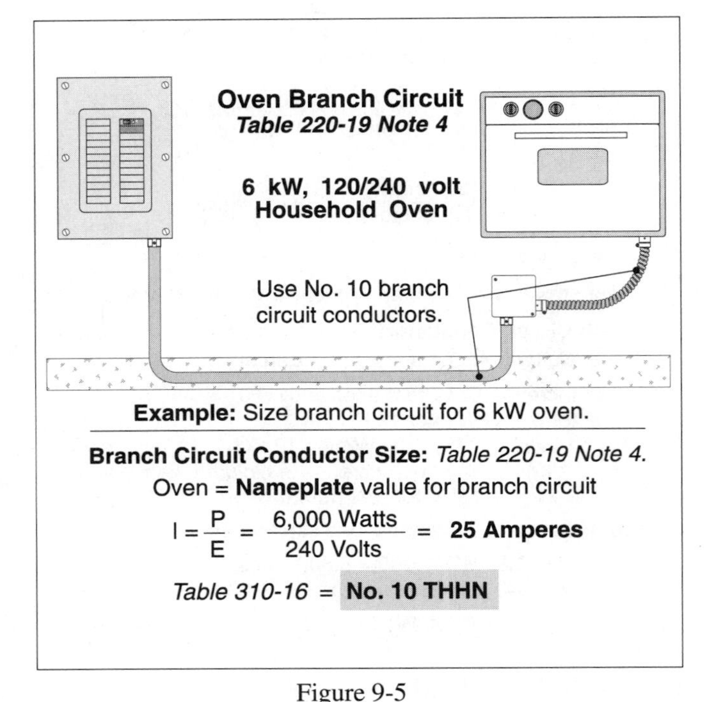

Figure 9-5

Branch Circuit One Oven Example

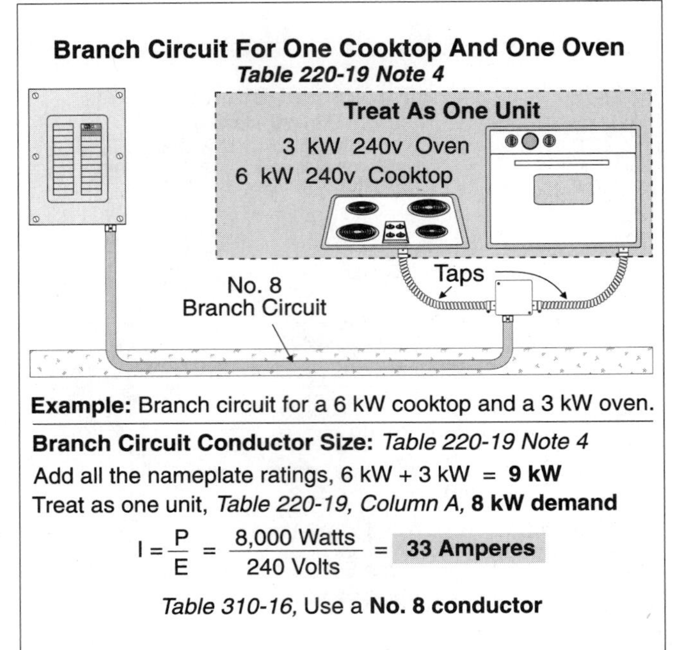

Figure 9-6

Branch Circuit One Cooktop And One Oven Example

One Counter-mounted Cooking Unit And Up To Two Ovens [220–19 Note 4]

To calculate the load for one counter-mounted cooking unit and up to two wall-mounted ovens, complete the following steps:

Step 1: → Total Load. Add the nameplate ratings of the cooking appliances and treat this total as one range.
Step 2: → Table 220-19 Demand. Determine the demand load for one unit from Table 220–19, Column A.
Step 3: → Net Computed Load. If the total nameplate rating exceeds 12 kW, increase Column A (8 kW) 5 percent for each kW, or major fraction (.5 kW), that the combined rating exceeds 12 kW.

❑ **Cooktop And One Oven Example**

What is the branch circuit load (in amperes) for one 6-kW counter-mounted cooking unit and one 3-kW wall-mounted oven, Fig. 9–6?

(a) 25 amperes (b) 38 amperes (c) 33 amperes (d) 42 amperes

- Answer: (c) 33 amperes

Step 1: → Total connected load. 6 kW + 3 kW = 9 kW.
Step 2: → Demand load for one range, Column A of Table 220–19, 8 kW.

$$I = \frac{P}{E},\ I = \frac{8{,}000\ VA}{240\ volts} = 33.3\ amperes$$

❑ **Cooktop And Two Ovens Example**

What is the branch circuit load (in amperes) for one 6 kW counter-mounted cooking unit and two 4 kW wall-mounted ovens, Fig. 9–7?

(a) 22 amperes (b) 27 amperes (c) 33 amperes (d) 37 amperes

- Answer: (d) 37 amperes

Step 1: → The total connected load:
6 kW + 4 kW + 4 kW = 14 kW
Step 2: → Demand load for one range, Column A or Table 220–19, = 8 kW
Step 3 → Since the combined total (14 kW) exceeds 12 kW, we must increase Column A (8 kW) 5 percent for each kW, or major fraction (.5 kW), that the range exceeds 12 kW. 14 kW exceeds 12 kW by 2 kW, which results in a 10 percent increase of the Column A value. 8 kW × 1.1 = 8.8 kW.

$$I = \frac{P}{E},\ I = \frac{8{,}800\ VA}{240\ volts} = 36.7\ amperes$$

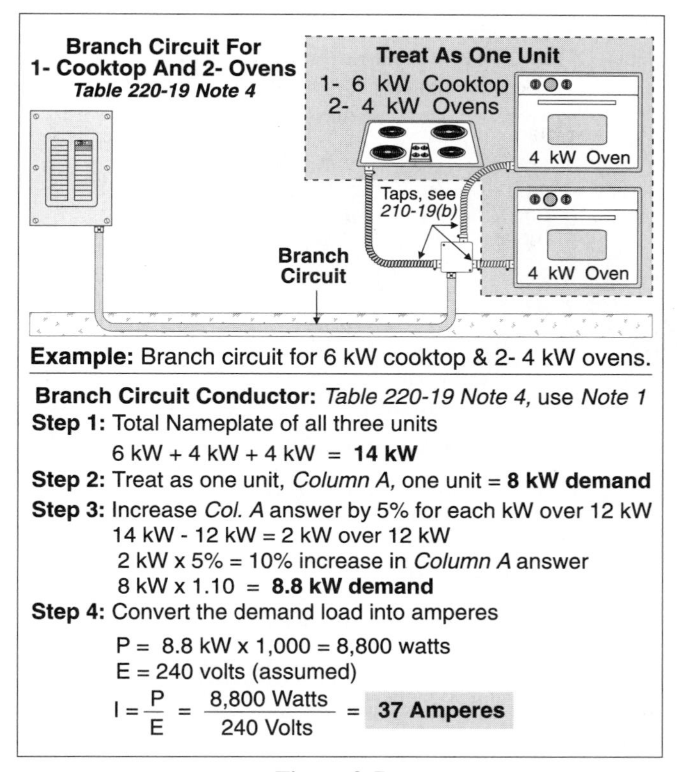

Figure 9-7

Branch Circuit, One Cooktop And Two Ovens Example

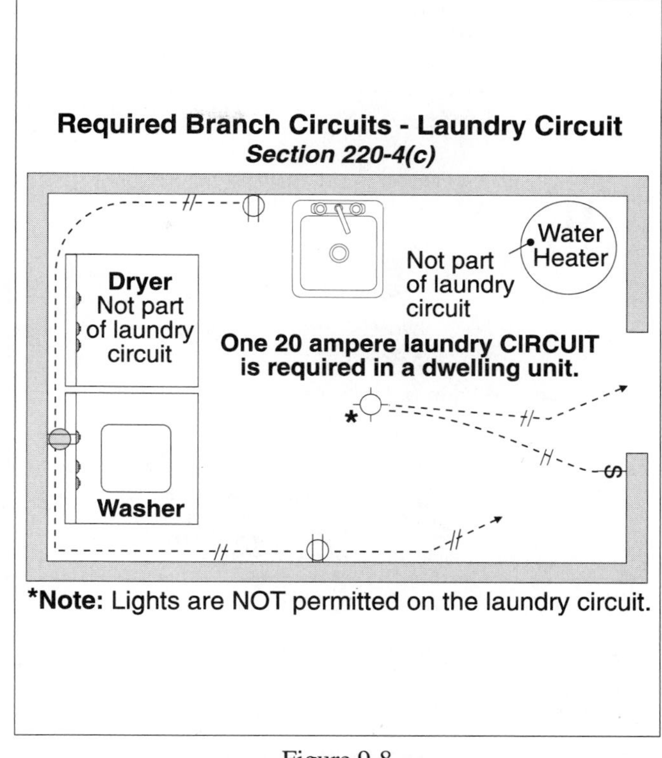

Figure 9-8

Laundry Room Circuit Requirement

9–6 *LAUNDRY RECEPTACLE(S) CIRCUIT [220–4(c)]*

One 20-ampere branch circuit for the laundry receptacle outlet (or outlets) is required and the laundry circuit cannot serve any other outlet, such as the laundry room lights [210–52(f)]. The NEC® does not require a separate circuit for the washing machine but does require a separate circuit for the laundry room receptacle or receptacles. If the washing machine receptacle(s) are located in the garage or basement, GFCI protection might be required, see Section 210–8(a)(2) and (4) and their exceptions, Fig. 9–8.

Feeder and Service

Each dwelling-unit shall have a feeder load consisting of 1,500 volt-amperes for the 20-ampere laundry circuit [220-16(b)].

Other related Code rules

A laundry outlet must be within 6 feet of washing machine [210–50(c)].

Laundry area receptacle outlet required [210–52(f)].

9–7 *LIGHTING AND RECEPTACLES*

General Lighting Volt-Amperes Load [220–3(b)]

The NEC® requires a minimum 3 volt-amperes per square-foot [Table 220–3(b)] for the *general lighting* and general-use receptacles. The dimensions for determining the area shall be computed from the outside of the building and shall not include open porches, garages, or spaces not adaptable for future use, Fig. 9–9.

Note: The 3 volt-ampere-per-square-foot rule for general lighting includes all 15- and 20-ampere general-use receptacles; but, it does not include the appliance or laundry circuit receptacles. See the Note at the Bottom of Table 220–3(b) for more details.

❑ **General Lighting VA Load Example**

What is the general lighting and receptacle load for a 2,100 square-foot dwelling-unit that has thirty four convenience receptacles and twelve lighting fixtures rated 100 watts each, Fig. 9–10?

(a) 2,100 VA (b) 4,200 VA (c) 6,300 VA (d) 8,400 VA

- Answer: (c) 6,300 VA.

2,100 square feet × 3 VA = 6,300 VA. No additional load is required for general-use receptacles and lighting outlets, see Note to Table 220–3(b).

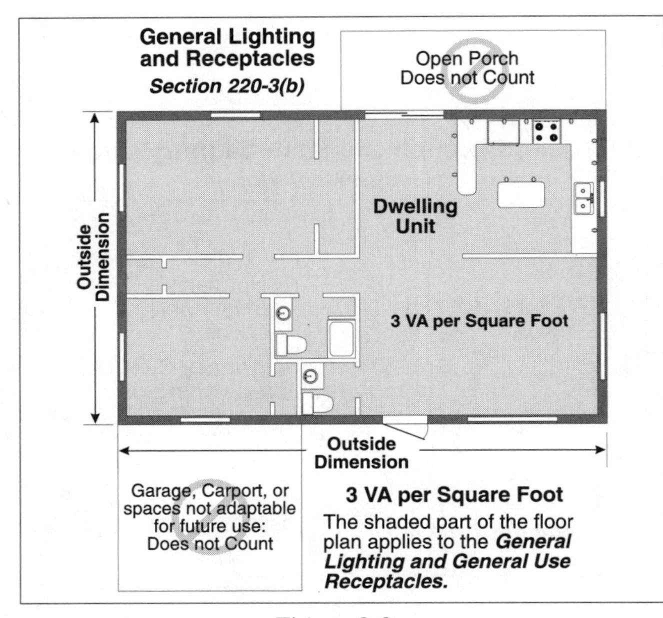

Figure. 9-9

General Lighting Requirement

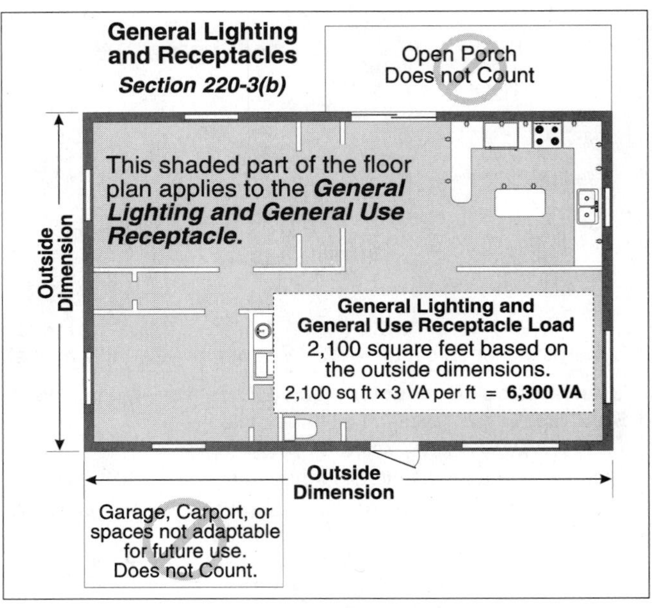

Figure. 9-10

General Lighting Example

Number Of Circuits Required [Chapter 9, Example No. 1(a)]

The number of branch circuits required for general lighting and receptacles shall be determined from the general lighting load and the rating of the circuits [220–4(a)].

To determine the number of branch circuits for general lighting and receptacles, follow these steps:

Step 1: → Determine the general lighting VA load:
Living area square footage × 3 VA.

Step 2: → Determine the general lighting ampere load:
Amperes = VA/E.

Step 3: → Determine the number of branch circuits:

$$\frac{\text{General Lighting Amperes (Step 2)}}{\text{Circuit Amperes (100 \%)}}$$

❑ Number Of 15 Ampere Circuits Example

How many 15-ampere circuits are required for a 2,100 square foot dwelling-unit, Fig. 9–11?

(a) 2 circuits (b) 3 circuits
(c) 4 circuits (d) 5 circuits

• Answer: (c) 4

Step 1: → General lighting VA = 2,100 square feet × 3 VA = 6,300 VA

Step 2: → General lighting ampere:

$$I = \frac{VA}{E},\ I = \frac{6{,}300\ VA}{120\ volts*}$$

I = 53 amperes

*Use 120 volts single-phase unless specified otherwise [220–2].

Step 3 → Determine the number of circuits:

$$\frac{\text{General Lighting Amperes}}{\text{Circuit Amperes}}$$

$$\text{Number of circuits} = \frac{53\ \text{amperes}}{15\ \text{amperes}}$$

Number of circuits = 3.53 or 4

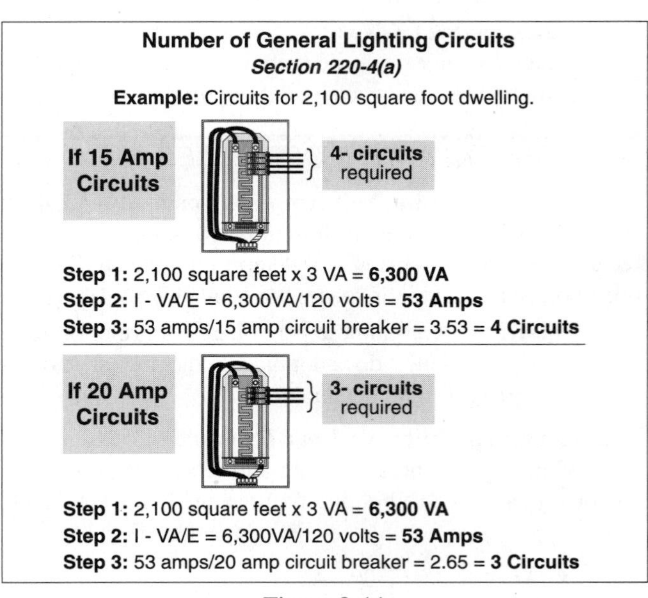

Figure 9-11

Number of General Lighting Circuits Example

❑ **Number Of 20-Ampere Circuits Example**

How many 20-ampere circuits are required for a 2,100 square foot home?

(a) 2 circuits (b) 3 circuits (c) 4 circuits (d) 5 circuits

- Answer: (b) 3

Step 1: → General lighting VA = 2,100 square foot × VA = 6,300 VA.

Step 2: → General lighting amperes: $I = \frac{VA}{E}$, $I = \frac{6{,}300\ VA}{120\ volts}$, I = 53 amperes

Step 3: → Determine the number of circuits: $\frac{\text{General Lighting Amperes}}{\text{Circuit Amperes}}$

Number of circuits = $\frac{53\ \text{amperes}}{20\ \text{amperes}}$ Numberof circuits= 2.65 or 3 circuits

PART B - *STANDARD METHOD – FEEDER/SERVICE LOAD CALCULATIONS*

9–8 *DWELLING-UNIT FEEDER/SERVICE LOAD CALCULATIONS (Part B Of Article 240)*

When determining a dwelling-unit feeder, or service using the standard method, the following steps should be used.

Step 1: → **General Lighting And Receptacles, Small Appliance And Laundry Circuits [220–11].** The NEC® recognizes that the general lighting and receptacles, small appliance, and laundry circuits will not all be on, or loaded, at the same time and permits a demand factor to be applied [220–16]. To determine the feeder demand load for these loads, follow these steps.

(a) **General Lighting.** Determine the total connected load for general lighting and receptacles (3 VA per square foot), two small appliance circuits (3,000 VA), and one laundry circuit (1,500 VA).

(b) **Demand Factor.** Apply Table 220–11 demand factors to the total connected load (Step 1).
First 3,000 VA @ 100 percent demand
Remainder @ 35 percent demand

Step 2: → **Air Conditioning versus Heat.** Because the air conditioning and heating loads are not on at the same time, it is permissible to omit the smaller of the two loads [220–21].
The A/C demand load is sized at 125 percent of the largest A/C VA, plus the sum of the other A/C VA's [440–34]. Fixed electric space-heating demand loads are calculated at 100 percent of the total heating load [220–15].

Step 3: → **Appliances [220–17]**
A demand factor of 75 percent is permitted to be applied when four or more appliances are fastened in place and are on the feeder, such as dishwasher, disposal, trash compactor, water heater, etc. This does not apply to motors [220–14], space-heating equipment [220–15], clothes dryers [220–18], cooking appliances [220–19], or air-conditioning equipment [440–34].

Step 4: → **Clothes Dryer [220–18]**
The feeder or service demand load for electric clothes dryers located in a dwelling-unit shall not be less than: (1) 5000 watts or (2) the nameplate rating if greater than 5,000 watts. A feeder or service dryer load is not required if the dwelling-unit does not contain an electric dryer!

Step 5: → **Cooking Equipment [220–19]**
Household cooking appliances rated over 1¾ kW can have the feeder and service loads calculated according to the demand factors of Section 220–19, Table and Notes 1, 2, and 3.

Step 6: → **Feeder And Service Conductor Size**
Four Hundred Amperes And Less. The feeder or service conductors are sized according to Note 3 of Table 310–16 for 3-wire, single-phase, 120/240-volt systems up to 400 amperes and the grounded conductor is sized to the maximum unbalanced load [220–22] using Table 310–16.
Over Four Hundred Amperes. The ungrounded and grounded (neutral) conductors are sized according to Table 310–16.

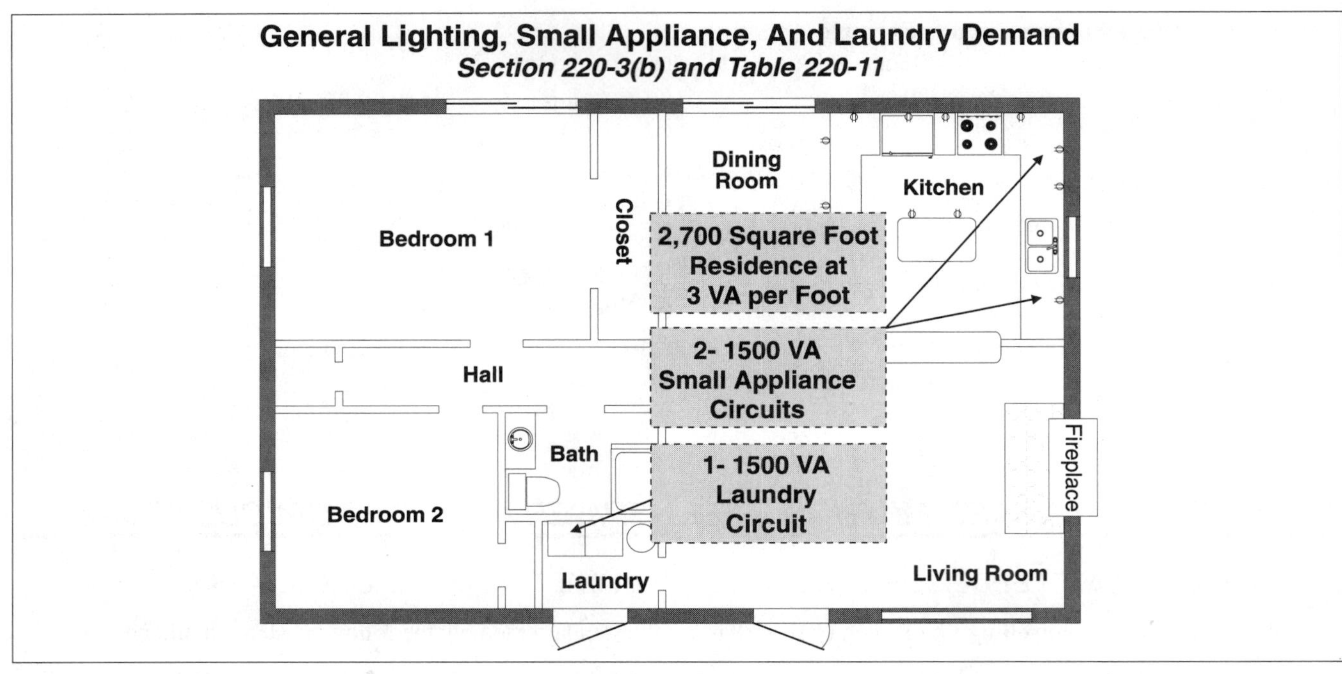

Figure 9-12

General Lighting, Small Appliance and Laundry Demand Example

9–9 *DWELLING-UNIT FEEDER/SERVICE CALCULATIONS EXAMPLES*

Step 1: General Lighting, Small Appliance And Laundry Demand [220–11]

➛ **General Lighting Question 1**

What is the general lighting, small appliance and laundry demand load for a 2,700 square foot dwelling-unit, Fig. 9–12?

(a) 8,100 VA (b) 12,600 VA (c) 2,700 VA (d) 6,360 VA

- Answer: (d) 6,360 VA

General Lighting and Receptacles	2,700 square feet × 3 VA =	8,100 VA		
Small Appliance Circuits	1,500 VA × 2 =	3,000 VA		
Laundry Circuit	1,500 VA × 1 =	+ 1,500 VA		
Total Connected Load		12,600 VA		
Demand Factors:				
Total Connected Load		12, 600 VA		
First 3,000 VA @ 100%		–3,000 VA	× 1.00 =	3,000 VA
Remainder @ 35%		9,600 VA	× .35 =	+ 3,360 VA
General lighting, small appliance, and laundry demand load =				6,360 VA

➛ **General Lighting Question 2**

What is the general lighting, small appliance, and laundry demand load for a 6,540 square foot dwelling-unit?

(a) 8,100 VA (b) 12,600 VA (c) 2,700 VA (d) 10,392 VA

- Answer: (d) 10,392 VA

General Lighting and Receptacles	6,540 square feet × 3 VA =	19,620 VA		
Small Appliance Circuits	1,500 VA × 2 =	3,000 VA		
Laundry Circuit	1,500 VA × 1 =	+ 1,500 VA		
Total connected load		24,120 VA		
Demand Factors:				
Total Connected Load		24,120 VA		
First 3,000 VA @ 100		–3,000 VA	× 1.00 =	3,000 VA
Remainder @ 35%		21,120 VA	× .35 =	+ 7,392 VA
General lighting, small appliance, and laundry demand load =				10,392 VA

Step 2: Air Conditioning Versus Heat [220–15]

When determining the cooling versus heating load, we must compare the air conditioning load at 125 percent against the heating load of 100 percent. Section 220–15 only refers to the heating load and Section 440–32 contains the requirement of the air conditioning load. We are permitted to omit the smaller of the two loads according to Section 220–21.

➛ **Air Conditioning versus Heat Question 1**

What is the service demand load for a 5-horsepower 230-volt A/C versus three 3-kW baseboard heaters, Fig. 9–13?

(a) 6,400 VA (b) 3,000 W

(c) 8,050 VA (d) 9,000 W

- Answer: (d) 9,000 VA is greater than the 8,050 VA air conditioning

  A/C [440–34] 230 volts × 28 amperes* = 6,440 × 1.25 = 8,050 VA, omit [220–21]

  Heat [220–15] 3,000 watts × 3 units = 9,000 watts, Table 430-148

* Table 430–148

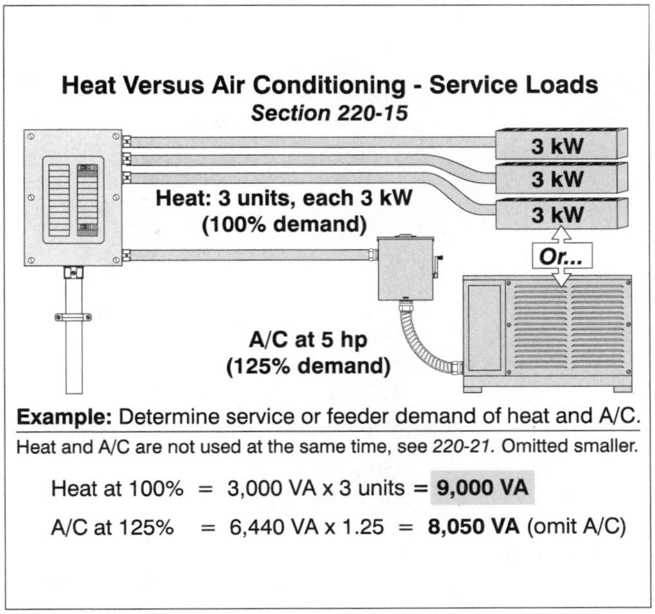

Figure 9-13

Air Conditioning versus Heat Example

➛ **Air Conditioning versus Heat Question 2**

What is the service or feeder demand load for a 4-horsepower 230-volt A/C unit and a 10-kW electric space heater?

(a) 10,000 VA (b) 3,910 VA (c) 6,440 VA (d) 10,350 VA

- Answer: (a) 10,000 VA heat is greater than the 6,469 VA air conditioning

  A/C: [440–34] Since a 4-horsepower 230 volt motor is not listed in Table 430–148, we must determine the approximate VA rating by averaging the values for a 3-hpP and 5-hp motor.

3-hp = 230 volts × 17 amperes* =	3,910 VA
5-hp = 230 volts × 28 amperes* =	+ 6,440 VA
Total VA for 3-hp + 5-hp (8-hp) =	10,350 VA

VA for 4-hp = 10,350 VA (8-hp)/2 = 5,175 VA × 1.25 = 6,469 VA, omit [220–21]

*Table 430–148

Step 3: Appliance Demand Load [220–17]

➛ **Appliance Question 1**

What is the demand load for a 940 VA disposal, a 1,250 VA dishwasher, and a 4,500 VA water heater?

(a) 5,018 VA (b) 6,690 VA (c) 8,363 VA (d) 6,272 VA

- Answer: (b) 6,690 VA, no demand factor for three units

Disposal	940 VA	
Dishwasher	1,250 VA	
Water Heater	+ 4,500 VA	
	6,690 VA	at 100%

➛ **Appliance Question 2**

What is the demand load for a 940 VA disposal, a 1,250 VA dishwasher, a 1,100 VA trash compactor, and a 4,500 VA water heater, Fig. 9–14?

(a) 7,790 VA (b) 5,843 VA (c) 7,303 VA (d) 9,738 VA

- Answer: (b) 5,843 VA

Disposal	940 VA	
Dishwasher	1,250 VA	
Trash Compactor	1,100 VA	
Water Heater	+ 4,500 VA	
	7,790 VA	× .75 = 5,843 VA

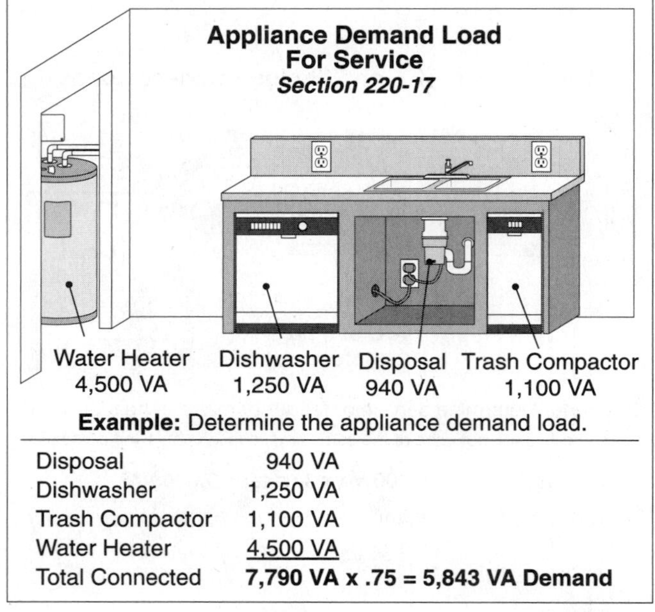

Figure 9-14

Appliance Demand Load Example

Dryer Demand Load For Service
Section 220-18

Dryer:
4 kW 120/240 volt
Demand = 5 kW

4 kW
120/240 volt

Service demand load is based on the nameplate value but **5,000 VA** is the ***minimum*** value permitted.

Figure 9-15

Dryer Demand Load Example

Step 4: Dryer Demand Load [220–18]

➛ **Dryer Question 1**

What is the service and feeder demand load for a 4-kW dryer, Fig. 9–15?

(a) 4,000 VA (b) 3,000 VA (c) 5,000 VA (d) 5,500 VA

- Answer: (c) 5,000 VA. The dryer load must not be less than 5,000 VA.

➛ **Dryer Question 2**

What is the service and feeder demand load for a 5.5-kW dryer?

(a) 4,000 VA (b) 3,000 VA (c) 5,000 VA (d) 5,500 VA

- Answer: (c) 5,500 VA. The dryer load must not be less than the nameplate rating if greater than 5 kW.

Step 5: Cooking Equipment Demand Load [220–19]

➛ **Column B Question 3**

What is the service and feeder demand load for two 3-kW cooking appliances in a dwelling-unit?

(a) 3 kW (b) 4.8 kW (c) 4.5 kW (d) 3.9 kW

- Answer: (c) 4.5 kW, Table 220–19, Column B

  3 kW × 2 units = 6 kW × 0.75 = 4.5 kW

➛ **Column C Question 4**

What is the service and feeder demand load for one 6-kW cooking appliance in a dwelling-unit?

(a) 6 kW (b) 4.8 kW (c) 4.5 kW (d) 3.9 kW

- Answer: (b) 4.8 kW, Table 220–19, Column C

  6 kW × 0.8 = 4.8 kW

➛ **Column B And C Question 5**

What is the service and feeder demand load for two 3-kW ovens and one 6-kW cooktop in a dwelling-unit, Fig. 9–16?

(a) 6 kW (b) 4.8 kW (c) 4.5 kW (d) 9.3 kW

- Answer: (d) 9.3 kW, Table 220–19, Column B and C

Column C demand 6 kW × 0.8 =	4.8 kW
Column B demand (3 kW × 2) = 6 kW × 0.75 =	+ 4.5 kW
Total demand	9.3 kW

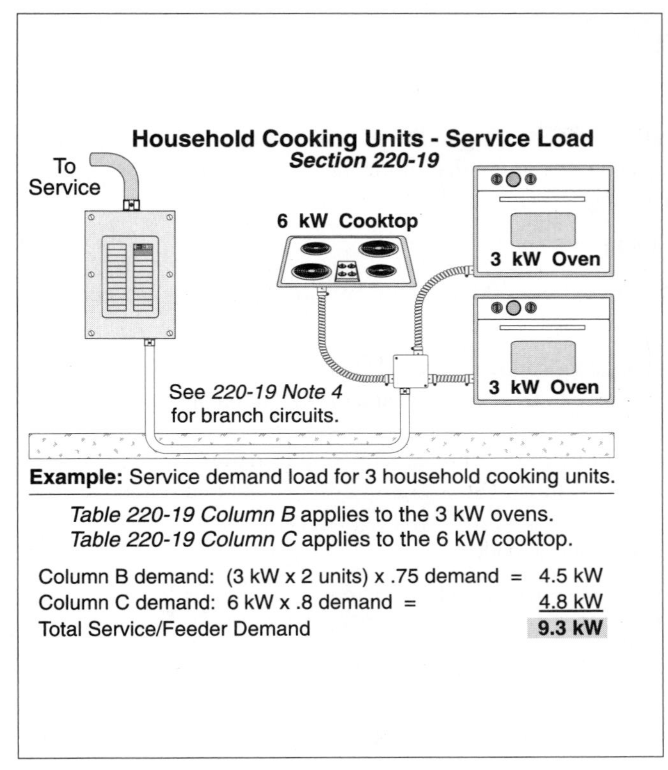

Figure 9-16

Cooking Equipment Demand Load Example

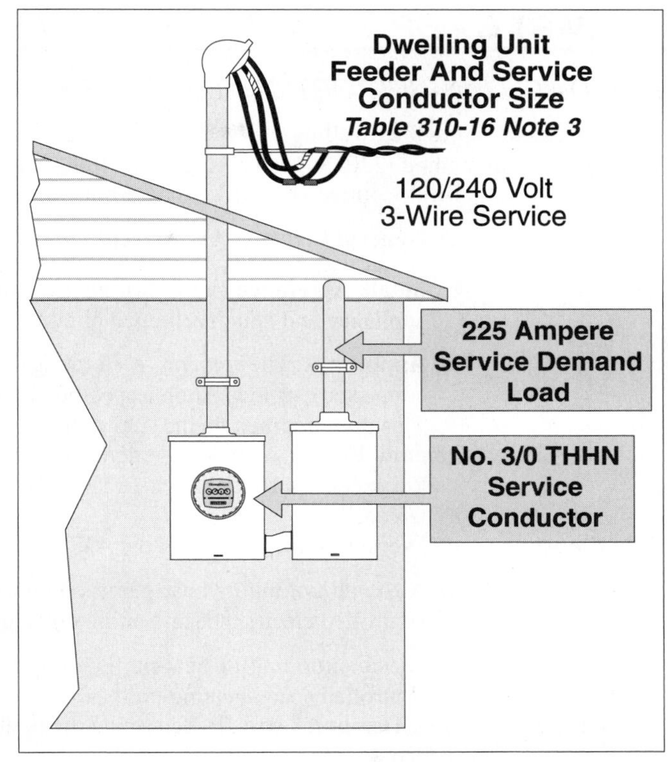

Figure 9-17

Feeder and Service Conductor Size – Note 3 of Table 310–16

➛ **Column A Question 1**

What is the service and feeder demand load for an 11.5-kW range in a dwelling-unit?

(a) 11.5 kW (b) 8 kW (c) 9.2 kW (d) 6 kW

- Answer: (b) 8 kW, Table 220–19, Column A

➛ **Note 1 Question 2**

What is the service and feeder demand load for a 13.6-kW range in a dwelling-unit?

(a) 8.8 kW (b) 8 kW (c) 9.2 kW (d) 6 kW

- Answer: (a) 8.8 kW,

  8 kW × 1.1* = 8.8 kW

* Column A value (8 kW) must be increased 5 percent for each kW or major fraction of a kW (.5 kW or larger) over 12 kW [Table 220–19, Note 1].

Step 6: Service Conductor Size [Note 3 of Table 310–16]

➛ **Service Conductor Question 1**

What size THHN feeder or service conductor (120/240 volt, single-phase) is required for a 225 ampere service demand load, Fig. 9–17?

(a) No. 0 (b) No. 00 (c) No. 000 (d) No. 0000

- Answer: (c) No. 3/0

➛ **Service Conductor Question 2**

What size THHN feeder or service conductor (208Y/120-volt, single-phase) is required for a 225-ampere service demand load?

(a) No. 1/0 (b) No. 2/0 (c) No. 3/0 (d) No. 4/0

- Answer: (d) 4/0, Table 310–16, Note 3, does not apply to 208Y/120-volt systems because the grounded conductor will carry current even if the loads are balanced.

PART C - *OPTIONAL METHOD – FEEDER/SERVICE LOAD CALCULATIONS*

9–10 *DWELLING-UNIT OPTIONAL FEEDER/SERVICE CALCULATIONS [220–30]*

Instead of sizing the dwelling-unit feeder and/or service conductors according to the standard method (Part B of Article 220), an optional method [220–30] can be used. The following steps can be used to determine the demand load:

Step 1: → **Total Connected Load.** To determine the total connected load, add the following loads:

(a) **General Lighting And Receptacles:** 3 volt-amperes per square-foot.

(b) **Small Appliance And Laundry Branch Circuits**: 1,500 volt-amperes for each 20 ampere small appliance and laundry branch circuit.

(c) **Appliances:** The *nameplate* VA rating of all appliances and motors that are fastened in place (permanently connected), or located on a specific circuit. Be sure to use the range and dryer nameplate rating! This does not include the A/C or heating loads.

Step 2: → **Demand Factor.** A 40 percent demand factor is permitted to be applied to that portion of the total connected load in excess of 10 kW. First 10 kW of total connected load (Step 1) is at 100 percent and remainder is at 40 percent.

Step 3: → **A/C versus Heat.** Determine the A/C versus Heat demand load:

(a) Air conditioning or heat-pump compressors at 100 percent vs. 65 percent of three or less separately controlled electric space-heating units, or

(b) Air conditioning or heat-pump compressors at 100 percent vs. 40 percent of four or more separately controlled space-heating units.

Step 4: → **Total Demand Load.** To determine the total demand load, add the demand load from Step 2 and the largest of Step 3.
I = VA/E to be used to convert the total demand VA load to amperes.

Step 5: → **Conductor Size**
Four Hundred Amperes And Less. The ungrounded conductors are permitted to be sized to Note 3 of Table 310–16 for 120/240 volt single-phase systems up to 400 amperes. The grounded neutral conductor must have an ampacity to carry the unbalanced load, sized to Table 310-16.
Over Four Hundred Amperes. The ungrounded and grounded neutral conductors are sized according to Table 310–16.

9–11 *DWELLING-UNIT OPTIONAL CALCULATION EXAMPLES*

❑ **Optional Load Calculation Example No. 1**

What size service conductor is required for a 1,500 square foot dwelling-unit containing the following loads:

Dishwasher at 1,200 VA, 120 volt
Disposal at 900 VA, 120 volt
Cook top at 6,000 VA, 120/240 volt
A/C at 5 horsepower, 240 volt
Water heater at 4,500 VA, 240 volt
Dryer at 4,000 VA, 240 volt
Two-ovens each at 3,000 VA, 120/240 volt
Heat Strip at 10 kW, 240 volt

(a) No. 5 (b) No. 4 (c) No. 3 (d) No. 2

- Answer: (c) No. 3, 105 amperes – Note 3 of Table 310–16.

STEP 1: Total Connected Load

				Totals
(a) General Lighting 1,500 square feet × 3 VA =		4,500 VA		(a) 4,500 VA
(b) Small Appliance Circuits 1,500 VA × 2		3,000 VA		
Laundry Circuit		+ 1,500 VA		
		4,500 VA	=	(b) 4,500 VA
(c) Appliances (nameplate)				
Dishwasher	1,200 VA			
Water Heater	4,500 VA			
Disposal	900 VA			
Dryer*	4,000 VA			
Cooktop**	6,000 VA			
Ovens 3,000 VA × 2 units	+ 6,000 VA			
	22,600 VA =			(c) 22,600 VA

* The dryer load is 4,000 VA, not 5,000 VA! Section 220–18 (Part B of Article) does not apply to Section 220–30 (Part C).
** Cooking equipment is also calculated at nameplate value.

STEP 2: The Demand Load

(a) General Lighting (and receptacles)	4,500 VA		
(b) Small Appliance and Laundry Circuits	4,500 VA		
(c) Appliance nameplate ratings	+ 22,600 VA		
Total Connected Load	31,600 VA		
Demand Factor			
Total Connected Load	31,600 VA		
First 10,000 @ 100%	– 10,000 VA	× 1.00 =	10,000 VA
Remainder @ 40%	21,600 VA	× 0.40 =	+ 8,640 VA
Demand load			18,640 VA

STEP 3: Air-Conditioning Versus Heat

Air-conditioning or heat-pump compressors at 100 percent versus 65 percent of three or less separately controlled electric space-heating units.

Air Conditioner 230* volts × 28 amperes** = 6,440 VA, omit, Heat Strip 10,000 VA × .65 = 6,500 VA
* Some exams use 240 volts, ** Table 430–148

STEP 4: Total Demand Load

Total of Step 2 = Other Loads	18,640 VA
Total of Step 3 = Heat Load	+ 6,500 VA
Total Demand Load =	25,140 VA

STEP 5: Service Conductors [Table 310–16, Note 3]

$I = \frac{VA}{E}$, $I = \frac{25{,}140\ VA}{240\ volts} = 105$ amperes

The feeder and service conductor is sized to 110 amperes, Note 3 of Table 310–16 = No. 3 AWG.

❑ **Optional Load Calculation Example No. 2**

What size service conductor is required for a 2,330 square-foot residence with a 300 square-foot porch and a 150 square-foot carport that contains the following:

Receptacles, 15 extra	Recessed Fixtures, 10 – 100 watts (high-hats)
Dishwasher @ 1.5 kVA, 120 volt	Disposal @ 1 kVA 120 volt
Trash compactor @1.5 kVA, 120 volt	Water heater @ 6 kW, 240 volt
Range @ 14 kW, 120/240 volt	Dryer @ 4.5 kVA 120/240 volt
Air conditioner @ 6 kVA, 240 volt	Baseboard heat, four each @ 2.5 kW, 240 volt

(a) No. 5 (b)No. 4 (c) No. 3 (d) No. 2

- Answer: (d) No. 2, Note 3 of Table 310–16

STEP 1: Total Connected Load

			Totals
(a) General lighting (and receptacles): 2,330 square feet* × 3 VA = 6,990 VA			(a) 6,990 VA
(b) Small appliance and laundry branch circuits:			
Small Appliance Circuit 1,500 VA × 2 circuits =	3,000 VA		
Laundry Circuit	+1,500 VA		
	4,500 VA	=	(b) 4,500 VA

(c) Appliance Nameplate VA:			
Dishwasher	1,500 VA		
Disposal	1,000 VA		
Trash Compactor	1,500 VA		
Water Heater	6,000 VA		
Range**	14,000 VA		
Dryer***	+ 4,500 VA		
	28,500 VA	=	(c) 28,500 VA

*The open porch and carport is not counted [220–3(b)]. The 15 receptacles and the 10 recessed fixtures are included in the 3 VA per square-foot general lighting and receptacle load [Table Notes to Table 220–3(b)].

**All appliances at nameplate rating

*** The dryer load is 4,500 VA, NOT 5,000 VA! Be careful, Section 220–18 (Part B of Article 220) does not apply to Section 220–30 (Part C).

STEP 2: The Demand Load

(a) General Lighting (and receptacles)	6,990 VA		
(b) Small Appliance and Laundry Circuit	4,500 VA		
(c) Appliance nameplate ratings	+ 28,500 VA		
	39,990 VA		
First 10,000 @ 100% =	– 10,000 VA	× 1.0 =	10,000 VA
Remainder @ 40% =	29,990 VA	× .40 =	+ 11,996 VA
Demand load			21,996 VA

STEP 3: Air Conditioning Versus Heat

Air conditioning or heat-pump compressors at 100 percent vs. 40 percent of four or more separately controlled electric space-heating units.

Air Conditioner 230* volts × 28 amperes** = 6,440 VA

Heat Strip 10,000 VA × .40 = 4,000 VA* omit 220–21

* Some exams use 240 volts, ** Table 430–148

STEP 4: The Total Demand Load

Step 2 = Other Loads	21,996 VA
Step 3 = A/C Load	+ 6,440 VA
Total Demand Load	28,436 VA

STEP 5: Feeder And Service Conductors [Table 310–16, Note 3]

$I = \frac{VA}{E}, \quad I = \frac{28{,}436 \text{ VA}}{240 \text{ volts}} = 118 \text{ amperes}$

The feeder and service conductor sizes are sized to 125 amperes, Note 3 of Table 310–16 = No. 2 AWG.

9–12 *NEUTRAL CALCULATIONS – GENERAL [220–22]*

The feeder and service neutral load is the maximum *unbalanced demand load* between the grounded conductor (neutral) and any one ungrounded conductor as determined by Article 220 Part B (standard calculations). This means that since 240 volt loads are not connected to the neutral conductor, they are not considered for sizing the neutral conductor.

❑ **Neutral Not Over 200 Amperes Example**

What size 120/240-volt single-phase ungrounded and grounded conductor is required for a 375-ampere two-family dwelling-unit demand load, of which 175 amperes consist of 240 volt loads, Fig. 9–18?

(a) Two 400 THHN and 1– 350 THHN
(b) Two 350 THHN and 1– 350 THHN
(c) Two 500 THHN and 1– 500 THHN
(d) Two 400 THHN and 1– No. 3/0 THHN

- Answer: (d) 400 kcmil THHN, and No. 3/0 THHN

The ungrounded conductor is sized at 375 amperes (400 kcmil THHN) according to Note 3 of Table 310-16. The grounded conductor (neutral) is sized to carry a maximum unbalanced load of 200 amperes (375 amperes less 175 amperes – 240 volt loads).

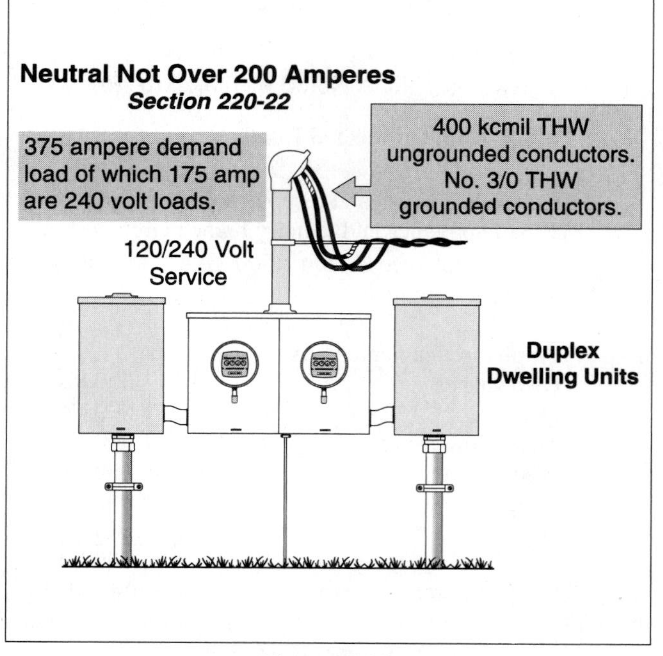

Figure. 9-18

Neutral Conductor Size Example

Cooking Appliance Neutral Load [220–22]

The feeder or service cooking appliance neutral load for household cooking appliances such as electric ranges, wall-mounted ovens, or counter-mounted cooking units, shall be calculated at 70 percent of the demand load as determined by Section 220–19.

❑ **Range Neutral Example**

What is the neutral load (in amperes) for one 14-kW range?

(a) 33 amperes (b) 26 amperes
(c) 45 amperes (d) 18 amperes

- Answer: (b) 26 amperes

36.7 amperes × 0.7 = 25.69 amperes

Step 1: → Since the range exceeds 12 kW, we must comply with Note 1 of Table 220–19. The first step is to determine the demand load as listed in Column A of Table 220–19 for one unit, = 8 kW

Step 2: → We must increase the Column A value (8 kW) by 5 percent for each kW that the average range (in this case 14 kW) exceeds 12 kW. This results is an increase of the Column A value (8 kW) by 10 percent.
8 kW x 1.1 = 8.8 kW or (8,000 watts + 800 watts)

Step 3: → Convert the demand load in kW to amperes.
$I = \frac{P}{E}$, $I = \frac{8{,}800 \text{ VA}}{240 \text{ volts}} = 36.7 \text{ amperes} \times 0.7 = 26 \text{ amperes}$

Dryer Neutral Load [220–22]

The feeder and service dryer neutral demand load for electric clothes dryers shall be calculated at 70 percent of the demand load as determined by Section 220–18.

❑ **Dryer Over 5-kW Example**
What is the neutral load (in amperes) for one 5.5-kVA dryer?

(a) 4 kW (b) 5 kW (c) 6 kW (d) None of these

• Answer: (a) 4 kW, 5.5 kVA × 0.7 = 3.85 kW

Unit 9 – One-Family Dwelling-Unit Load Calculations Summary Questions

Part A – General Requirements

9–2 Voltages [220–2]

1. Unless other voltages are specified, branch-circuit, feeder and service loads shall be computed at a nominal system voltages of ______.
(a) 120/240 (b) 120/240 (c) 208Y/120 (d) any of these

9–3 Fraction Of An Ampere

2. There are no NEC® rules for rounding when a calculation results in a fraction of an ampere, but it does contains the following note, except where the computations result in a ______ of an ampere or larger, such fractions may be dropped.
(a) .05 (b) .5 (c) .49 (d) .51

9–4 Appliance (Small) Circuits [220–4(b)]

3. A minimum of ______ 20-ampere small appliance branch circuits are required for receptacle outlets in the kitchen, dining room, breakfast room, pantry, or similar areas of dining.
(a) one (b) two (c) three (d) none of these

4. When sizing the feeder or service, each dwelling-unit shall have a minimum feeder load of ______ VA for the two small appliance branch circuits.
(a) 1,500 (b) 2,000 (c) 3,000 (d) 4,500

9–5 Cooking Equipment – Branch Circuit [Table 220–19, Note 4]

Table 220–19 Column A

5. What is the branch circuit demand load (in amperes) for one 11-kW range?
(a) 21 amperes (b) 27 amperes (c) 38 amperes (d) 33 amperes

6. What is the branch circuit load (in amperes) for one 13.5-kW range?
(a) 33 amperes (b) 37 amperes (c) 50 amperes (d) 58 amperes

Wall Mounted Oven And/Or Counter-Mounted Cooking Unit

7. • What is the branch circuit load (in amperes) for one 6-kW wall-mounted oven?
(a) 20 amperes (b) 25 amperes (c) 30 amperes (d) 40 amperes

8. • What is the branch circuit load (in amperes) for one 4.8-kW counter-mounted cooking unit?
(a) 15 amperes (b) 20 amperes (c) 25 amperes (d) 30 amperes

9. What is the branch circuit load (in amperes) for one 6-kW counter-mounted cooking unit and one 3-kW wall-mounted oven?
(a) 25 amperes (b) 38 amperes (c) 33 amperes (d) 42 amperes

10. What is the branch circuit load (in amperes) for one 6-kW counter-mounted cooking unit and two 4-kW wall-mounted ovens?
(a) 22 amperes (b) 27 amperes (c) 33 amperes (d) 37 amperes

9–6 Laundry Receptacle(s) Circuit [220–4(c)]

11. The NEC® does not require a separate circuit for the washing machine, but requires is a separate circuit for the laundry room receptacle or receptacles, of which one can be for the washing machine.
(a) True (b) False

12. Each dwelling-unit shall have a feeder load consisting of ________ VA volt-ampere for the 20 ampere laundry circuit.
(a) 1,500 (b) 2,000 (c) 3,000 (d) 4,500

9–7 Lighting And Receptacles

13. The NEC® requires a minimum 3 volt-ampere per square foot [Table 220–3(b)] for the required general lighting [210–70(a)] and receptacles [210–52(a)]. The dimensions for determining the area shall be computed from the outside of the building and shall not include ________.
(a) open porches
(b) garages
(c) spaces not adaptable for future use
(d) all of these

14. The 3 volt-ampere per square foot for general lighting includes all 15- and 20-ampere general use receptacles, but not the appliance or laundry circuit receptacles.
(a) True (b) False

15. What is the general lighting load and receptacle load for a 2,100-square foot home that has fourteen extra convenience receptacles and nine extra recessed fixtures rated 100 watts each?
(a) 2,100 VA (b) 4,200 VA (c) 6,300 VA (d) 8,400 VA

Number of General Lighting Circuits [Chapter 9, Example No. 1(a)]

16. The number of branch circuits required for general lighting and receptacles shall be determined from the general lighting load and the rating of the circuits.
(a) True (b) False

17. How many 15-ampere general lighting circuits are required for a 2,340-square foot home?
(a) two (b) three (c) four (d) five

18. How many 20-ampere general lighting circuits are required for a 2,100-square foot home?
(a) two (b) three (c) four (d) five

19. How many 20 ampere circuits are required for the general lighting and receptacles for a 1,500-square foot dwelling-unit?
(a) two (b) three (c) four (d) five

Part B – Standard Method – Load Calculation Questions

9–8 Dwelling Unit Load Calculations (Part B Of Article 240)

20. The NEC® recognizes that the general lighting and receptacles' small appliance and laundry circuits will not all be on or loaded at the same time and permits a ________ to be applied to the total of these loads.
(a) demand factor
(b) adjustment factor
(c) correction factor
(d) correction

21. Because the A/C and heating loads are not on at the same time (simultaneously), it is permissible to omit the smaller of the two loads when determining the A/C versus heat demand load.
(a) True (b) False

22. A load demand factor of 75 percent is permitted for ________ or more appliances fastened in place, such as dishwashers, disposals, trash compactors, water heaters, etc.
(a) one (b) two (c) three (d) four

23. The feeder or service demand load for electric clothes dryers located in dwelling-units shall not be less than ________.
(a) 5000 watts
(b) nameplate rating
(c) a or b
(d) the greater of a or b

24. • A feeder or service dryer load is required, even if the dwelling-unit does not contain an electric dryer.
(a) True (b) False

25. Household cooking appliances rated 1¾-kW can have the feeder and service loads calculated according to the demand factors of Section 220–19, Table, and Notes.
(a) True (b) False

26. Dwelling-unit feeder and service ungrounded conductors are sized according to Note 3 of Table 310–16 for ________.
(a) 3-wire single phase 120/240 volt systems up to 400 amperes
(b) 3-wire single phase 120/208 volt systems up to 400 amperes
(c) 4-wire single phase 120/240 volt systems up to 400 amperes
(d) none of these

27. What is the general lighting load for a 2,700-square foot dwelling-unit?
(a) 8,100 VA (b) 12,600 VA (c) 2,700 VA (d) 6,840 VA

28. What is the total connected load for general lighting and receptacles, small appliance and laundry circuits for a 6,540-square foot dwelling-unit?
(a) 8,100 VA (b) 12,600 VA (c) 2,700 VA (d) 24,120 VA

29. What is the feeder or service demand load for a 5-horsepower 230-volt A/C and three 3-kW baseboard heaters?
(a) 6,400 VA (b) 3,000 VA (c) 8,050 VA (d) 9,000 VA

30. What is the feeder or service demand load for a 4-horsepower 230-volt A/C and a 10-kW electric space heater?
(a) 10,000 VA (b) 3,910 VA (c) 6,440 VA (d) 10,350 VA

31. What is the feeder or service demand load for a 940 VA disposal, 1,250 VA dishwasher and a 4,500 VA water heater?
(a) 5,018 VA (b) 6,690 VA (c) 8,363 VA (d) 6,272 VA

32. • What is the feeder or service demand load for a 940 VA disposal, 1,250 VA dishwasher, 1,100 VA trash compactor and a 4,500 VA water heater?
(a) 7,790 VA (b) 5,843 VA (c) 7,303 VA (d) 9,738 VA

33. What is the feeder or service demand load for a 4-kW dryer?
(a) 4,000 VA (b) 3,000 VA (c) 5,000 VA (d) 5,500 VA

34. What is the feeder or service demand load for a 5.5-kW dryer?
(a) 4,000 VA (b) 3,000 VA (c) 5,000 VA (d) 5,500 VA

35. • What is the feeder or service demand load for two 3-kW cooking appliances in a dwelling-unit?
(a) 3 kW (b) 4.8 kW (c) 4.5 kW (d) 3.9 kW

36. What is the feeder or service demand load for one 6-kW cooking appliance in a dwelling-unit?
(a) 6 kW (b) 4.8 kW (c) 4.5 kW (d) 3.9 kW

37. What is the feeder or service demand load for one 6-kW and two 3-kW cooking appliances in a dwelling-unit?
(a) 6 kW (b) 4.8 kW (c) 4.5 kW (d) 9.3 kW

38. What is the feeder or service demand load for an 11.5-kW range in a dwelling-unit?
(a) 11.5 kW (b) 8 kW (c) 9.2 kW (d) 6 kW

39. What is the feeder or service demand load for a 13.6-kW range in a dwelling-unit?
(a) 8.8 kW (b) 8 kW (c) 9.2 kW (d) 6 kW

40. What size 120/240-volt single-phase service or feeder THHN conductors are required for a dwelling-unit that has a 190-ampere service demand load?
(a) No. 1/0 (b) No. 2/0 (c) No. 3/0 (d) No. 4/0

41. • What is the net computed demand load for the general lighting, small appliance and laundry circuits for a 1,500-square foot dwelling-unit?
(a) 4,500 VA (b) 9,000 VA (c) 5,100 VA (d) none of these

42. What is the feeder or service demand load for a 3-horsepower, 230-volt air conditioner with a ⅛-horsepower blower versus 4-kW heat?
(a) 3,910 VA (b) 5,222 VA (c) 4,000 VA (d) None of these

43. What is the feeder or service demand load for a dwelling-unit that has one water heater at 4,000 VA and one dishwasher at 1,500 VA?
(a) 4 kW (b) 5.5 kW (c) 4.13 kW (d) none of these

44. What is the net computed load for a 4.5-kW dryer?
(a) 4.5 kW (b) 5 kW (c) 5.5 kW (d) none of these

45. What is the total net computed demand load for a 14-kW range?
(a) 8 kW (b) 8.4 kW (c) 8.8 kW (d) 14 kW

46. If the total demand load is 30-kW for a 120/240-volt dwelling-unit, what is the service and feeder conductor size?
(a) No. 4 (b) No. 3 (c) No. 2 (d) No. 1

47. If the service raceway contains No. 2 service conductors, what size bonding jumper is required for the service raceway?
(a) No. 8 (b) No. 6 (c) No. 4 (d) No. 3

48. If the service contains No. 2 conductors, what is the minimum size grounding electrode conductor required?
(a) No. 8 (b) No. 6 (c) No. 4 (d) No. 3

Part C - Optional Method – Load Calculation Questions

9–10 Dwelling Unit Optional Feeder/Service Calculations (220–30)

49. When sizing the feeder or service conductor according to the optional method we must:
(a) Determine the total connected load of the general lighting and receptacles, small appliance and laundry branch circuits, and the nameplate VA rating of all appliances and motors.
(b) Determine the demand load by applying the following demand factor to the total connected load. First 10 kVA 100 percent, remainder 40 percent.
(c) Determine the A/C versus heat demand load.
(d) All of these.

Use this information for Question 50.
A 1,500-square foot dwelling-unit containing the following loads:

Dishwasher at 1,500 VA
Disposal at 1,000 VA
Cooktop at 6,000 VA
A/C at 3 horsepower
Water heater at 5,000 VA
Dryer at 5,500 VA
Two ovens each at 3,000 VA
Heat strip at 10 kW

50. • Using the optional method, what size aluminum conductor is required for the service for the described loads?
(a) No. 1 (b) No. 1/0 (c) No. 2/0 (d) No. 3/0

A 2,330-square foot residence with a 300-square foot porch and 150-square foot carport contains the following:

15 extra general-use receptacles
Dishwasher at 1.5 kVA, 120 volt
Trash compactor at 1.5 kVA, 120 volt
Range at 14 kW, 120/240 volt
Air conditioner at 5 hp, 240 volt
30 – 100 watt recessed fixtures (high hats)
Disposal at 1 kVA, 120 volt
Water heater at 6 kW, 240 volt
Dryer at 4.5 kW 120/240 volt
Baseboard heat four each at 2.5 kW, 240 volt

51. Using the optional method, what size conductor is required for the service, using the described loads?
(a) No. 5 (b) No. 4 (c) No. 3 (d) No. 2

9–12 Neutral Calculations General [220–22]

52. The feeder and service neutral load is the maximum unbalanced demand load between the grounded conductor (neutral) and any one ungrounded conductor as determined by Article 220 Part B. Since 240 volt loads cannot be connected to the neutral conductor, 240 volt loads are not considered for sizing the feeder neutral conductor.
(a) True (b) False

53. • What size single-phase 3-wire feeder is required for a 475 ampere demand load of which 275 amperes consists of 240 volt loads?
(a) 2– 750 THHN and 1– 350 THHN
(b) 2– 500 THHN and 1– 350 THHN
(c) 2– 750 THHN and 1– 500 THHN
(d) 2– 750 THHN and 1– No. 3/0 THHN

54. The feeder or service neutral load for household cooking appliances such as electric ranges, wall-mounted ovens, or counter-mounted cooking units shall be calculated at ____ percent of the demand load as determined by Section 220–19.
(a) 50 (b) 60 (c) 70 (d) 80

55. What is the dwelling-unit service neutral load for one 9-kW range?
(a) 5.6 kW (b) 6.5 kW (c) 3.5 kW (d) 12 kW

56. The service neutral demand load for household electric clothes dryers shall be calculated at ____ percent of the demand load as determined by Section 220–18.
(a) 50 (b) 60 (c) 70 (d) 80

57. What is the feeder neutral load for one 6-kW household dryers?
(a) 4.2 kW (b) 4.7 kW (c) 5.4 kW (d) 6.6 kW

Challenge Questions

9–5 Cooking Equipment – Branch Circuit [Table 220–19 Note 4]

58. The branch-circuit demand load for one 18-kW range is ______ kW.
(a) 12 (b) 8 (c) 10.4 (d) 18

59. A dwelling-unit kitchen will have the following appliances: One 9-kW cooktop and one wall-mounted oven rated 5.3-kW. The branch-circuit demand load for these appliances will be ______ kW.
(a) 14.3 (b) 12 (c) 8.8 (d) 8

9–8 Standard Method – Load Calculation Questions

General Lighting [Table 220-11]

60. An apartment building contains twenty units, and each unit has 840-square feet. What is the general lighting feeder demand load for each dwelling-unit? Note: The laundry facilities are provided on the premises for all tenants and no laundry circuit is required in each unit, see 210–52(f) Exception No. 1.
(a) 3,520 VA (b) 3,882 VA (c) 4,220 VA (d) 6,300 VA

61. An 1,800-square foot residence has a 300-square foot open porch and a 450-square foot carport with the following loads: 45 general use receptacles; 19 high-hats; and six 150 VA flood lights. What is the demand load for the general lighting, receptacles, small appliance circuits, and laundry circuits?
(a) 5,400 VA (b) 7,900 VA (c) 5,415 VA (d) 6,600 VA

62. How many 15-ampere, 120-volt branch circuits are required for general use receptacle and lighting for a 1,800-square foot dwelling-unit?
(a) 1 circuit (b) 2 circuits (c) 3 circuits (d) 4 circuits

63. How many 20-ampere, 120-volt branch circuits are required for general use receptacle and lighting for a 2,800-square foot dwelling-unit?
(a) 2 circuit (b) 6 circuits (c) 7 circuits (d) 4 circuits

Appliance Demand Factors [Section 220–17]

64. A dwelling-unit contains the following: A 1.2-kW washing machine; a 4-kW water heater; a 1.2-kW dishwasher; and a 1.5-kW trash compactor. The appliance demand load added to the service is ______.
(a) 5.9 kW (b) 6.7 kW (c) 7.7 kW (d) 8.8 kW

65. • A dwelling-unit contains the following: A water heater (4-kW); a dishwasher (½-horsepower); a dryer; a pool pump (¾-horsepower); a cooktop (6-kW); an oven (6-kW); A/C (4-horsepower); and heat (6-kW). What is the appliance demand load for the dwelling-unit?
(a) 6.7 kW (b) 4 kW (c) 9 kW (d) 11 kW

Dryer Calculation [Section 220–18]

66. A dwelling-unit contains a 5.5-kW electric clothes dryer. What is the feeder demand load for the dryer?
(a) 3.38 kW (b) 4.5 kW (c) 5 kW (d) 5.5 kW

Cooking Equipment Demand Load [220–19]
General Calculations

67. The feeder load for twelve 1¾-kW cooktops is ______ .
(a) 21 kW (b) 10.5 kW (c) 13.5 kW (d) 12.5 kW

68. The minimum neutral for a branch-circuit to an 8¾-kW household range is ______ . See Article 210.
(a) No. 10 (b) No. 12 (c) No. 8 (d) No. 6

69. What is the maximum dwelling-unit feeder or service demand load for fifteen 8-kW cooking units?
(a) 38.4 kW (b) 30 kW (c) 110 kW (d) 27 kW

Note 2 of Table 220–19

70. What is the feeder demand load for five 10-kW, five 14-kW, and five 16-kW household ranges?
(a) 70 kW (b) 23 kW (c) 14 kW (d) 33 kW

Note 3 to Table 220–19

71. A dwelling-unit has one 6-kW cooktop and one 6-kW oven. What is the minimum feeder demand load for the cooking appliances?
(a) 13 kW (b) 8.8 kW (c) 7.8 kW (d) 8.2 kW

72. The maximum feeder demand load for an 8½-kW range is ______.
(a) 8.5 kW (b) 8 kW (c) 6.3 kW (d) 12 kW

73. Given: Five 5-kW ranges; two 4-kW ovens; and four 7-kW cooking units. The minimum feeder demand load is ______.
(a) 61 kW (b) 22 kW (c) 19.5 kW (d) 18 kW

74. Given: Two 3-kW wall mounted ovens and one 6-kW cooktop. The feeder demand load would be ______.
(a) 9.3 kW (b) 11 kW (c) 8.4 kW (d) 9.6 kW

9–9 Service Questions

75. • Both units of a duplex apartment require a 100 ampere main; the resulting 200 ampere service would require ______ THHN.
(a) No. 1/0 (b) No. 2/0 (c) No. 3/0 (d) No. 4

76. After all demand factors have been taken into consideration, the demand load for a dwelling-unit is 21,560 VA. The minimum service size for this residence would be ______ if the optional method of service calculations were used. Be sure to read 220–30 carefully!
(a) 90 amperes (b) 100 amperes (c) 110 amperes (d) 125 amperes

77. After all demand factors have been taken into consideration, the net computed load for a service is 24,221 VA. The minimum service size is ______.
(a) 150 amperes (b) 175 amperes (c) 125 amperes (d) 110 amperes

9–10 Optional Method – Load Calculation Questions [220–30]

78. Using the optional calculations method, the demand load for a 6-kW central space heating and a 4-kW air-conditioning unit would be ______.
(a) 9,000 watts (b) 3,900 watts (c) 4,000 watts (d) 5,000 watts

79. The total connected load of a dwelling-unit is 25 kVA, not including heat or air conditioning. If the heat is separately controlled in five rooms (10 kW), and the air conditioning is 6 kW, then the total service demand load would be ______ if the optional method is used. Note: Don't forget to apply optional method demand factors.
(a) 22 kVA (b) 29 kVA (c) 35 kVA (d) 36 kVA

80. Using the optional method, the service size would be ______ amperes for the following: 1,200-square feet first floor, plus 600-square feet on the second floor, and a 200-square foot open porch; a 4-kW, 230 volt water heater; a ½-horsepower, 115 volt dishwasher; a 4-kW clothes dryer; 4-horsepower A/C, 6-kW heat strips; a ¾-horsepower 115 volt pool pump; one 6-kW oven; one 6-kW Cooktop; 20 receptacles; and 10 lights. The service source voltage is 230/115 volt single phase.
(a) 150 (b) 175 (c) 125 (d) 110

81. An 1,800-square foot residence contains the following: a 300-square foot porch; a 150-square foot carport; a water heater of 4-kW; 10-kW heat separated in five rooms; a 1.5-kW dishwasher; a 6-kW range; a 4.5-kW dryer; two 3-kW ovens; a 6-kW air conditioner; 35 general use receptacles; and sixteen 75 watt lighting fixtures. The service size for the loads would be ______ amperes. The optional method of service calculations were used.
(a) 175 (b) 110 (c) 125 (d) 150

82. A dwelling-unit has 1200-square feet on the first floor, 600-square feet upstairs (unfinished but adaptable for future use), and a 200-square foot open porch with the following loads: a pool pump ¾-horsepower; range – 13.9-kW; dishwasher – 1.2-kW; water heater – 4-kW; dryer – 4-kW; A/C 5 horsepower; and heat 6-kW. Using the optional calculation, what is the demand load for the service?
(a) No. 4 (b) No. 3 (c) No. 2 (d) No. 1

CHAPTER 3
Advanced NEC® Calculations And Code Questions

Scope of Chapter 3

Unit 10

Multifamily Dwelling-Unit Load Calculations

OBJECTIVES

After reading this unit, the student should be able to explain the following concepts:

- Air conditioning versus heat
- Appliance (small) circuits
- Appliance demand load
- Clothes dryer demand load
- Cooking equipment calculations
- Dwelling-unit calculations – Optional method
- Dwelling-unit calculations – Standard method
- Laundry circuit
- Lighting and receptacles

After reading this unit, the student should be able to explain the following terms:

- General lighting
- General–use receptacles
- Optional method
- Rounding
- Standard method
- Unbalanced demand load
- Voltages

10–1 *MULTIFAMILY DWELLING-UNIT CALCULATIONS – STANDARD METHOD*

When determining the ungrounded and grounded (neutral) conductor for multifamily dwelling-units (Fig. 10–1), apply the following multifamily standard method steps:

Step 1: → **General Lighting And Receptacles, Small Appliance And Laundry Circuits [220–11].** The NEC® recognizes that the general lighting and receptacles, small appliance, and laundry circuits will not all be on, or loaded, at the same time. The NEC® permits the following demand factor to be applied to these loads [220–16].

(a) Total Connected Load. Determine the total connected general lighting and receptacles (3 VA), small appliance (3,000 VA), and the laundry (1,500 VA) circuit load of all dwelling-units. The laundry load (1,500 VA) can be omitted if laundry facilities are provided on the premises [210–52(f) Exception 1].

(b) Demand Factor. Apply Table 220–11 demand factors to the total connected load (Step 1a).
First 3,000 VA at 100% demand
Next 117,000 VA at 35% demand
Remainder at 25% demand

Step 2: → **A/C versus Heat [220–15, 220–21, and 440–34].** Because the A/C and heating loads are not on at the same time (simultaneously), it is permissible to omit the smaller of the two loads.
Air conditioning: The A/C demand load shall be calculated at 125 percent of the largest A/C motor VA, plus the sum of the other A/C motor VA's [440–34].
Heat: Fixed electric space-heating loads shall be computed at 100 percent of the total connected load [220–15].

Step 3: → **Appliances [220–17].** A demand factor of 75 percent is permitted for four or more appliances fastened in place such as a dishwasher, disposal, trash compactor, water heater, etc. This does not apply to space-heating equipment [220–15], clothes dryers [220–18], cooking appliances [220–19], or air-conditioning equipment [440–34].

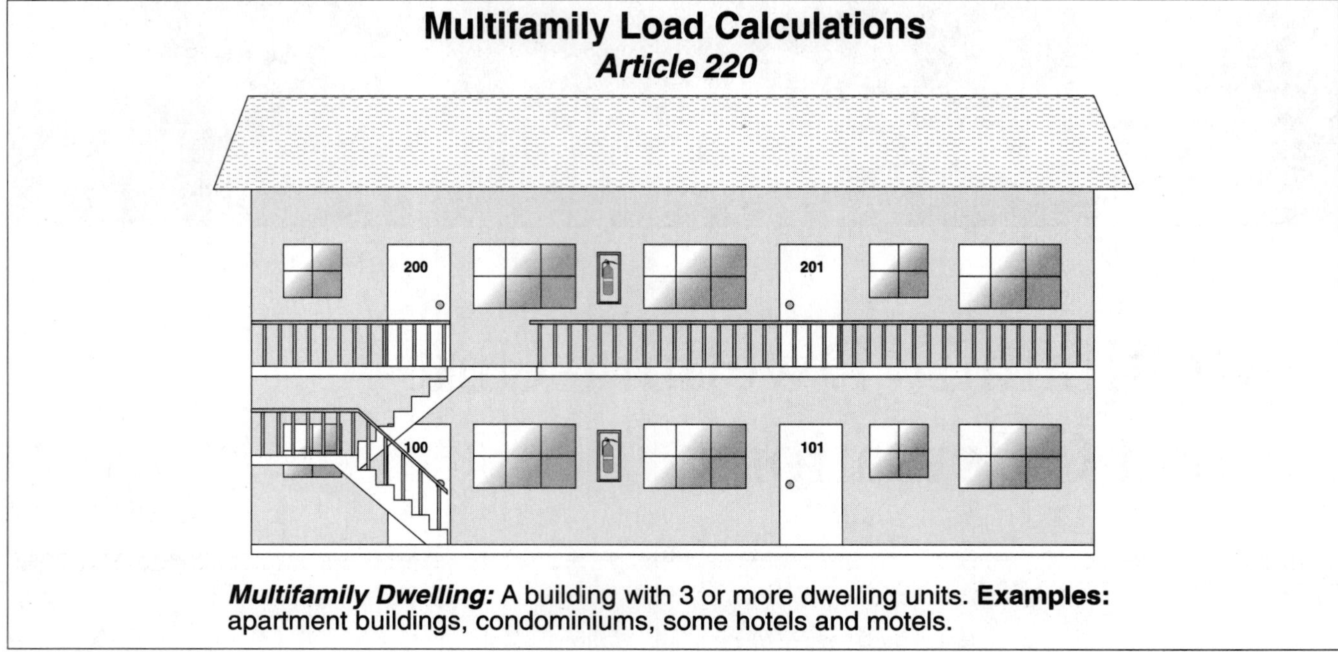

Figure 10-1

Multifamily Load Calculations

Step 4: → **Clothes Dryers [220–18].** The feeder or service demand load for electric clothes dryers located in dwelling-units shall not be less than: (1) 5,000 watts, or (2) the nameplate rating (whichever is greater) adjusted according to the demand factors listed in Table 220–18.
Note: A dryer load is not required if the dwelling-unit does not contain an electric dryer. Laundry room dryers shall not have their loads calculated according to this method. This is covered in Unit 11.

Step 5: → **Cooking Equipment [220–19].** Household cooking appliances rated over 1¾ kW can have their feeder and service loads calculated according to the demand factors of Section 220–19, Table and Notes.

Step 6: → **Feeder And Service Conductor Size**
Four hundred amperes and less. The ungrounded conductors are sized according to Note 3 of Table 310–16 for 120/240-volt single-phase systems.
Over four hundred amperes. The ungrounded conductors are sized according to Table 310–16 to the calculated unbalanced demand load.

10–2 *MULTIFAMILY DWELLING-UNITS CALCULATION EXAMPLES – STANDARD METHOD*

Step 1. General Lighting, Small Appliance, And Laundry Demand [220–11]

➛ **Question 1.**

What is the demand load for an apartment building that contains twenty units? Each apartment is 840 square feet.
Note: Laundry facilities are provided on the premises for all tenants [210-52(f)].

(a) 5,200 VA (b) 40,590 VA (c) 110,400 VA (d) none of these

- Answer: (b) 40,590 VA

General Lighting	840 square feet × 3 VA	2,520 VA		
Small Appliance Circuits	1,500 VA × 2 circuits	3,000 VA		
Laundry Circuit (none)		+ 0 VA		
Total connected load × units		5,520 VA	× 20 =	110,400 VA
Demand factor		110,400 VA		
First 3,000 VA at 100%		– 3,000 VA	× 1.00 =	3,000 VA
Next 117,000 VA at 35%		107,400 VA	× .35 =	+37,590 VA
Total demand load =				40,590 VA

➛ **Question 2.**

What is the general lighting net computed demand load for a twenty unit apartment? Each unit is 990 square feet.

(a) 74,700 VA (b) 149,400 VA (c) 51,300 VA (d) 105,600 VA

• Answer: (c) 51,300 VA

General lighting	990 square feet × 3 VA	2,970 VA		
Small appliance	1,500 VA × 2 circuits	3,000 VA		
Laundry circuit	1,500 VA × 1 circuits	+1,500 VA		
Total connected load × units		7,470 VA	× 20 =	149,400 VA
Demand factor		149,400 VA		
First 3,000 VA at 100%		–3,000 VA	× 1.00 =	3,000 VA
		146,400 VA		
Next 117,000 VA at 35%		–117,000 VA	× .35 =	40,950 VA
Remainder VA at 25%		29,400 VA	× .25 =	+ 7,350 VA
Total demand load =				51,300 VA

Step 2. Air Conditioning Versus Heat [220–15 and 440-34]

➛ **Question 1.**

What is the net computed A/C versus heat load for a forty unit multifamily building that has a 3-horsepower 230-volt A/C unit and two 3-kW baseboard heaters in each unit?

(a) 160 kW (b) 240 kW (c) 60 kW (d) 50 kW

• Answer: (b) 240 kW

A/C [440-34] – 230 volts × 17 amperes* = 3,910 VA, 3,910 VA × 1.25 = 4,888 VA

4,888 VA + (3,910 VA × 39 units) = 157,378 VA, Omit smaller than heat [220-21].

Heat [220–15] 3,000 watts × 2 units = 6,000 watts, 6,000 × 40 units = 240,000 watts/1,000 = 240 kW

➛ **Question 2.**

What is the A/C versus heat demand load for a twenty-five-unit multifamily building that has a 3-horsepower 230 volt A/C unit and a 5-kW heat strip?

(a) 160 kVA (b) 125 kVA (c) 6 kVA (d) 5 kVA

• Answer: (b) 125 kVA

Air Conditioning [220–14] – 230 volts × 17 amperes* = 3,910 VA, 3,910 VA × 1.25 = 4,888 VA

4,888 VA + (3,910 × 24 units) = 98,728 VA, Omit smaller than heat [220-21].

Heat [220–15] 5,000 watts × 25 units = 125,000 watts/1,000 = 125 kW

* Table 430–148

Step 3. Appliance Demand Load [220–17]

➛ **Question 1.**

What is the appliance demand load for a twenty-unit multifamily building that contains a 940 VA disposal, a 1,250 VA dishwasher, and a 4,500 VA water heater?

(a) 100 kVA (b) 134 kVA (c) 7 kVA (d) 5 kVA

• Answer: (a) 100 kVA

Disposal	940 VA	
Dishwasher	1,250 VA	
Water heater	+ 4,500 VA	
	6,690 VA	× 20 units = 133,800 VA
Demand factor at 75%		× .75
Total demand load =		100,350 VA

➛ Question 2.

What is the appliance demand load for a thirty-five-unit multifamily building that contains a 900 VA disposal, a 1,200 VA dishwasher, and a 5,000 VA water heater.

(a) 71 kVA (b) 142 kVA (c) 107 kVA (d) 186 kVA

- Answer: (d) 186 kVA

Disposal	900 VA	
Dishwasher	1,200 VA	
Water heater	+ 5,000 VA	
Unit total connected load	7,100 VA × 35 Units =	248,500 VA
Demand factor at 75%		× .75
Total demand load =		186,375 VA/1,000 = 186 VA

Step 4. Dryer Demand Load [220–18]

➛ Question 1.

A multifamily dwelling (twelve-unit building) contains a 4.5-kVA electric clothes dryer in each unit. What is the feeder and service dryer demand load for the building?

(a) 5 kVA (b) 27 kVA (c) 60 kVA (d) none of these

- Answer: (b) 27 kVA

5 kVA* × 12 units = 60 kVA × .45 = 27 kVA

* The minimum load is 5 kVA, for standard calculations.

➛ Question 2.

What is the demand load for twenty 5.25-kW dryers installed in dwelling-units of a multifamily building?

(a) 5 kVA (b) 27 kVA (c) 60 kVA (d) 37 kVA

- Answer: (d) 37 kVA

5.25 kVA × 20 units = 105 kVA × .35 = 36.75 kVA

Step 5. Cooking Equipment Demand Load [220–19]

➛ Column A Question

What is the feeder and service demand load for five 9-kW ranges?

(a) 9 kW (b) 45 kW (c) 20 kW (d) none of these

- Answer: (c) 20 kW [Table 220–19, Column A]

➛ Note 1 Question

What is the feeder and service demand load for three ranges rated 16 kW each?

(a) 15 kW (b) 14 kW (c) 17 kW (d) 21 kW

- Answer: (c) 17 kW [220–19 Note 1] (closest answer)

Step 1. → "Column A" demand load = 14 kW (3 units).

Step 2. → The average range (16 kW) exceeds 12 kW by 4 kW, increase "Column A" demand load (14 kW) by 20 percent, 14 kW × 1.2 = 16.8 kW or 14 kW + 2.8 kW =16.8 kW

➛ Note 2 Question

What is the feeder and service demand load for three ranges rated 9-kW and three ranges rated 14-kW?

(a) 36 kW (b) 42 kW (c) 78 kW (d) 22 kW

- Answer: (d) 22 kW [220–19, Note: 2]

Step 1: → Determine the total connected load.

9 kW (use minimum 12 kW)	3 × 12 kW =	36 kW
14 kW	3 × 14 kW =	+ 42 kW
Total connected load		78 kW

Step 2. → Determine the average range rating, 78 kW/6 units = 13 kW average rating.

Step 3. → Demand load Table 220–19 "Column A", 6 ranges = 21 kW.

Step 4. → The average range (13 kW) exceeds 12 kW by 1 kW. Increase "Column A" demand load (21 kW) by 5 percent, 21 kW × 1.05 = 22.05 kW, or 21 kW + 1.05 kW = 22.05 kW.

➛ Note 3 Column B Question

What is the feeder and service demand load for ten 3-kW ovens?

(a) 10 kW (b) 30 kW
(c) 15 kW (d) 20 kW

- Answer: (c) 15 kW [220–19, Column B]
  3 kW × 10 units = 30 kW × .49 = 14.70 kW demand load.

➛ Note 3 Column C Question

What is the feeder and service demand load for eight 6-kW cooktops?

(a) 10 kW (b) 17 kW
(c) 14.7 kW (d) 48 kW

- Answer: (b) 17 kW [220–19, Column C]
  6 kW × 8 units = 48 kW × .36 = 17.28 kW demand load.

Step 6. Service Conductor Size

➛ Question 1.

What size aluminum service conductors are required for a 120/240-volt, single-phase multifamily building that has a total demand load of 93 kVA, Fig. 10–2?

(a) 300 THHN kcmil (b) 350 THHN kcmil
(c) 500 THHN kcmil (d) 600 THHN kcmil

- Answer: (d) 600 kcmil aluminum
  I = VA/E
  I = 93,000 VA/240 volts
  I = 388 amperes [Note 3 of Table 310–16]

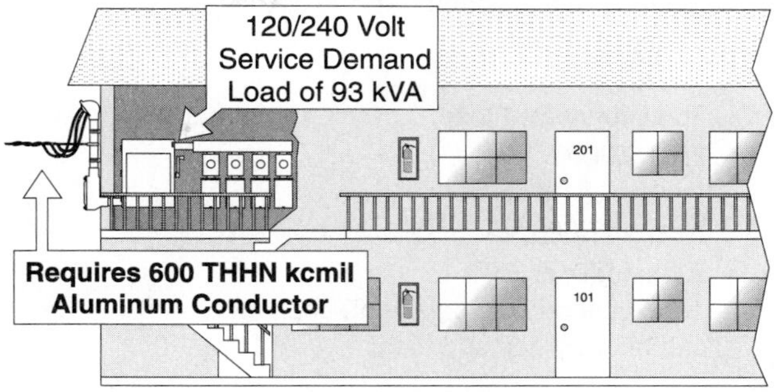

Figure 10-2
Conductor Sizing – Multifamily Dwelling

➛ Question 2.

What size service conductors are required for a multifamily building that has a total demand load of 270 kVA for a 120/208-volt three-phase system?

Note: Service conductors are run in parallel.

(a) 2 – 300 THHN kcmil per phase (b) 2 – 350 THHN kcmil per phase
(c) 2 – 500 THHN kcmil per phase (d) 2 – 600 THHN kcmil per phase

- Answer: (c) 2 – 500 THHN kcmil per phase
  I = VA/(E × 1.732)
  I = 270,000 VA/(208 volts × 1.732)
  I = 750 amperes
  Amperes per parallel set = 750 amperes/2 raceways, = 375 amperes per phase.
  500 THHN kcmil conductor has an ampacity of 380 amperes [Table 310–16 at 75°C]. These parallel conductors rated 760 amperes (380 amperes × 2) are permitted to be protected by a 800 ampere protection device [240–3(b)].

10–3 *MULTIFAMILY DWELLING-UNITS CALCULATION EXAMPLES – STANDARD METHOD*

❑ Multifamily Load Calculation Example

What is the demand load for an apartment building that contains twenty-five units? Each apartment is 1,000 square feet. Air conditioning 5-horsepower, heat 7.5-kW, dishwasher 1.2-kVA, disposal 1.5-kVA, water heater 4.5-kW, dryer 4.5-kW, and range 15.5-kW. System voltage 208/120-volt wye three-phase.

Step 1. General Lighting, Small Appliance, And Laundry Demand [220–11]

Twenty units each containing:			
General lighting 1,000 square feet × 3 VA	3,0000 VA		
Small appliance circuits 1,500 VA × 2	3,000 VA		
Laundry circuit	+ 1,500 VA		
	7,500 VA		
Total connected load, 7,500 VA × 25 units =	187,500 VA		
First 3,000 VA at 100% =	– 3,000 VA	× 1.00 =	3,000 VA
	184,000 VA		
Next 117,000 VA at 35%	– 117,000 VA	× .35 =	40,950 VA
Remainder at 25%	67,000 VA	× .25 =	+ 16,750 VA
Total general lighting, small appliance, and laundry demand load			60,700 VA

Step 2. Air Conditioning Versus Heat [220–15 And 440–34]

Twenty five units each containing A/C 5-horsepower versus 7.5-kW heat:
A/C 5 horsepower single-phase [440-34], 208 volts × 30.8 amperes* = 6,406 VA
6,406 VA × 1.25 = 8,008 VA, 8,008 VA + (6,406 × 24 units) = 161,752 VA
Omit smaller than heat [220–21]
Total heat demand load [220–15] 7,500 VA × 25 units = 187,500 VA
* The minimum dryer load for standard load calculations is 5 kW

Step 3. Appliance Demand Load [220–17]

Twenty five units each containing:		
Disposal	1,200 VA	
Dishwasher	1,500 VA	
Water heater	+ 4,500 VA	
Total connected load	7,200 VA	× 25 units = 180,000 VA
Demand factor at 75%		× .75
Total appliance demand load		135,000 VA

Step 4. Dryer Demand Load [220–18]

Twenty five 4.5-kW dryers:
Total connected load = 5 kVA* × 25 units = 125 kW
Demand factor for 25 units = 32.5%
Total demand dryer load = 125 kW × .325 = 40.625 kW
* The minimum dryer load for standard load calculations is 5 kW

Step 5. Cooking Equipment Demand Load [220–19]

Twenty five 13.5-kW ranges:
Step 1. Column "A" demand load for 25 units = 40 kW.
Step 2. The average range (13.5 kW) exceeds 12 kW by 3.5 kW, increase Column "A" demand load (40 kW) by 20%.
Range demand load = 40 kW × 1.2 = 48 kW or 40 kW + 8 kW = 48 kW

Step 6. Service Conductor Size [Note 3 Of Table 310–16]

Step 1. Total general lighting, small appliance, laundry demand load	60,700 VA
Step 2. Total heat demand load [220–15] 7,500 VA × 25	187,500 VA
Step 3. Total appliance demand load	135,000 VA
Step 4. Total demand dryer load	40,625 W
Step 5. Range demand load	+ 48,000 W
Total demand load	471,825 VA

Service conductor amperes = VA/(E × 1.732)
I = 471,825 VA/(208 volts × 1.732), I = 1,310 amperes

10–4 *MULTIFAMILY DWELLING-UNIT CALCULATIONS [220–32] – OPTIONAL METHOD*

Instead of sizing the ungrounded conductors according to the standard method (Part B of Article 220), the optional method can be used for feeders and service conductors in multifamily dwelling-units. Follow the following steps for determining the demand load.

Step 1: → **Determine the total connected load.** Add the following loads:

(a) General Lighting. 3 volt-amperes per square-foot.

(b) Small appliance and laundry branch circuit. 1,500 volt-amperes for each 20 ampere small appliance and laundry branch circuit.

(c) Appliances. The *nameplate* VA rating of all appliances and motors fastened in place (permanently connected), but not the A/C or heating load. Be sure to use the range and dryer nameplate rating!

(d) A/C versus Heat. Determine the largest of the A/C versus heat.

(1) Air conditioning or heat-pump compressors at 100 percent.
(2) Heat at 100 percent.

Step 2. → **Determine the demand load.** The net computed demand load is determined by applying the demand factor from Table 220–32 to the total connected load (Step 1). The net computed demand load (kVA) can be converted to amperes by:

Single Phase $I = \frac{VA}{E}$ **Three Phase** $I = \frac{VA}{(E \times 1.732)}$

Step 3. → **Feeder and Service conductor**

Four hundred amperes and less. The ungrounded conductors are sized according to Note 3 of Table 310–16 for 120/240-volt, single-phase systems up to 400 amperes.

Over four hundred amperes. The ungrounded conductors are sized according to Table 310–16 based on the calculated demand load.

When do you use the standard method verses the optional method? For the purpose of exam preparation, always use the standard load calculations unless the question specifies the optional method. In the field, you will probably want to use the optional method because it provides for a smaller service.

10–5 *MULTIFAMILY DWELLING-UNIT EXAMPLE QUESTIONS [220–32] – OPTIONAL METHOD*

A multifamily building has twelve units on each phase, each is 1,500 square feet and contains the following:

25 extra receptacles	Dryer at 4.5 kVA
Washing machine at 1.2 kVA	Range at 14.4 kW
Dishwasher at 1.5 kVA	Water heater at 4 kW
Heat at 5 kW	A/C 3-horsepower with 1/8-horsepower compressor fan

➛ **General Lighting Demand Question**

Using the optional calculations, what is the net computed load for the building general lighting and general-use receptacles, small appliance, and laundry circuits? The system voltage is 120/208-volts.

(a) 45 kVA (b) 90 kVA (c) 108 kVA (d) 60 kVA

- Answer: (a) 45 kVA [Table 220–32]

(a) General Lighting, 1,500 square feet × 3 VA =	4,500 VA
(b) Two Small Appliance Circuits	3,000 VA
(c) Laundry Circuit	+ 1,500 VA
	9,000 VA

Total Demand Load for 12 units × 9,000 VA = 108,000 VA × .41 = 44,280 VA

Note: The washing machine is calculated as part of the 1,500 VA laundry circuit.

➛ **Air Conditioning Versus Heat Question**

Using the optional calculations, what is the demand load for the building air conditioning versus heat?

(a) 52 kVA (b) 60 kVA (c) 30 kVA (d) 25 kVA

- Answer: (d) 25 kVA [220–32]

A/C = 3-horsepower, 208 volts × 18.7 amperes =	3,889 VA
⅛-horsepower, 208 volts × 1.6 amperes =	+ 333 VA
	4,222 VA

4,222 VA × 12 units × .41 = 20,772 VA *

Heat = 5,000 watts × 12 units = 60,000 VA × .41 = 24,600 watts.

* Omit the air conditioner load because it is smaller than the heat [220–21].

➛ Appliance Demand Question

Using the optional calculations, what is the twelve-unit building net demand load for the water heaters and dishwashers?

(a) 41 kVA (b) 53 kVA
(c) 33 kVA (d) 27 kVA

- Answer: (d) 27 kVA [220–32]

Water heater	4,000 VA
Dishwasher	+ 1,500 VA
	5,500 VA

Total Appliance Demand Load for the 12 units,
5,500 VA × 12 units = 66,000 VA × .41 = 27,060 VA

Note: The washing machine is calculated as part of the laundry circuit.

Dryer Demand Load Question

Using the optional calculations, what is the twelve-unit building net computed load for the 4.5-kVA dryer?

(a) 60 kVA (b) 25 kVA
(c) 55 kVA (d) 22 kVA

- Answer: (d) 22 kVA

4.5 kW* × 12 units × .41 = 22.14 kVA

* Be sure to use the nameplate rating, not 5,000 VA.

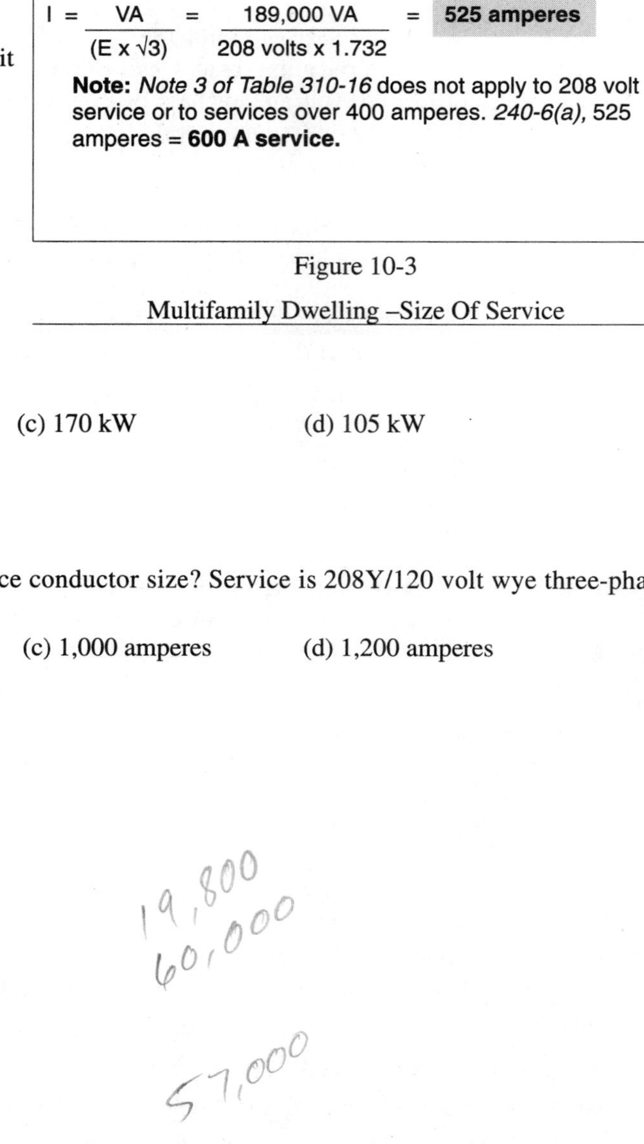

Figure 10-3

Multifamily Dwelling –Size Of Service

Range Demand Load Question

Using optional calculations, what is the twelve-unit building demand load for the 14.4-kW ranges?

(a) 70 kW (b) 40 kW (c) 170 kW (d) 105 kW

- Answer: (a) 70 kW

14.4 kW × 12 units × .41 = 70.85 kW

Service Conductor Size Question

If the total demand load is 189 kVA, what is the service conductor size? Service is 208Y/120 volt wye three-phase, Fig. 10–3.

(a) 600 amperes (b) 800 amperes (c) 1,000 amperes (d) 1,200 amperes

- Answer: (a) 600 amperes

$$I = \frac{VA}{(E \times 1.732)}$$

$$I = \frac{189{,}00\ VA}{(208\ volts \times 1.732)} = 525\ amperes$$

19,800
60,000
19950
57,000

Unit 10 – Multifamily Dwelling-Unit Load Calculations Summary Questions

10–1 Multifamily Dwelling Calculations – Standard Method

1. • An apartment building contains twenty dwelling-units, 840 square feet each. What is the general lighting feeder demand load for the building? *Note:* Laundry facilities are provided on the premises for all tenants.
(a) 5,200 VA (b) 40,590 VA (c) 110,400 VA (d) None of these

2. An apartment building contains twenty dwelling-units, 990 square feet each. What is the general lighting net computed demand load for the building?
(a) 74,700 VA (b) 149,400 VA (c) 51,300 VA (d) 105,600 VA

3. A forty unit multifamily building has a 3-horsepower, 230-volt A/C and two 3-kW baseboard heaters in each unit. What is the A/C versus heat net computed demand load?
(a) 160 kW (b) 240 kW (c) 60 kW (d) 50 kW

4. A twenty-five unit multifamily building has a 3-horsepower, 230-volt A/C and a 5-kW heat strip. What is the A/C versus heat net computed demand load?
(a) 160 kW (b) 125 kW (c) 6 kW (d) 5 kW

5. A sixteen unit multifamily building contains a 940 VA disposal, a 1,250 VA dishwasher, and a 4,500 VA water heater. What is the service demand load for these appliances?
(a) 100 kVA (b) 134 kVA (c) 80 kVA (d) 5 kVA

6. An apartment building contains twenty-eight dwelling-units, each unit contains a 900 VA disposal, 1,200 VA dishwasher and a 5,000 VA water heater. What is the feeder and service demand load for the appliances?
(a) 149 kVA (b) 142 kVA (c) 107 kVA (d) 186 kVA

7. A multifamily dwelling (forty unit building) contains a 4.5-kW electric clothes dryer in each unit. What is the feeder and service demand load for all the dryers?
(a) 50 kW (b) 27 kW (c) 60 kW (d) none of these

8. What is the demand load for ten 5.25-kW dryers installed in dwelling-units of a multifamily building?
(a) 26 kW (b) 37 kW (c) 60 kW (d) 37 kW

9. What is the demand load for twelve 3.25-kW ovens?
(a) 10 kW (b) 18 kW (c) 15 kW (d) 20 kW

10. What is the demand load for eight 7-kW cooktops?
(a) 20 kW (b) 17 kW (c) 14.7 kW (d) 48 kW

11. • What is the demand load for five 12.4-kW ranges?
(a) 9 kW (b) 45 kW (c) 20 kW (d) None of these

12. What is the demand load for three ranges rated 15.5-kW?
(a) 15 kW (b) 14 kW (c) 17 kW (d) 21 kW

13. What is the feeder and service demand load for 3 ranges rated 11-kW and 3 ranges rated 14-kW?
(a) 36 kW (b) 42 kW (c) 78 kW (d) 22 kW

14. • What size aluminum service conductors are required for a 115/230-volt, single-phase multifamily building that has a total demand load of 90-kW?
(a) 500 THHN kcmil (b) 600 THHN kcmil (c) 700 THHN kcmil (d) 800 THHN kcmil

15. What size service conductors are required for a multifamily building that has a total demand load of 260-kW for a 120/208-volt three phase system?
(a) 2 – 300 THHN kcmil (b) 2 – 350 THHN kcmil
(c) 2 – 500 THHN kcmil (d) 2 – 600 THHN kcmil

Use this information for The Next Five Questions.
A multifamily building has 12 units. Each is 1500 square feet and contains the following:
System voltage 120/240-volt, single-phase.

25 extra receptacles	Dryer at 4.5-kW
Washing machine at 1.2 kVA	Range at 14.45-kW
Dishwasher at 1.5 kVA	Water heater at 4-kW
Heat at 5-kW	A/C 3 HP with ⅛-horsepower compressor fan

16. • What is the building's net computed load for the general lighting, small appliance and laundry in VA?
(a) 40 kVA (b) 108 kVA (c) 105 kVA (d) 90 kVA

17. What is the building's demand load for 3-horsepower air conditioning with a ⅛-horsepower compressor fan versus 5-kW heat in VA?
(a) 52 kVA (b) 60 kVA (c) 30 kVA (d) 105 kVA

18. What is the building demand load for the appliances in VA?
(a) 40 kVA (b) 55 kVA (c) 43 kVA (d) 50 kVA

19. What is the building demand load for the 4.5-kW dryer?
(a) 60 kVA (b) 27 kVA (c) 55 kVA (d) 101 kVA

20. What is the building demand load for the 14.45-kW range?
(a) 27 kW (b) 35 kW (c) 168 kW (d) 30 kW

21. If the total demand load of a multifamily dwelling-unit is 206 kVA, what is the service and feeder conductor size? The system voltage is 120/240, single-phase.
(a) 600 amperes (b) 800 amperes (c) 1,000 amperes (d) 1,200 amperes

22. • If a service contains three sets of parallel 400 kcmil conductors, what size bonding jumper is required for each service raceway?
(a) No. 1 (b) No. 1/0 (c) No. 2/0 (d) No. 3/0

23. • If 400 kcmil service conductors are in parallel in three raceways, what is the minimum size grounding electrode conductor required?
(a) No. 1/0 (b) No. 2/0 (c) No. 3/0 (d) No. 4/0

10–3 Multifamily Dwelling Calculations [220–32] – Optional Method

24. • When determining the service (using optional calculations) for a multifamily dwelling, the total connected load shall have the demand factors of Table 220–32 applied. When determining the total connected load, the largest of the _________ shall be used.
(a) 100 percent of the air conditioning
(b) 125 percent of the air conditioning
(c) 100 percent of the heat
(d) a or c

25. A building has sixty units. Each unit is 1,500 square feet. Using the optional dwelling-unit calculations, what is the net computed load for the building general lighting and general use receptacles, small appliance and laundry circuits?
(a) 145 kVA (b) 190 kVA (c) 108 kVA (d) 130 kVA

26. A 60 unit multifamily dwelling has a 3-horsepower air conditioner with a ⅛-horsepower blower and 5-kW heat. Using the optional method, what is the A/C versus heat demand load?
(a) 50 kVA (b) 61 kVA (c) 30 kVA (d) 72 kVA

27. Using the optional method for dwelling-unit calculations, what is the sixty unit multifamily building net demand load for a 4-kW water heater and a 1.5-kVA dishwasher?
(a) 80 kVA (b) 50 kVA (c) 30 kVA (d) 60 kVA

28. Each unit of a sixty unit apartment building has a 4-kW dryer. Using the optional method, the demand load that would be added to the service is ______ kW.
(a) 75 (b) 240 (c) 72 (d) 58

29. Using the optional method for dwelling-unit calculations, what is the sixty unit multifamily building net computed load for a 4.5-kW dryer?
(a) 65 kW (b) 25 kW (c) 55 kW (d) 75 kW

30. Using the optional method for dwelling-unit calculations, what is a sixty unit multifamily building demand load for the 14-kW range?
(a) 150 kW (b) 50 kW (c) 100 kW (d) 200 kW

31. If the total demand load is 270 kVA, what is the service and feeder conductor size? The service is 208/120-volt wye three phase.
(a) 600 amperes (b) 800 amperes (c) 1,000 amperes (d) 1,200 amperes

32. Each unit of a twenty unit multifamily dwelling has 900 square feet of living space. The air conditioning is 5-horsepower, heat 5-kW, water heater 5-kW, and range 14-kW. The service for this apartment building is approximately ______ kVA if the optional method calculation is used.
(a) 200 (b) 250 (c) 280 (d) 320

33. • If the service conductors are in parallel in two raceways (500 THHN kcmil), what size bonding jumper is required for each service raceway?
(a) No. 2 (b) No. 1 (c) No. 1/0 (d) No. 2/0

34. • If the service conductors are in parallel in two raceways (500 THHN kcmil), what is the minimum size grounding electrode conductor required?
(a) No. 1/0 (b) No. 2/0 (c) No. 3/0 (d) No. 4/0

Challenge Questions

General Lighting and Receptacle Calculations [Table 220– 11]

35. Each dwelling-unit of a twenty unit multifamily building has 900 square feet of living space. What is the general lighting load for the multifamily dwelling-unit apartment building?
(a) 45 kVA (b) 60 kVA (c) 37 kVA (d) 54 kVA

36. An apartment building contains twenty dwelling-units and each dwelling-unit has 840 square feet of living space. What is the general lighting feeder demand load for the multifamily building, if laundry facilities are provided on the premises for all tenants and no laundry circuit is installed in each unit?
(a) 35 kVA (b) 41 kVA (c) 45 kVA (d) 63 kVA

Appliance Demand Factors [Section 220–17].

37. An apartment building contains twenty dwelling-units. Each unit contains a 900 VA disposal, a 1,200 VA dishwasher, and a 5,000 VA water heater. What is the feeder demand load for the appliances in this building?
(a) 106,500 VA (b) 117,100 VA (c) 137,000 VA (d) 60,000

Dryer Calculation [Section 220–18]

38. The nameplate rating for each household dryer in a ten unit apartment building is 4-kW. This would add ______ kW to the service size.
(a) 20 (b) 25 (c) 40 (d) 50

Ranges – Note 1 of Table 220–19

39. • The demand load for thirty 15.8-kW household ranges is ______ kW.
(a) 31 (b) 47 (c) 54 (d) 33

Ranges – Note 2 of Table 220–19

40. What is the feeder demand load for five 10-kW, five 14-kW, and five 16-kW household ranges?
(a) 210 kW (b) 30 kW (c) 14 kW (d) 33 kW

41. What is the kW to be added to service loads for ten 12-kW, eight 14-kW, and two 9-kW household ranges?
(a) 33 kW (b) 35 kW (c) 36.75 kW (d) 29.35 kW

Ranges – Note 3 to Table 220–19

42. What is the minimum demand load for five 5-kW cooktops, two 4-kW ovens, and four 7-kW ranges?
(a) 15.5 kW (b) 8.8 kW (c) 19.5 kW (d) 18.2 kW

43. • What is the maximum dwelling-unit feeder or service demand load for fifteen 8-kW cooking units?
(a) 38.4 kW (b) 30 kW
(c) 120 kW (d) none of these

Ranges – Note 5 of Table 220–19

44. A school has twenty 10-kW ranges installed in the home economics class. The minimum load this would add to the service is ______ kW.
(a) 35 (b) 44.8 (c) 56 (d) 160

10–12 Neutral Calculation [220–22]

45. The service neutral demand load for household electric clothes dryers shall be calculated at ______ percent of the demand load as determined by Section 220–18.
(a) 50 (b) 60 (c) 70 (d) 80

46. The feeder neutral demand for fifteen 8-kW cooking units would be ______ kW.
(a) 38.4 (b) 30 (c) 120 (d) 21

47. What is the feeder neutral load (in amperes) for fifteen 6-kW household dryers?
(a) 105 amperes (b) 125 amperes
(c) 150 amperes (d) none of these

48. A twelve unit multifamily dwelling contains a 4 kVA electric clothes dryer in each unit. What is the feeder or service neutral demand load (in amperes)?
(a) 115 amperes (b) 225 amperes
(c) 80 amperes (d) 55 amperes

49. • The feeder and service neutral demand load can be reduced 70 percent for that portion of the unbalanced load over 200 amperes. This applies to ________ systems.
(a) 3-wire single phase 120/240-volt
(b) 3-wire single phase 120/208-volt
(c) 4-wire three phase 120/208-volt
(d) a and c

50. • What is the dwelling-unit service neutral load (in amperes) for ten 9-kW ranges?
(a) 40 amperes
(b) 55 amperes
(c) 75 amperes
(d) 105 amperes

Unit 11

Commercial Load Calculations

OBJECTIVES

After reading this unit, the student should be able to briefly explain the following concepts:

Part A – General
Conductor overcurrent protection [240–3]
Conductor ampacity
Fraction of an ampere
General requirements

Part B – Loads
Air conditioning
Dryers
Electric heat
Kitchen equipment
Laundry circuit
Lighting demand factors
Lighting without demand factors
Lighting miscellaneous
Multioutlet assembly [220–3(c) Exception No. 1]
Receptacles [220–3(b)(7)]
Banks and offices general lighting and receptacles
Signs [600–6(c)]

Part C – Load Calculations
Marina [555–5]
Mobile home park [550–22]
Recreational vehicle park [551–73]
Restaurant – Optional method Section 220–36
School – Optional method Section 220–34

After reading this unit, the student should be able to briefly explain the following terms:

General lighting
General lighting demand factors
Nonlinear loads
VA rating
Ampacity
Continuous load
Next size up protection device
Overcurrent protection
Rounding
Standard ampere ratings for overcurrent protection devices
Voltages

PART A – *GENERAL*

11–1 *GENERAL REQUIREMENTS*

Article 220 provides the requirement for branch circuits, feeders and services. In addition to this Article, other Articles are applicable such as Branch Circuits–210, Feeders–215, Services–230, Overcurrent Protection–240, Wiring Methods–300, Conductors–310, Appliances–422, Electric Space-Heating Equipment–424, Motors–430, and Air-Conditioning–440.

11–2 *CONDUCTOR AMPACITY [ARTICLE 100]*

The ampacity of a conductor is the rating, in amperes, that a conductor can carry continuously without exceeding its insulation temperature rating [NEC® Definition – Article 100]. The allowable ampacities as listed in Table 310–16 are affected by ambient temperature, conductor insulation, and conductor bunching or bundling [310–10], Fig. 11–1.

Continuous Loads

Conductors are sized at 125 percent of the continuous load before any derating factor, and the overcurrent protection devices are sized at 125 percent of the continuous loads [210–22(c), 220–3(a), 220–10(b), and 384–16(c)].

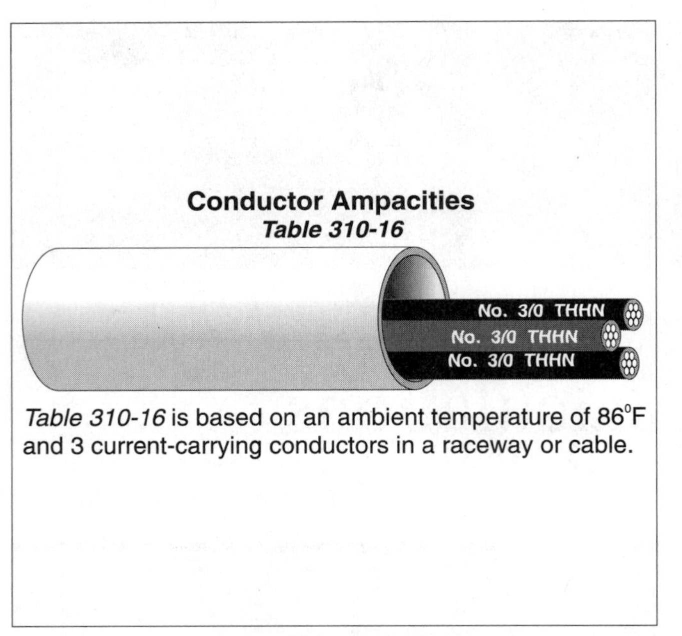

Figure 11-1

Conductor Ampacities – Table 310–16

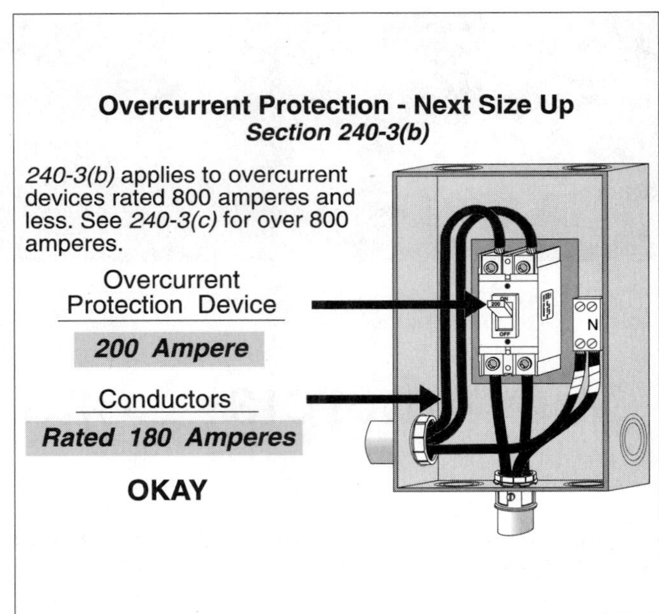

Figure 11-2

Overcurrent Protection – Next Size Up

11–3 *CONDUCTOR OVERCURRENT PROTECTION [240–3]*

The purpose of overcurrent protection devices is to protect conductors and equipment against excessive or dangerous temperatures [240–1 FPN]. There are many rules in the National Electrical Code for conductor protection and there are many different installation applications where the general rule of protecting the conductor at its ampacity does not apply. Examples would be motor circuits, air-conditioning, tap conductors, etc. See Section 240–3 for specific rules on conductor overcurrent protection.

Next Size Up Okay [240–3(b)]

If the ampacity of a conductor does not correspond with the standard ampere rating of a fuse or circuit breaker, as listed in Section 240–6(a), the next size up protection device is permitted. This practice only applies if the conductors do not supply multioutlet receptacles and if the next size up overcurrent protection device does not exceed 800 amperes [Note 9 of Table 310–16], Fig. 11–2.

Standard Size Overcurrent Devices [240–6(a)]

The following is a list of some of the standard ampere ratings for fuses and inverse time circuit breakers: 15, 20, 25, 30, 35, 40, 45, 50, 60, 70, 80, 90, 100, 110, 125, 150, 175, 200, 225, 250, 300, 350, 400, 500, 600, 800, 1,000, 1,200, and 1,600 amperes.

11–4 *VOLTAGES [220–2]*

Unless other voltages are specified, branch-circuit, feeder, and service loads shall be computed at a nominal system voltage of 120, 120/240, 208Y/120, 240, 480Y/277, 480 volts, or 600Y/347, Fig. 11–3.

11–5 *FRACTION OF AN AMPERE*

There is no specific NEC® rule for rounding fractions of an ampere but Chapter 9, Part B, Examples, contains the following note, "except where the computations result in a major fraction of an ampere (0.5 or larger), such fractions may be dropped." So for all practical purposes, we should follow this practice.

PART B – *LOADS*

11–6 *AIR CONDITIONING*

Branch Circuit [440–22(a) And 440–32]

Branch circuit conductors that supply air-conditioning equipment must be sized no less than 125 percent of the air conditioner rating [440–32], and the protection is sized between 175 percent and up to 225 percent of the air- conditioning rating.

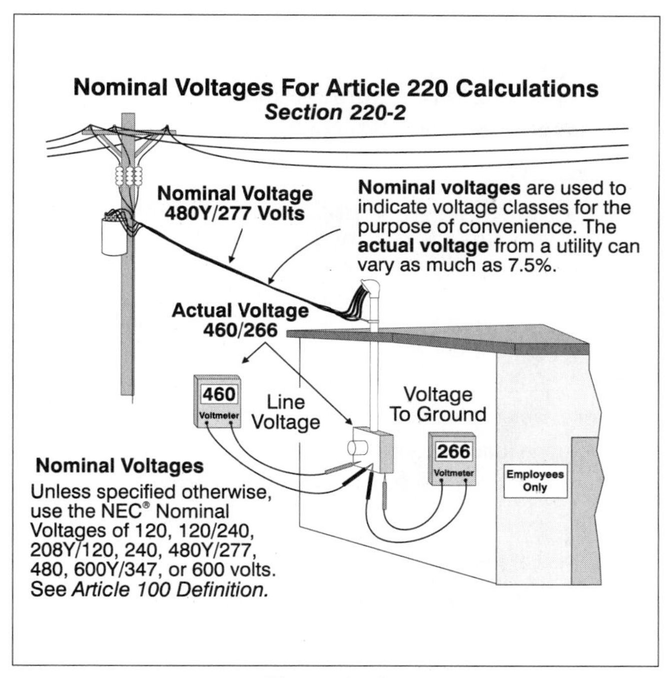

Figure 11-3

Nominal Voltage

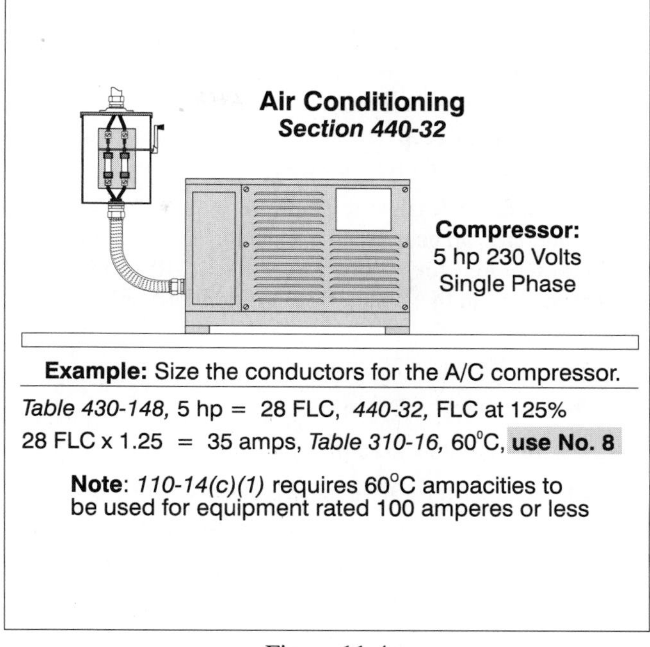

Figure 11-4

Air Conditioning – Branch Circuit Sizing Example

❑ Air Conditioning Branch Circuit Example

What size branch circuit is required for a 5-horsepower, 230 volt air conditioner that has a current rating of 28 amperes, Fig. 11–4?

(a) No. 10 (b) No. 8 (c) No. 14 (d) none of these

- Answer: (b) No. 8

Branch Circuit Conductor [440–32]: The ampere rating of a single-phase 5-horsepower motor is 28 amperes. Always assume single-phase unless the question specifies three-phase. The branch circuit conductors are sized no less than 125 percent of 28 amperes, 28 amperes × 1.25 = 35 amperes, which requires a No. 8 at 60ºC [Table 310–16 and 110–14(c)].

Branch Circuit Protection [440–22]: The branch circuit protection is sized from 175 percent up to 225 percent of the ampere rating, 28 amperes × 1.75 = 49 amperes up to 28 amperes × 2.25 = 63 amperes.

Feeder or Service Conductors [440–34]

Feeder circuit conductors that supply air-conditioning equipment must be sized no less than 125 percent of the largest air conditioner VA rating plus 100 percent of the other air conditioners' VA ratings.

VA Rating – The VA rating of an air conditioner is determined by multiplying the voltage rating of the unit by its ampere rating. For the purpose of this book, we will determine the unit ampere rating by using motor full-load current ratings as listed in Article 430, Table 430–148, and Table 430–150 (three-phase).

❑ Air Conditioning Feeder/Service Conductor Example

What is the demand load required for the air-conditioning equipment of a twelve-unit office building where each unit contains one air conditioner rated 5-horsepower, 230 volts?

(a) 77 kVA (b) 79 kVA (c) 45 kVA (d) none of these

- Answer: (b) 79 kVA

The VA of a single-phase 5-horsepower motor rated 230 volts is calculated as volts times amperes, 230 volts × 28 amperes = 6,440 VA . The feeder demand load for air-conditioning is calculated at no less than 125 percent of the largest air conditioner plus 100 percent of all other units. [(6,440 VA × 1.25) + (6,440 VA × 11 units)] = 78,890 VA/1,000 = 79 kVA.

Note: Section 220–21 of the NEC® permits the smaller of the air-conditioning or heat load to be omitted for calculation purposes.

11–7 *DRYERS*

The branch circuit conductor and overcurrent protection device for commercial dryers is sized to the appliance nameplate rating. The feeder demand load for dryers is calculated at 100 percent of the appliance rating. Section 220–18 demand factors do not apply to commercial dryers.

❑ **Dryer Branch Circuit Example**

What size branch circuit conductor and overcurrent protection is required for a 7-kW dryer rated 240 volts when the dryer is located in the laundry room of a multifamily dwelling, Figure 11–5?

(a) No. 12 with a 20-ampere breaker

(b) No. 10 with a 20-ampere breaker

(c) No. 12 with a 30-ampere breaker

(d) No. 10 with a 30-ampere breaker

- Answer: (d) No. 10 with a 30-ampere breaker.
  The ampere rating of the dryer is: I = VA/E
  I = 7,000 VA/240 volts = 29 amperes.

The ampacity of the conductor and overcurrent device must not be less than 29 amperes [240–3]

Table 310–16, No. 10 conductor at 60ºC is rated 30-amperes protected with a 30-ampere protection device [240–6(a)].

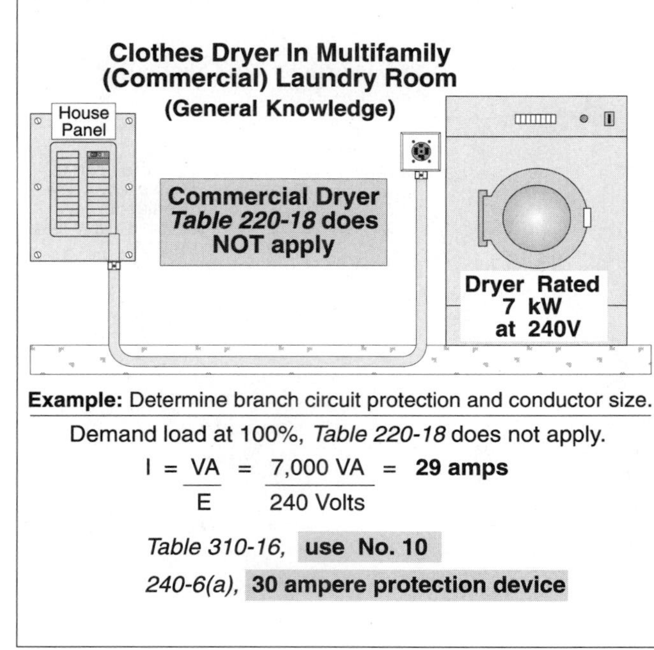

Figure 11-5

Clothes Dryer – Branch Circuit Sizing Example

❑ **Dryer Feeder Example**

What is the service demand load for ten 7-kW dryers located in a laundry room?

(a) 70 kVA (b) 52.5 kVA (c) 35 kVA (d) none of these

- Answer: (a) 70 kVA
  The NEC® does not permit a demand factor for commercial dryers, therefore the dryer demand load must be calculated at 100 percent; 7 kW × 10 units = 70 kVA.

Note: If the dryers are on continuously, the conductor and protection device must be sized at 125 percent of the load [210–22(c), 220–3(a), 220–10(b), and 384–16(c)].

11–8 *ELECTRIC HEAT*

Branch Circuit Sizing [424–3(b)]

Branch circuit conductors and the overcurrent protection device for electric heating shall be sized not be less than 125 percent of the total heating load, including blower motors.

❑ **Heating Branch Circuit Example**

What size conductor and protection is required for a three-phase 240-volt, 15-kW heat strip with a 5.4-ampere blower motor, Fig. 11–6?

(a) No. 10 with 30-ampere protection (b) No. 6 with 50-ampere protection

(c) No. 8 with 40-ampere protection (d) No. 6 with 60-ampere protection

- Answer: (d) No. 6 with 60-ampere protection [424–3(b)]
  $I = VA/(E \times \sqrt{3})$, I = 15,000 VA/(240 volts × 1.732) = 36 amperes. The conductors and protection must not be less than 125 percent of [36 amperes (heat) + 5.4 amperes (motor)] = 41.4 amperes × 1.25 = 51.75 amperes. The conductors are sized to the 60ºC column ampacities of Table 310–16 [110–14(c)(1)], which is a No. 6 rated 55 amperes. The overcurrent protection device must be sized no less than 52 amperes, which is a 60-ampere device [240–6(a)].

Heating Feeder/Service Demand Load [220–15]

The feeder and service demand load for electric heating equipment is calculated at 100 percent of the total heating load.

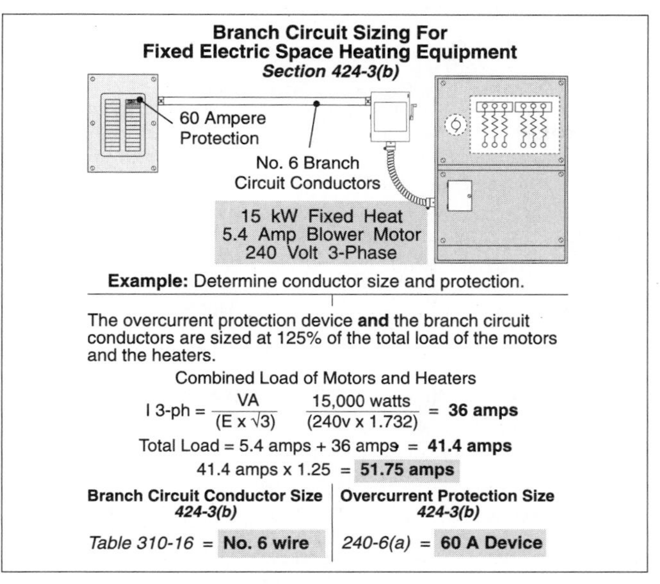

Figure 11-6

Fixed Electric Heat – Branch Circuit Sizing Example

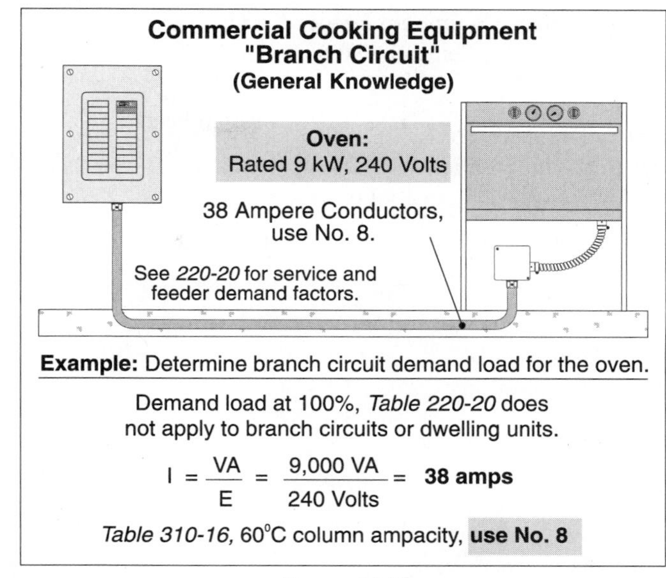

Figure 11-7

Commercial Cooking Equipment – Branch Circuit Sizing Example

❑ **Heating Feeder/ Service Demand Load Example**

What is the feeder and service demand load for a building that has seven three-phase 208 volt, 10-kW heat strips with a 5.4-ampere blower motor (1,945 VA) for each unit?

(a) 84 kVA (b) 53 kVA (c) 129 kVA (d) 154 kVA

- Answer: (a) 84 kVA

  10,000 VA + 1,945 VA = 11,945 VA × 7 units = 83,615 VA/1,000 = 83.615 kVA.

Note: Section 220–21 of the NEC® permits the smaller of the air-conditioning or heat load to be omitted for feeder and service calculations.

11–9 *KITCHEN EQUIPMENT*

Branch Circuit

Branch circuit conductors and overcurrent protection for commercial kitchen equipment are sized according to the appliance nameplate rating.

❑ **Kitchen Equipment Branch Circuit Example No. 1**

What is the branch circuit demand load (in amperes) for one 9-kW oven rated 240 volts, Fig. 11–7?

(a) 38 amperes (b) 27 amperes (c) 32 amperes (d) 33 amperes

- Answer: (a) 38 amperes

  I = VA/E

  I = 9,000 VA/240 volts = 38 amperes

❑ **Kitchen Equipment Branch Circuit Example No. 2**

What is the branch circuit load for one 14.47-kW range rated 208 volts, three-phase?

(a) 60 amperes (b) 40 amperes (c) 50 amperes (d) 30 amperes

- Answer: (b) 40 amperes

  I = VA/(E × 1.732)

  I = 14,470 VA/(208 volts × 1.732) = 40 amperes

Kitchen Equipment Feeder/Service Demand Load [220–20]

The service demand load for thermostatic control or intermittent use commercial kitchen equipment is determined by applying the demand factors from Table 220–20, to the total connected kitchen equipment load. *The feeder or service demand load cannot be less than the two largest appliances!*

Note: The demand factors of Table 220–20 do not apply to space heating, ventilating, or air–conditioning equipment.

❑ Kitchen Equipment Feeder/Service Example No. 1

What is the demand load for the following kitchen equipment loads, Fig. 11–8?

Water heater	5.0 kW	Booster heater	7.5 kW
Mixer	3.0 kW	Oven	5.0 kW
Dishwasher	1.5 kW	Disposal	1.0 kW

(a) 15 kW (b) 23 kW
(c) 12.5 kW (d) none of these

- Answer: (a) 15 kW

Water heater	5 kW
Booster heater	7.5 kW
Mixer	3 kW
Oven	5 kW
Dishwasher	1.5 kW
Disposal	+ 1 kW
Total connected	23 kW

The demand factor for six loads is 65 percent

23 kW × .65 = 14.95 kW.

The demand load cannot be less than the two largest appliances; 5 kW + 7.5 kW = 12.5 kW.

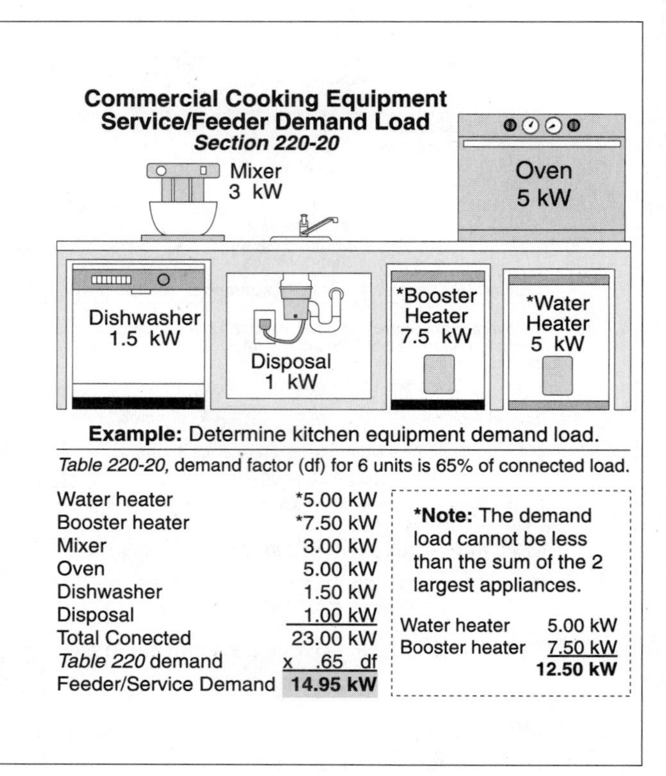

Figure 11-8

Commercial Cooking Equipment – Service Demand Load Example

❑ Kitchen Equipment Feeder/Service Example No. 2

What is the demand load for the following kitchen equipment loads?

Water heater	10 kW
Booster heater	15 kW
Mixer	4 kW
Oven	6 kW
Dishwasher	1.5 kW
Disposal	1 kW

(a) 24.4 kW (b) 38.2 kW (c) 25.0 kW (d) 18.9

- Answer: (c) 25 kW*

Water heater	10 kW
Booster heater	15 kW
Mixer	4 kW
Oven	6 kW
Dishwasher	1.5 kW
Disposal	+ 1 kW
Total connected	37.5 kW

The demand factor for six appliances is 65 percent; 37.5 kW × .65 = 24.4 kW

* The demand load cannot be less than the two largest appliances; 10 kW + 15 kW = 25 kW

11–10 *LAUNDRY EQUIPMENT*

Laundry equipment circuits are sized to the appliance nameplate rating. For exam purposes, it is generally accepted that a laundry circuit is not considered a continuous load. Assume all commercial laundry circuits to be rated 1,500 VA unless noted otherwise in the question.

❑ Laundry Equipment Example

What is the demand load for ten washing machines located in a laundry room, Fig. 11–9?

(a) 1,500 VA (b) 15,000 VA (c) 1,125 VA (d) none of these

- Answer: (b) 15,000 VA

1,500 VA × 10 units = 15,000 VA

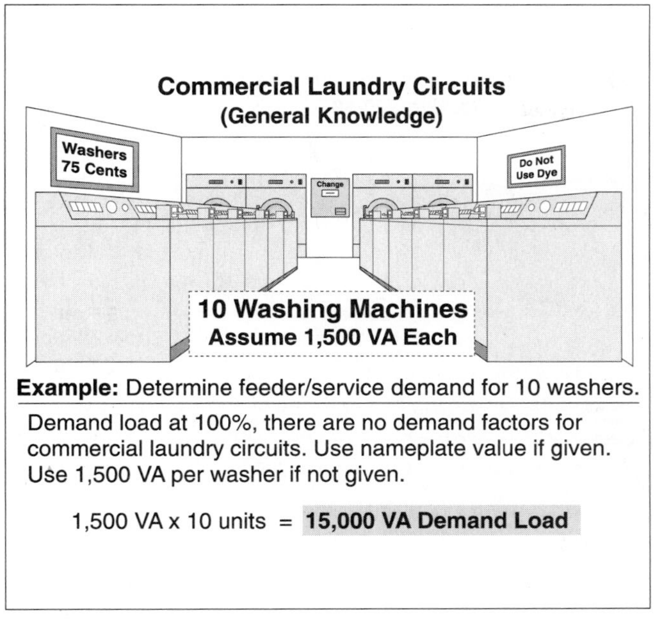

Figure 11-9

Commercial Laundry Service Demand Load Example

General Lighting Demand Load - Hotel/Motel
Tables 220-3(b) and 220-21

Hotel - 40 Units
600 square feet per unit

Example: Determine general lighting demand load.

Table 220-3(b), lighting load is 2 VA per square foot.

600 square feet x 40 units x 2 VA per foot = 48,000 VA

Table 220-11, First 20,000 VA at 50% - 20,000 VA x .5 = 10,000 VA

Next 80,000 VA at 40% 28,000 VA x .4 = 11,200 VA

Demand Load = 21,200 VA

Figure 11-10

General Lighting Demand Factor – Hotel Example

11–11 *LIGHTING – DEMAND FACTORS [TABLES 220–3(b) and 220–11]*

The NEC® requires a minimum service load per square foot for general lighting depending on the type of occupancy [Table 220–3(b)]. For the guest rooms of hotels and motels, hospitals, and storage warehouses, the general lighting demand factors of Table 220–11 can be applied to the general lighting load.

Hotel or Motel Guest Rooms – General Lighting

The general lighting demand load of 2-VA per square foot [Table 220–3(b)] for the guest rooms of hotels and motels is permitted to be reduced according to the demand factors listed in Table 220–11:

General Lighting Demand Factors

First 20,000 VA at 50 percent demand factor

Next 80,000 VA at 40 percent demand factor

Remainder VA at 30 percent demand factor

❑ **Hotel General Lighting Demand Example**

What is the general lighting demand load for a forty room hotel? Each unit contains 600 square feet of living area, Fig. 11–10.

(a) 48 kVA (b) 24 kVA (c) 20 kVA (d) 21 kVA

- Answer: (d) 21 kVA [Table 220–3(b) and Table 220–11]

40 units × 600 square feet =	24,000 square feet × 2 VA =	48,000 VA		
First 20,000 VA at 50 %		– 20,000 VA	× .5 =	10,000 VA
Next 80,000 VA at 40 %		28,000 VA	× .4 =	+ 11,200 VA
Total demand load				21,200 VA

11–12 *LIGHTING WITHOUT DEMAND FACTORS [TABLES 220–3(b) and 220–10(b)].*

The general lighting load for commercial occupancies other than guest rooms of motels and hotels, hospitals, and storage warehouses is assumed continuous and shall be calculated at 125 percent [220–10(b)] of the general lighting load as listed in Table 220–3(b).

❑ **Store General Lighting Example**

What is the general lighting load for a 21,000 square foot store, Fig. 11–11?

(a) 40 kVA (b) 63 kVA (c) 79 kVA (d) 81 kVA

- Answer: (c) 79 kVA

  21,000 square feet × 3 VA × 1.25 = 78,750 VA

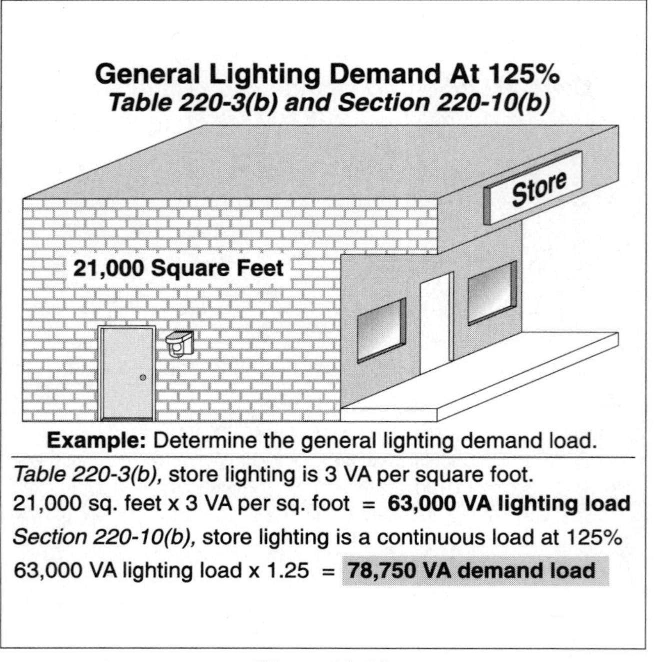

Figure 11-11

General Lighting – No Demand Factor – Store Example

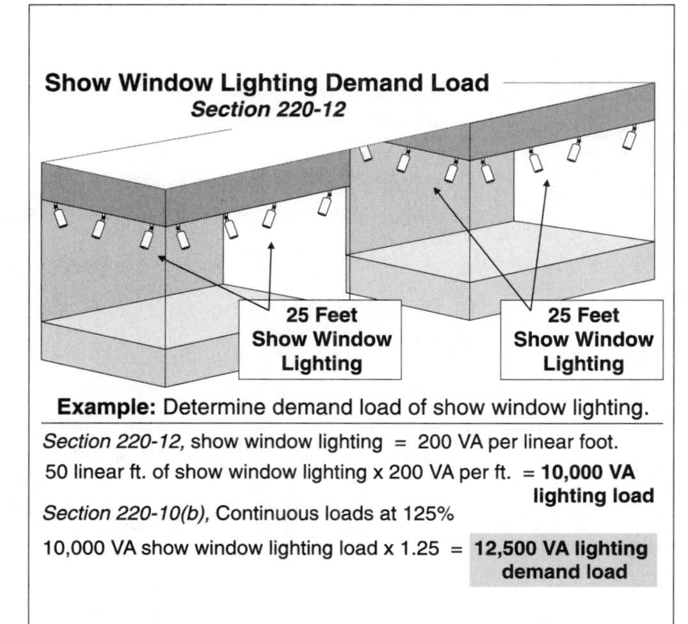

Figure 11-12

Show Window Demand Load Example

❑ **Club General Lighting Example**

What is the general lighting load for a 4,700 square foot dance club?

(a) 4,700 VA (b) 9,400 VA (c) 11,750 VA (d) 250 kVA

- Answer: (c) 11,750 VA

  4,700 square feet × 2 VA × 1.25 = 11,750 VA

❑ **School General Lighting Example**

What is the general lighting load for a 125,000 square foot school?

(a) 125 kVA (b) 375 kVA (c) 475 kVA (d) 550 kVA

- Answer: (c) 475 kVA

  125,000 square feet × 3 VA × 1.25 = 468,750 VA

11–13 *LIGHTING – MISCELLANEOUS*

Show-Window Lighting [220–12]

The demand load for each linear foot of show-window lighting shall be calculated at 200 VA per foot. Show-window lighting is assumed to be a continuous load. See Example 3 in the back of the NEC®, also Section 220–3(c) Exception No. 3 for the requirements for show-window branch circuits.

❑ **Show-Window Load Example**

What is the demand load in kVA for 50 feet of show-window lighting, Fig. 11–12?

(a) 6 kVA (b) 7.5 kVA (c) 9 kVA (d) 12.5 kVA

- Answer: (d) 12.5 kVA

  50 feet × 200 VA per foot = 10,000 VA × 1.25 = 12,500 VA

11–14 *MULTIOUTLET RECEPTACLE ASSEMBLY [220–3(c) EXCEPTION No. 1]*

Each 5 feet, or fraction of a foot, of multioutlet receptacle assembly shall be considered to be 180 VA for service calculations. When a multioutlet receptacle assembly is expected to have a number of appliances used simultaneously, each foot, or fraction of a foot, shall be considered as 180 VA for service calculations. A multioutlet receptacle assembly is not generally considered to be a continuous load.

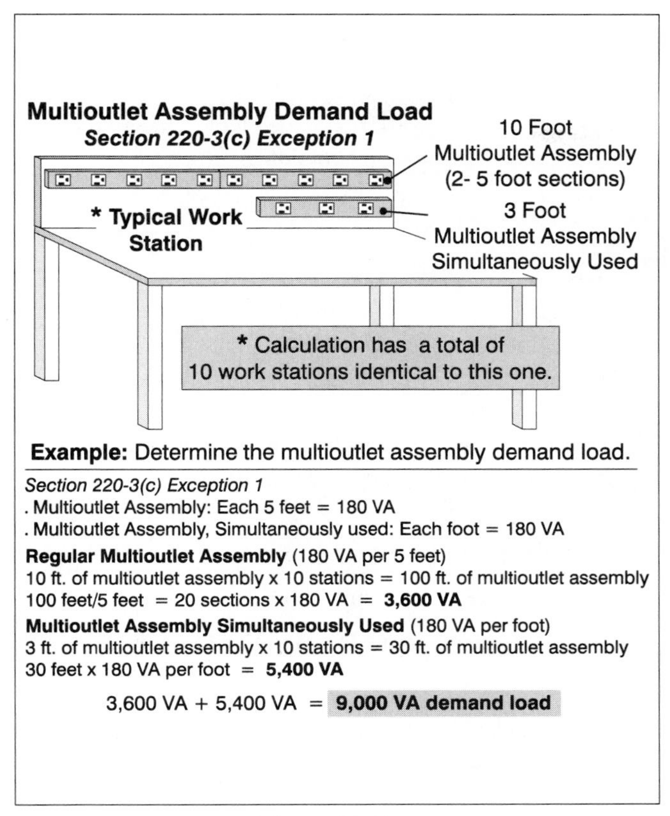

Figure 11-13

Multioutlet Assembly Demand Load Example

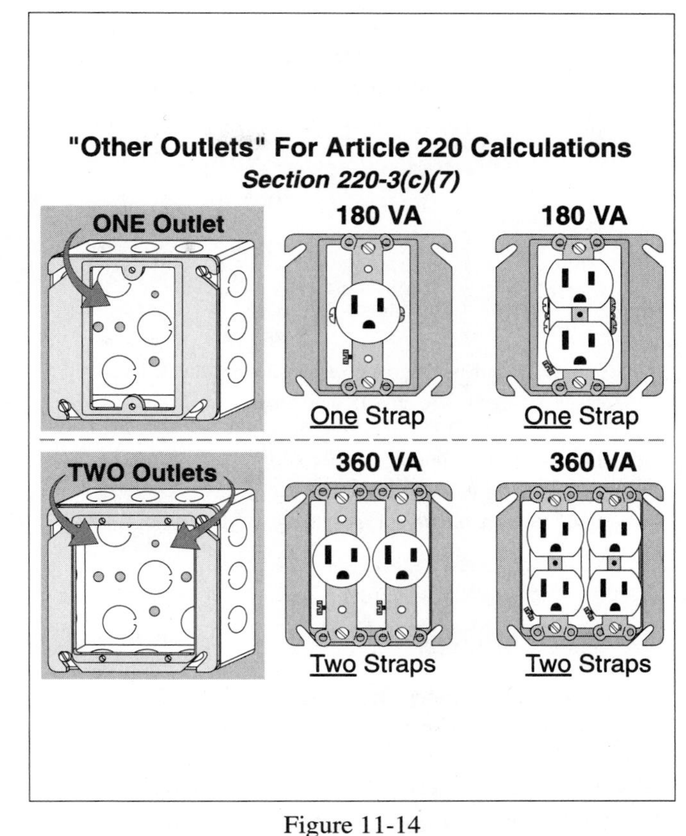

Figure 11-14

Receptacle Outlet – 180 VA

❑ Multioutlet Receptacle Assembly Example

What is the demand load for ten workstations that have 10 feet of multioutlet receptacle assembly and 3 feet of multioutlet receptacle assembly simultaneously used, Fig. 11–13?

(a) 5 kVA (b) 6 kVA (c) 7 kVA (d) 9 kVA

- Answer: (d) 9 kVA

100 feet/ 5 = 20 sections at 180 VA =	3,600 VA
30 feet/ 1 = 30 sections at 180 VA =	5,400 VA
	9,000 VA /1,000 = 9 kVA

11–15 *RECEPTACLES VA LOAD [220–3(c)(7) and 220–13]*

Receptacle VA Load

The minimum load for each commercial or industrial general-use receptacle outlet shall be 180 volt-amperes. Receptacles are generally not considered to be a continuous load, Fig. 11–14.

Number Of Receptacles Permitted On A Circuit

The maximum number of receptacle outlets permitted on a commercial or industrial circuit is dependent on the circuit ampacity. The number of receptacles per circuit is calculated by dividing the VA rating of the circuit by 180 VA for each receptacle strap.

❑ Receptacles Per Circuit Example

How many receptacle outlets are permitted on a 15-ampere, 120-volt circuit, Fig. 11–15?

(a) 10 (b) 13 (c) 15 (d) 20

- Answer: (a) 10 Receptacles [220–3(c)(7)]

  The total circuit VA load for a 15-ampere circuit is 120 volts × 15 amperes = 1,800 VA.

  The number of receptacle outlets per circuit = 1,800 VA/180 VA = 10 receptacles.

Note. Fifteen ampere circuits are permitted for commercial and industrial occupancies according to the National Electrical Code, but some local codes require a minimum 20-ampere rating for commercial and industrial circuits [310–5].

Number Of Receptacles Per Circuit
Section 220-3(c)(7)

15 ampere, 120 Volt Circuit Breaker

360 VA 360 VA 360 VA

180 VA per Receptacle (strap)

360 VA 360 VA

Example: Determine how many receptacles on 15 amp breaker.

Total circuit load in VA, on a 15 amp breaker:
120 volt circuit breaker = Volts x Amps

120 volts x 15 amperes = **1,800 VA load permitted**

220-3(c)(7), each receptacle (strap) = 180 VA

1,800 VA load/180 VA per receptacle = **10 receptacles permitted**

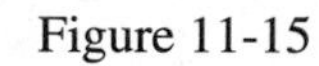

Figure 11-15

Number Of Receptacles Per Circuit Example

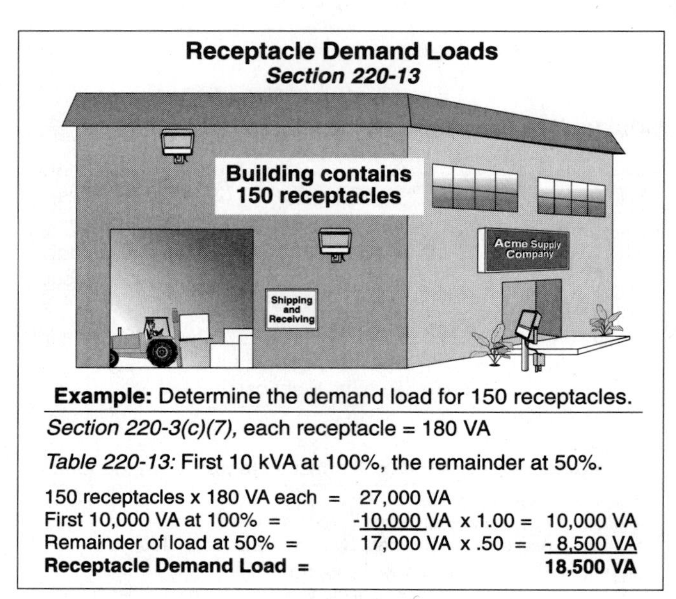

Figure 11-16

Receptacle Service Demand Load Example

Receptacle Service Demand Load [220–13]

The feeder and service demand load for commercial receptacles is calculated at 180 volt-amperes per receptacle strap [220–3(c)(7)]. The demand factors of Table 220–13 can be used for that portion of the receptacle load in excess of 10 kVA. Receptacle loads are generally not considered to be continuous loads.

Table 220–13 Receptacle Demand Factors:

First 10 kVA at 100 percent demand factor

Remainder kVA at 50 percent demand factor

❑ **Receptacle Service Demand Load Example**

What is the service demand load for one hundred and fifty, 20-ampere, 120 volt general-use receptacles in a commercial building, Fig. 11–16?

(a) 27 kVA (b) 14 kVA (c) 4 kVA (d) none of these

- Answer: (d) none of these [220–13]

Total receptacle load	150 receptacles × 180 VA =	27,000 VA	
First 10 kVA at 100%		– 10,000 VA × 1.00 =	10,000 VA
Remainder at 50%		17,000 VA × .50 =	+ 8,500 VA
Total receptacle demand load			18,500 VA/1,000 = 18.5 kVA

11–16 *BANKS AND OFFICES GENERAL LIGHTING AND RECEPTACLES*

Some testing agencies include the receptacle demand load to be part of the general lighting load for banks and offices. If that is the case, the general lighting demand load for banks and offices would be calculated as 3.5-VA per square foot times 125 percent for continuous lighting load, plus the receptacle demand load after applying Table 220–13 demand factors.

Receptacle Demand [Table 220–13]. The receptacle demand load is calculated at 180 volt-amperes for each receptacle outlet [220–3(c)(7)] if the number of receptacles are known, or 1 VA per square foot if the number of receptacles are unknown.

❑ **Bank General Lighting And Receptacle Example**

What is the general lighting demand load (including receptacles) for an 18,000 square foot bank? The number of receptacles are unknown, Fig. 11–17.

(a) 68 kVA (b) 110 kVA (c) 84 kVA (d) 93 kVA

- Answer: (d) 93 kVA

Lighting demand	78,750 VA*
Receptacle demand	+ 14,000 VA*
Total demand load	92,750 VA/1,000 = 92.75 kVA

* See next page for details.

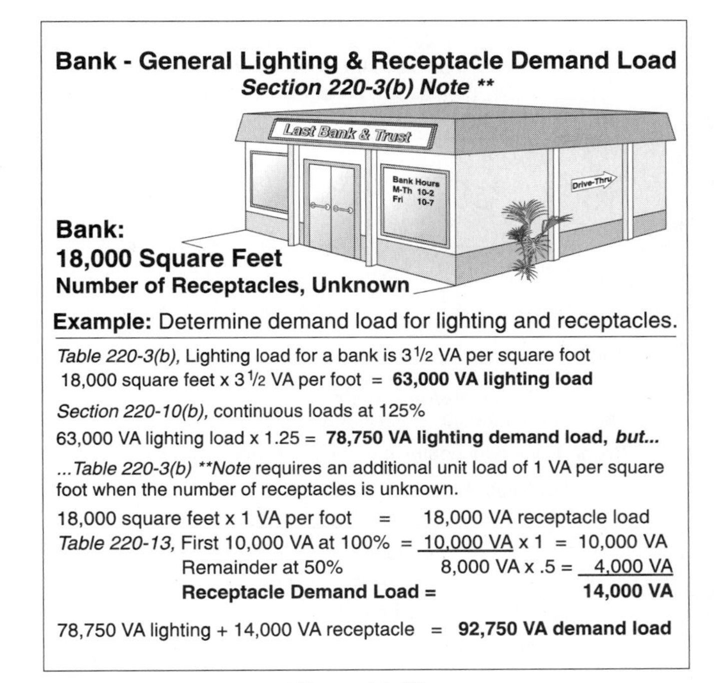

Figure 11-17

General Lighting Demand Load – Bank Example

Signs - Commercial Buildings, Pedestrian Access
Section 220-3(a) and 600-5

Section 600-5(a): Load for exterior outling lighting is 1,200 VA.

Last Bank & Trust

Bank Hours M-Th 10-2 Fri 10-7

Drive-Thru

Commercial Building With Pedestrial Access

Example: Determine service demand load for exterior sign.

600-5(a) requires a 20 ampere branch circuit for an exterior sign.
600-5(b) states that the sign is computed at 1,200 VA.
220-10(b), continuous loads on feeders are calculated at 125%.

1,200 VA sign outlet x 1.25 = **1,500 VA demand load**

Figure 11-18

Sign Demand Load Example

General Lighting Load: 18,000 square feet × 3.5 VA × 1.25 = 78,750 VA

Receptacle Load: 1 VA per square foot and apply Table 220–13 demand factors.

18,000 square feet × 1 VA =	18,000 VA		
First 10 kVA at 100%	–10,000 VA	× 1.00 =	10,000 VA
Remainder at 50%	8,000 VA	× .50 =	+ 4,000 VA
Receptacle demand load			14,000 VA

❑ **Office General Lighting And Receptacle Example**

What is the general lighting demand load for a 28,000 square foot office building that has 160 receptacles?

(a) 111 kVA (b) 110 kVA (c) 128 kVA (d) 142 kVA

- Answer: (d) 142 kVA

General lighting	122,500 VA
Receptacles	+ 19,200 VA
Total load	141,700 VA/1,000 = 142 kVA

General Lighting Load: 28,000 square feet × 3.5 VA × 1.25 = 122,500 VA.

Receptacle Load:

160 receptacles × 180 VA =	28,800 VA		
First 10, 000 VA at 100%	– 10,000 VA	× 1.00 =	10,000 VA
Remainder at 50%	18,800 VA	× .50 =	+ 9,200 VA
Receptacle demand load			19,200 VA

11–17 *SIGNS [220-3(c)(6) AND 600–5(b)]*

The NEC® requires each commercial occupancy that is accessible to pedestrians to be provided with at least one 20-ampere branch circuit for a sign [600–5(b)(1)]. The load for the required exterior signs or outline lighting shall be a minimum of 1,200 VA [220–3(c)(6) and 600-5(b)(3)]. A sign outlet is considered to be a continuous load and the feeder load must be sized at 125 percent ot the continuous load [210–22(c), 220–3(a), 220–10(b), and 384–16(c)].

❑ **Sign Demand Load Example**

What is the demand load for one electric sign, Fig. 11–18?

(a) 1,200 VA (b) 1,500 VA (c) 1,920 VA (d) 2,400 VA

- Answer: (b) 1,500 VA

  1,200 VA × 1.25 = 1,500 VA

11–18 *NEUTRAL CALCULATIONS [220–22]*

The neutral load is considered the maximum unbalanced demand load between the grounded (neutral) conductor and any one ungrounded (hot) conductor, as determined by the calculations in Article 220, Part B. This means that line-to-line loads are not considered when sizing the neutral conductor.

Reduction Over 200 Amperes

For balanced 3-wire, single-phase and 4-wire, 3-phase wye systems, the neutral demand load can be reduced 70 percent for that portion of the unbalanced load over 200 amperes.

Reduction Not Permitted

The neutral demand load shall not be permitted to be reduced for 3-wire, single-phase 208Y/120-, or 480Y/277-volt circuits consisting of two line wires and the common conductor (neutral) of a 4-wire, 3-phase wye system. This is because the common (neutral) conductor of a 3-wire circuit connected to a 4-wire, 3-phase wye system carries approximately the same current as the phase conductors; see Note 10(b) of Table 310–16. This can be proven with the following formula:

$$I_n = \sqrt{L_1^2 + L_2^2 - (L_1 \times L_2)}$$

L_1 = Current of Line$_1$, Line$_2$ = Current of Line$_2$

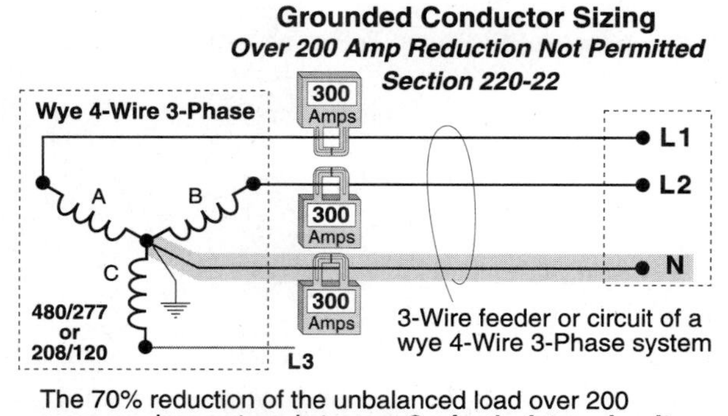

Figure 11-19

Grounded Conductor Not Permitted To Be Reduced Example

❑ **Three-Wire Wye Neutral Current Example**

What is the neutral current for a balanced 300-ampere 3-wire 208Y/120-volt feeder, Fig. 11–19?

(a) 100 amperes (b) 200 amperes (c) 300 amperes (d) none of these

- Answer: (c) 300 amperes

  $I_n = \sqrt{L_1^2 + L_2^2 - (L_1 \times L_2)}$

  $I_n = \sqrt{(300^2 + 300^2) - (300 \times 300)}$

  $I_n = \sqrt{180{,}000 - 90{,}000}$ $I_n = \sqrt{90{,}000}$ $I_n = 300$ amperes

Nonlinear Loads

The neutral demand load cannot be reduced for electric-discharge lighting, electronic ballasts, dimmers, controls, computers, laboratory test equipment, medical test equipment, recording studio equipment, or other *nonlinear loads*. This restriction only applies to circuits that are supplied from a 4-wire, wye-connected, 3-phase system, such as a 208Y/120- or 480Y/277-volt system. Nonlinear loads can cause triplen *harmonic currents* that add on the neutral conductor, which can require the neutral conductor to be larger than the ungrounded conductor load. See Section 220–22 FPN No. 2.

PART C – *LOAD CALCULATIONS*

MARINA [555–5]

The National Electrical Code permits a demand factor to apply to the receptacle outlets for boat slips at a marina. The demand factors of Table 555–5 are based on the number of receptacles. The receptacles must also be balanced between the lines to determine the number of receptacles on any given line.

❑ **Marina Receptacle Outlet Demand Example**

What size 120/240 volt, single-phase service is required for a marina that has twenty 20-ampere, 120 volt receptacles and twenty 30-ampere, 240 volt receptacles, Fig. 11–20?

(a) 200 amperes (b) 400 amperes (c) 600 amperes (d) 800 amperes

- Answer: (c) 600 amperes

  Section 555–5 permits a demand factor according to the number of receptacles. The receptacles must be balanced to determine the number of receptacles on any given line. Ten 20-ampere, 120-volt receptacles are on line 1, ten 20-ampere, 120-volt receptacles are on line 2. Twenty 30-ampere, 240-volt receptacles are on lines 1 and 2.

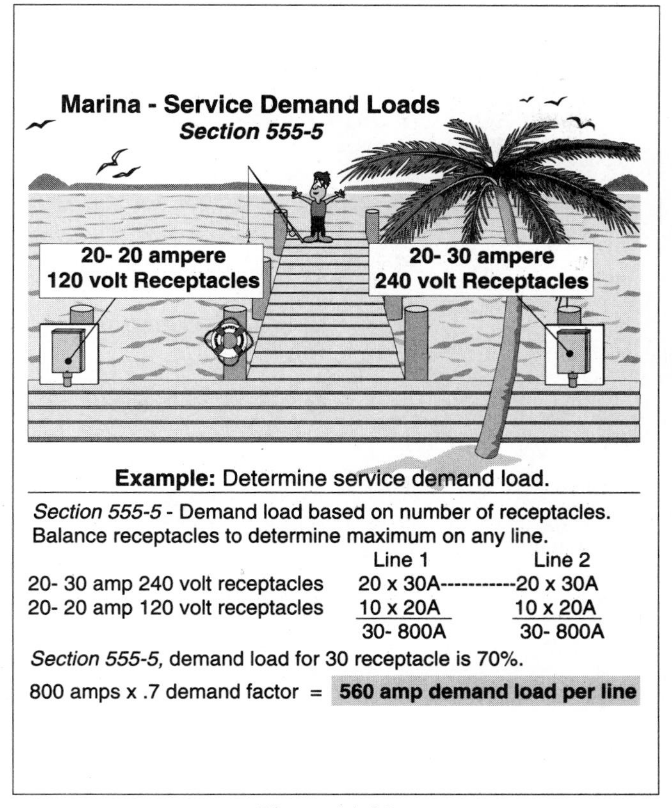

Figure 11-20

Marina Service Demand Load Example

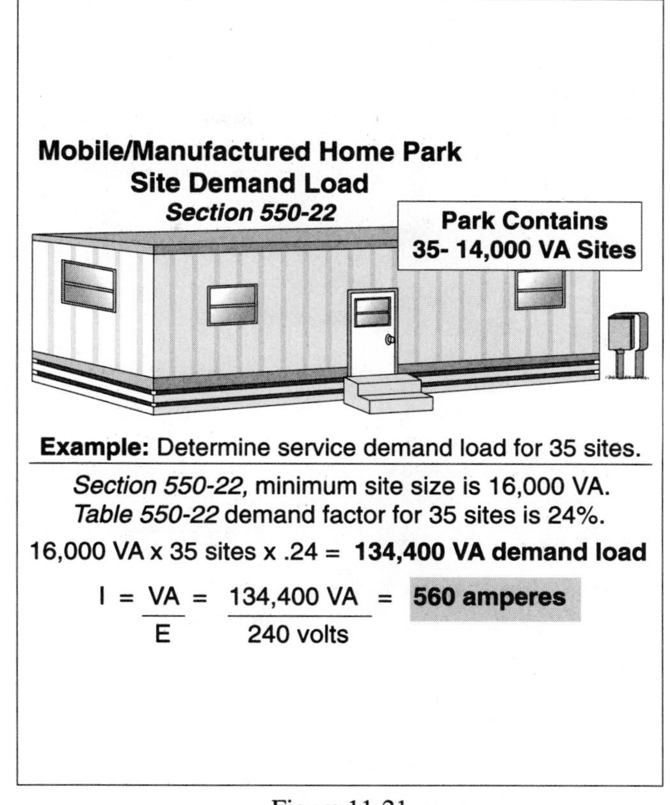

Figure 11-21

Mobile Home Park Service Demand Load Example

The total load on each line is [(10 receptacles × 20 amperes) + (20 receptacles × 30 ampere)] = 800 amperes per line. The demand factor for 30 receptacles (per line) is 70 percent; 800 amperes × 0.7 = 560 amperes per line.

MOBILE/MANUFACTURED HOME PARK [550–22]

The service demand load for a mobile/manufactured home park is sized according to the demand factors of Table 550–22 to the larger of 16,000 VA for each mobile/manufactured home lot, or the calculated load for each mobile/manufactured home site according to Section 550–13.

❑ **Mobile/Maunfactured Home Park Example**

What is the demand load for a mobile/manufactured home park that has facilities for 35 sites? The system is 120/240-volts single-phase, Fig. 11–21.

(a) 400 amperes (b) 600 amperes (c) 800 amperes (d) 1,000 amperes

- Answer: (b) 600 amperes

  16,000 VA × 35 sites × .24 = 134,400 VA

  I = VA/E

  I = 134,400 VA/240 volts = 560 amperes.

RECREATIONAL VEHICLE PARK [551–73]

Recreational vehicle parks are calculated according to the demand factor of Table 551–73. The total calculated load is based on:

2,400 VA for each 20-ampere supply facilities site,

3,600 VA for each 20- and 30-ampere supply facilities site, and

9,600 VA for each 50-ampere, 120/240-volt supply facilities site.

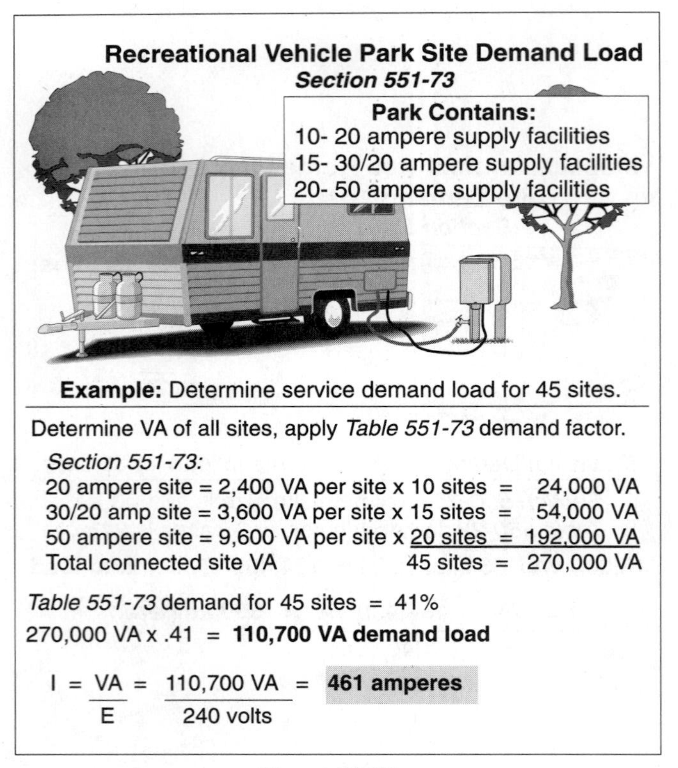

Figure 11-22

Recreational Vehicle Park Site Demand Load Example

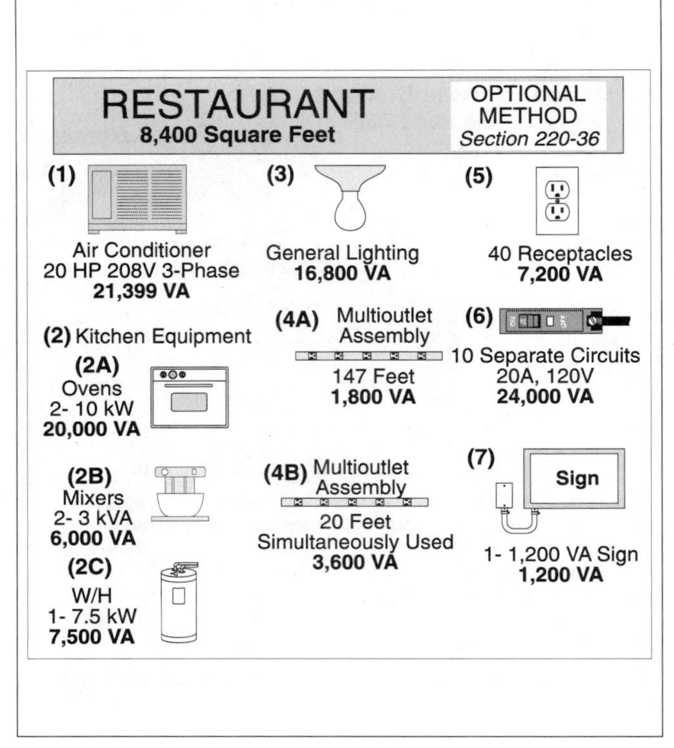

Figure 11-23

Restaurant Demand Load Optional Method – Example

❑ **Recreational Vehicle Park Example**

What is the demand load for a recreational vehicle park that has ten 20-ampere supply facilities, fifteen 20- and 30-ampere supply facilities, and twenty 50-ampere supply facilities? The system is 120/240-volts single-phase, Fig. 11–22.

(a) 400 amperes (b) 500 amperes (c) 600 amperes (d) 700 amperes

• Answer: (b) 500 amperes

The service for the recreational vehicle park is sized according to the demand factors of Table 551–73

20-ampere supply facilities, 2,400 VA × 10 sites =	24,000 VA
20- and 30-ampere supply facilities, 3,600 VA × 15 sites =	54,000 VA
50-ampere supply facilities, 9,600 VA × 20 sites =	+ 192,000 VA
Total connected load	270,000 VA

Demand factor [Table 551–73] for 45 sites, 41%

Demand load = 270,000 VA × .41 = 110,700 VA

I = VA/E

I = 110,700 VA/240 volts = 461 amperes

RESTAURANT – OPTIONAL METHOD [220–36]

An optional method of calculating the demand service load for a restaurant is permitted. The following steps can be used to determine the service size.

Step 1: → Determine the total connected load.
Add the nameplate rating of all loads at 100 percent and include both the air conditioner and heat load.

Step 2: → Apply the demand factors from Table 220–36 to the total connected load (Step 1).

All Electric Restaurant Demand Factors

0 - 250 kVA	*80%*
251- 280 kVA	*70%*
281 - 325 kVA	*60%*
Over 326 kVA	*50%*

Not All Electric Restaurant Demand Factors

0 - 250 kVA	*100%*
251- 280 kVA	*90%*
281 - 325 kVA	*80%*
326 - 375 kVA	*70%*
376 - 800 kVA	*65%*
Over 800 kVA	*50%*

❑ Restaurant Optional Method Example

Using the optional method, what is the service size for the loads listed in Figure 11–23?

Step 1: → Determine the total connected load.

(1) Air conditioning demand [Table 430–150 and 440–34]

VA = 208 Volts × 59.4 amperes × 1.732 = 21,399 VA

(2) Kitchen equipment

Ovens	10 kVA × 2 units =	20.0 kVA
Mixers	3 kVA × 2 units =	6.0 kVA
Water heaters	7.5 kVA × 1 unit =	+ 7.5 kVA
Total connected load		33.5 kVA

(3) Lighting [Table 220–3(b)]

General lighting, 8,400 square feet × 2 VA per square foot = 16,800 VA

(4) Multioutlet assembly [220–3(c) Exception]

147 feet/5 feet = thirty sections at 5-feet, 180 VA × 30 = 5,400 VA

Simultaneous used 180 VA × 20 feet = 3,600 VA

(5) Receptacles [220–3(c)(7)]

180 VA per receptacle × 40 receptacles = 7,200 VA

(6) Separate circuits (noncontinuous)

120 volts × 20 amperes = 2,400 VA × 10 circuits = 24,000 VA

Sign [220–3(c)(6)and 600-5] = 1,200 VA

(1) Air conditioning	21,399 VA
(2) Kitchen equipment	33,500 VA
(3) Lighting	16,800 VA
(5) Multioutlet assembly	5,400 VA
(6) Receptacles	9,000 VA
(7) Separate circuits	24,000 VA
(8) Sign	+ 1,200 VA
Total connected load	113,099 VA

Step 2: → Apply the demand factors from Table 220–36 to the total connected load.

All Electric – If the restaurant is all electric, a demand factor of 80 percent is permitted to be applied to the first 250 kVA.

113,099 VA × 0.8 = 90,479 VA

$I = VA/(E \times \sqrt{3})$

I = 90,479 VA/(208 volts × 1.732) = 251 amperes

Not All Electric – If the restaurant is not all electric, then a demand factor of 100 percent applies to the first 250 kVA.

113,099 VA at 100% = 113,099 VA

$I = VA/(E \times \sqrt{3})$

I = 113,099 VA/(208 volts × 1.732) = 314 amperes

SCHOOL – OPTIONAL METHOD [220–34]

An optional method of calculating the demand service load for a school is permitted. The following steps can be used to determine the service size.

Step 1: → Determine the total connected load. Add the nameplate rating of all loads at 100 percent and select the larger of the air-conditioning versus heat load.

Step 2: → Determine the average VA per square foot by dividing the total connected load (Step 1) by the square feet of the building.

Step 3: → Determine the demand VA per square foot by applying the demand factors of Table 220–34 to the average VA per square foot (Step 2).

Step 4: → Determine the school net VA, by multiplying the demand VA per square foot (Step 3) by the square foot of the school building.

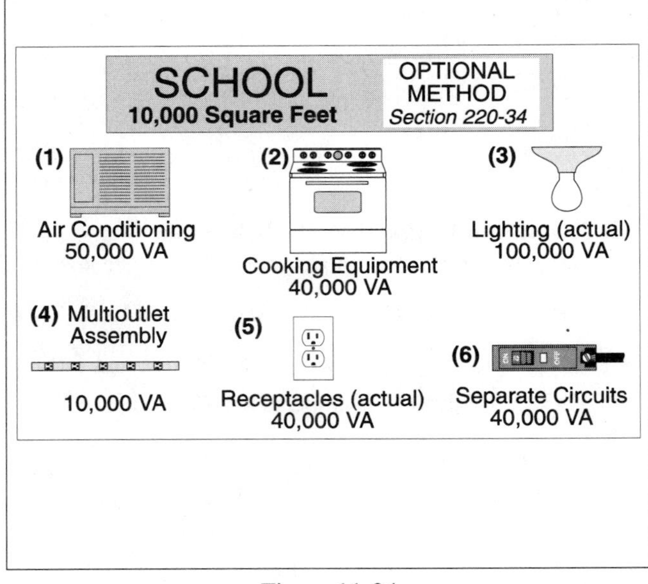

Figure 11-24

School Demand Load Optional Method – Example

❑ School Optional Method Example

What is the service size for the following loads? The system voltage is 208Y/120-volt three-phase, Fig. 11–24.

(1) Air conditioning	50,000 VA
(2) Cooking equipment	40,000 VA
(3) Lighting	100,000 VA
(4) Multioutlet assembly	10,000 VA
(5) Receptacles	40,000 VA
(6) Separate circuits	+ 40,000 VA
	280,000 VA

(a) 300 amperes (b) 400 amperes (c) 500 amperes (d) 600 amperes

• Answer: (c) 500 amperes

Step 1: → Determine the average VA per square foot.
(Total Connected Load/Square Feet Area): 280,000 VA/ 10,000 square feet = 28 VA per square foot.

Step 2: → Apply the demand factors from Table 220–34 to the average VA per square foot.

Average VA per square foot	28 VA		
First 3 VA at 100%	– 3 VA	× 1.00 =	3.00 VA
	25 VA		
Next 17 VA at 75%	– 17 VA	× .75 =	12.75 VA
Remainder at 25%	8 VA	× .25 =	+ 2.00 VA
Net VA per square foot			17.75 VA

Total demand load = 17.75 VA per square foot × 10,000 square feet = 175,000 VA

Service Size – $I = VA/(E \times \sqrt{3})$

$I = 175{,}000\ VA/(208\ volts \times 1.732) = 486$ amperes

Service Demand Load Using the Standard Method

For exam preparation purposes, you are not expected to calculate the total demand load for a commercial building. However, you are expected to know how to determine the demand load for the individual loads (Steps 1 through 10 below). For your own personal knowledge, you can use the following steps to determine the total demand load for a commercial building for the purpose of sizing the service.

Part A – Determine The Demand Load For Each Type Of Load

Step 1: → Determine the general lighting load Table 220–3(b) and Table 220–11.

Step 2: → Determine the receptacle demand load. [220–13].

Step 3: → Determine the appliance demand load at 100 percent.

Step 4: → Determine the demand load for show windows at 125 percent, [220–12].

Step 5: → Determine the demand load for multioutlet assembly, [220–3(c) Exception No. 1].

Step 6: → Determine the larger of :
Air Conditioner, largest 125 percent plus all others at 100 percent, [440–33] vs. heat at 100 percent, [220–15].

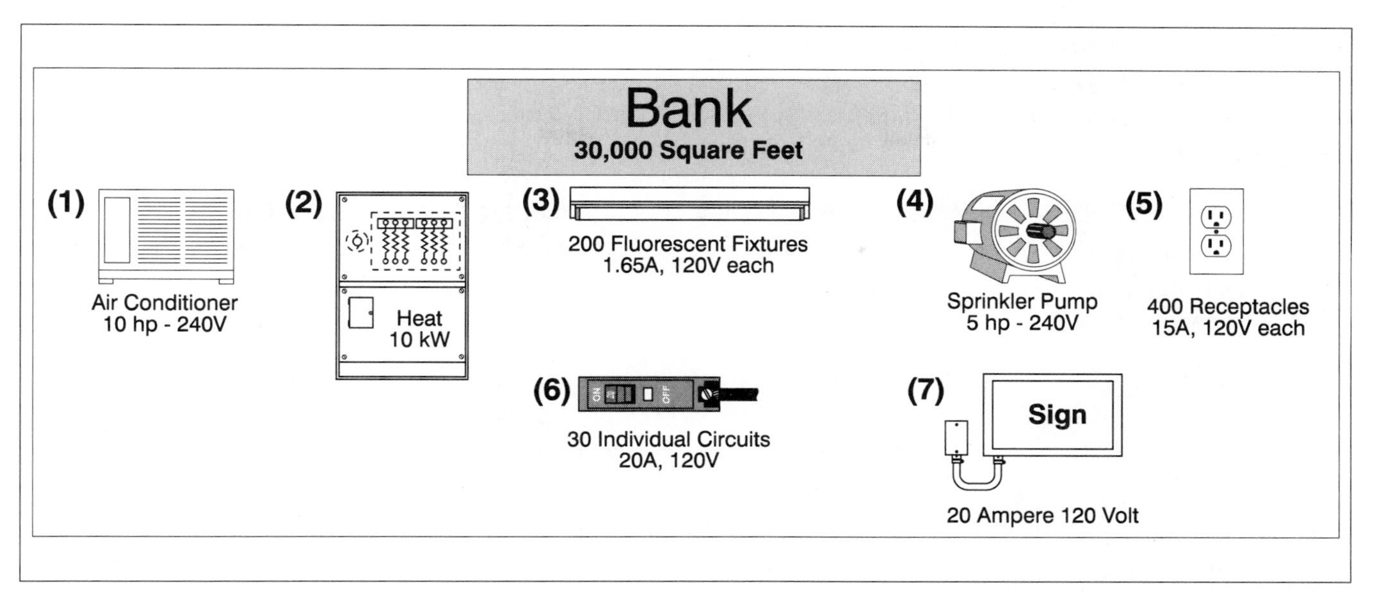

Figure 11-25

Bank Example

Step 7: → Determine the demand load for educational cooking equipment, [220–19 Note 5].
Step 8: → Determine the kitchen equipment demand load, [Table 220–20].
Step 9: → All other noncontinuous loads at 100 percent, [220–10(b)].
Step 10: → All other continuous loads at 125 percent, [220–10(b)].

Part B – Determine the Total Demand Load
Add up the individual demand loads of Steps 1 through 10.

Part C – Determine the Service Size
Divide the total demand load (Part B) by the system voltage.
Single-Phase = I = VA/E   Three-Phase = I = VA/(E × 1.732)

PART D – *LOAD CALCULATION EXAMPLES*

BANK (120/240-VOLT SINGLE-PHASE), Fig. 11–25.

(1) Air Conditioning [Article 440, 430–24, and Table 430–148]
10-horsepower, 230 volts × 50 amperes × 1.25 = 14,375 VA

(2) Heat [220–15 and 220–21] 10 kW omit 220–21

(3) Lighting [Table 220–3(c), and 220–10(b)]
30,000 square foot × 3.5 VA = 105,000 VA × 1.25 = 131,250 VA
Actual Lighting, Omit
200 units × 1.65 ampere × 120 volts × 1.25 = 49,500 VA

(4) Motor [Table 430–148]
5-horsepower, 230 volts × 28 amperes = 6,440 VA
(Some exams use 240 volts × 28 amperes = 6,720 VA)

(5) Receptacles [220–13]
Actual Load 400 receptacles × 180 VA = 72,000 VA

	72,000 VA		
	– 10,000 VA	× 1.00 =	10,000 VA
	62,000 VA	× .50 =	+ 31,000 VA
Net computed load			41,000 VA

(6) Separate Circuits (non-continuous)

30 circuits × 20 amperes × 120 volts = 72,000 VA

(7) Sign [220–3(c)(6), 220–10(b) and 600–5(b)(3)]

1,200 VA × 1.25 = 1,500 VA

SUMMARY	Overcurrent Protection	Conductor	Neutral
(1) Air conditioner	14,375 VA	17,969 VA *1	0 VA
(2) Lighting	131,250 VA	131,250 VA	105,000 VA *2
(4) Motor	16,100 VA *3	6,440 VA	0 VA
(5) Receptacles	41,000 VA	41,000 VA	41,000 VA
(6) Separate Circuits	72,000 VA	72,000 VA	72,000 VA
(7) Sign	+ 1,500 VA	+ 1,500 VA	+ 1,200 VA *2
	276,225 VA	**270,159 VA**	**219,200 VA**

*1 Includes 25% for the largest motors (14,375 × 1.25)

*2 The neutral reflects the feeder conductor load at 100 percent, not 125 percent!

*3 Feeder protection is sized according to the largest motor short-circuit ground-fault protection device [430–63].

28 amperes × 2.5 = 70 amperes, 70 amperes × 230 volts = 16,100 VA

1. Feeder/service protection device

I = VA/E, I = 276,225 VA /240 volts = 1,151 amperes

2. Feeder and service conductor

I = VA/E, I = 270,159 VA/240 volts = 1,125 amperes

3. Neutral Amperes

I = VA/E, I = 219,200/240 volts = 913 amperes

The neutral conductor is permitted to be reduced according to the requirements of Section 220–22.

Neutral demand load =	913 amperes	
First 200 amperes	– 200 amperes × 1.00 =	200 amperes
Remainder	713 amperes × .70 =	+ 499 amperes
		699 amperes

❑ **Summary Example**

The service overcurrent protection device must be sized no less than 1,151 amperes. The service conductors must have an ampacity no less than the overcurrent protection device rating [240–3(c)]. The grounded (neutral) conductor must not be less than 699 amperes.

➛ What size service overcurrent protection device is required?

(a) 800 amperes (b) 1,000 amperes (c) 1,200 amperes (d) 1,600 amperes

- Answer: (c) 1,200 amperes [240–6(a)].

➛ What size service THHN conductors are required in each raceway if the service is parallel in four raceways?

(a) 250 kcmil (b) 300 kcmil (c) 350 kcmil (d) 400 kcmil

- Answer: (c) 350 kcmil

  1,200 amperes/4 raceways = 300 amperes per raceway, sized based on 75ºC terminal rating [110–14(c)(2)]

  350 kcmil rated 310 amperes × 4 = 1,240 amperes [240–3(c) and Table 310–16].

Note: 300 kcmil THHN has an ampacity of 320 amperes at 90ºC, but we must size the conductors at 75ºC, not 90ºC!

➛ What size service grounded (neutral) conductor is required in each of the four raceways?

(a) No. 1/0 (b) No. 2/0 (c) 250 kcmil (d) 350 kcmil

- Answer: (b) No. 2/0

  The grounded conductor must be sized no less than:

  (1) 12½ percent area of the line conductor, 350,000 × 4 = 1,400,000 × .125 = 175,000 [250–23(b)], 175,000/4 = 43,750 circular mil, No. 3 per raceway [Chapter 9, Table 8]. **(2)** When paralleling conductors, no conductor smaller than No. 1/0 is permitted [310–4]. **(3)** The service neutral conductor must have an ampacity of at least 699 amperes, 699 amperes/4 = 175 amperes, Table 310–16, No. 2/0 is required at 75ºC [110–14(c)].

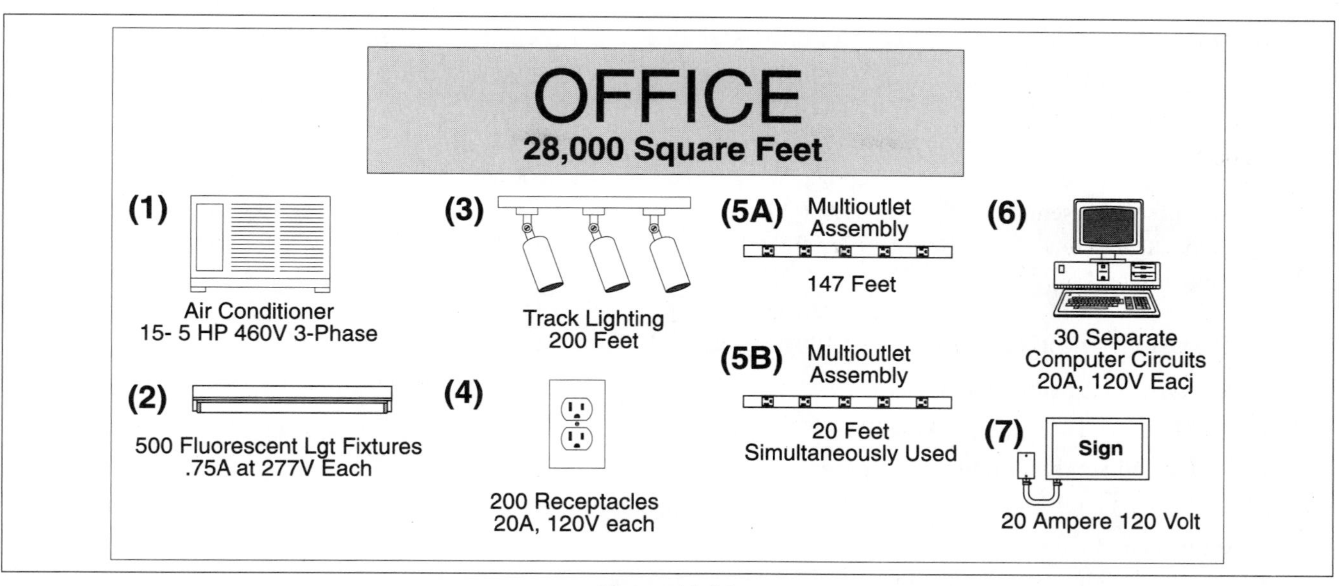

Figure 11-26

Office Example

➛ What size grounding electrode conductor is required to a concrete encased electrode?

(a) No. 4 (b) No. 1/0 (c) No. 2/0 (d) No. 3/0

- Answer: (a) No. 4 [250–94 Exception No. 1b]

OFFICE BUILDING (480Y/277-VOLT THREE-PHASE), Fig. 11–26.

(1) Air Conditioning (Table 430–150 and 440–34)

460 volts × 7.6 amperes × 1.732 =	6,055 VA × 15 units =	90,826 VA
25 percent of the largest + VA = 6,055 × .25 =		1,514 VA

(2) Lighting (NEC®) [Table 220–3(b) and 220–10(b)]

28,000 square feet × 3.5 VA = 98,000 VA × 1.25 = 122,500 VA (omit)

(2) Lighting Electric Discharge (Actual) [220–10(b)]

500 lights × .75 amperes × 277 volts =	103,875 VA × 1.25 =	129,844 VA

(3) Lighting, Track (Actual) [220–3(c)(5), 220–10(b) and 410–102] 150 VA per 2 feet

200 feet/2 feet = 100 sections x 150 VA =	15,000 VA × 1.25 =	18,750 VA

(4) Receptacles [220-13](Actual)

200 receptacles × 180 VA =	36,000 VA	
	– 10,000 VA × 100% =	10,000 VA
	26,000 VA × 50% =	+ 13,000 VA
		23,000 VA

(5) Multioutlet Assembly [220–3(c), Exception]

(A) 147 feet/5 feet = thirty 5-foot sections, 30 × 180 VA =	5,400 VA
(B) Simultaneously used 20 feet × 180 VA =	3,600 VA

(6) Separate Circuits (noncontinuous)

120 volts × 20 amperes = 2,400 VA × 30 circuits =	72,000 VA

(7) Sign [220–10(b) and 600–6]

1,200 VA × 1.25 =	1,500 VA

SUMMARY	Overcurrent Protection	Conductor	Neutral	
(1) Air conditioning	92,339 VA	92,339 VA	0 VA	
(2) Lighting (actual)	129,844 VA	129,844 VA	103,875 VA	*1
(3) Lighting (track)	18,750 VA	18,750 VA	15,000 VA	*1
(4) Receptacles (actual)	23,000 VA	23,000 VA	23,000 VA	
(5) Multioutlet Assembly	9,000 VA	9,000 VA	9,000 VA	
(6) Separate Circuits	72,000 VA	72,000 VA	72,000 VA	
(7) Sign	+ 1,500 VA	+ 1,500 VA	+ 1,200 VA	*1
	346,434 VA	**346,434 VA**	**234,875 VA**	

*1 The neutral reflects the feeder load at 100 percent, not 125 percent!

1. Feeder/service protection device

$I = VA/(E \times \sqrt{3})$, $I = 346{,}434\ VA/(480 \times 1.732) = 416$ amperes

2. Feeder and service conductor amperes

$I = VA/(E \times \sqrt{3})$, $I = 346{,}434\ VA/(480 \times 1.732) = 416$ amperes

3. Neutral conductor

$I = VA/(E \times \sqrt{3})$, $I = 234{,}875\ VA/(480\ volts \times 1.732) = 283$ amperes*.

❑ **Summary Example**

The service overcurrent protection device and conductor must be sized no less than 416 amperes, and the grounded (neutral) conductor must not be less than 283 amperes.

➛ What size service overcurrent protection device is required?

(a) 300 amperes (b) 350 amperes (c) 450 amperes (d) 500 amperes

- Answer: (c) 450 amperes [240–6(a)]. Since the total demand load for the overcurrent device is 416 amperes, the minimum service permitted is 450 amperes.

➛ What size service conductors are required if the total calculated service demand load is 416 amperes?

(a) 300 kcmil (b) 400 kcmil (c) 500 kcmil (d) 600 kcmil

- Answer: (d) 600 kcmil rated 420 amperes [240–3(b) and Table 310–16]
  The service conductors must have an ampacity of at least 416 amperes, protected by a 450 ampere protection device [240–3(b)], and the conductors must be selected based on 75°C insulation rating [110–14(c)(2)].

➛ What size service grounded (neutral) conductor is required for this service?

(a) No. 1/0 (b) 3/0 (c) 250 kcmil (d) 350 kcmil

- Answer: (d) 350 kcmil
  The grounded conductor must be sized no less than:
  (1) The required grounding electrode conductor [250–23(b) and Table 250–94], = No. 1/0. **(2)** An ampacity of at least 283 amperes, Table 310–16, 350 kcmil is rated 310 amperes at 75°C [110–14(c)].

Note: Because of 125 amperes of electric discharge lighting [103,875 VA/(480 × 1.732)], the neutral conductor is not permitted to be reduced for that portion of the load in excess of 200 amperes [220-22].

RESTAURANT (STANDARD) (208Y/120 VOLT, THREE-PHASE SYSTEM), Fig. 11–27.

(1) Air conditioning [Table 430–150, and 440–34]

208 volts × 59.4 amperes × 1.732 × 1.25 = 26,749 VA

(2) Kitchen equipment [220–20]

(A) Ovens	10 kW × 2 units =	20.0 kW		
(B) Mixer	3 kW × 2 units =	6.0 kW		
(C) Water heaters	7.5 kW × 1 unit =	+ 7.5 kW		
Total connected		33.5 kW	× .7 =	23.45 kW

(3a) Lighting NEC® [Table 220–3(b) and 220–10(b)]

8,400 square feet × 2 VA =16,800 VA × 1.25 = 21,000 VA, omit - less than actual.

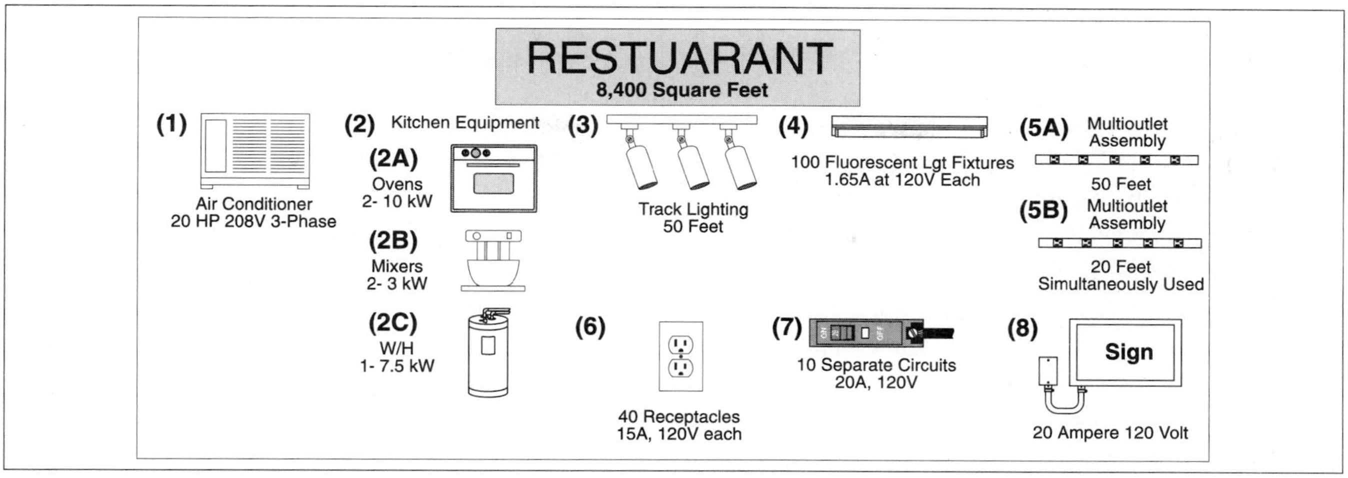

Figure. 11-27

Restaurant Standard Method Example

(3b) Lighting, Track [220–3(c)(5), 220–10(b) and 410–102]
150 VA per two feet, 50 feet/2 = 25 sections
25 sections × 150 VA = 3,750 VA × 1.25 = 4,688 VA

(4) Lighting (actual) [220–10(b)]
100 lights × 1.65 amperes × 120 volts = 19,800 × 1.25 = 24,750 VA

(5) Multioutlet Assembly [220–3(c), Exception]
5,400 VA (1,800 VA + 3,600 VA)
(A) 50/5 = ten 5-foot sections, 10 sections × 180 VA = 1,800 VA
(B) Simultaneously used 20 feet × 180 VA = 3,600 VA

(6) Receptacles (Actual) [220–13] 40 receptacles × 180 VA = 7,200 VA

(7) Separate Circuits [220–10(b)]
120 volts × 20 amperes = 2,400 VA × 10 circuits = 24,000 VA

(8) Sign [220–10(b) and 600–6] 1,200 VA × 1.25 = 1,500 VA

SUMMARY	Overcurrent Protection	Conductor
(1) Air Conditioning	26,749 VA	26,749 VA
(2) Kitchen equipment	23,450 VA	23,450 VA
(3) Lighting (track)	4,688 VA	4,688 VA
(4) Lighting (actual)	24,750 VA	24,750 VA
(5) Multioutlet Assembly	5,400 VA	5,400 VA
(6) Receptacles (actual)	7,200 VA	7,200 VA
(7) Separate Circuits	24,000 VA	24,000 VA
(8) Sign	+ 1,500 VA	+ 1,500 VA
	117,737 VA	**117,737 VA**

1. Feeder and service protection device

$I = VA/(E \times \sqrt{3})$

I = 117,737 VA/(208 × 1.732) = 327 amperes

2. Feeder and service conductor

$I = VA/(E \times \sqrt{3})$

I = 117,737 VA/(208 × 1.732) = 327 amperes

❑ **Summary Example**

The service overcurrent protection device must be sized no less than 327 amperes. The service conductors must be no less than 327 amperes.

➤ What size service overcurrent protection device is required?

(a) 300 amperes (b) 350 amperes (c) 450 amperes (d) 500 amperes

- Answer: (b) 350 amperes [240–6(a)]
  Since the total demand load for the overcurrent device is 327 amperes, the minimum service permitted is 350 amperes [240–6(a)].

➤ What size service conductors are required?

(a) No. 4/0 (b) 250 kcmil (c) 300 kcmil (d) 400 kcmil

- Answer: (d) 400 kcmil
  The service conductors must have an ampacity of at least 327 amperes, and be protected by a 350 ampere protection device [240–3(b)] next size up is OK, and the conductors must be selected according to Table 310–16, 400 kcmil based on 75°C insulation rating [110–14(c)(2)] is rated 335 amperes at 75°C.

Unit 11 – Commercial Load Calculations Summary Questions

Part A – General

11–2 Conductor Ampacity [Article 100]

1. The ________ of a conductor is the rating in amperes a conductor can carry continuously without exceeding its insulation temperature rating. The allowable ampacities as listed in Table 310–16 are affected by ambient temperature, current flow, conductor insulation and conductor bunching or bundling [310–10].
   (a) load rating (b) ampacity (c) demand load (d) continuous factor

11–3 Conductor Overcurrent Protection [240–3]

2. The purpose of conductor ________ is to protect the conductors against excessive or dangerous temperatures. If the ampacity of a conductor does not correspond with the standard ampere rating of a fuse or circuit breaker, the next size up protection device is permitted. This applies only if the conductors do not supply multioutlet receptacles and if the next size overcurrent protection device does not exceed 800-amperes.
   (a) short-circuit protection (b) ground-fault protection
   (c) overload protection (d) overcurrent protection

3. The following is a list of some of the standard ampere ratings for overcurrent protection devices (fuses and inverse time circuit breakers): 15, 25, 35, 45, 80, 90, 110, 175, 250 and 350-amperes.
   (a) True (b) False

11–4 Voltages [220–2]

4. Unless other voltages are specified, branch-circuit, feeder, and service loads shall be computed at a nominal system voltage of ________.
   (a) 600/347 (b) 240/120 (c) 208Y/120 (d) Any of these

11–5 Fraction Of An Ampere

5. There are no NEC® rules for rounding when a calculation results in a fraction of an ampere, but it does contain the following note, "except where the computations result in a ________ of an ampere or larger, such fractions may be dropped."
   (a) .05 (b) .5 (c) .49 (d) .51

Part B – Loads

11–6 Air Conditioning

6. What is the feeder or service demand load required for the air conditioning of a six unit office building? Each unit contains one air conditioner rated 3-horsepower, 230 volts.
   (a) 13 kVA (b) 24 kVA (c) 45 kVA (d) 51 kVA

11–7 Dryers

7. What size branch circuit conductor and overcurrent protection is required for a 6-kW dryer rated 240 volts, located in the laundry room of a multifamily dwelling?
   (a) No. 12 with a 20-ampere breaker
   (b) No. 10 with a 20-ampere breaker
   (c) No. 12 with a 30-ampere breaker
   (d) No. 10 with a 30-ampere breaker

8. What is the feeder and service demand for eight 6.75-kW dryers?
   (a) 70 kW (b) 54 kW (c) 35 kW (d) 27 kW

11–8 Electric Heat

9. • What size conductor and protection is required for a single-phase 15-kW, 480-volt heat strip that has a 1.6-ampere blower motor?
   (a) No. 10 THHN with 30-amperes protection
   (b) No. 6 THHN with 45-amperes protection
   (c) No. 10 THHN with 40-amperes protection
   (d) No. 4 THHN with 70-amperes protection

10. What is the feeder and service demand load for a building that has four 20-kW, 230-volt single-phase heat strips that have a 5.4-ampere blower motor for each unit?
    (a) 100 kVA (b) 50 kVA (c) 125 kVA (d) 85 kVA

11–9 Kitchen Equipment

11. What is the branch circuit demand load (in amperes) for one 11.4-kW oven rated 240 volts?
    (a) 60 amperes (b) 27 amperes (c) 48 amperes (d) 33 amperes

12. • What is the feeder and service demand load for the following?
    Water heater .... 9 kW
    Booster heater .... 12 kW
    Mixer.......................... 3 kW
    Oven........................... 2 kW
    Dishwasher................ 1.5 kW
    Disposal .... 1 kW
    (a) 19 kW (b) 21 kW (c) 29 kW (d) 12 kW

13. What is the feeder and service demand load for the following?
    Water heater .... 14 kW
    Booster heater .... 11 kW
    Mixer........................... 7 kW
    Oven............................ 9 kW
    Dishwasher................ 1.5 kW
    Disposal .... 3 kW
    (a) 25 kW (b) 30 kW (c) 20 kW (d) 45 kW

11–10 Laundry Equipment

14. What is the feeder and service demand load for seven washing machines at 1,500 VA?
    (a) 10,500 VA (b) 15,000 VA (c) 1,125 VA (d) none of these

11–11 Lighting – Demand Factors [220–11]

15. What is the general lighting feeder or service demand load for a 24 room motel, 685 square feet each?
    (a) 10 kVA (b) 15 kVA (c) 20 kVA (d) 25 kVA

16. What is the general lighting and receptacle feeder or service demand load for a 250,000 square foot storage warehouse that has 200 receptacles?
    (a) 25 kVA (b) 55 kVA (c) 95 kVA (d) 105 kVA

11–12 Lighting Without Demand Factors [Table 220–3(b) and 220–10(b)]

17. What is the feeder and service (protection device) general lighting demand load for a 3,200 square foot dance club?
    (a) 3,200 VA (b) 6,400 VA (c) 8,000 VA (d) 12,000 VA

18. What is the feeder and service general lighting load for a 90,000 square foot school?
    (a) 238 kVA (b) 338 kVA (c) 90 kVA (d) 270 kVA

11–13 Lighting – Miscellaneous

19. How many 2 x 4 fluorescent fixtures, each rated 277 volts, .8-amperes, can be connected to a 20-ampere circuit? The four lamps are rated 40 watts each and the fixture is to be on for more than 3 hours.
(a) 5 (b) 7 (c) 9 (d) 20

20. How many 360 watt, 120-volt incandescent fixtures can be connected to a 20-ampere 120-volt circuit continuously?
(a) 5 (b) 7 (c) 9 (d) 12

21. • What is the VA demand feeder and service load for 130 feet of show window lighting?
(a) 26 kVA (b) 33 kVA (c) 9 kVA (d) 11 kVA

11–14 Multioutlet Receptacle Assembly [220–3(c) Exception No. 1]

22. What is the feeder demand load for 50 feet of multioutlet assembly and 10 feet of multioutlet assembly simultaneously used?
(a) 3,600 VA (b) 12,000 VA (c) 7,200 VA (d) 5,500 VA

11–15 Receptacle VA Load [220–3(c)(6)]

23. How many receptacle outlets are permitted on a 20-ampere 120-volt circuit?
(a) 10 (b) 13 (c) 15 (d) 20

24. What is the service demand load for one-hundred-ten 15- or 20-ampere general use receptacles in a commercial building?
(a) 5 kVA (b) 10 kVA (c) 15 kVA (d) 20 kVA

25. What is the service demand load for the general lighting and receptacles, for a 30,000 square foot bank?
(a) 100 kVA (b) 150 kVA (c) 200 kVA (d) 250 kVA

26. • What is the service general lighting demand and receptacle load for a 10,000 square foot office building with 75 receptacles?
(a) 25 kVA (b) 35 kVA (c) 45 kVA (d) 55 kVA

11–17 Signs [220–(c)(6), 600–5(b)(3), And 220–10(b)]

27. What is the feeder demand load for sizing the overcurrent protection device for one electric sign?
(a) 1,400 VA (b) 1,500 VA (c) 1,920 VA (d) 2,400 VA

11–18 Neutral Calculations [220–22]

28. What is the neutral current for a balanced 150-ampere 3-wire 208Y/120-volt feeder?
(a) 0 amperes (b) 150 amperes (c) 250 amperes (d) 300 amperes

Part C – Load Calculations

Marina [555–5]

29. A marina has the following: Twenty-four slips (20-amperes 240-volt receptacles) and thirty slips designed for boats over 20 feet (30-ampere 240-volt receptacles). What size service is required for the marina?
(a) 400 amperes (b) 550 amperes (c) 900 amperes (d) 1,000 amperes

30. What size conductor is required for the service if the total demand load per phase is 570-amperes and the service is parallel in two raceways?
(a) 4/0 (b) 250 kcmil (c) 300 kcmil (d) 350 kcmil

Mobile Home Park [550–22]

31. What is the feeder and service load for a mobile home park that has the facilities for forty-two sites? The system is 120/240-volt single-phase.
(a) 650 amperes (b) 510 amperes (c) 660 amperes (d) 730 amperes

Recreational Vehicle Park [551–73]

32. A recreational vehicle park has seventeen sites with only 20-ampere 240-volt receptacles, thirty-five sites with 20- and 30-ampere 240-volt receptacles, and ten sites with 50-ampere 240-volt receptacles. The feeder and service demand load is approximately ________ amperes.
(a) 355 (b) 450 (c) 789 (d) 1,114

Restaurant – Optional Method [220–36]

33. A restaurant has a total connected load of 400 kVA and is all electric. What is the demand load for the service?
(a) 363 kVA (b) 325 kVA (c) 200 kVA (d) 275 kVA

34. A restaurant has a total connected load of 400 kVA and is *not* all electric. What is the demand load for the service?
(a) 260 kVA (b) 325 kVA (c) 300 kVA (d) 275 kVA

School – Optional Method [220–34]

35. A 28,000 square foot school has a total connected load of 590,000 VA. What is the total demand load?
(a) 350 kVA (b) 400 kVA (c) 450 kVA (d) 500 kVA

Challenge Questions

General Commercial Calculations

36. • If the service high leg conductor only supplies three-phase loads, what size service conductor would be required for the high leg? Three-phase loads: 10-horsepower, 230-volt 3-phase A/C and 10-kW of heat, 230-volt 3-phase.
(a) 10 TW (b) 12 THW (c) 10 THHN (d) No. 8 THW

37. • A commercial office building has forty-six 250 watt, 120-volt lights installed on a 208Y/120-volt three–phase system. If the fixtures are on continuously, how many 20 ampere, 120-volt circuits would be required?
(a) 9 circuits (b) 7 circuits (c) 5 circuits (d) 4 circuits

Conductor Sizing

38. • Three parallel service raceways are installed. Each raceway contains three 500-kcmil THHN conductors. What size bonding jumper is required for each service raceway?
(a) 250,000 (b) No. 2/0 (c) No. 1/0 (d) No. 4/0

39. • If each service raceway contains 300-kcmil conductors, what size bonding jumper is required for the raceway?
(a) No. 1 (b) No. 2 (c) No. 1/0 (d) No. 4

11–9 Kitchen Equipment [220–20]

40. What is the kitchen equipment demand load for: one 14-kW range, one 1.5-kW water heater, one .75-kW mixer, one 2-kW dishwasher, one 3-kW booster, and one 3-kW coffee machine?
(a) 17 kW (b) 14 kW (c) 15 kW (d) 24 kW

11–11 Lighting – Demand Factors [Table 220–3(b) And 220–11]

41. • Each unit of a one-hundred unit hotel is 12 × 15 feet. In addition, there is an office that is 60 × 20 feet, and there are hallways of 120 square feet. What is the general lighting demand load?
(a) 29 kVA (b) 37 kVA (c) 23 kVA (d) 25 kVA

11–15 Receptacle VA Load [220–3(c)(7) And 220–13]

42. If, in the hall and other areas of a motel, there are one hundred receptacle outlets (not in the motel rooms), the demand load added to the service for these receptacles would be ______ VA.
(a) 0 (b) 10,000 (c) 14,000 (d) 20,000

43. What is the receptacle demand load for a 20,000 square foot office building?
(a) 10,000 VA (b) 40,000 VA (c) 30,000 VA (d) 15,000 VA

11–16 Banks And Offices – General Lighting And Receptacle

44. What is the general lighting and general use receptacle load for a 30,000 square foot bank?
(a) 220 kVA (b) 200 kVA (c) 175 kVA (d) 150 kVA

11–18 Neutral Calculations [220–22]

A 208Y/120-volt 3-phase service has a total connected load of 1,450-amperes.
600-amperes of these are phase-to-phase loads
300-amperes are balanced 120-volt fluorescent lighting
550-amperes of other 120-volt loads

45. • The neutral demand for this service is ______ amperes.
(a) 1,150 (b) 650 (c) 745 (d) 420

Part C – Load Calculations

Marina Calculations [555]

46. A marina shore power facility has twenty 20-ampere, 240-volt receptacles, seventeen 30-ampere 240-volt receptacles, and seven 50-ampere 240-volt receptacles. After applying demand factors, the service demand load for the shore power boxes is ______ amperes.
(a) 1,260 (b) 630 (c) 1,160 (d) 625

Mobile/manufactured Home Parks [550]

47. A seventy-five site mobile home park is designed to contain mobile homes that have a 14,000 VA load per site. The service demand load for the park is ______ kVA.
(a) 1,200 (b) 264 (c) 222 (d) 201

Motel

48. A forty unit motel (300 square feet in each unit) includes 3-kVA air conditioning, 4-kW heat, and every two units share one 1.5-kW water heater. What is the demand load for the motel?
(a) 281 kVA (b) 202 kVA (c) 226 kVA (d) 251 kVA

Recreational Vehicle Park [Article 551]

49. A recreational vehicle park has forty-two sites. The minimum feeder demand load for these sites would be ______ kVA.
(a) 151 (b) 101 (c) 139 (d) 56

Restaurant Calculations

50. • The branch circuit rating required for a 12-kW range in a restaurant would be ______ amperes.
(a) 50 (b) 45 (c) 35 (d) 30

51. • A new restaurant has a total connected lighting load of 30 kVA. The kitchen equipment includes two gas stoves, one gas grill, three gas ovens, one 75-gallon gas water heater, one 5-kW dishwasher, two 2-kW coffee makers, five 2-kW kitchen appliances on their own circuit, and ten 1.5-kVA small appliance circuits. Using the optional method, the service demand load would be closest to ______ kVA.
(a) 64 (b) 50 (c) 45 (d) 38

School – Optional Method [220–34]

52. • Using the optional method, what is the demand load (VA per square foot) for a 10,000 square foot school that has a total connected load of 320 kVA?
(a) 15.75 VA per square foot
(b) 18.75 VA per square foot
(c) 29.00 VA per square foot
(d) 12.75 VA per square foot

53. Using the optional method, what is the demand load (kVA) for a 20,000 square foot school that has a total connected load of 160 kVA?
(a) 135 kVA (b) 106 kVA (c) 120 kVA (d) 112 kVA

Unit 12

Delta/Delta And Delta/Wye Transformer Calculations

OBJECTIVES

After reading this unit, the student should be able to briefly explain the following concepts:

- Current flow
- Delta transformer voltage
- Delta high leg
- Delta primary and secondary
- Line currents
- Delta primary or secondary phase Currents
- Delta phase versus line
- Delta current triangle
- Delta transformer balancing
- Delta transformer sizing
- Delta panel schedule in kVA
- Delta panelboard and conductor sizing
- Delta neutral current
- Delta maximum unbalanced load
- Delta/delta example
- Wye transformer voltage
- Wye voltage triangle
- Wye transformer current
- Line current
- Phase current
- Wye phase versus line
- Wye transformer loading and balancing
- Wye transformer sizing
- Wye panel schedule in kVA
- Wye panelboard and conductor sizing
- Wye neutral current
- Wye maximum unbalanced load
- Delta/wye example
- Delta versus wye

After reading this unit, the student should be able to briefly explain the following terms:

- Delta connected
- kVA rating
- Line
- Line current
- Line voltage
- Phase
- Phase current
- Phase load – delta
- Phase load – wye
- Phase voltage
- Ratio
- Balanced load
- Unbalanced load
- Delta secondary
- Wye secondary
- Winding
- Wye connected
- Primary/secondary voltage – amperes
- Three-phase load
- Single-phase load

INTRODUCTION

This unit deals with three-phase delta/delta and delta/wye transformer sizing and balancing. The system voltages used in this unit were selected because of their common industry configuration.

DEFINITIONS

Delta Connected

Delta-connected means the windings of three single-phase transformers are connected in series to form a closed circuit. A line can be traced from one point of the delta system, through all transformers, then back to the original starting point (series). A delta-connected transformer is represented by the Greek letter delta Δ. Many call it a delta high leg system because the voltage from one conductor to ground is 208 volts, Fig. 12–1.

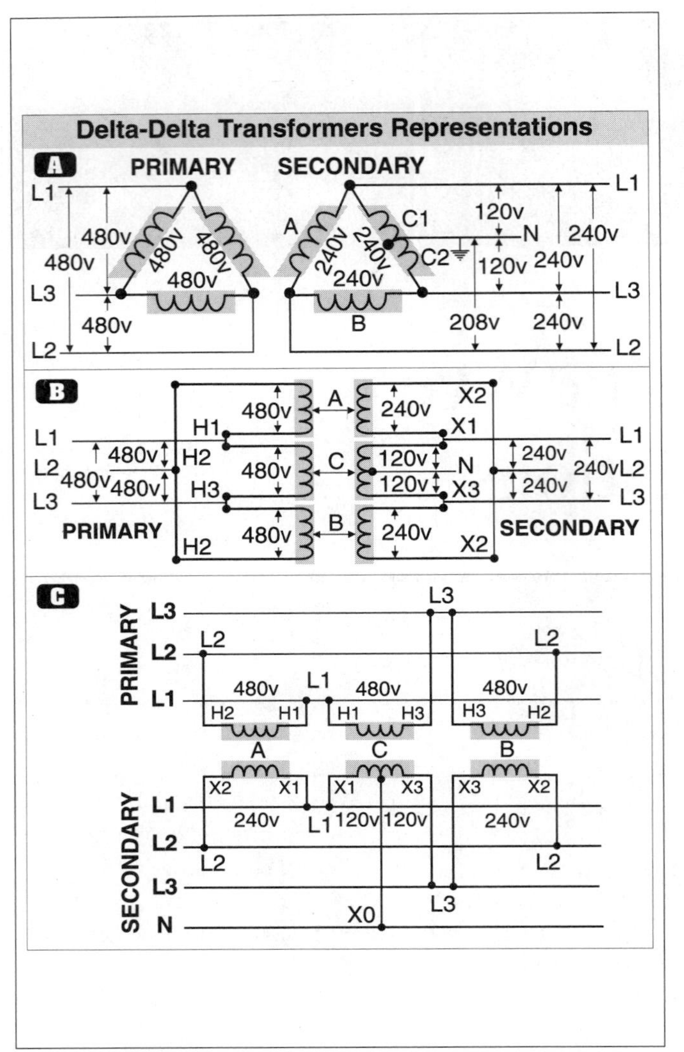

Figure 12-1

Different Ways of Drawing Delta Transformers

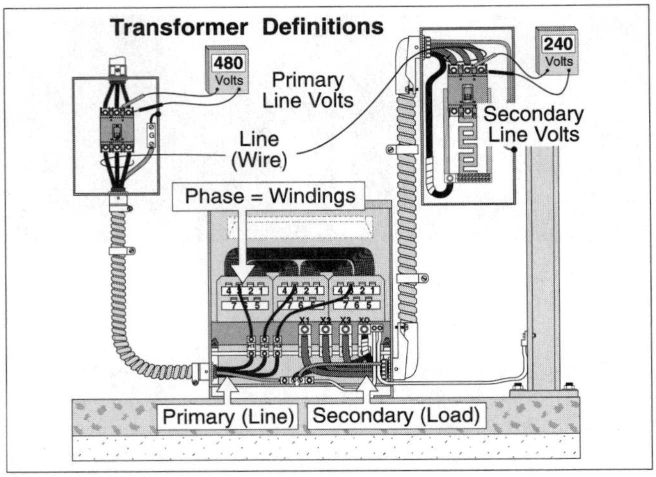

Figure 12-2

Transformer Definitions

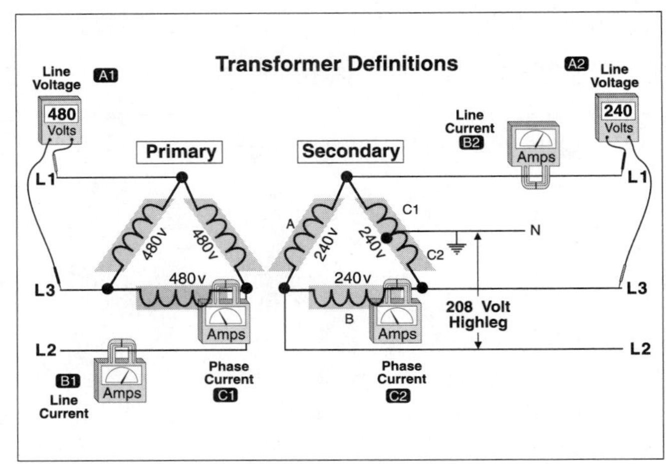

Figure 12-3

Transformer Definitions

kVA Rating

Transformers are rated in *kilovolt-amperes* (kVA) and are sized according to the VA rating of the loads.

Note: One kilovolt-ampere is equal to 1,000 volt-amperes.

Line

The *line* is considered the electrical system supply (hot, ungrounded) conductors, Fig. 12–2 and Fig. 12– 3.

Line Current

The *line current* for both delta and wye systems is the current on the line (hot, ungrounded) conductors, calculated according to the following formulas, Fig. 12–3, B1 and B2.

Single-Phase Line Current = VA_{Line}/E_{Line}

Three-Phase Line Current = $VA_{Line}/(E_{Line} \times \sqrt{3})$

Line Voltage

The line voltage is the voltage that is measured between any two line (hot ungrounded) conductors. It is also called line-to-line voltage. Line voltage is greater than phase voltage for wye systems and line voltage is the same as *phase voltage* for delta systems, Fig. 12–3, A1 and A2.

Phase

The *phase* winding is the coil shaped conductors that serve as the primary or secondary of the transformer.

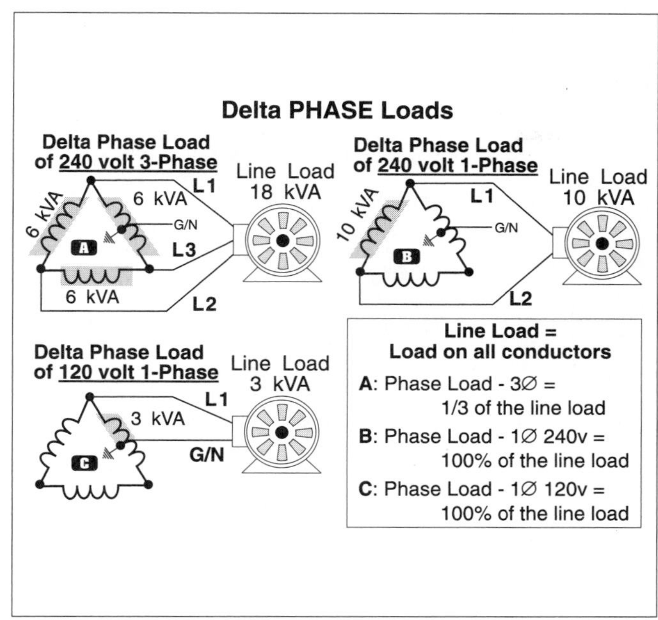

Figure 12-4

Delta Phase Loads

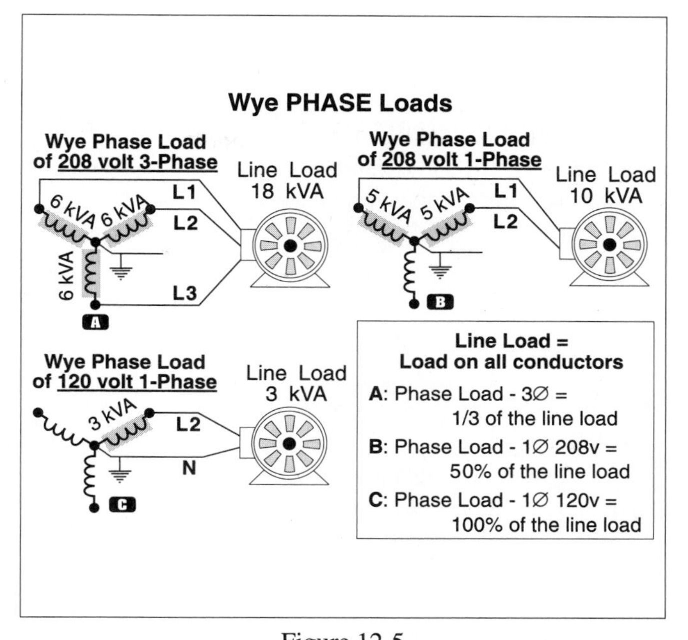

Figure 12-5

Wye Phase Loads

Phase Current

The *phase current* is the current of the transformer winding. For delta systems, the phase current is less than the line current. For wye systems, the phase current is the same as the line current, Fig. 12–3, C1 and C2.

Phase Load Delta

The *phase load* is the load on the transformer winding, Fig. 12–4.

The phase load of a three-phase, 240-volt load = line load/3.

The phase load of a single-phase, 240-volt load = line load.

The phase load of a single-phase, 120-volt load = line load.

Phase Load Wye

The phase load is the load on the transformer winding, Fig. 12–5.

The phase load of a three-phase, 208-volt load = line load/3.

The phase load of a single-phase, 208-volt load = line load/2.

The phase load of a single-phase, 120-volt load = line load.

Phase Voltage

The *phase voltage* is the internal transformer voltage generated across any one winding of a transformer. For a delta secondary, the phase voltage is equal to the line voltage. For wye secondaries, the phase voltage is less than the line voltage, Fig. 12–6.

Ratio (Voltage)

The *ratio* is the relationship between the number of primary winding turns as compared to the number of secondary winding turns. The ratio is a comparison between the primary phase voltage to the secondary phase voltage. For typical delta/delta systems, the ratio is 2:1. For typical wye systems, the ratio is 4:1, Fig. 12–7.

Unbalanced Load (Neutral Current)

The *unbalanced load* is the load on the secondary grounded (neutral) conductors.

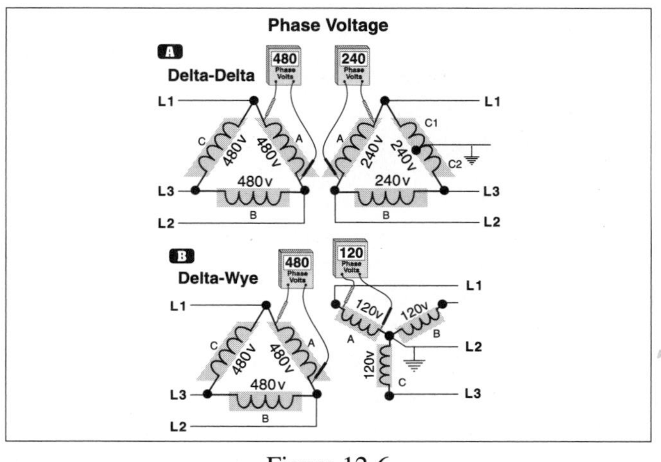

Figure 12-6

Phase Voltage

Delta Secondary – Unbalanced Current
The unbalanced load is calculated using:
$I_{Line\ 1} - I_{Line\ 3}$

Note: L_2 is called the high leg and its voltage to ground is approximately 208 volts, therefore no neutral loads are connected to this line.

Wye Secondary – Unbalanced Current

The unbalanced load is calculated using:

$$\sqrt{L_1^2 + L_2^2 + L_3^2 - (L_1 \times L_2 + L_2 \times L_3 + L_1 \times L_3)}$$

Winding

The *primary winding* is the winding(s) on the input side of a transformer and the *secondary winding* is the output side of a transformer.

Wye Connected

Wye-connected means a connection of three single-phase transformers of the same rated voltage to a common point (neutral) and the other ends are connected to the line conductors, Fig. 12–8.

12–1 *CURRENT FLOW*

When a load is connected to the secondary of a transformer, current will flow through the secondary conductor windings. The current flow in the secondary creates an electromagnetic field that opposes the primary electromagnetic field. The secondary flux lines effectively reduce the strength of the primary flux lines. As a result, less *counter-electromotive force* is generated in the primary winding conductors. With less *CEMF* to oppose the primary applied voltage, the primary current automatically increases in direct proportion to the secondary current, Fig. 12–9.

Note: The primary and secondary line currents are inversely proportional to the voltage ratio of the transformer. This means that the winding with the most number of turns will have a higher voltage and lower current as compared to the winding with the least number of turns, which will have a lower voltage and higher current, Fig. 12–10.

The following Tables show the current relationship between kVA and voltage for common size transformers.

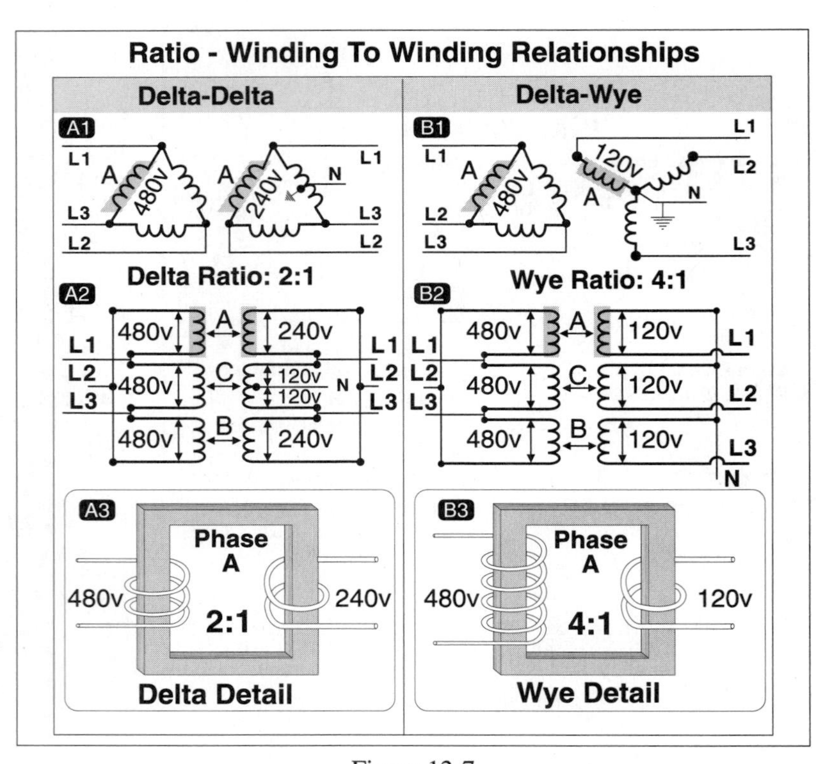

Figure 12-7

Ratio – Winding to Winding Relationships

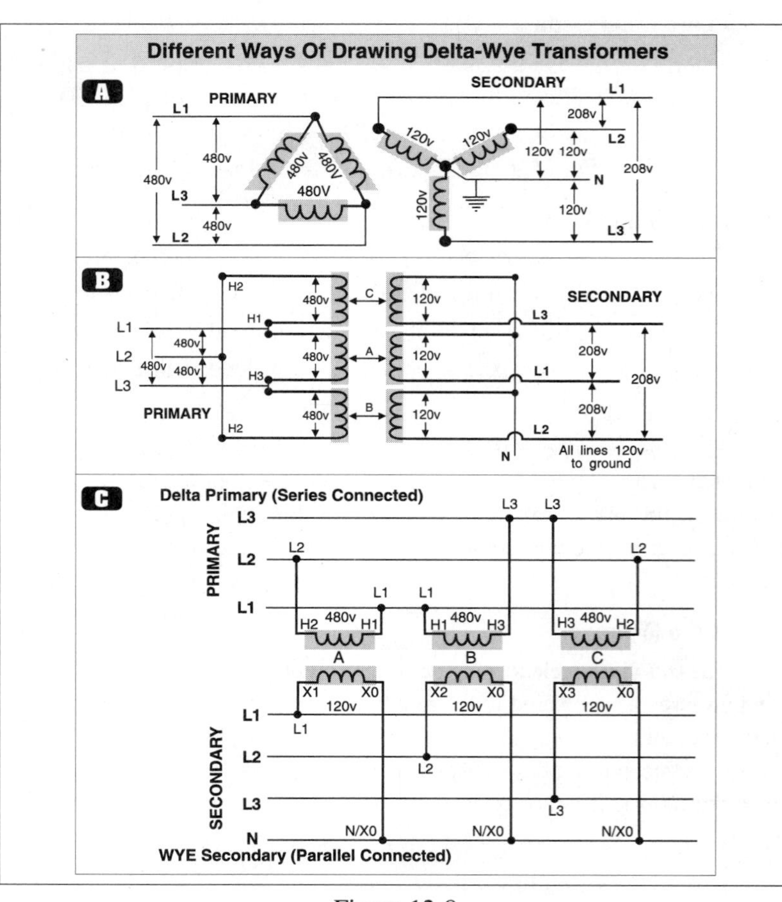

Figure 12-8

Different Ways of Drawing Delta/Wye Transformers

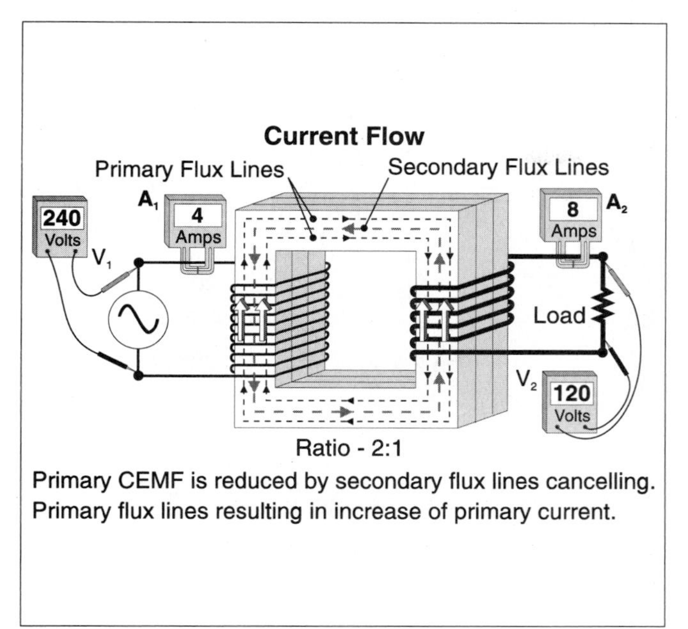

Figure 12-9

Current Flow

Step-Down Transformer Relationships

Primary

MORE WINDINGS (Turns)

- Higher Voltage
- Lower Current
- Smaller Wire

Secondary

LESS WINDINGS (Turns)

- Lower Voltage
- Higher Current
- Larger Wire

Figure 12-10

Step-Down Transformer Relationships

Single-Phase Transformers I=kVA/E			
kVA Rating	**Current at 208 Volts**	**Current at 240 Volts**	**Current at 480 Volts**
7.5	63 amperes	31 amperes	16 amperes
10	83 amperes	42 amperes	21 amperes
15	125 amperes	63 amperes	31 amperes
25	208 amperes	104 amperes	52 amperes
37.5	313 amperes	156 amperes	78 amperes

Three-Phase Transformers I=kVA/(E x √3)			
kVA Rating	**Current at 208 Volts**	**Current at 240 Volts**	**Current at 480 Volts**
15	42 amperes	36 amperes	18 amperes
22.5	63 amperes	54 amperes	27 amperes
30	83 amperes	72 amperes	36 amperes
37.5	104 amperes	90 amperes	45 amperes
45	125 amperes	108 amperes	54 amperes
50	139 amperes	120 amperes	60 amperes
75	208 amperes	180 amperes	90 amperes
112.5	313 amperes	271 amperes	135 amperes

PART A – *DELTA/DELTA TRANSFORMERS*

12–2 *DELTA TRANSFORMER VOLTAGE*

In a delta configured transformer, the *line voltage* equals the *phase voltage*, Fig. 12–11.

$E_{Line} = E_{Phase}$

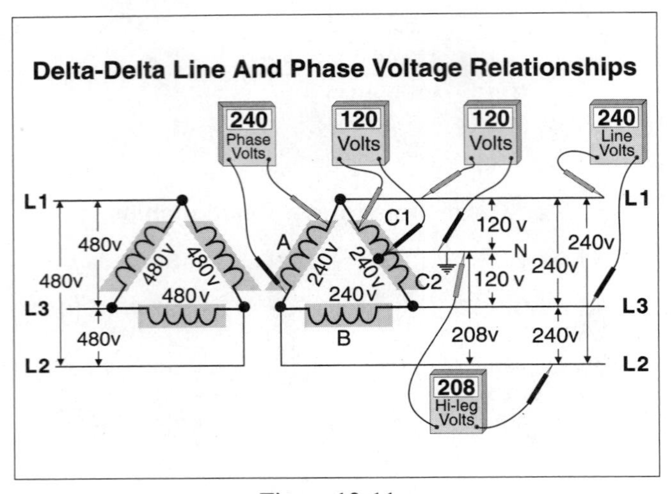

Figure 12-11

Delta/Delta Line/Phase Relationship

Delta High-Leg

L1 L3 L2 A B C1 C2 N 208v 240v 240v 120 v 208v 240v

Line 2 (L2) is the **High-Leg** conductor.

208 Hi-leg Volts

Figure 12-12

Delta High leg

Primary Delta Voltage

LINE Voltage	PHASE Voltage
L_1 to L_2 = 480 volts	Phase Winding A = 480 volts
L_2 to L_3 = 480 volts	Phase Winding B = 480 volts
L_3 to L_1 = 480 volts	Phase Winding C = 480 volts

Secondary Delta Voltage

LINE Voltage	PHASE Voltage	NEUTRAL Voltage
L_1 to L_2 = 240 volts	Phase Winding A = 240 volts	Neutral to L_1 = 120 volts
L_2 to L_3 = 240 volts	Phase Winding B = 240 volts	Neutral to L_2 = 208 volts
L_3 to L_1 = 240 volts	Phase Winding C = 240 volts	Neutral to L_3 = 120 volts

12–3 *DELTA HIGH LEG*

The term *high leg* (bastard leg, identified with an orange cover [384–3(e)]) is used to identify the conductor that has a higher voltage to ground. The high leg voltage is the vector sum of the voltage of transformer "A" and "C_1", or transformers "B" and "C_2," which equals 120 volts × 1.732 = 208 volts (for a 120/240 volt secondary). If the secondary voltage is 230/115 volts, the high leg voltage to ground (or neutral) would be 115 volts × 1.732 = 199.18 volts.

Note: The actual voltage is often less than the nominal system voltage because of voltage drop, Fig. 12–12.

12–4 *DELTA PRIMARY AND SECONDARY LINE CURRENTS*

In a delta configured transformer, the line current does not equal the phase current.

The primary or secondary *line current* of a transformer can be calculated by the formula:

$$I_{Line} = \frac{VA_{Line}}{E_{Line} \times \sqrt{3}}$$

❑ **Delta Primary Line Current Example**

What is the primary line current for a 150-kVA, 480- to 240/120-volt three-phase transformer, Fig. 12–13, Part A?

(a) 416 amperes (b) 360 amperes (c) 180 amperes (d) 144 amperes

- Answer: (c) 180 amperes

  $I_{Line} = VA_{Line}/(E_{Line} \times \sqrt{3})$

  I_{Line} = 150,000 VA/(480 volts × 1.732)

  I_{Line} = 180 amperes

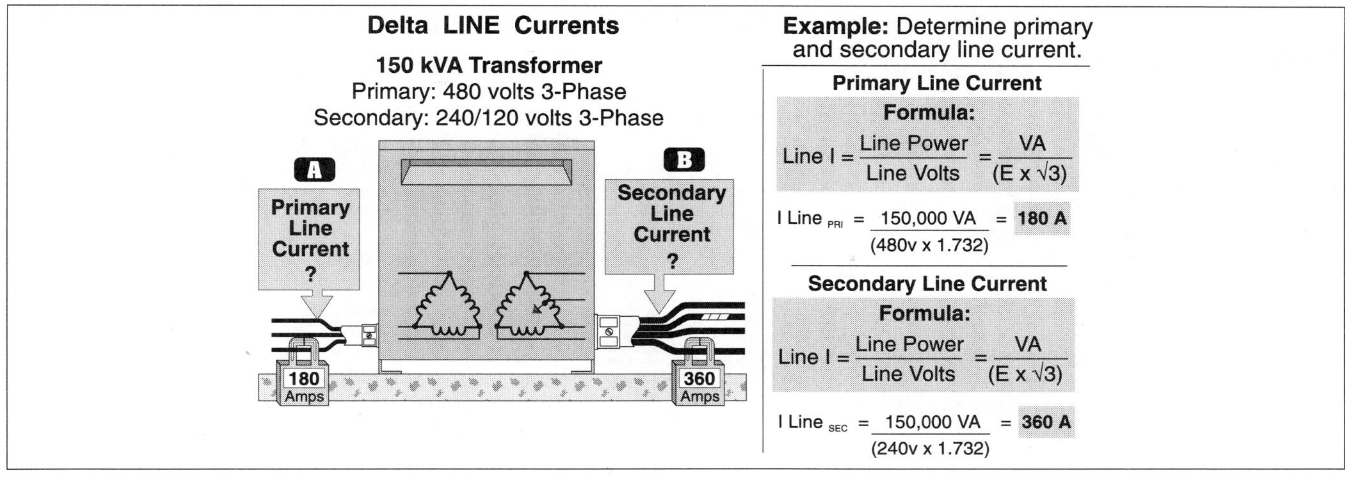

Figure 12-13

Delta Primary/Secondary Line Currents

❑ **Delta Secondary Line Current Example**

What is the secondary line current for a 150-kVA, 480- to 240-volt three-phase transformer, Fig. 12–13, Part B?

(a) 416 amperes (b) 360 amperes (c) 180 amperes (d) 144 amperes

- Answer: (b) 360 amperes

$I_{Line} = VA_{Line}/(E_{Line} \times \sqrt{3})$

$I_{Line} = 150{,}000 \text{ VA}/(240 \text{ volts} \times 1.732)$

$I_{Line} = 360 \text{ amperes}$

12–5 *DELTA PRIMARY OR SECONDARY PHASE CURRENTS*

The *phase current* of a transformer winding is calculated by dividing the *phase VA** by the *phase volts:*

$\mathbf{I_{Phase} = VA_{Phase}/E_{Phase}}$

The phase load of a three-phase 240-volt load = line load/3.
The phase load of a single-phase 240-volt load = line load.
The phase load of a single-phase 120-volt load = line load.

Note: Remember that it takes 3 phases or windings to create a 3-phase system.

❑ **Delta Primary Phase Current Example**

What is the primary phase current for a 150-kVA, 480- to 240/120-volt three-phase transformer, Fig. 12–14, Part A?

(a) 416 amperes (b) 360 amperes (c) 180 amperes (d) 104 amperes

- Answer: (d) 104 amperes

$I_{Phase} = \frac{VA_{Phase}}{E_{Phase}} \quad VA_{Phase} = \frac{150{,}000 \text{ VA}}{3 \text{ Phases}} = 50{,}000 \text{ VA} \quad E_{Phase} = 480 \text{ volts}$

$I_{Phase} = 50{,}000 \text{ VA}/480 \text{ volts}$

$I_{Phase} = 104 \text{ amperes}$

❑ **Delta Secondary Phase Current Example**

What is the secondary phase current for a 150-kVA, 480 to 240/120 volt three-phase transformer, Fig. 12–14, Part B?

(a) 416 amperes (b) 360 amperes (c) 208 amperes (d) 104 amperes

- Answer: (c) 208 amperes

$I_{Phase} = \frac{VA_{Phase}}{E_{Phase}}$

Phase power = 150,000 VA/3 = 50,000 VA

$I_{Phase} = 50{,}000 \text{ VA}/240 \text{ volts}$

$I_{Phase} = 208 \text{ amperes}$

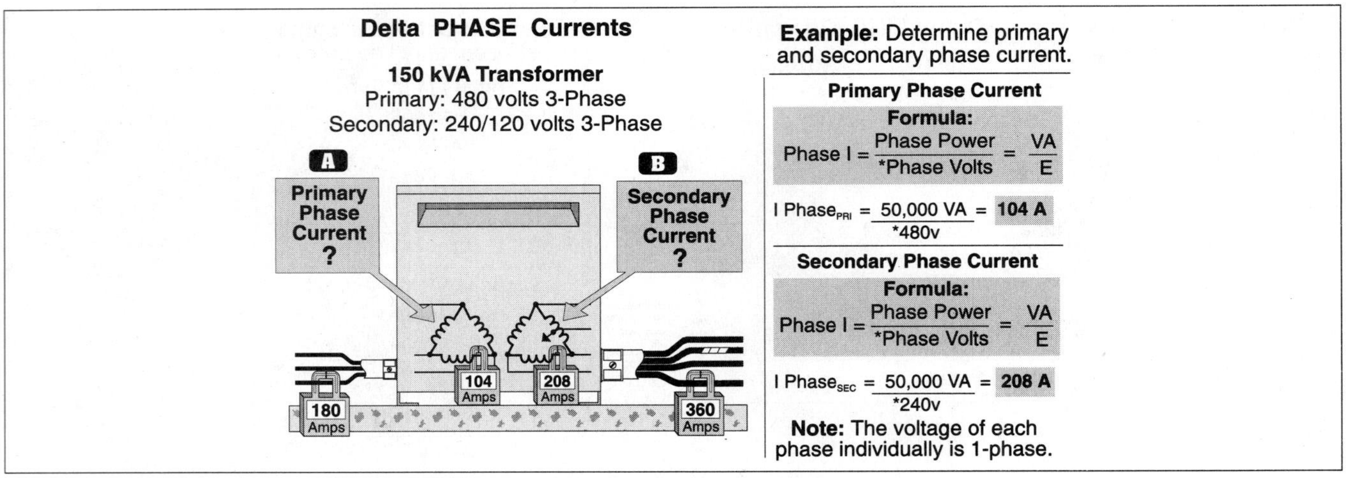

Delta PHASE Currents

150 kVA Transformer
Primary: 480 volts 3-Phase
Secondary: 240/120 volts 3-Phase

A — Primary Phase Current ?
B — Secondary Phase Current ?

180 Amps | 104 Amps | 208 Amps | 360 Amps

Example: Determine primary and secondary phase current.

Primary Phase Current

Formula:
Phase I = Phase Power / *Phase Volts = VA / E

I Phase$_{PRI}$ = 50,000 VA / *480v = **104 A**

Secondary Phase Current

Formula:
Phase I = Phase Power / *Phase Volts = VA / E

I Phase$_{SEC}$ = 50,000 VA / *240v = **208 A**

Note: The voltage of each phase individually is 1-phase.

Figure 12-14

Primary/Secondary Phase Relationships

12–6 *DELTA PHASE VERSUS LINE*

Since each line conductor from a delta transformer is actually connected to two transformer windings (phases), the effects of loading on the line (conductors) can be different than on the phase (winding).

❑ **Delta Phase VA, Three-Phase Example**

A 36-kVA, 240-volt three-phase load has the following effect on a delta system, Fig. 12–15:

LINE: Total line power = 36 kVA
Line current = $I_L = VA_{Line}/(E_{Line} \times \sqrt{3})$
I_L = 36,000 VA/(240 volts × $\sqrt{3}$)
I_L = 87 amperes, or
$I_L = I_P \times \sqrt{3}$
I_L = 50 amperes × 1.732
I_L = 87 amperes

PHASE: Phase power = 12 kVA (winding)
Phase current = $I_P = VA_{Phase}/E_{Phase}$
I_P = 12,000 VA/240 volts = 50 amperes, or
$I_P = I_L/\sqrt{3}$
I_P = 87 amperes/1.732 = 50 amperes

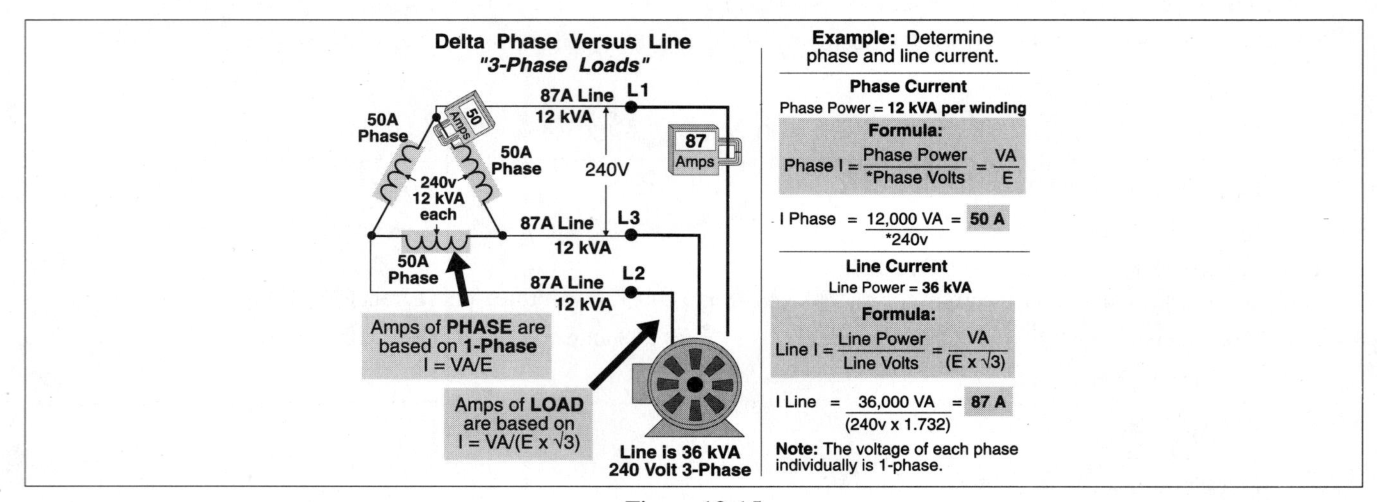

Figure 12-15

Delta Secondary Phase versus Line

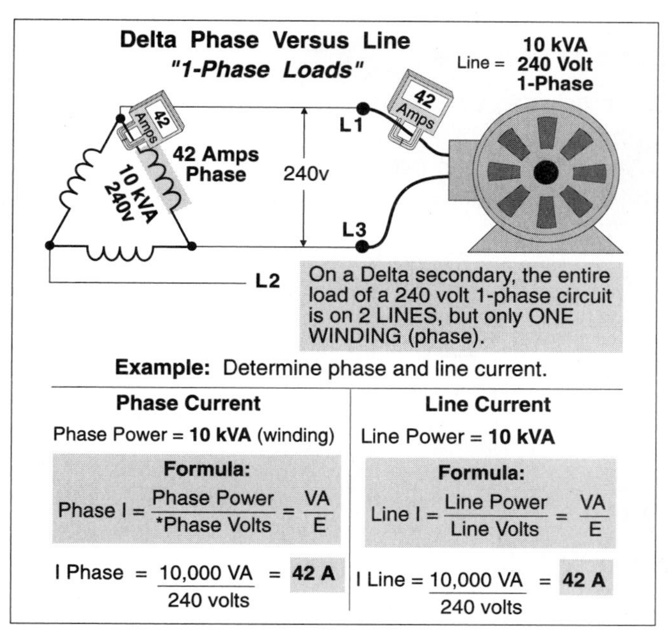

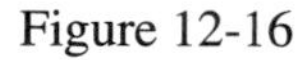

Figure 12-16

Delta Secondary Phase versus Line

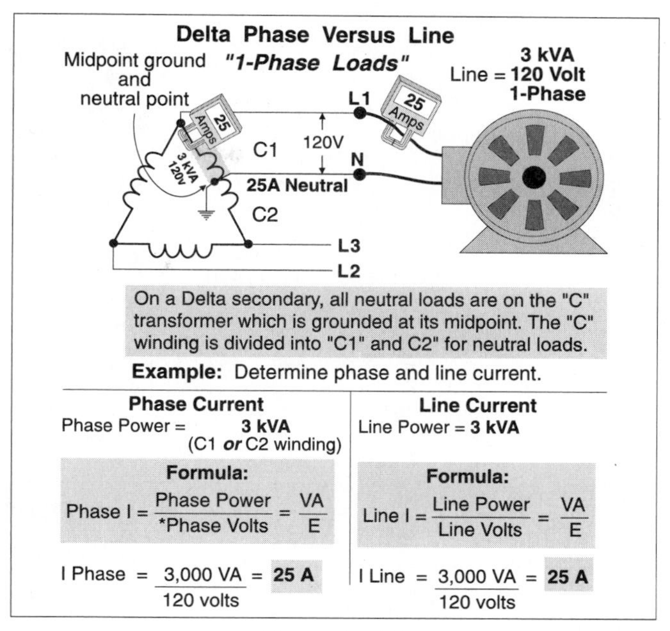

Figure 12-17

Delta Secondary Phase versus Line

- ❑ **Delta Phase VA, Single-Phase 240 Volt Example**

A 10-kVA, 240-volt single-phase load has the following effect on a delta/delta system, Fig. 12–16:

LINE: Total line power = 10 kVA
Line current = $I_L = VA_{Line}/E_{Line}$ VA
I_L = 10,000 VA/240 volts
I_L = 42 amperes

PHASE: Phase power = 10 kVA (winding)
Phase current = $I_{P} = VA_{Phase}/E_{Phase}$
I_P= 10,000 VA/240 volts = 42 amperes

- ❑ **Delta Phase VA, Single-phase 120 Volt Example**

A 3-kVA, 120 volt single-phase load has the following effect on the system, Fig. 12–17:

LINE: Line power = 3 kVA
Line current = $I_L = VA_{Line}/E_{Line}$ VA
I_L = 3,000 VA/120 volts
I_L = 25 amperes

PHASE: Phase power = 3 kVA (C_1 or C_2 winding)
Phase current = $I_{P} = VA_{Phase}/E_{Phase}$
I_P= 3,000 VA/120 volts = 25 amperes

12–7 *DELTA CURRENT TRIANGLE*

The three-phase line and phase current of a delta system are not equal, the difference is the square root of 3 ($\sqrt{3}$).

$\mathbf{I_L = I_P \times \sqrt{3}}$ $\mathbf{I_P = I_L / \sqrt{3}}$

The delta triangle (Fig. 12–18) can be used to calculate delta three-phase line and phase currents. Place your finger over the desired item and the remaining items show the formula to use.

12–8 *DELTA TRANSFORMER BALANCING*

To properly size a delta/delta transformer, the transformer phases (windings) must be *balanced.* The following steps are used to balance the transformer.

Step 1 → Determine the VA rating of all loads.
Step 2 → Balance three-phase loads: ⅓ on Phase A, ⅓ on Phase B, and ⅓ on Phase C.

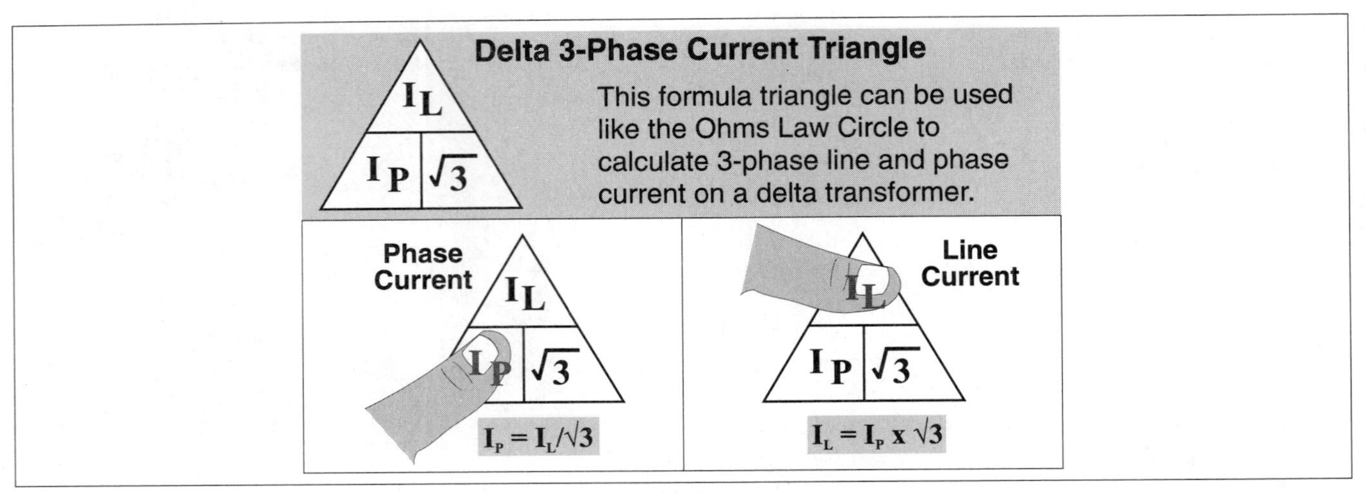

Figure 12-18

Delta Current Triangle

Step 3 → Balance single-phase 240 volt loads (largest to smallest): 100 percent on Phase A or Phase B. It is permissible to place some 240 volt single-phase load on Phase C when necessary for balance.

Step 4 → Balance the 120 volt loads (largest to smallest): 100 percent on C_1 or 100 percent on C_2.

❑ **Delta Transformer Balancing Example**

Balance and size a 480- to 240/120-volt three-phase transformer for the following loads: One 36 kVA three-phase heat strip, two 10 kVA single-phase 240-volt loads, and three 3 kVA 120-volt loads, Fig. 12–19.

	Phase A (L_1 and L_2)	Phase B (L_2 and L_3)	C_1 (L_1)	C_2 (L_3)	Line Total
(1) 36 kVA 240 volt, 3Ø	12 kVA	12 kVA	6 kVA	6 kVA	36,000 VA
(2) 10 kVA 240 volt, 1Ø	10 kVA	10 kVA			20,000 VA
(3) 3 kVA 120 volt, 1Ø			6 kVA *	3 kVA *	9,000 VA
	22 kVA	22 kVA	12 kVA	9 kVA	65,000 VA

*Indicates neutral load.

12–9 *DELTA TRANSFORMER SIZING*

Once you balance the transformer, you can size the transformer according to the load of each phase. The "C" transformer must be sized using two times the highest of "C_1" or "C_2." The "C" transformer is actually a single unit. If one side has a larger load, that side determines the transformer size.

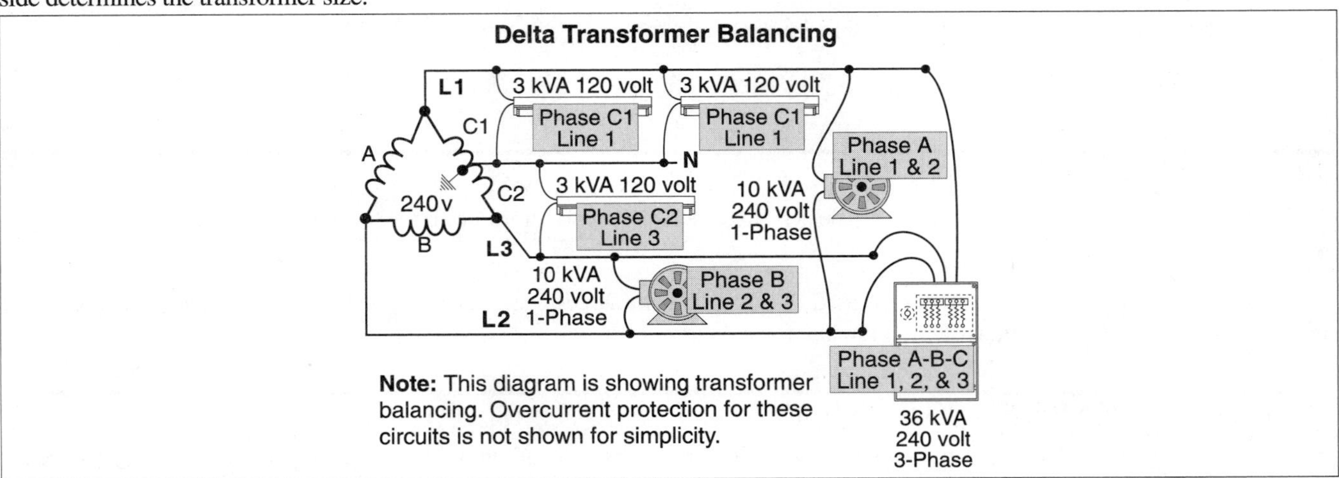

Figure 12-19

Delta Transformer Balancing

❑ **Delta Transformer Sizing Example**

What size 480- to 240/120-volt three-phase transformer is required for the following loads: One 36 kVA three-phase heat strip, two 10 kVA single-phase 240-volt loads, and three 3 kVA 120-volt loads?

(a) three 1Ø 25 kVA transformers
(b) one 3Ø 75 kVA transformer
(c) a or b
(d) none of these

- Answer: (c) a or b

  Phase winding A = 22 kVA

  Phase winding B = 22 kVA

  Phase winding C = (12 kVA of $C_1 \times 2$) = 24 kVA

12–10 *DELTA PANEL SCHEDULE IN kVA*

When balancing a panelboard in kVA be sure that three-phase loads are balanced ⅓ on each line and 240-volt single-phase loads are balanced ½ on each line.

❑ **Delta Panel Schedule in kVA Example**

Balance a 240/120-volt three-phase panelboard for the following loads in kVA: one 36 kVA three-phase heat strip, two 10 kVA single-phase 240-volt loads, and three 3 kVA 120-volt loads.

	Line 1	Line 2	Line 3	Line Total
36 kVA 240 volt, 3Ø	12 kVA	12 kVA	12 kVA	36,000 VA
10 kVA 240 volt, 1Ø	5 kVA	5 kVA		10,000 VA
10 kVA 240 volt, 1Ø		5 kVA	5 kVA	10,000 VA
(3) 3 kVA 120 volt, 1Ø	+ 6 kVA *		3 kVA *	9,000 VA
	23 kVA	22 kVA	20 kVA	65,000 VA

*Indicates neutral load.

12–11 *DELTA PANELBOARD AND CONDUCTOR SIZING*

When selecting the sizing of *panelboards* and *conductors,* we must balance the line loads in amperes.

❑ **Delta Conductors Sizing Example**

Balance a 240/120-volt three-phase panelboard for the following loads in amperes: one 36 kVA three-phase heat strip, two 10 kVA single-phase 240-volt loads, and three 3 kVA 120-volt loads.

	Line 1	Line 2	Line 3	Line Amperes
36 kVA 240 volt, 3Ø	87 Amps	87 Amps	87 Amps	36,000 VA/(240 volts × 1.732)
10 kVA 240 volt, 1Ø	42 Amps	42 Amps		10,000 VA/240 volts
10 kVA 240 volt, 1Ø		42 Amps	42 Amps	10,000 VA/240 volts
3 – 3 kVA 120 volt, 1Ø	+ 50 Amps *		25 Amps *	3,000 VA/120 volts
	179 Amps	171 Amps	154 Amps	

*Indicates neutral load.

Why balance the panel in amperes? Why not take the VA per phase and divide by phase voltage?

- Answer: Line current of a three-phase load is calculated by the formula I_L = VA/(VA Line × $\sqrt{3}$). In our examples 36,000 VA/(240 volts × 1.732) = 87 amperes per line. If we took the per line power of 12,000 VA and divided by one line voltage of 120 volts, we come up with an incorrect line current of 12,000 VA/120 volts = 100 amperes.

12–12 *DELTA NEUTRAL CURRENT*

The *neutral current* for a delta secondary is calculated as $Line_1$ neutral current less $Line_3$ neutral current.

❑ **Delta Neutral Current Example**

What is the neutral current for the following loads: one 36 kVA three-phase heat strip, two 10 kVA, single-phase 240-volt loads, and three 3 kVA 120-volt loads?

In our continuous example, $Line_1$ neutral current = 50 amperes and $Line_3$ neutral current = 25 amperes.

(a) 0 amperes
(b) 25 amperes
(c) 50 amperes
(d) none of these

- Answer: (b) 25 amperes.

  Neutral current = 50 amperes – 25 amperes = 25 amperes

	Line 1	Line 2	Line 3	Ampere Calculation
36 kVA 240 volts, 3Ø	87 amps	87 amps	87 amps	36,000 VA/(240 volts × 1.732)
10 kVA 240 volts, 1Ø	42 amps	42 amps		10,000 VA/240 volts
10 kVA 240 volts, 1Ø		42 amps	42 amps	10,000 VA/240 volts
3 kVA 120 volts, 1Ø	**50 amps** *		**25 amps** *	3,000 VA/120 volts

* Indicates neutral (120-volt) loads

12–13 *DELTA MAXIMUM UNBALANCED LOAD*

The *maximum unbalanced load* (neutral) is the largest phase 120 volt neutral current. The neutral current is the actual current on the neutral conductor.

❑ **Delta Maximum Unbalanced Load Example**

What is the maximum unbalanced load for the following loads: one 36-kVA three-phase heat strip, two 10-kVA single-phase 240-volt loads, and three 3-kVA 120-volt loads?

(a) 0 amperes (b) 25 amperes (c) 50 amperes (d) none of these

- Answer: (c) 50 amperes

  Maximum unbalanced current equals the largest line neutral current.

	Line 1	Line 2	Line 3	Ampere Calculation
36 kVA 240 volt, 3Ø	87 amps	87 amps	87 amps	36,000 VA/(240 volts × 1.732)
10 kVA 240 volt, 1Ø	42 amps	42 amps		10,000 VA/240 volts
10 kVA 240 volt, 1Ø		42 amps	42 amps	10,000 VA/240 volts
3 kVA 120 volt, 1Ø	**50 amps** *		**25 amps** *	3,000 VA/120 volts

*Indicates neutral (120-volt) loads

12–14 *DELTA/DELTA EXAMPLE*

One – Dishwasher 4.5 kW, 120v

One – 10 HP A/C motor 240-volt, 3Ø

One – 18 kW 240-volt 3Ø heat strip

Two – 14 kW ranges, 240-volt 1Ø

Two – 3 HP motors, 240-volt, 1Ø

One – 10 kW water heater, 240-volt, 1Ø

Eight – 1.5 kW 120-volt lighting circuits

Motor VA

VA 1Ø = Table Volts x Table Amperes, Table 430–148

3 HP, VA = E × I

VA = 230* volts × 17 amperes = 3,910 VA, Table 430–150

10 HP, VA = 230 volts* × 28 amperes × 1.732 = 11,154

VA 3Ø = Table Volts × Table Amperes × $\sqrt{3}$, Table 430–150

Note: Some exam testing agencies use actual volts (240 volts) instead of table volts.

	Phase A (L_1 & L_2)	Phase B (L_2 & L_3)	C_1 (L_1)	C_2 (L_3)	Line Total
Heat 18 kW (A/C omitted)	6,000 VA	6,000 VA	3,000 VA	3,000 VA	18,000 VA
14 kW Ranges 240 volt, 1Ø	14,000 VA	14,000 VA			28,000 VA
10 kW Water Heater 240 volt, 1Ø	10,000 VA				10,000 VA
3 HP 240 volt, 1Ø		3,910 VA			3,910 VA
3 HP 240 volt, 1Ø		3,910 VA			3,910 VA
Dishwasher 4.5 kW, 120 volt			4,500 VA *		4,500 VA
Lighting (8 – 1.5 kW), 120 volt (3 on L_1, 5 on L_3)			4,500 VA *	7,500 VA*	12,000 VA
	30,000 VA	27,820 VA	12,000 VA	10,500 VA	80,320 VA

*Indicates neutral (120 volt) loads.

Note: The phase totals (30,000 VA, 27,820 VA, 22,500 VA) should add up to the Line total (80,320 VA). This is done as a check to make sure all items have been accounted for and added correctly.

10 horsepower, VA = Table Volts × Table Amperes × 1.732

VA = 230 volts × 28 amperes × 1.732 = 11,154 VA (omit)

➛ **Transformer Size Question**

What size transformers are required?

(a) three 30 kVA single-phase transformers (b) one 90 kVA three-phase transformer

(c) a or b (d) none of these

- Answer: (c) a or b

 Phase A = 30 kVA

 Phase B = 28 kVA

 Phase C = 24 kVA (12 kVA × 2)

➛ **High leg Voltage Question**

What is the high leg voltage to the ground?

(a) 120 volts (b) 208 volts (c) 240 volts (d) None of these

- Answer: (b) 208 volts

 120 volts × 1.732 = 208 volts, in effect, the vector sum of the voltage of transformer "A" and "C_1" or transformers "B" and "C_2." 240 volts + 120 volts vectorial = 208 volts

➛ **Neutral kVA Question**

What is the maximum kVA on the neutral?

(a) 3 kVA (b) 6 kVA (c) 7.5 kVA (d) 9 kVA

- Answer: (d) 9 kVA

 If Phase C_2 loads are not on then phase C_1 neutral loads would impose a 9 kVA load to neutral (4.5 kW + 4.5 kW). This does not include the 6 kVA on transformer "C" from the heat load since there are no neutrals involved.

➛ **Maximum Unbalanced Current Question**

What is the maximum unbalanced load on the neutral?

(a) 25 amperes (b) 50 amperes (c) 75 amperes (d) 100 amperes

- Answer: (c) 75 amperes

 I = P/E = 9,000 VA/120 volts = 75 amperes.

 The neutral must be sized to the carry maximum unbalanced neutral current, which in this case is 75 amperes.

➛ **Neutral Current Question**

What is the current on the neutral?

(a) 0 amperes (b) 13 amperes (c) 25 amperes (d) none of these

- Answer: b) 13 amperes

 C_1 = 9,000 VA/120 volts = 75 amperes

 C_2 = 7,500 VA/120 volts = 62.5 amperes

 Neutral current = 75 amperes less 62.5 amperes = 12.5 amperes

➛ **Phase VA Load, Single-phase Question**

What is the phase load of each 3-horsepower 240-volt single-phase motor?

(a) 1,855 VA (b) 3,910 VA (c) 978 VA (d) none of these

- Answer: (b) 3,910 VA

 VA = 230 volts × 28 amperes. On a delta system, a 240-volt single-phase load is entirely on one transformer phase.

➛ **Phase VA Load, Three-phase Question**

What is the phase load for the three-phase, 10-horsepower motor?

(a) 11,154 VA (b) 3,718 VA (c) 5,577 VA (d) none of these

- Answer: (b) 3,718 VA

 VA = Volts × Amperes × 1.732

 VA = 230 volts × 28 amperes × 1.732, VA = 11,154 VA

 11,154 VA/3 phases = 3,718 VA per phase

➛ **High leg Conductor Size Question**

If only the three-phase 18 kW load is on the high leg, what is the current on the high leg conductor?

(a) 25 amperes (b) 43 amperes (c) 73 amperes (d) 97 amperes

- Answer: (b) 43 amperes

  To calculate current on any conductor:

  $I = P/(E \times \sqrt{3})$ (three-phase)

  $I = 18{,}000 \text{ VA}/(240 \text{ volts} \times 1.732) = 43$ amperes

➛ **Voltage Ratio Question**

What is the phase voltage ratio of the transformer?

(a) 4:1 (b) 1:4 (c) 1:2 (d) 2:1

- Answer: (d) 2:1

  480 primary phase volts to 240 secondary phase volts

❑ **Panel Loading VA Example**

Balance the loads on the panelboard in VA.

	Line 1	Line 2	Line 3	Line Total
18 kW Heat 3Ø	6,000 VA	6,000 VA	6,000 VA	18,000 VA
14 kW Ranges 240 volts, 1Ø	7,000 VA	7,000 VA		14,000 VA
14 kW Ranges 240 volts, 1Ø		7,000 VA	7,000 VA	14,000 VA
Water Heater, 240 volts, 1Ø	5,000 VA	5,000 VA		10,000 VA
3 hp 240 volts, 1Ø Motor		1,955 VA	1,955 VA	3,910 VA
3 hp 240 volts, 1Ø Motor		1,955 VA	1,955 VA	3,910 VA
Dishwasher 1Ø 120V	4,500 VA*			4,500 VA
Lighting 1Ø 120V	+ 4,500 VA*		7,500 VA *	4,500 VA
	27,000 VA	28,910 VA	24,410 VA	80,320 VA

* Indicates neutral (120 volt) loads.

❑ **Panel Sizing, Amperes Example**

Balance the loads from the previous example on the panelboard in amperes.

	Line 1	Line 2	Line 3	Ampere Calculation
18 kW Heat 3Ø	43 amps	43 amps	43 amps	18,000 VA/(240 volts × 1.732)
14 kW Ranges 240 volts, 1Ø	58 amps	58 amps		14,000 VA/240 volts
14 kW Ranges 240 volts, 1Ø		58 amps	58 amps	14,000 VA/240 volts
Water Heater, 240 volts, 1Ø	42 amps	42 amps		10,000 VA/240 volts
3 hp 240 volts, 1Ø Motor		17 amps	17 amps	FLC
3 hp 240 volts, 1Ø Motor		17 amps	17 amps	FLC
Dishwasher 1Ø 120 volts	38 amps *			4,500 VA/120 volts
Lighting 1Ø 120 volts	+ 38 amps *		63 amps *	7,500 VA/120 volts
(3 loads on L_1 and 5 loads on L_3)	219 amps	235 amperes	198 amps	

*Indicates neutral (120 volt) loads.

PART B – *DELTA/WYE TRANSFORMERS*

12–15 *WYE TRANSFORMER VOLTAGE*

In a wye configured transformer, the *line voltage* does not equal the *phase voltage*, Fig. 12–20.

> **Reminder:** In a delta configured transformer, line current does not equal phase current.

Primary Voltage (Delta)

LINE Voltage	PHASE Voltage
L_1 to L_2 = 480 volts	Phase Winding A = 480 volts
L_2 to L_3 = 480 volts	Phase Winding B = 480 volts
L_3 to L_1 = 480 volts	Phase Winding C = 480 volts

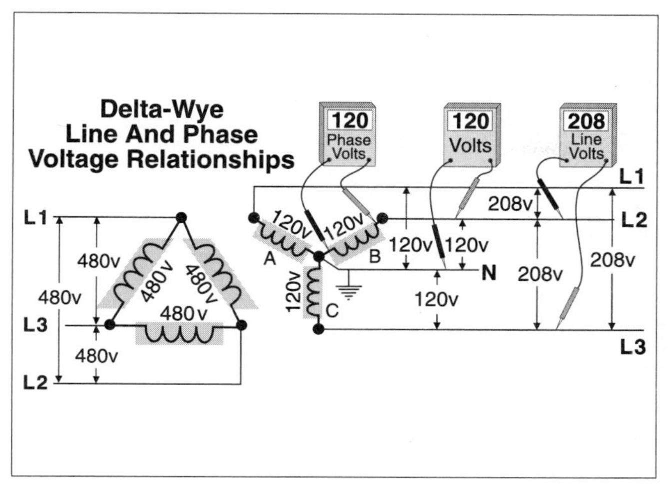

Figure 12-20

Delta/Wye Voltage Relationships

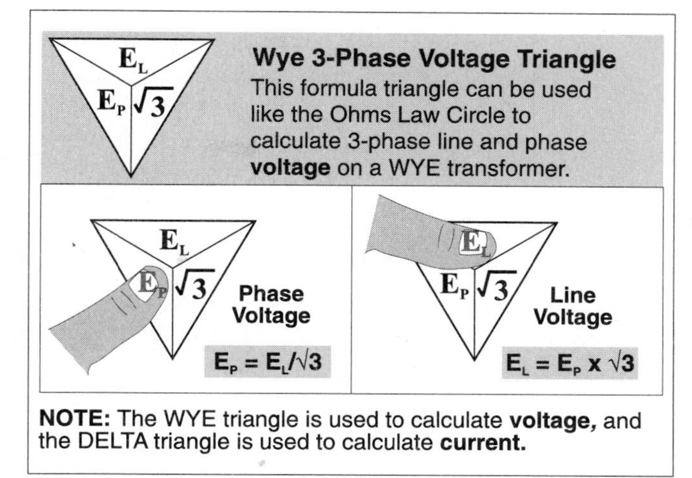

Figure 12-21

Wye Voltage Triangle

Secondary Voltage (Wye)

LINE Voltage	PHASE Voltage	NEUTRAL Voltage
L_1 to L_2 = 208 volts	Phase Winding A = 120 volts	Neutral to L_1 = 120 volts
L_2 to L_3 = 208 volts	Phase Winding B = 120 volts	Neutral to L_2 = 120 volts
L_3 to L_1 = 208 volts	Phase Winding C = 120 volts	Neutral to L_3 = 120 volts

12–16 *WYE LINE AND PHASE VOLTAGE TRIANGLE*

The *line voltage* and *phase voltage* of a wye system are not the same. The difference is a factor of the $\sqrt{3}$.

$\mathbf{E_{Line} = E_{Phase} \times \sqrt{3}}$ **or** $\mathbf{E_{Phase} = E_{Line}/\sqrt{3}}$

The wye voltage triangle (Fig. 12–21) can be used to calculate wye three-phase line and phase voltage. Place your finger over the desired item, and the remaining items show the formula to use. These formulas can be used when either one of the phase or line voltages is known.

12–17 *WYE TRANSFORMERS CURRENT*

In a wye configured transformer, the three-phase and single-phase 120 volt *line current* equals the *phase current*, Fig. 12–22.

$\mathbf{I_{Phase} = I_{Line}}$

12–18 *WYE LINE CURRENT*

The **line current** of both delta and wye transformers can be calculated by the following formula:

$$\mathbf{I_{Line} = VA_{Line}/(E_{Line} \times \sqrt{3})}$$

❑ **Primary Line Current Question**

What is the primary line current for a 150 kVA 480- to 208Y/120-volt three-phase transformer, Fig. 12–23?

(a) 416 amperes (b) 360 amperes
(c) 180 amperes (d) 144 amperes

- Answer: (c) 180 amperes.

$I_{Line} = VA_{Line}/(E_{Line} \times \sqrt{3})$

I_{Line} = 150,000 VA/(480 volts × 1.732)

I_{Line} = 180 amperes

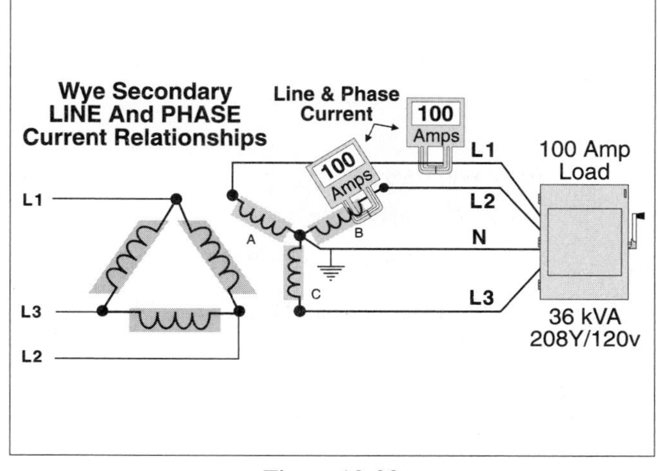

Figure 12-22

Wye Secondary Line and Phase Current

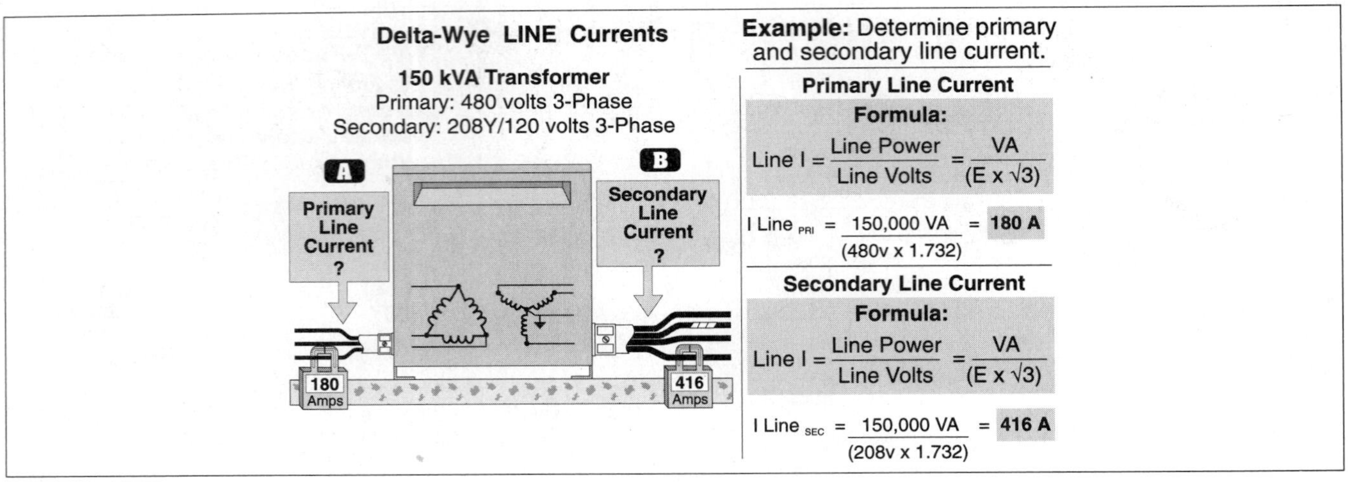

Delta-Wye LINE Currents

150 kVA Transformer
Primary: 480 volts 3-Phase
Secondary: 208Y/120 volts 3-Phase

A Primary Line Current ?
B Secondary Line Current ?
180 Amps
416 Amps

Example: Determine primary and secondary line current.

Primary Line Current

Formula:

$$\text{Line I} = \frac{\text{Line Power}}{\text{Line Volts}} = \frac{\text{VA}}{(\text{E} \times \sqrt{3})}$$

$$\text{I Line}_{PRI} = \frac{150{,}000 \text{ VA}}{(480\text{v} \times 1.732)} = \textbf{180 A}$$

Secondary Line Current

Formula:

$$\text{Line I} = \frac{\text{Line Power}}{\text{Line Volts}} = \frac{\text{VA}}{(\text{E} \times \sqrt{3})}$$

$$\text{I Line}_{SEC} = \frac{150{,}000 \text{ VA}}{(208\text{v} \times 1.732)} = \textbf{416 A}$$

Figure 12-23

Delta/Wye Primary/Secondary Line Current

❑ **Secondary Line Current Question**

What is the secondary line current for a 150-kVA, 480- to 208Y/120-volt, three-phase transformer, Fig. 12–23?

(a) 416 amperes (b) 360 amperes (c) 180 amperes (d) 144 amperes

- Answer: (a) 416 amperes

$I_{Line} = VA_{Line}/(E_{Line} \times \sqrt{3})$

I_{Line} = 150,000 VA/(208 volts × 1.732)

I_{Line} = 416 amperes

12–19 *WYE PHASE CURRENT*

The phase current of a wye configured transformer winding is the same as the line current. The phase current of a delta configured transformer winding is less than the line current by the $\sqrt{3}$.

❑ **Primary Phase Current Question**

What is the primary (delta) phase current for a 150-kVA, 480- to 208Y/120-volt, three-phase transformer, Fig. 12–24, Part A?

(a) 416 amperes (b) 360 amperes (c) 180 amperes (d) 104 amperes

- Answer: (d) 104 amperes

$$I_{Line} = \frac{VA_{Line}}{E_{Line} \times \sqrt{3}}$$

I_{Line} = 150,000 VA/(480 volts × 1.732)

I_{Line} = 180 amperes

$$I_{Phase} = \frac{VA_{Phase}}{E_{Phase}}$$

I = 50,000 VA/480 volts

I_{Phase} = 104 amperes, or

$I_{Phase} = I_{Line}/\sqrt{3} = 108/1.732$ = 104 amperes

❑ **Secondary Phase Current Question**

What is the secondary (wye) phase current for a 150-kVA, 480- to 208Y/120-volt, three-phase transformer, Fig. 13-24, Part B?

(a) 416 amperes (b) 360 amperes (c) 208 amperes (d) 104 amperes

- Answer: (a) 416 amperes

$$I_{Line} = \frac{VA_{Line}}{(E_{Line} \times \sqrt{3})}$$

I_{Line} = 150,000 VA/(208 volts × 1.732)

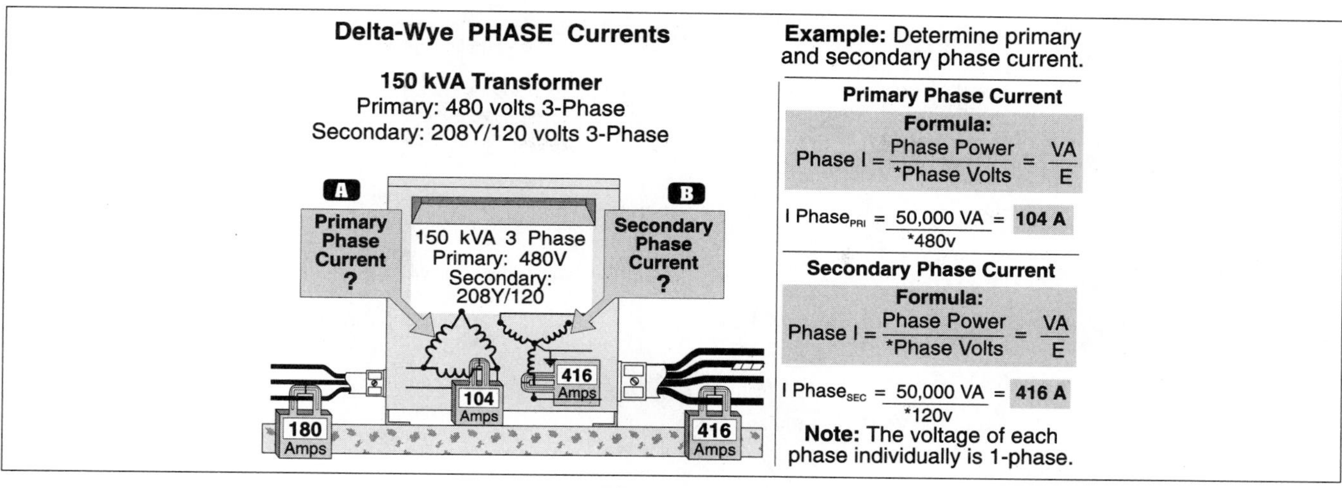

Figure 12-24

Delta/Wye Primary/Secondary Phase Current

I_{Line} = 416 amperes, or

$$I_{Phase} = \frac{VA_{Phase}}{E_{Phase}}$$

I_{Phase} = 50,000 VA/120 volts

I_{Phase} = 416 ampere

12–20 *WYE PHASE VERSUS LINE CURRENT*

Since each line conductor from a wye transformer is connected to a different transformer winding (phase), the effects of three-phase loading on the line are the same as the phase, Fig. 12–25.

❑ **Phase VA Load, Three-Phase Example**

A 36-kVA, 208 volt, three-phase load has the following effect on the system:

LINE: Line power = 36 kVA

Line current = I_L = VA/(E × $\sqrt{3}$)

I_L = 36,000 VA/(208 volts × $\sqrt{3}$)

I_L = 100 amperes

PHASE: Phase power = 12 kVA (any winding)

Phase current = I_P = VA_{Phase}/E_{Phase}

I_P = 12,000 VA/120 volts = 100 amperes

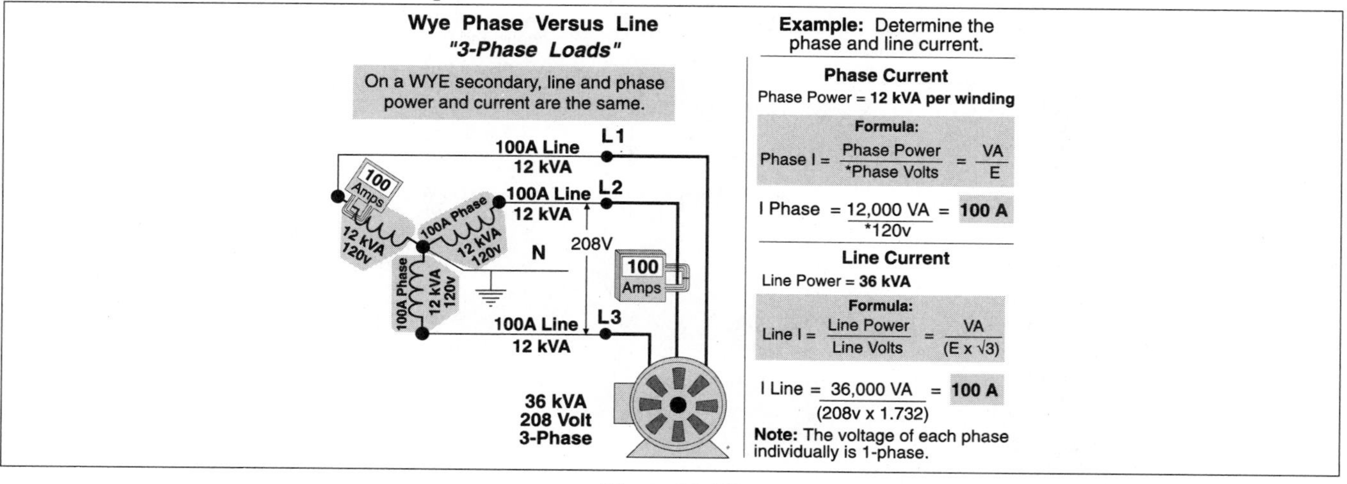

Figure 12-25

Wye Secondary Phase vs. Line

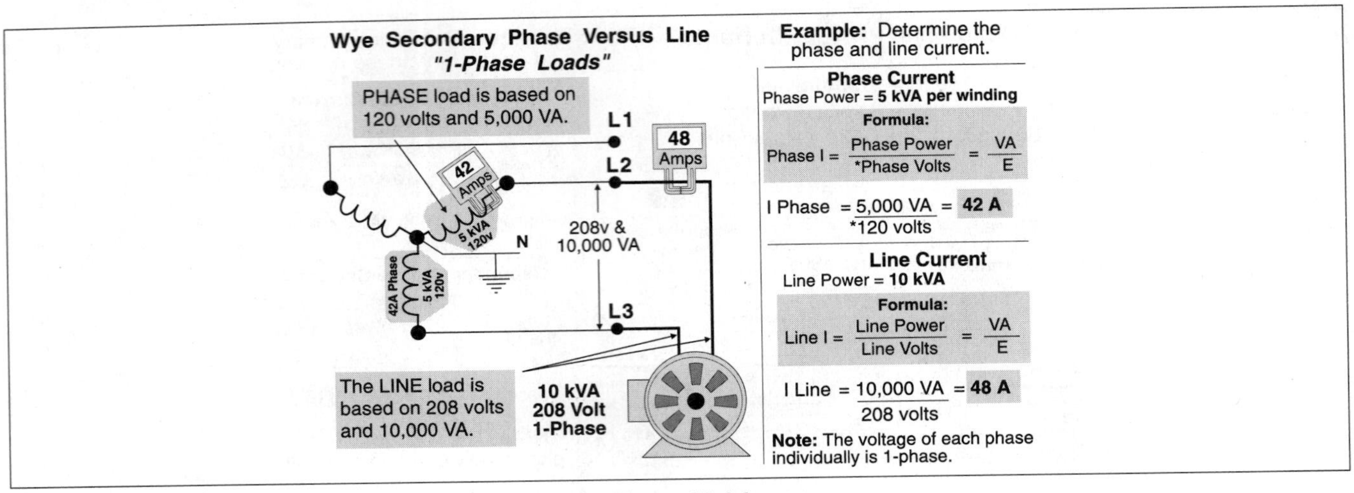

Wye Secondary Phase Versus Line
"1-Phase Loads"

PHASE load is based on 120 volts and 5,000 VA.

L1, L2, L3, N — 42 Amps — 48 Amps — 5 kVA 120v — 42A Phase — 208v & 10,000 VA

The LINE load is based on 208 volts and 10,000 VA.

10 kVA 208 Volt 1-Phase

Example: Determine the phase and line current.

Phase Current
Phase Power = **5 kVA per winding**
Formula: Phase I = Phase Power / *Phase Volts = VA / E
I Phase = 5,000 VA / *120 volts = **42 A**

Line Current
Line Power = **10 kVA**
Formula: Line I = Line Power / Line Volts = VA / E
I Line = 10,000 VA / 208 volts = **48 A**

Note: The voltage of each phase individually is 1-phase.

Figure 12-26

Wye Secondary Phase vs. Line

❑ Phase VA Load, Single-Phase 208 Volt Example

A 10-kVA, 208-volt, single-phase load has the following effect on the system, Fig. 13-26:

LINE: Line power = 10 kVA
Line current = I_L = VA/E
I_L = 10,000 VA/208 volts
I_L = 48 amperes
PHASE: Phase power = 10 kVA (winding)
Phase current = I_P = VA_{Phase}/E_{Phase}
I_P= 5,000 VA/120 volts
I_P= 42 amperes

❑ Phase VA Load, Single-phase 120 Volt Example

A 3-kVA, 120-volt single-phase load has the following effect on the system, Fig. 12–27:

LINE: Line power = 3 kVA
Line current = I_L = VA/E
I_L = 3,000 VA/120 volts
I_L = 25 amperes
PHASE: Phase power = 3 kVA (any winding)
Phase current = I_P = VA_{Phase}/E_{Phase}
I_P = 3,000 VA/120 volts
I_P=25 amperes

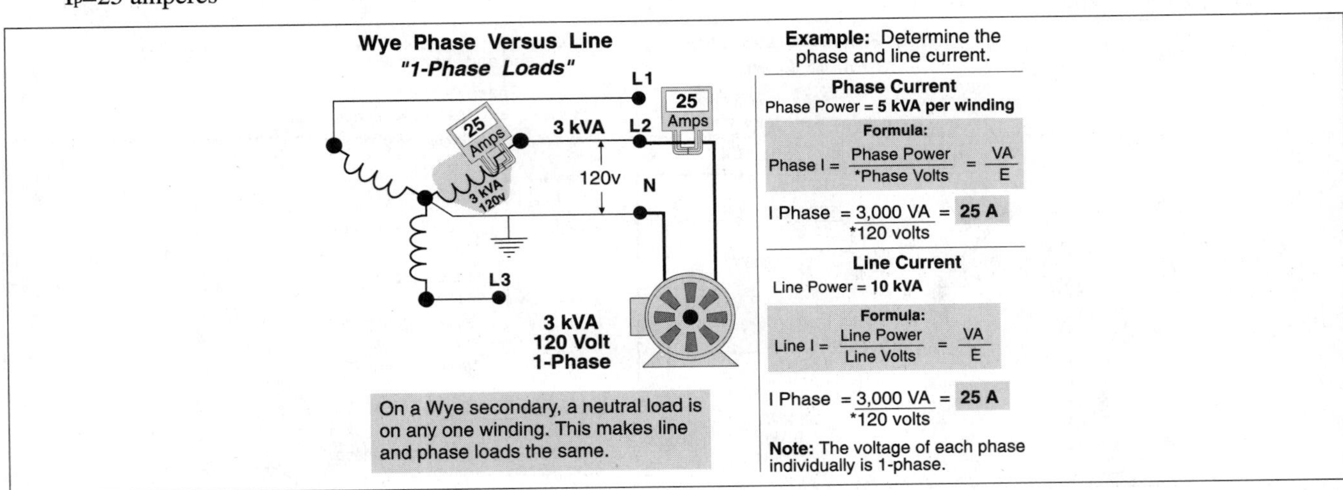

Figure 12-27

Wye Secondary Phase vs. Line

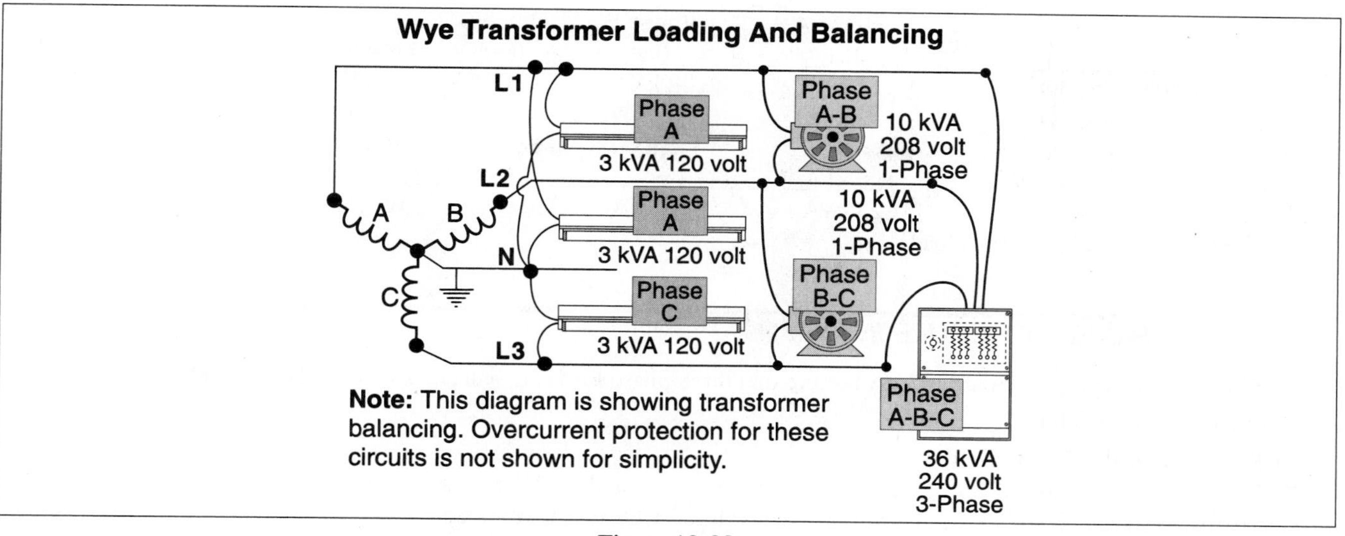

Figure 12-28

Wye Transformer Loading and Balancing

12–21 *WYE TRANSFORMER LOADING AND BALANCING*

To properly size a delta/wye transformer, the secondary transformer phases (windings) or the line conductors must be balanced. Balancing the panel (line conductors) is identical to balancing the transformer for wye configured transformers! The following steps should be helpful to balance the transformer.

Step 1: → Determine the VA rating of all loads.
Step 2: → Balance 3Ø loads: ⅓ on Phase A, ⅓ on Phase B, and ⅓ on Phase C.
Step 3: → Balance single-phase 208-volt loads (largest to smallest): 50 percent on each phase (A to B, B to C, or A to C).
Step 4: → Balance the 120-volt loads (largest to smallest): 100 percent on any phase.

❑ **Transformer Loading and Balancing Example**

Balance and size a 480- to 208Y/120-volt three-phase transformer for the following loads: One 36 kVA, three-phase heat strip, two 10-kVA, single-phase 208-volt loads, and three 3 kVA, 120-volt loads, Fig. 12–28.

	Phase A (L1)	Phase B (L2)	Phase C (L3)	Line Total
36 kVA 208 volt, 3Ø	12 kVA	12 kVA	12 kVA	36 kVA
10 kVA 208 volt, 1Ø	5 kVA	5 kVA		10 kVA
10 kVA 208 volt, 1Ø		5 kVA	5 kVA	10 kVA
(3) 3 kVA 120 volt, 1Ø	+ 6 kVA *		3 kVA *	9 kVA
	23 kVA	22 kVA	20 kVA	65 kVA

* Indicates neutral (120 volt) loads.

12–22 *WYE TRANSFORMER SIZING*

Once you balance the transformer, you can size the transformer according to the load of each phase.

❑ **Transformer Sizing Example**

What size 480- to 208Y/120-volt, three-phase phase transformer is required for the following loads: one 36-kVA, three-phase heat strip, two 10-kVA single-phase 208-volt loads, and three 3-kVA 120-volt loads?

(a) three 1Ø 25 kVA transformer
(b) one 3Ø 75 kVA transformer
(c) a or b
(d) none of these

- Answer: (c) a or b

  Phase A = 23 kVA

  Phase B = 22 kVA

  Phase C = 20 kVA

	Phase A (L_1)	Phase B (L_2)	Phase C (L_3)	Line Total
36 kVA 208 volts, 3Ø	12 kVA	12 kVA	12 kVA	36 kVA
10 kVA 208 volts, 1Ø	5 kVA	5 kVA		10 kVA
10 kVA 208 volts, 1Ø		5 kVA	5 kVA	10 kVA
(3) 3 kVA 120 volts, 1Ø	+ 6 kVA *		3 kVA *	9 kVA
	23 kVA	22 kVA	20 kVA	65 kVA

* Indicates neutral (120 volt) loads.

12–23 *WYE PANEL SCHEDULE IN kVA*

When balancing a panelboard in kVA be sure that three-phase loads are balanced ⅓ on each line and 208 volt 1Ø loads are balanced ½ on each line.

❑ **Panel Schedule, kVA Example**

Balance and size a 208Y/120-volt, three-phase phase panelboard in kVA for the following loads: one 36-kVA, three-phase heat strip, two 10-kVA, single-phase 208-volt loads, and three 3-kVA, 120-volt loads.

	Line 1	Line 2	Line 3	Line Total
36 kVA 208 volt, 3Ø	12 kVA	12 kVA	12 kVA	36,000 VA
10 kVA 208 volt, 1Ø	5 kVA	5 kVA		10,000 VA
10 kVA 208 volt, 1Ø		5 kVA	5 kVA	10,000 VA
3 kVA 120 volt, 1Ø	6 kVA *			6,000 VA
3 kVA 120 volt, 1Ø			3 kVA *	3,000 VA
	23 kVA	22 kVA	20 kVA	65,000 VA

* Indicates neutral (120 volt) loads.

12–24 *WYE PANELBOARD AND CONDUCTOR SIZING*

When selecting and sizing the panelboard and conductors, we must balance the line loads in amperes.

❑ **Panelboard Conductors Sizing Example**

Balance and size a 208Y/120-volt, three-phase panelboard in amperes for the following loads: one 36-kVA, three-phase heat strip, two 10-kVA, single-phase 208-volt loads, and three 3-kVA, 120-volt loads.

	Line 1	Line 2	Line 3	Ampere Calculation
36 kVA 208 volt, 3Ø	100 amps	100 amps	100 amps	36,000/(208 volts × 1.732)
10 kVA 208 volt, 1Ø	48 amps	48 amps		10,000 VA/208 volts
10 kVA 208 volt, 1Ø		48 amps	48 amps	10,000 VA/208 volts
2 – 3 kVA 120 volt, 1Ø	50 amps *			6,000 VA/120 volts
3 kVA 120 volt, 1Ø	+		25 amps *	3,000 VA/120 volts
	198 amps	196 amps	173 amps	

Why balance the panel in amperes? Why not take wattage or VA per phase and divide by one phase voltage?

- Answer: Line current of a single-phase load is calculated by the formula, I_L = VA Line/ELine.

  In our examples, I_L = 10,000 VA/208 volts = 48 amperes per line.

  If we took the per line power of 5,000 and divided by one line voltage of 120 volts, we come up with an incorrect line current of 5,000 VA/120 volts = 42 amperes.

12–25 *WYE NEUTRAL CURRENT*

To determine the neutral current of a wye system, we use the following formula:

I neutral = $\sqrt{L_1^2 + L_2^2 + L_3^2 - (L_1 \times L_2 + L_2 \times L_3 + L_1 \times L_3)}$

❑ **Neutral Current Example**

Balance and size the neutral current for the following loads: one 36-kVA three-phase heat strip, two 10-kVA single-phase 208-volt loads, and three 3-kVA 120-volt loads. L_1 = 50 amperes, L_2 = 0 amperes, L_3 = 25 amperes.

(a) 0 amperes (b) 25 amperes (c) 35 amperes (d) 50 amperes

- Answer: (c) 35 amperes

  Based on the previous example

$I_{Neutral} = \sqrt{L_1^2 + L_2^2 + L_3^2 - (L_1 \times L_2 + L_2 \times L_3 + L_1 \times L_3)}$

$I_{Neutral} = \sqrt{(50^2 + 0^2 + 25^2) - [(50 \times 0) + (0 \times 25) + (50 \times 25)]}$

$I_{Neutral} = \sqrt{2{,}500 + 625 - 1250} = \sqrt{1{,}875} = 43.3$ amperes neutral current

12–26 *WYE MAXIMUM UNBALANCED LOAD*

The maximum unbalanced load (neutral) is the largest phase 120 volt neutral current with other lines off.

❑ **Maximum Unbalanced Load Example**

Balance and size the maximum unbalanced load for the following loads: one 36-kVA, three-phase heat strip, two 10-kVA, single-phase 208-volt loads, and three 3-kVA, 120-volt loads. L1 = 50 amperes, L2 = 0 amperes, L3 = 25 amperes.

(a) 0 amperes (b) 25 amperes (c) 50 amperes (d) none of these

- Answer: (c) 50 amperes

  The maximum unbalanced current equals the largest line neutral current.

	Line 1	Line 2	Line 3	Ampere Calculation
36 kVA 208 volts, 3Ø	100 amps	100 amps	100 amps	36,000 VA/(208 volts × 1.732)
10 kVA 208 volts, 1Ø	48 amps	48 amps		10,000 VA/208 volts
10 kVA 208 volts, 1Ø		48 amps	48 amps	10,000 VA/208 volts
2 – 3 kVA 120 volts, 1Ø	**50 amps** *			3,000 VA/120 volts
3 kVA 120 volts, 1Ø			25 amps *	3,000 VA/120 volts
	198 amps	196 amps	173 amps	

* Indicates neutral (120 volt) loads.

12–27 *DELTA/WYE EXAMPLE*

One – Dishwasher 4.5 kW, 120-volts

One – 10 hp A/C motor 208-volts, 3Ø

One – 18 kW 208-volts, 3Ø heat strip

Two – 14 kW ranges 208-volts, 1Ø

Two – 3 hp motors, 208-volts, 1Ø

One 10 kW water heater 208-volts, 1Ø

Eight 1.5 kW, 120-volts, lighting circuits

	Phase A (L1)	Phase B (L2)	Phase C (L3)	Line Total
Heat (A/C omitted *)	6,000 VA	6,000 VA	6,000 VA	18.0 kVA
14 kW Ranges 208 volt, 1Ø	7,000 VA	7,000 VA		14.0 kVA
14 kW Ranges 208 volt, 1Ø		7,000 VA	7,000 VA	14.0 kVA
10 kW Water Heater 208 volt, 1Ø	5,000 VA		5,000 VA	10.0 kVA
3 hpP 208 volt, 1Ø Motor	1,945 VA	1,945 VA		3.89 kVA
3 hp 208 volt, 1Ø Motor		1,945 VA	1,945 VA	3.89 kVA
Dishwasher 4.5 kW, 120 volt			4,500 VA *	4.5 kVA
Lighting (8 – 1.5 kW circuits)	6,000 VA *	3,000 VA *	3,000 VA *	12.0 kVA
	25,945 VA	26,890 VA	27,445 VA	80.28 kVA

Note: The phase totals (26 kVA, 27 kVA and 27.5 kVA) should add up to the line total (80.5 kVA). This method is used as a check to make sure all items have been accounted for and added correctly.

Motor VA

VA 1Ø = Table Volts × Table Amperes

3 hp, VA = E × I, VA = 208 volts × 18.7 amperes = 3,890 VA/2 = 1,1945 VA per phase

VA 3Ø = Table Volts × Table Amperes × $\sqrt{3}$

10 hp A/C, VA = Table Volts × Table Amperes × 1.732

10 hp A/C, VA = 208 volts × 30.8 amperes × 1.732 = 11,096 VA (omitted - smaller than 18 kW heat)

➛ Transformer Sizing Question
What size transformers are required?

(a) three 30 kVA single-phase transformers
(b) one 90 kVA three-phase transformer
(c) a or b
(d) none of these

- Answer: (c) a or b
  Phase A = 26 kVA
  Phase B = 27 kVA
  Phase C = 28 kVA

➛ Maximum kVA on Neutral Question
What is the maximum kVA on the neutral?

(a) 3 kVA (b) 6 kVA (c) 7.5 kVA (d) 9 kVA

- Answer: (c) 7.5 kVA Phase "C"
  Count 120-volt loads only, do not count phase to phase (208-volt) loads

➛ Maximum Neutral Current Question
What is the maximum current on the neutral?

(a) 50 ampere (b) 63 ampere (c) 75 amperes (d) 100 amperes

- Answer: (b) 63 amperes
  I = P/E = 7,500 VA/120 volts = 63 amperes. The neutral must be sized to carry the maximum unbalanced neutral current which in this case is 63 amperes.

➛ Neutral Current Question
What is the current on the neutral?

(a) 25 amperes (b) 33 amperes (c) 50 amperes (d) 63 amperes

- Answer: (b) 33 amperes

Formula

$$I_{Neutral} = \sqrt{L_1^2 + L_2^2 + L_3^2 - (L_1 \times L_2 + L_2 \times L_3 + L_1 \times L_3)}$$

L_1 = 6,000 VA/120 volts = 50 amperes
L_2 = 3,000 VA/120 volts = 25 amperes
L_3 = 7,500 VA/120 volts = 63 amperes

$$\sqrt{(50^2 + 25^2 + 63^2) - [\,(50 \times 25) + (25 \times 63) + (50 \times 63)\,]}$$

$$\sqrt{(2{,}500 + 625 + 3{,}969) - (1{,}250 + 1{,}575 + 3{,}150)}$$

$$\sqrt{7{,}094 - 5{,}975} \quad \sqrt{1{,}119} = 33 \text{ amperes}$$

➛ Phase VA, Three-Phase Question
What is the per phase load of each 3-horsepower 208-volt, single-phase motor?

(a) 1,945 VA (b) 3,910 VA (c) 978 VA (d) none of these

- Answer: (a) 1,945 VA
  VA 1Ø = Table Volts × Table Amperes
  3 HP, VA = E × I, VA = 208 volts × 18.7 amperes = 3,890 VA
  On a wye system, a 208 volt 1Ø load is on two transformer phases
  3,890 VA/2 phases = 1,945 VA

➛ Phase VA, Single-Phase Question
What is the per phase load for the three-phase, 10-horsepower A/C motor?

(a) 11,154 VA (b) 3,698 VA (c) 5,577 VA (d) None of these

- Answer: (b) 3,698 VA
  10-horsepower, VA = Table volts × Table amperes x 1.732
  VA = 208 volts × 30.8 amperes × 1.732 = 11,095 VA
  11,095 VA/3 phases = 3,698 VA per phase

➛ **Voltage Ratio Question**

What is the phase voltage ratio of the transformer?

(a) 4:1 (b) 1:4 (c) 1:2 (d) 2:1

- Answer: (a) 4:1

  480 primary phase volts to 120 secondary phase volts

❑ **Balance Panel VA Example**

Balance the loads on the panelboard in VA.

	Line 1	Line 2	Line 3	Line Total
Heat 18 kW(A/C 11 kW/ omitted)	6,000 VA	6,000 VA	6,000 VA	18.0 kVA
14 kW Ranges 208 volts, 1Ø	7,000 VA	7,000 VA		14.0 kVA
14 kW Ranges 208 volts, 1Ø		7,000 VA	7,000 VA	14.0 kVA
10 kW Water Heater 208 volts, 1Ø	5,000 VA		5,000 VA	10.0 kVA
3 hp 208 volts, 1Ø Motor	2,000 VA	2,000 VA		4.0 kVA
3 hp 208 volts, 1Ø Motor		2,000 VA	2,000 VA	4.0 kVA
Dishwasher 4.5 kW 120 volt			4,500 VA *	4.5 kVA
Lighting (8 – 1.5 kW circuits)	+ 6,000 VA *	3,000 VA *	3,000 VA *	12 .0 kVA
(4 on L_1, 2 on L_2, 2 on L_3)	26,000 VA	27,000 VA	27,500 VA	80.5 kVA

❑ **Panelboard Balancing and Sizing Example**

Balance the previous example loads on the panelboard in amperes.

	Line 1	Line 2	Line 3	Ampere Calculations
18 kW Heat 3Ø	50.0 amps	50.0 amps	50.0 amps	18,000 VA/(208 volts × 1.732)
14 kW Ranges 208 volts, 1Ø	67.0 amps	67.0 amps		14,000 VA/208 volts
14 kW Ranges 208 volts, 1Ø		67.0 amps	67.0 amps	14,000 VA/208 volts
Water Heater, 208 volts, 1Ø	48.0 amps		48.0 amps	10,000 VA/208 volts
3 hp 208 volts, 1Ø Motor	18.7 amps	18.7 amps		Motor FLC
3 hp 208 volts, 1Ø Motor		18.7 amps	18.7 amps	Motor FLC
Dishwasher 1Ø 120 volt			38.0 amps	4,500 VA/120 volts
Lighting 1Ø 120 volt	+ 50.0 amps	25.0 amps	25.0 amps	1,500 VA/120 volts
	233.7 amps	246.4 amps	246.7 amps	

12–28 *DELTA VERSUS WYE*

Wye Truisms:

Phase **CURRENT** is the **SAME** as line current.

$I_{Phase} = I_{Line}$

$I_{Line} = I_{Phase}$

Wye Phase **VOLTAGE** is **DIFFERENT** than line voltage.

Line Voltage is greater than Phase Voltage by the square root of 3 ($\sqrt{3}$).

$E_{Line} = E_{Phase} \times \sqrt{3}$

$E_{Phase} = E_{Line} / \sqrt{3}$

Delta Truisms:

Phase **VOLTAGE** is the **SAME** as line voltage.

$E_{Phase} = E_{Line}$

$E_{Line} = E_{Phase}$

Delta phase **CURRENT** is **DIFFERENT** than line current.

Line current is greater than Phase Current by the square root of 3 ($\sqrt{3}$).

$I_{Line} = I_{Phase} \times \sqrt{3}$

$I_{Phase} = I_{Line} / \sqrt{3}$

Unit 12 – Delta/Delta And Delta/Wye Transformers Summary Questions

Definitions

1. Delta connected means the windings of three single-phase transformers (same rated voltage) are connected in ________ . A delta-connected transformer is represented by the greek letter delta Δ.
(a) series (b) parallel (c) series-parallel (d) a and b

2. Which system is called the high leg system because the voltage from one conductor to ground is between 190 and 208-volts-to-ground?
(a) delta (b) wye

3. The ________ is the electrical system ungrounded conductors.
(a) load (b) line (c) phase (d) system

4. The__________ voltage is the voltage measured between any ungrounded conductors.,
(a) load (b) line (c) phase (d) system

5. • Line voltage of a ________ system is greater than phase voltage and line voltage of a _______ system is the same as phase voltage.
(a) delta, delta (b) wye, wye
(c) delta, wye (d) wye, delta

6. The________ is the coil shape conductors that serve as the primary or secondary of the transformer.
(a) line (b) load (c) phase (d) none of these

7. The phase current is the current of the transformer winding. For ______ systems, the phase current is less than the line current. For ______ systems, the phase current is the same as the line current.
(a) wye, wye (b) delta, delta
(c) wye, delta (d) delta, wye

8. The phase load is the load on the transformer winding. For delta systems the phase load is _______.
(a) 3Ø 240-volt load = line load/3
(b) 1Ø 240-volt load = line load
(c) 1Ø 120-volt load = line load
(d) all of these

9. The phase load is the load on the transformer winding. For wye systems the phase load is ________.
(a) 3Ø 208-volt load = line load/3
(b) 1Ø 208-volt load = line load/2
(c) 1Ø 120-volt load = line load
(d) all of these

10. The phase voltage is the internal transformer voltage generated across any one "winding" of a transformer. For _______ systems, the phase voltage is equal to the line voltage. For ______ systems the phase voltage is less than the line voltage.
(a) wye, wye (b) delta, delta
(c) wye, delta (d) delta, wye

11. The ratio is the relationship between the number of primary winding turns compared to the number of secondary winding turns. The ratio is a comparison between the primary phase voltage and the secondary phase voltage. For typical delta/delta systems, the ratio is _____, and for typical wye systems the ratio is ________.
(a) 1:2, 1:4 (b) 2:1, 4:1 (c) 4:1, 2:1 (d) none of these

12. What is the turns ratio of a 480- to 208Y/120-volt transformer?
(a) 4:1   (b) 1:4   (c) 1:2   (d) 2:1

13. • The unbalanced load is the load on the secondary grounded conductors. For ________ systems the formula is:
$\sqrt{L_1^2 + L_2^2 + L_3^2 - (L_1 \times L_2 + L_2 \times L_3 + L_1 \times L_3)}$
(a) delta   (b) wye   (c) a or b   (d) none of these

14. The unbalanced (neutral) current for a ________ system can be calculated by: $I_{Line\ 1} - I_{Line3}$.
(a) delta   (b) wye   (c) a or b   (d) none of these

15. ________-connected means a connection of three single-phase transformers to a common point (neutral) and the other end is connected to the line conductors.
(a) delta   (b) wye   (c) a or b   (d) none of these

12–1 Current Flow

16. When a load is connected to the secondary of a transformer, current will flow through the secondary conductor winding. The current flow in the secondary creates an electromagnetic field that opposes the primary electromagnetic field. The secondary flux lines ________.
(a) effectively reduce the strength of the primary flux lines
(b) less CEMF is generated in the primary winding conductors
(c) primary current automatically increases in direct proportion to the secondary current
(d) all of these

17. The primary and secondary line currents are directly proportional to the ratio of the transformer.
(a) True   (b) False

Part A – Delta/Delta Transformer

12–2 Delta Transformers Voltages

18. In a delta configured transformer, the Line Voltage equals the Phase Voltage.
(a) True   (b) False

12–3 Delta High Leg

19. If the secondary voltage of a delta/delta transformer is 220/110-volts, the high leg voltage to ground (or neutral) would be ______ volts.
(a) 190   (b) 196   (c) 202   (d) 208

12–4 Delta Primary And Secondary Line Currents

20. In a delta configured transformer, the line current equals the phase current.
(a) True   (b) False

21. What is the primary line current for a 45-kVA, 480- to 240/120-volt 3Ø transformer?
(a) 124 amperes   (b) 108 amperes   (c) 54 amperes   (d) 43 amperes

22. What is the secondary line current for a 45-kVA, 480- to 240-volt 3Ø transformer?
(a) 124 amperes   (b) 108 amperes   (c) 54 amperes   (d) 43 amperes

12–5 Delta Primary Or Secondary Phase Currents

23. The phase current of a transformer winding is calculated by dividing the phase load by the phase volts: $I_{Phase} = VA_{Phase}/E_{Phase}$.
(a) True   (b) False

24. What is the primary phase current for a 15-kVA, 480- to 240/120-volt 1Ø transformer?
(a) 124 amperes   (b) 62 amperes   (c) 45 amperes   (d) 31 amperes

25. What is the secondary phase current for a 15-kVA, 480- to 240/120-volt 1Ø transformer?
(a) 124 amperes (b) 62 amperes (c) 45 amperes (d) 31 amperes

12–6 Delta Phase Versus Line

26. A 15-kVA 240-volt 3Ø load has the following effect on a delta system:
(a) $I_L = 15{,}000/(240 \times \sqrt{3})$, $I_L = 36$ amperes (b) $I_P = 5{,}000/240$, $I_P = 21$ amperes
(c) $I_L = I_P \times \sqrt{3}$, $I_L = 21 \times 1.732$, $I_L = 36$ amperes (d) All of the above

27. • A 5-kVA 240-volt 1Ø load has the following effect on a delta/delta system:
(a) Line Power = 5 kVA (b) Phase Power = 5 kVA
(c) I_P = 5,000 VA/240-volts = 21 amperes (d) All of the above

28. • A 2-kVA 120-volt 1Ø load has the following effect on a delta system:
(a) Line Power = 2 kVA (b) I_L = 2,000 VA/120-volts, I_L = 17 amperes
(c) Phase Power = 3 kVA (C1 or C2 winding) (d) a and b

12–7 Delta Current Triangle

29. The line and phase current of a delta system are not equal. The difference between line and phase current can be described by which of the following equations?
(a) $I_L = I_P \times \sqrt{3}$ (b) $I_P = I_L/\sqrt{3}$ (c) a and b (d) none of these

12–8 And 12–9 Delta Transformer Balancing And Sizing

30. To properly size a delta/delta transformer, the transformer must be balanced in kVA. Which of the following steps should be used to balance the transformer?
(a) Balance 3Ø loads, ⅓ on Phase A, 1/3 on Phase B, and 1/3 on Phase C.
(b) Balance 1Ø 240-volt loads, 100 percent on Phase A or Phase B.
(c) Balance the 120-volt loads, 100 percent on Phase C1 or C2.
(d) All of these

31. • Balance and size a 480- to 240/120-volt 3Ø delta transformer for the following loads: One 18-kVA 3Ø heat strip, two 5-kVA 1Ø 240-volt loads, and three 2-kVA 120-volt loads. What size transformer is required?
(a) Three 1Ø 12.5-kVA transformers (b) One 3Ø 37.5-kVA transformer
(c) a or b (d) None of these

12–10 Delta Panel Schedule In kVA

32. When balancing a panelboard in kVA be sure that 3Ø loads are balanced 1/3 on each line and 240-volt 1Ø loads are balanced ½ on each line.
(a) True (b) False

33. • Balance a 120/240-volt delta 3Ø panelboard in kVA for the following loads: One 18-kVA 3Ø heat strip, two 5-kVA 1Ø 240-volt loads, and three 2-kVA 120-volt loads.
(a) The largest line load is 13 kVA. (b) The smallest line load is 10 kVA.
(c) Total line load is equal to 34 kVA. (d) All of these.

12–11 Delta Panelboard And Conductor Sizing

34. When selecting and sizing panelboards and conductors, we must balance the line loads in amperes. Balance a 120/240-volt 3Ø panelboard in amperes for the following loads: One 18-kVA 3Ø heat strip, two 5-kVA 1Ø 240-volt loads, and three 2-kVA 120-volt loads.
(a) The largest line is 98 amperes. (b) The smallest line is 81 amperes.
(c) Each line equals to 82 amperes. (d) a and b.

12–12 Delta Neutral Current

35. What is the neutral current for: One 18-kVA, 3Ø heat strip, two 5-kVA 1Ø 240-volt loads, and three 2-kVA 120-volt loads?
(a) 17 amperes (b) 34 amperes (c) 0 amperes (d) A and B

12–13 Delta Maximum Unbalanced Load

36. • What is the maximum unbalanced load in amperes for: One 18-kVA, 3Ø heat strip, two 5-kVA 1Ø 240-volt loads, and three 2-kVA, 120-volt loads?
(a) 25 amperes (b) 34 amperes (c) 0 amperes (d) None of these

37. What is the phase load for a 18-kVA, 3Ø heat strip?
(a) 18 kVA (b) 9 kVA (c) 6 kVA (d) 3 kVA

38. • What is the transformer winding phase load for a 5-kVA, 1Ø 240-volt load?
(a) 5 kVA (b) 2.5 kVA (c) 1.25 kVA (d) None of these

39. If only a 3Ø, 12-kW load is on the high leg, what is the current on the high leg conductor?
(a) 29 amperes (b) 43 amperes (c) 73 amperes (d) 97 amperes

40. What is the ratio of a 480- to 240/120-volt transformer?
(a) 4:1 (b) 1:4 (c) 1:2 (d) 2:1

Part B Delta/Wye Transformers

12–15 Wye Transformers Voltages

41. In a wye configured transformer, the line voltage equals the phase voltage.
(a) True (b) False

12–17 Wye Transformers Current

42. In a wye configured transformer, the line current equals the phase current.
(a) True (b) False

12–18 Line Current

43. The line current of both delta and wye 3-phase transformers can be calculated by the formula:
$I_{Line} = VA_{Line}/(E_{Line} \times \sqrt{3})$.
(a) True (b) False

44. What is the primary line current for a 22-kVA, 480- to 208Y/120-volt 3Ø transformer?
(a) 61 amperes (b) 54 amperes (c) 26 amperes (d) 22 amperes

45. What is the secondary line current for a 22-kVA, 480- to 208Y/120-volt 3Ø transformer?
(a) 61 amperes (b) 54 amperes (c) 27 amperes (d) 22 amperes

12–19 Phase Current

46. The phase current of a wye configured transformer winding is the same as the line current. The phase current of a delta configured transformer winding is less than the line current by the $\sqrt{3}$.
(a) True (b) False

47. • What is the primary phase current for a 22-kVA, 480- to 208Y/120-volt 3Ø transformer?
(a) 15 amperes (b) 22 amperes (c) 36 amperes (d) 61 amperes

48. • What is the secondary phase current for a 22-kVA, 480- to 208Y/120-volt 3Ø transformer?
(a) 15 amperes (b) 22 amperes (c) 36 amperes (d) 61 amperes

12–20 Wye Phase Versus Line

49. Since each line conductor from a wye transformer is connected to a different transformer winding (phase), the effects of loading on the line are the same as the phase. A 12-kVA, 208-volt 3Ø load has the following effect on the system:
(a) Line Current = I_L = VA/(E x $\sqrt{3}$), I_L = 12,000/(208 x $\sqrt{3}$), I_L = 33 amperes
(b) Phase Power = 4 kVA
(c) Phase Current = I_P = VA_{Phase}/E_{Phase}, I_P = 4,000/120 = 30 amperes
(d) All of these

50. A 5-kVA, 208-volt 1Ø load has the following effect on the system:
(a) Line Power = 5 kVA
(b) Line Current = IL = VA/E, I_L = 5,000/208, I_L = 24 amperes
(c) Phase Power = 5 kVA
(d) Phase Current = I_P = VA_{Phase}/E_{Phase}, I_P= 2,500/120 = 21 amperes
(e) All of these

51. A 2-kVA, 120-volt 1Ø load has the following effect on the system:
(a) Line Power = 2 kVA
(b) Line Current = I_L = VA/E, I_L = 2,000/120, I_L = 17 amperes
(c) Phase Power = 2 kVA
(d) Phase Current = I_P = VA_{Phase}/E_{Phase}, I_P= 2,000/120 = 17 amperes
(e) All of these

52. • What is the phase load of a 3-horsepower, 208- volt, 1Ø motor?
(a) 1,945 VA (b) 3,890 VA (c) 978 VA (d) none of these

53. • What is the per phase load for a 3Ø, 7.5-horsepower, 208-volt motor?
(a) 2,906 VA (b) 8,718 VA (c) 4.359 VA (d) none of these

12–21 And 12–22 Wye Transformer Balancing And Sizing

54. To properly size a delta/wye transformer, the secondary transformer phases (windings) or the line conductors must be balanced. Balancing the panel (line conductors) is identical to balancing the transformer for wye configured transformers. Which of the following steps should be used to balance the transformer?
(a) balance 3Ø loads, 1/3 on each phase
(b) balance 1Ø 208-volt loads 50 percent on each phase
(c) balance the 120-volt loads 100 percent on any phase
(d) all of these

55. • Balance and size a 480- to 208Y/120-volt 3Ø transformer for the following loads: One 18-kVA, 3Ø heat strip, two 5-kVA, 1Ø 208-volt loads, and three 2-kVA, 120-volt loads. What size transformer is required?
(a) three 1Ø 12.5-kVA transformers (b) one 3Ø 37.5-kVA transformer
(c) a or b (d) none of these

12–23 Wye Panel Schedule In kVA

56. When balancing a panelboard in kVA be sure that 3Ø loads are balanced 1/3 on each line and 208-volt 1Ø loads are balanced 1/2 on each line. Balance a 208Y/120-volt 3Ø panelboard in kVA for the following loads: One 18-kVA, 3Ø heat strip, two 5-kVA, 1Ø 208-volt loads, and three 2-kVA, 120-volt loads.
(a) The largest line is 12.5 kVA.
(b) The smallest line is 10.5 kVA.
(c) Each line equals to 11.33 kVA.
(d) all of these.

12–24 Wye Panelboard And Conductor Sizing

57. • When selecting and sizing panelboards and conductors, we must balance the line loads in amperes. Balance a 208Y/120-volt 3Ø panelboard in amperes for the following loads: One 18-kVA, 208-volt 3Ø heat strip, two 5-kVA, 1Ø 208-volt loads and three 2-kVA, 120-volt loads.
(a) The largest line current is 108 amperes (b) The smallest line is 91 amperes
(c) a and b (d) None of these

12–25 Wye Neutral Current

58. To determine the neutral current for a 3-wire wye system we must use the following formula:
$I \text{ neutral} = \sqrt{(L_1^2 + L_2^2 + L_3^2) - (L_1 \times L_2 + L_2 \times L_3 + L_1 \times L_3)}$
(a) True (b) False

59. • What is the neutral current for: One 18-kVA, 3Ø heat strip, two 5-kVA, 1Ø 208-volt loads and three 2-kVA, 120-volt loads.
(a) 47 amperes (b) 29 amperes (c) 34 amperes (d) A and B

12–26 Wye Maximum Unbalanced Load

60. • What is the maximum unbalanced load in amperes for: One 18-kVA, 3Ø heat strip, two 5-kVA, 1Ø 208-volt loads, and three 2-kVA, 120-volt loads.
(a) 17 amperes (b) 34 amperes (c) 0 amperes (d) a and b

12–27 Delta Versus Wye

61. Which of the following statements are true for wye systems?
(a) Phase current is the same as line current: $I_P = I_L$.
(b) Phase voltage is the same as the line voltage.
(c) $E_L = E_P \times \sqrt{3}$.
(d) a and c

62. Which of the following statements are true for delta systems?
(a) Phase voltage is the same as line voltage: $E_P = E_L$.
(b) Phase current is the same as the line current.
(c) Line current is greater than Phase Current: $I_L = I_P \text{ x } \sqrt{3}$.
(d) a and c

Challenge Questions

Delta/Delta Questions

The following statement applies to the next seven questions.

The following loads are to be connected to a three-phase delta/delta 480-volt to 240-volt transformer.
1 – 25-horsepower, 240-volt, three-phase synchronous motor, air conditioning
2 – 1½-horsepower, 240-volt, single-phase motors
2 – 2-horsepower, 120-volt motors
3 – 1900 watt, 120-volt lighting loads
18 kW, 240-volt, 3-phase heat
Balance the above loads as closely as possible, then answer the next seven questions.

63. • Three single-phase transformers are connected in series. What size transformers are required for the above loads?
(a) 10/10/10 kVA (b) 15/15/15 kVA (c) 10/10/20 kVA (d) 15/15/10 kVA

64. • The line VA load of the 25-horsepower three-phase synchronous motor would be ______ VA.
(a) 9,700 (b) 21,113 (c) 32,031 (d) 28,090

65. • After balancing the loads, the maximum unbalanced neutral current is ______ amperes.
(a) 27 (b) 1.5 (c) 48 (d) 148

66. • The transformer phase load of one 1½-horsepower motor would be ______ VA.
(a) 1,380 (b) 2,760 (c) 2,300 (d) 1,150

67. • The line load of one of the 1½-horsepower motors would be ______ VA.
(a) 1,380 (b) 2,760 (c) 2,300 (d) 1,150

68. • The transformer phase VA load of the 25-horsepower three-phase synchronous motor is closest to ______ VA.
(a) 3,520 (b) 7,038 (c) 21,100 (d) 32,100

69. • If the only load on the high leg is the three-phase 25-horsepower motor, the high leg conductor would have to be sized a minimum of ______ amperes.
(a) 13 (b) 66 (c) 18 (d) 22

The following statement applies to the next two questions.

A three-phase 37.5-kVA delta/delta transformer has a primary of 480-volts and a secondary of 240/120-volts.

70. • The primary line current is closest to ______ amperes.
(a) 45 (b) 90 (c) 50 (d) 65

71. • The maximum overcurrent device size for the 37.5-kVA delta/delta transformer is ______ amperes.
(a) 40 (b) 45 (c) 50 (d) 60

The following statement applies to the next question.

An office building contains the following loads:
1 – 5-horsepower 240-volt three-phase motor
7 – 1,500 watt, 120-volt heaters
2 – 3-horsepower 240-volt motors
5 – 3 kW 240-volt, single-phase heaters
30 kW 120-volt small appliance
10 – 5 kW 240-volt, single-phase machines
40 kW 120-volt lighting

72. • If the only loads on the center tapped transformer (C_1 and C_2) are the 120-volt loads, what is the total kW load on this transformer?
(a) 30–40 kW (b) 50–60 kW (c) 80–90 kW (d) over 100 kW

73. • The total load on the center tapped 240/120-volt transformer is 100 kW, which consists of 80 kW of 120-volt balanced loads, and 20 kW of 240-volt loads. What is the maximum unbalanced neutral current?
(a) 167 amperes (b) 0 amperes (c) 192 amperes (d) 333 amperes

74. • On a delta/delta three-phase transformer, a 30,000 VA, 240-volt three-phase load would have a phase current of ______ amperes on the secondary winding.
(a) 125 (b) 72 (c) 42 (d) none of these

75. • What is the secondary line current of a three-phase, 45-kVA, delta/delta transformer that has a turn ratio of 2:1 and a primary voltage of 460?
(a) 113 amperes (b) 55 amperes
(c) 36 amperes (d) cannot be determined

76. • A 2,200/220-volt three-phase transformer is connected delta/delta. The secondary phase current is 300 amperes and the secondary line current is 240 volts. What is the primary line current?
(a) 40 amperes (b) 20 amperes (c) 50 amperes (d) 30 amperes

77. • For a given three-phase transformer, the connection giving the highest secondary voltage is ______.
(a) wye primary, delta secondary (b) delta primary, delta secondary
(c) wye primary, wye secondary (d) delta primary, wye secondary

Part B – Wye Transformers

78. • A delta/wye, 480/208, 120-volt three-phase transformer has a secondary line current of 200 amperes. What is the primary line current?
(a) 31 amperes (b) 72 amperes (c) 87 amperes (d) 53 amperes

79. • If the secondary phase voltage of a delta/wye connected transformer is 277 volts, and a 10-kVA three-phase load was connected to that secondary, the line current would be ______ amperes.
(a) 7 (b) 12 (c) 32 (d) none of these

The following statement applies to the next two questions.

1 – 15-horsepower, 208-volt three-phase induction motor
2 – 1.5-kW, 208-volt dryers
4 – 3-kW, 120-volt lighting banks
1 – 8-kW, 208-volt cooktops
5 – 2.5-kW, 208-volt heat strips

80. • Each 2.5-kW heat strip has a line load of ______ .
(a) 1,250 watts (b) 12,500 watts (c) 2,500 watts (d) no load

81. • The load per line of the three-phase motor can be determined by which of these formulas?
(a) 3 × V × I (b) (V × I)/3 (c) V × I (d) no phase load

82. • If a three-phase motor has a total load of 12,000 VA, and is connected to a delta/wye transformer, the load on one phase of the secondary will be ______ VA.
(a) 12,000 (b) 4,000 (c) 3,000 (d) 1,500

83. • A three-phase 480Y/208-120-volt transformer provides power to nine 2,000 VA, 120-volt lights and nine 2,000 watt, 208-volt heaters. The maximum neutral current is ______ amperes.
(a) 50 (b) 75 (c) 87 (d) 95

84. • There are five 2,500 watt, 208-volt heat strips on the secondary of a three-phase 208Y/120-volt transformer. What is the load per phase for each heat strip?
(a) 1,403 watts (b) 1,250 watts (c) 2,500 watts (d) 12,500 watts

The following statement applies to the next question.

The system is 208Y/120-volt, three-phase. The loads are:
1 – 10-horsepower, three-phase induction motor
4 – 1,500 VA, 120-volt circuits
5 – 1,500 VA, 120-volt lights
1 – 6,000 VA, three-phase water heater
1 – 3,000 watt, 208-volt single-phase industrial oven
1 – ¾-horsepower, 208-volt single-phase motor

85. • The phase load of the 10-horsepower three-phase motor is closest to ______ VA.
(a) 3,700 (b) 5,500
(c) 4,300 (d) 11,100

The following statement applies to the next question.

The system is three-phase 208Y/120-volt and the loads are:
9 – 2-kVA 120-volt lighting loads
5 – 3-kVA 120-volt loads
2 – 2.5 kW 208-volt heating units

86. • The neutral conductor must be sized to carry ______ amperes.
(a) 10 (b) 25
(c) 90 (d) 100

Index

P

R

S

T

U

V